Ophiolites and Oceanic Lithosphere

Ophiolites and Oceanic Lithosphere

edited by

I. G. Gass, S. J. Lippard & A. W. Shelton
Department of Earth Sciences, The Open University
Milton Keynes

1984

Published for
The Geological Society
by Blackwell Scientific Publications
Oxford London Edinburgh
Boston Melbourne

Published by

Blackwell Scientific Publications
Osney Mead, Oxford OX2 0EL
8 John Street, London WC1N 2ES
9 Forrest Road, Edinburgh EH1 2QH
52 Beacon Street, Boston, Massachusetts 02108, USA
99 Barry Street, Carlton, Victoria 3053, Australia

First published 1984

DISTRIBUTORS

USA
Blackwell-Mosby Book Distributors
11830 Westline Industrial Drive
St Louis, Missouri 63141

Canada
Blackwell-Mosby Book Distributors
120 Melford Drive, Scarborough
Ontario M1B 2X4

Australia
Blackwell Scientific Book Distributors
31 Advantage Road, Highett
Victoria 3190

British Library Cataloguing in Publication Data

Ophiolites and oceanic lithosphere.—(Geological Society special publications, ISSN 0305-8719; No. 13)
1. Submarine geology
I. Gass, I.G. II. Lippard, S.J. III. Shelton, A.W.
IV. Series
551.46'08 QE39

ISBN 0-632-01219-6

Printed in Great Britain by
Alden Press Ltd, Oxford

Contents

II. EMPLACEMENT (OBDUCTION) OF OPHIOLITES

OPHIOLITE EMPLACEMENT AND OBDUCTION

REGIONAL STUDIES

Preface

The proposal that ophiolites are on-land fragments of oceanic lithosphere has been with us for over 20 years and has been widely accepted by the Earth Sciences' community. Despite this, many oceanographers are reluctant to use ophiolite data. Their argument (and it comes largely from the geological rather than geophysical oceanographers) is that 'Even if ophiolites *are* on-land fragments of oceanic lithosphere, they must be atypical for otherwise they would have been subducted and not obducted'. So they conclude that, although the study of ophiolites is perfectly acceptable in its own right, its results should not be used in the investigation of present-day, *in-situ* oceanic lithosphere—i.e. the invoking of 'reversed uniformitarianism' is not acceptable. However using uniformitarian principles to interpret ophiolites from present-day oceanic lithosphere studies is equally problematical for (i) petrological/geochemical studies of occan-floor rocks are based on widely spaced boreholes of limited depth and dredge-haul sampling, (ii) geophysical data can only detect large-scale phenomena and are open to various interpretations, and (iii) direct observations from submersibles are still very limited. Such circumstances should have drawn the two groups of workers closer together.

We believe that the deliberate ignoring of ophiolite evidence has, at best, slowed down our understanding of processes by which oceanic lithosphere is produced. Two examples to illustrate this will suffice: in the early 1970s, studies on ophiolite metamorphism indicated that hot seawater passed through the uppermost 4 km of the oceanic crust soon after it was generated at a constructive plate margin. These hot brines enriched in transition metals emerged from the oceanic crust to react with cold seawater to produce metal-enriched muds that with time becamc massive sulphide deposits. It was not until later in the decade when the thermal budget imbalance was identified by heat-flow studies on the East Pacific Rise, that oceanographic investigations into the possibility of seawater circulation through oceanic crust adjacent to present-day constructive margins took place with the subsequent location of black smokers where metal-enriched brines emerge from the ocean crust. Similarly, there has been much nonsense talked about magma chambers beneath constructive margins. The presence of a crustal magma chamber is implicit in all ophiolite-based models and most workers accept that magma chambers exist in the present-day oceanic crust; indeed they have been identified geophysically beneath the fast-spreading East Pacific Rise (see Orcutt *et al.*). However, because none have been seismologically detected beneath the slow-spreading Mid-Atlantic Ridge, serious (*sic*) proposals on how oceanic crust could be produced without involving magmatic processes have been made at oceanographic conferences despite the fact that no ophiolite has yet been described that could be produced by anything other than magmatic processes.

The notable exception to this negative attitude was the Ophiolite Conference held in Nicosia, Cyprus, during April 1979 when many oceanographers took the opportunity to study the outcrops on the classic Troodos ophiolite; much invaluable discussion and productive research collaboration resulted. Our proposal to the Geological Society to hold an international conference entitled 'Ophiolites and Oceanic Lithosphere' was specifically intended to encourage and stimulate oceanographic-ophiolite research recognition and collaboration. By and large it failed, for although some 200 participants from eighteen countrics attended the conference held in the apartments of the Geological Society in Burlington House, London on the 17–19 November 1982, only seven of the fifty-seven papers presented were on oceanic studies, three compared oceanic and ophiolitic data and the remaining forty-seven were entirely concerned with ophiolites. Most papers were well presented and lively discussions, regrettably not published here, ensued. The ophiolite-oceanographic imbalance occurred despite canvassing by the organizers to encourage greater oceanographic participation and is reflected in these proceedings. (Of the thirty-three papers presented here twenty-six are on ophiolites, two compare oceanographic or ophiolitic data and only five are oceanographic.)

The papers in this volume fall into two categories. In the first section the nature and formation of oceanic lithosphere is discussed from integrated and individual studies of the present ocean crust and ophiolite complexes. The second section contains both review and original articles on the emplacement (obduction) of ophiolites and the use to which they can be put in understanding the processes of plate collision. The nineteen papers in Section I can be further subdivided into five groups on (i) magma chambers, their products and processes, (ii) fracture zones, (iii) mantle structures, (iv) lavas and sediments, and (v) isotope studies and metamorphism.

Magma chambers: products and processes: Orcutt, McClain & Burnett, in reviewing and updating seismic studies across the East Pacific Rise, confirm the presence of extensive shallow magma chambers but dispute the commonly held view that oceanic crust thickens with increasing age. *Fisk* uses petrological arguments to limit the depths and temperatures of magma chambers beneath constructive margins and concludes that for the Galapagos Rise and Mid-Atlantic Ridge basalts, the reservoirs were located at about 3 km below the ridge crest and had temperatures of 1150–1270°C. *Flower* discusses the dependence of basalt petrology and geochemistry on spreading rates, postulating that slow-spreading ridges are characterized by polybaric fractionation and fast-spreading ones by low-pressure isobaric systems. Peculiarities of magma compositions of intermediate-rate spreading ridges are attributed to a combination of slow- (plagioclase accumulation) and fast-spreading (open magma supply) characteristics.

In contributions to magma-chamber products and processes from ophiolites, *Smewing* et al. describe the cumulate sequence gabbros of the northern Oman ophiolite and suggest that these were formed in a series of elongate magma chambers along the constructive margin at which the ophiolite developed. In separate papers, *Gregory, Pallister* and *Browning* discuss mantle-melt reactions, parental magma compositions and crystallization sequences in the Oman ophiolite. Variations in these features over short along-strike distances indicate that melts of varying composition and degree of fractionation are transferred into the crust during spreading processes. *Elthon, Casey & Komor* identify the early crystallization of Mg-rich pyroxenes in the basal cumulates of the Bay of Islands, Newfoundland ophiolite and propose that they crystallized at high pressures ($>$10 kb) and were tectonically transported to the base of the crust. This interpretation is in conflict with Browning's Oman evidence which suggests that the early crystallization of clinopyroxene results from magma compositions that differ from MORB-type liquids.

Fracture zones: White, quoting seismic data, describes the anomalously thin and deformed oceanic crust in and adjacent to the Mid-Atlantic Ridge fracture zones and discusses its influence on the structure of slow-spreading ridges. Atlantic fracture-zone rocks and structures studied from submersibles and dredge hauls are described by the CYAGOR II group and *Honnorez, Mével & Montigny*. From ophiolite studies comes *Karson*'s paper which describes a fossil transform fault in the Coastal Complex of Newfoundland.

Mantle structures: This topic has only two contributions from ophiolite studies. *Smewing* et al. mentioned earlier under magma chambers, describe the mantle foliations and lineations and the layering in gabbros from the northern part of the Semail ophiolite. They show that the stress field, which produced the mantle foliation, also affected the lowermost

2 km of the crustal sequence indicating an effective coupling between crust and mantle during sea-floor spreading processes. *Nicolas & Rabinowicz* present a model for asthenospheric flow beneath ocean ridges based on their study of mantle tectonite fabrics.

Lavas and sediments: Two papers, both based on studies of the classic Troodos ophiolite, fall in this category. *Malpas & Langdon* describe the komatiitic mafic and ultramafic lavas that occur near the top of the Troodos lava sequence and conclude that they are products of off-axis magmatism. *Boyle & Robertson* studying the metallogenic sediments occurring above and within the Troodos eruptive sequence, compare them to those associated with present-day ridges and conclude that the Troodos ophiolite was produced at a fairly fast-spreading constructive margin.

Isotope studies and metamorphism: Five papers fall into this category. *Elthon* et al. discuss the processes of hydrothermal metamorphism of the oceanic crust based on a study of the Sarmiento ophiolite in S Chile, whereas *Stakes, Taylor & Fisher* describe oxygen-isotope and geochemical characterization based on the study of specimens from both the present oceanic crust and the Oman ophiolite. In particular, they show how water/rock ratios and magmatic *v.* seawater influences can be calculated. *Thirwell & Bluck* present an Sr–Nd-isotope study on the eruptive rocks of the Caledonian ophiolite at Ballantrae, SW Scotland which, they suggest, were formed in a variety of tectonic settings and not in a simple, single ophiolite sequence. *Menzies* shows that the majority of diopsides separated from orogenic and ophiolitic tectonite peridotites have Sr and Nd isotope composition close to MORB. He suggests that the orogenic lherzolites represent a source of MORB liquids whereas the ophiolite harzburgites represent a depleted residium after extraction of such a liquid. *Ahmed & Hall* provide evidence suggesting that serpentinization and rodingitization in the Sakhakot-Qila ophiolite of Pakistan took place whilst the ophiolite was still an *in-situ* part of the oceanic crust.

We have divided the papers in the second part of the volume into two categories. Those concerned with (i) ophiolite emplacement and obduction and (ii) regional studies.

Ophiolite emplacement and obduction: Of the papers in this section, two (*Spray* and *Casey & Dewey*) deal with the detachment of the potential ophiolite from its *in-situ* oceanic lithosphere setting, whereas *Ogawa & Naka, Searle & Stevens* and *Woodcock & Robertson* are concerned primarily with the emplacement of oceanic crust onto arc or continental margins. *Spray*, drawing attention to the elongate form of most ophiolites (length $>$ breadth $>$ thickness) and the fact that the igneous crystalline ages are very close to those given by the metamorphic soles that mark the initial detachment of the ophiolite from its oceanic setting, concludes that this happened to young, hot oceanic lithosphere close to its associated constructive margin. *Casey & Dewey* discuss the relation between the initiation of subduction and ophiolite obduction as a consequence of change in plate movement. *Ogawa & Naka* describe forearc ophiolites in Japan and the Western Pacific and conclude that much of their mélangic structure results from deformation in a fracture zone prior to emplacement on a forearc margin. *Searle & Stevens* are concerned with the origin and emplacement of the Newfoundland, Oman and Spontang (Himalayan) ophiolites as forearc continent collisions and identify modern analogues for this process from the Western Pacific. The account of the Spontang ophiolite is particularly interesting as it is, so far as we are aware, the first account of this remote complex to appear in Western literature. *Woodcock & Robertson* discuss contrasting styles of ophiolite nappe emplacement in various Tethyan ophiolites.

Regional studies: Seven papers are included under this heading although the article by

Rothery, describing the use of satellite imagery in the investigation of the Oman ophiolite, does not realistically fall in this, or any other, category in this volume. *Colley*'s account of a possible ophiolite from Fiji gives a new interpretation to this essentially volcanic complex. *Davies & Jaques* and *Milsom* provide new geological and geophysical data on Papua–New Guinea ophiolites whilst *Coleman, Wadge, Draper & Lewis* and *Sturt, Roberts & Furnes* review ophiolite complexes in Arabia, the Caribbean and the Scandinavian Calidonides respectively. *Hall*, in a controversial paper, claims that the well-known ophiolites of the Middle East, and notably the Oman, are virtually autochthonous being formed by intracontinental rifting of a passive continental margin.

ACKNOWLEDGMENTS: We gratefully acknowledge financial assistance from the Geological and Royal Societies of London enabling some overseas speakers to attend the conference. We also thank our contributors for an interesting set of papers and by and large for dealing with editorial requests courteously and promptly, the referees for their prompt and effective treatment of manuscripts, and the secretarial staff of the Department of Earth Sciences at the Open University for their invaluable assistance throughout.

I. G. GASS, S. J. LIPPARD & A. W. SHELTON, Department of Earth Sciences, The Open University, Milton Keynes MK7 6AA, UK.

I. NATURE AND FORMATION OF OCEANIC LITHOSPHERE

MAGMA CHAMBERS: PRODUCTS AND PROCESSES

Evolution of the ocean crust: results from recent seismic experiments

J. A. Orcutt, J. S. McClain & M. Burnett

SUMMARY: We present results from recent analyses of seismic refraction and sea-floor microseismicity studies in the Pacific and Atlantic oceans which lend support to the hypothesis that processes responsible for the construction of ophiolite suites are similar to phenomena extant at mid-ocean ridges. Seismicity at fast-spreading ridges is characterized by very low magnitude (0–1) and shallow (<2–3 km) microearthquakes and long-term oscillations or harmonic tremor. A detailed seismic-refraction experiment on a fast-spreading portion of the East Pacific Rise supports the hypothesis of the existence of a crustal magma chamber. Analyses of these data indicate that the chamber is largely unperturbed by the presence of conjugate spreading centres and the inverted delta-shaped zone of partial melt is characterized by a half width in excess of 6 km. Finally, a new approach for obtaining upper-crustal velocities when applied to data characterized by the presence of shear waves provides several counter-examples to the hypothesis that the shallow crust evolves with time.

During the past few years, marine seismology has proved to be an effective tool in studying the detailed elastic structure of the oceanic crust. Concomitant advances in the quality and quantity of physical properties' data from ophiolite suites has permitted detailed comparisons between these allochthonous terranes and the oceanic lithosphere. A properly testable 'ophiolite hypothesis' for the genesis, evolution and structure of the oceanic crust has arisen from this work. We present two new data sets which, in general, support this hypothesis. These consist of ocean-bottom seismograph seismicity and refraction studies of the East Pacific Rise which require a substantial crustal magma chamber. This chamber, by implication, is responsible for the bulk of the fractionation processes which construct a vertically heterogeneous crustal column. Finally, a recently developed method for accurately determining the compressional velocity of the shallow crust casts some doubt on the evolutionary systematics attributed to 'Layer 2A' and the oceanic crust in general.

Microearthquakes on the East Pacific Rise at 21°N

The fast-spreading East Pacific Rise is strikingly quiescent teleseismically and observed events are clearly confined to transform faults. Microearthquake surveys conducted near the rise crest have been no more successful than the teleseismic studies in detecting events which clearly lie along the spreading axis (Reid *et al.* 1977; Lilwall *et al.* 1981; Prothero & Reid 1982). This inactivity contrasts dramatically with the teleseismic and local observations along the Mid-Atlantic Ridge (e.g. Forsyth 1975; Weidner & Aki 1973; Lilwall *et al.* 1981). Riedesel *et al.* (1982) have recently exploited the growing body of knowledge of spreading-centre fault lengths obtained by detailed surveys on slow- and fast-spreading ridges to place upper bounds on the depth extent of the earthquake fault planes. Whereas faults along the slow-spreading Mid-Atlantic Ridge likely extend through the entire oceanic crust, East Pacific Rise events which escape detection on the worldwide seismic network must be characterized by faults less than 2 km in vertical extent.

Following the successes of the Rivera Submersible Experiment (RISE Project Group 1981) and the discovery of an extraordinarily active hydrothermal field at 21°N on the East Pacific Rise, we planned and executed a microseismicity survey of the rise axis in the vicinity of the hydrothermal vents (Riedesel *et al.* 1982). Five ocean-bottom seismographs (OBS) were deployed within and near (<2 km) the neovolcanic zone and located with an absolute accuracy of better than 100 m. Altogether more than 100 distinct events were recorded in the period 16 July 1980–8 August 1980. These events consisted of regional and local earthquakes as well as phenomena interpreted as volcanic tremor caused by the movement of magma or hydrothermal fluids.

The recorded microearthquakes allowed a geologically interesting estimate of the extreme depth of occurrence. The microearthquakes were extremely impulsive and the measurement of S-P times was easily performed. The earthquakes were quite small, the seismic moments were order 10^{17} dyne-cm corresponding to a magnitude 0–1. For these moments and a stress drop of one bar (Prothero & Reid 1982), the source area was on the order of 35 m^2. Eleven of these microearthquakes were recorded on three or more OBSs and their locations were determined to accuracies from 1 to several km. Even though many of the events appear to be close to the hydrothermal vents within the neovolcanic zone, the probable errors do not permit an unambiguous location at

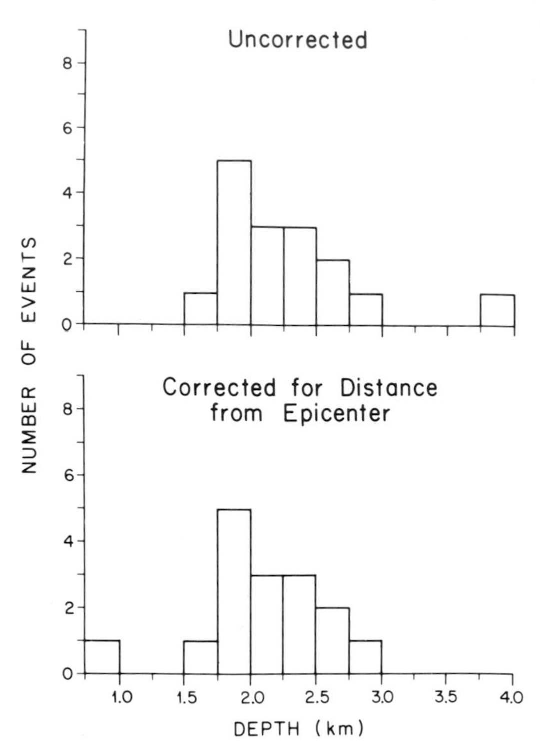

FIG. 1. A histogram of the depths for some earthquakes close to the vents. The S-P time at a given instrument is used to calculate the distance of the event. The top graph assumes that every event recorded at a given seismograph is directly beneath that OBS. The bottom graph corrects this for those few earthquakes with well-determined hypocentral depths.

the vents vis-à-vis the normal faults which bound the axis.

The S-P times alone however, require that the earthquakes be quite shallow. Figure 1a plots the maximum depth for the microearthquakes computed by assuming each event occurred beneath the recording station. Since the locations are not necessary, a large number of events recorded on only one or two instruments can be included in the study. The accurate epicentral locations of a number of these events were known so that their maximum hypocentral depths can be more accurately estimated. When these locations are employed, the corrected plot in Fig. 1b results. A maximum depth of 2–3 km appears to be reasonable for the microearthquakes although the results should not be interpreted to indicate a clustering between depths of 2 and 3 km.

Observed microearthquakes and brittle fracture clearly occurred at depths shallower than 3 km; this may be construed as a maximum depth of hydrothermal circulation and crustal embrittlement. This maximum depth coincides with the depth to the top of the crustal magma chamber at 21°N earlier reported by Reid *et al* (1977). We have closely examined the excitation of Stoneley or interface waves and conclude that the microearthquakes not only occur at depths of less than 3 km, but must occur very near the sea floor.

Two thirds of the events recorded apparently correspond to events interpreted as volcanic tremor. These events were very emergent, rising above the microseismic noise level and decaying, after a few minutes, below this noise level. The spectra were peaked between 3 and 5 Hz, frequencies somewhat higher than observed at subareal volcanoes such as Mt St Helens. The events are devoid of identifiable phases although the seismograms for a given source region were extremely repeatable from day to day. Furthermore, the temporal behaviour of individual events bore a strong resemblance to recordings obtained at Mt St Helens prior to the 18 May 1980 eruption.

This detailed experiment has provided strong evidence for considerable small-scale seismic activity at spreading centres with morphologies characteristic of fast-spreading rates. The seismicity is quite shallow, in contrast with the Mid-Atlantic Ridge, and seismic activity formerly associated only with subareal volcanoes is quite common on the sea floor. These observations are, furthermore, consistent with the presence of a shallow crustal magma chamber previously outlined by seismic-refraction studies and required by the 'ophiolite model' of the oceanic crust.

A detailed rise-crest refraction experiment—the MAGMA expedition

A seismic refraction experiment conducted on the East Pacific Rise at 8°N provided the first geophysical data which could be interpreted in terms of a crustal magma chamber associated with the genesis of the oceanic crust (Orcutt *et al.* 1975). Subsequently, a number of experiments on the East Pacific Rise provided further evidence for such a magma body (Reid *et al.* 1977; Bibee 1979; Herron *et al.* 1980; McClain & Lewis 1980; Hale *et al.* 1982) although similar experiments in the Atlantic were largely negative (Keen & Tramontini 1970; Poehls 1974; Whitmarsh 1975; Francis *et al.* 1977). Bibee (1979), although accumulating strong evidence for a rise-crest magma chamber, interpreted the data as requiring a sub-crustal rather than a crustal magma chamber. Such a structure was not only quite different from previous interpretations, but was not consistent with ophiolite models requiring a

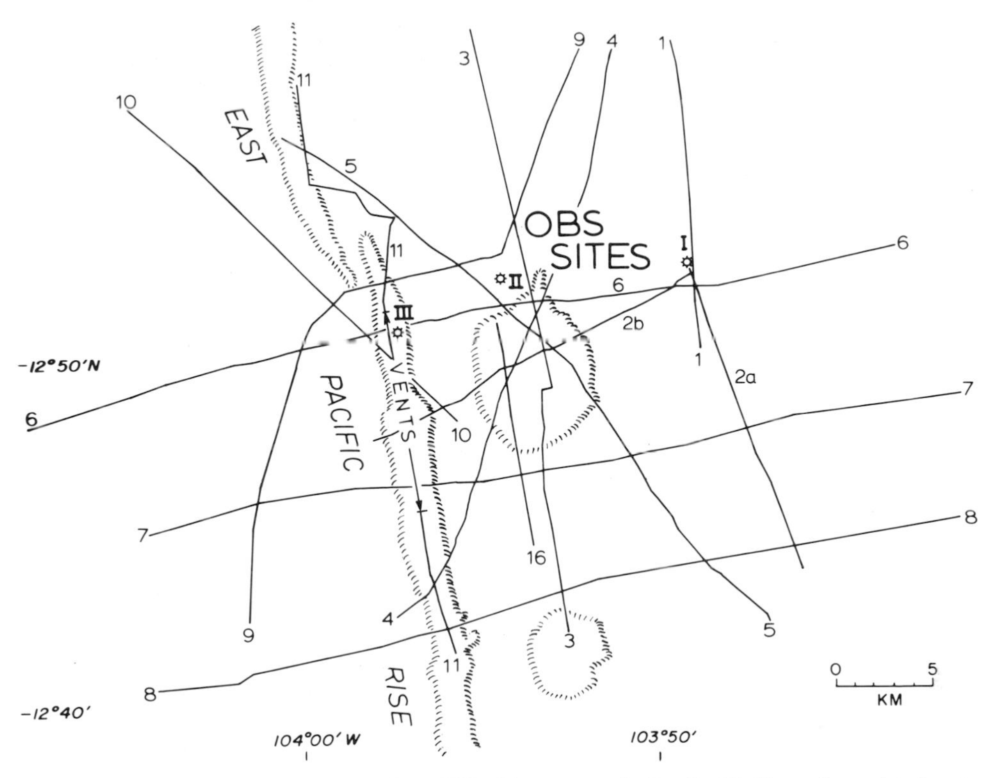

FIG. 2. Location of three OBS sites (I, II, and III) with respect to the East Pacific Rise and associated vent or 'Black Smoker' field. Seamounts are small-scale features less than 150 m in height. The conjugate-spreading centres are outlined near 12°50'N. Topography is adapted from SEABEAM chart provided by R. Hekinian and mapping conducted by R/V Melville during MAGMA expedition. Short lines up to 18 km in length consist of small shots (1–3 lb) fired every 230 m. Lines, originally intended to radiate from the OBS sites, suffer from the usual navigation problems encountered at low latitudes at sea.

crustal magma chamber for fractionation processes. In order to provide a data set capable of unambiguously constraining the depth and lateral extent of the crustal magma chamber, we conducted an extensive seismic-refraction experiment on the East Pacific Rise at 12°50'N in June and July 1982.

The East Pacific Rise near 13°N is characterized by a central high between 1 and 2 km in width. Hydrothermal vents have been observed using bottom photography within this central high between 12°45'N and 12°53'N (R. Hekinian, pers. comm. 1982) and SEABEAM profiling in the area reveals a small discontinuity in the rise axis (Fig. 2). The eastern–southern ridge dies out to the north of the western–northern branch. The overlapping nature belies a classical fracture zone, although the region is roughly in line with the O'Gorman 'Fracture Zone' described by Klitgord & Mammerickx (1982). A number of these small offsets have recently been observed along the East Pacific Rise (Macdonald & Fox 1983) and have been given the name 'conjugate-spreading centres' or 'degenerate transform faults'. The feature of 13°N is the smallest of the degenerate transform faults with an offset of less than 2.5 km. In fact, the feature was noted during the MAGMA expedition while conducting refraction profiles along the rise axis. The nature of the bathymetry was relayed to Macdonald & Fox on board the SEABEAM-equipped R/V Washington and the original French charts were extended to the north to include this sea-floor feature.

The MAGMA experiment utilized OBSs at three sites to record near-surface explosive sources. The sites were located on the ridge (age = 0), 6 km to the east (age = 0.11 Ma) and 16 km east (age = 0.30 Ma) (see Fig. 2). The OBSs were microprocessor-controlled digital instruments which record four components (vertical, two horizontals and a hydrophone) at 128 samples s^{-1} (Moore *et al.* 1981). A total of nine (eight successful) OBS deployments were made at the three sites. The single OBS to fail suffered from an erratic oscillator which caused time to

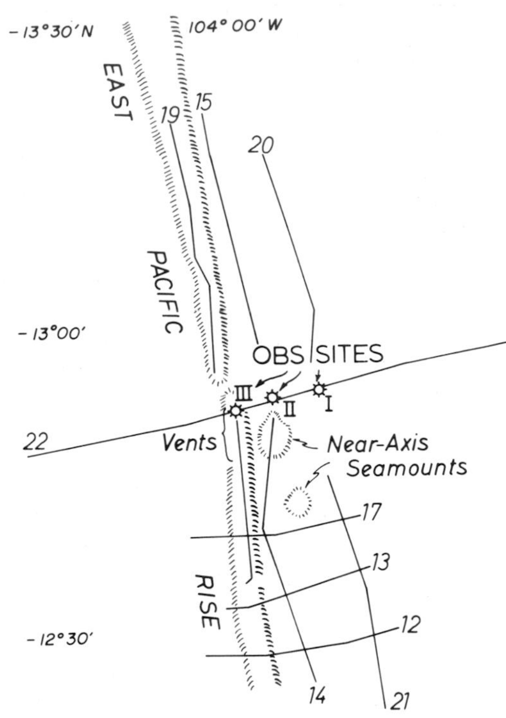

FIG. 3. Long refraction lines using conventional shots to 300 lb in size fired approximately every 0.9 km. The degenerate transform fault is sketched without the observed overlap (see Fig. 2).

advance too rapidly; the OBS was, however, successfully recovered. The redundancy of multiple OBSs at each site was required to store the large volume of data and to provide multiple coverage of the most essential lines. Not only were the OBSs turned on to record lines shot through the instruments, but OBSs at the other sites were frequently turned on to provide a broad three dimensional arrangement of sources and receivers. Each bottom seismograph was programmed to record 30 or 45 s of data at predetermined times. As many as 698 windows were used for an individual OBS, requiring that a very detailed shooting plan be worked out in advance. A total of 2003 shots were fired yielding some 4633 shot receiver pairs or 18,532 individual seismograms.

The refraction experiment, which extended over a 23-day period on station, was broken into two phases. In the first phase, short (15–17 km) lines were shot using small (1–3 lb) charges suspended from floating balloons at a depth of 10 m. This shooting technique provided a strikingly repeatable source. A grid of lines was shot in this manner (Fig. 2) with a shot spacing along each profile of about 230 m. The shots provided a dense areal coverage of the upper crust at the ridge axis and adjacent regions. The second phase of the refraction programme consisted of larger sources (15–300 lb) shot over an expanded two-dimensional grid (Fig. 3) with a lower shot density (about one shot every 900 m).

Data interpretation is in a very early stage, but a number of inferences can already be made. Profiles parallel to the ridge axis (Figs 4 and 5) exhibit dramatic differences between the ages of 0 and 0.30 Ma. The profiles over the older crust (bottom frames in Figs 4 and 5) are characterized by clear, impulsive arrivals to ranges in excess of 50 km. On the other hand, the axial profiles reveal weak emergent arrivals beyond ranges as small as

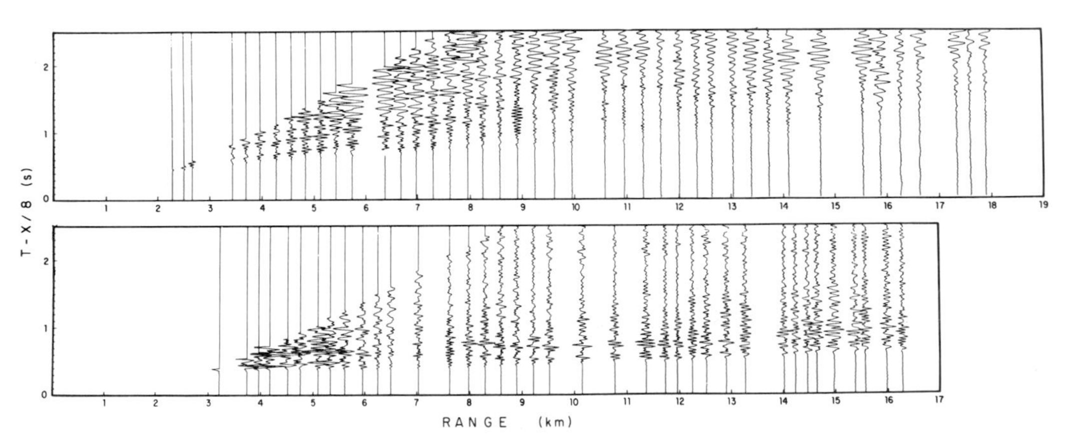

FIG. 4. Short refraction lines extending to a range of 16 km on 0.30-Ma crust and 18 km on zero-age crust. Profiles are reduced at 8 km s^{-1} and amplitudes are scaled as $(x/10)^{1.75}$. No topographic corrections have been applied. When the arrivals are impulsive the bubble pulse of the source is clearly seen. Note the extreme attenuation and delay of arrivals beyond a range of less than 10 km on the rise crest profile when compared to the generally impulsive 0.30-Ma crust.

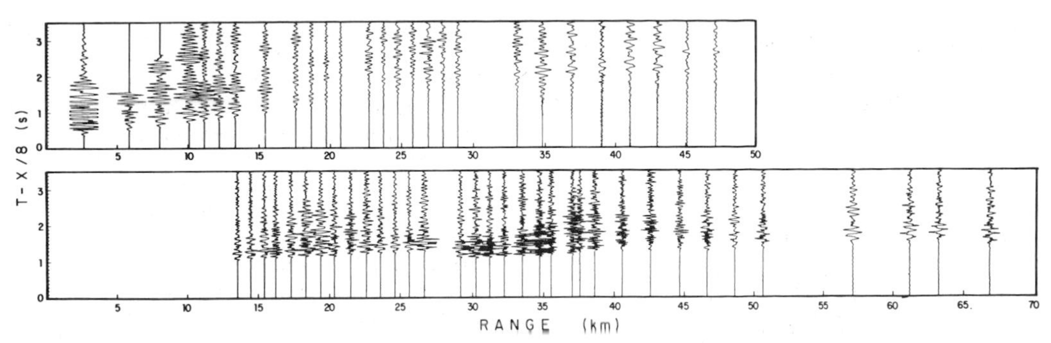

FIG. 5. Long refraction lines extending to a range of 64 km on 0.30-Ma crust and 45 km on zero-age crust. Profiles are plotted as in Fig. 4. Again note the significant attenuation of the zero-age profile vis-à-vis the profile collected on crust with an age of 0.30 Ma. The frequency content of the rise-crest line is restricted to quite low frequencies.

7 km. The contrast could not be the result of attenuation exclusively, but requires a low-velocity medium at a fairly shallow depth beneath the sea floor.

The effect of a crustal low-velocity zone is shown in Figs 6 and 7. The velocity–depth profile in Fig. 6 is the FF2 model of Spudich & Orcutt (1980). The ray-trace diagram and accompanying travel-time curve illustrate the effects of high-velocity gradients in the shallow and deep crust. Few rays turn within the fairly homogeneous 'plutonic' lower crust so we would expect the arrivals from this portion of the crust to be relatively weak. Fig. 7 illustrates the effect of introducing a shallow low-velocity zone into the structure. Now, as rays penetrate deeper than approximately 1 km into the crust, the wavefronts are refracted downward away from the sea floor and a substantial 'shadow zone' is generated. The expected effect would be a significant attenuation of seismograms beyond some cut-off distance. We propose this effect is responsible for the rapid attenuation of amplitudes in Figs 4 and 5 beyond ranges of 7 km. This extremely short range

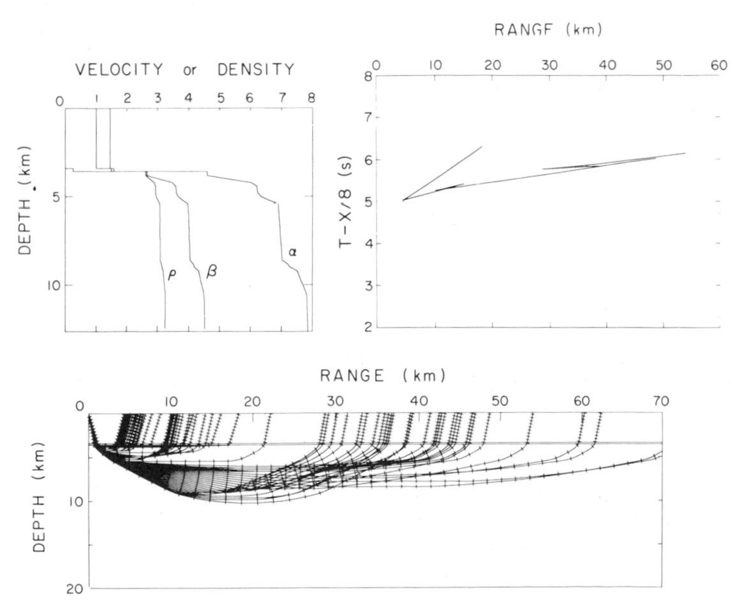

FIG. 6. Ray-trace diagram corresponding to the elastic profiles plotted in upper left-hand corner. The velocity–depth profile is adapted from the work of Spudich & Orcutt (1980). The large triplication in the travel-time curve in the upper right-hand corner is caused by the Moho transition. Note rays return to the surface along the entire length of the profile to produce profiles similar to the 0.30 Ma crustal sections in Figs 4 and 5.

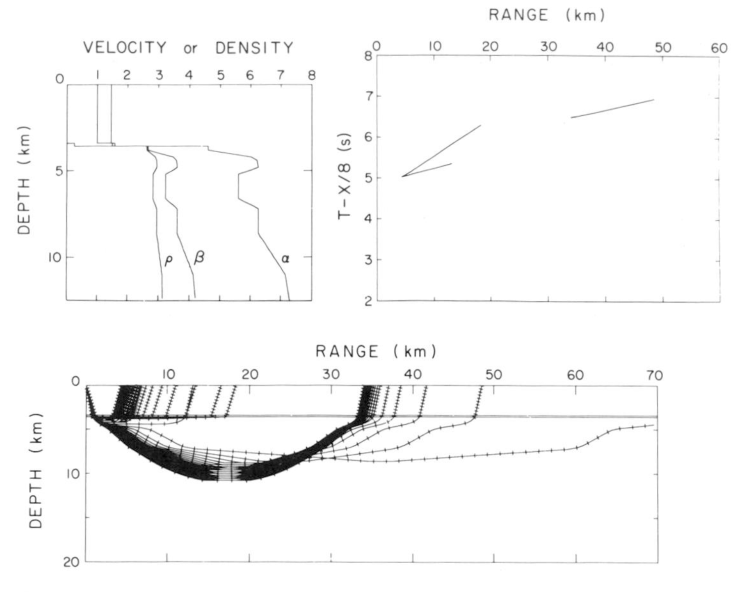

FIG. 7. Figure plotted as in Fig. 6 except the velocity–depth profile includes a pronounced low-velocity zone. The ray-trace diagram illustrates the downward refraction of rays as they touch and enter the low-velocity zone. Note the 'shadow zone' and delay introduced in the travel-time curve. The cut-off of clear, impulsive arrivals at the shadow is diagnostic of the depth to the top of the low-velocity zone.

requires that the magma chamber responsible for the low-velocity zone be very shallow in the crust, perhaps no deeper than 1, at most 2 km.

The data collected on lines parallel to the rise axis at a distance of 6 km most closely resemble the 0.30-Ma profiles in that distant arrivals are quite clear and impulsive. However, as illustrated in Fig. 8, a clear shadow zone and arrival delay are encountered at ranges on the order of 7–11 km. We currently interpret this to indicate

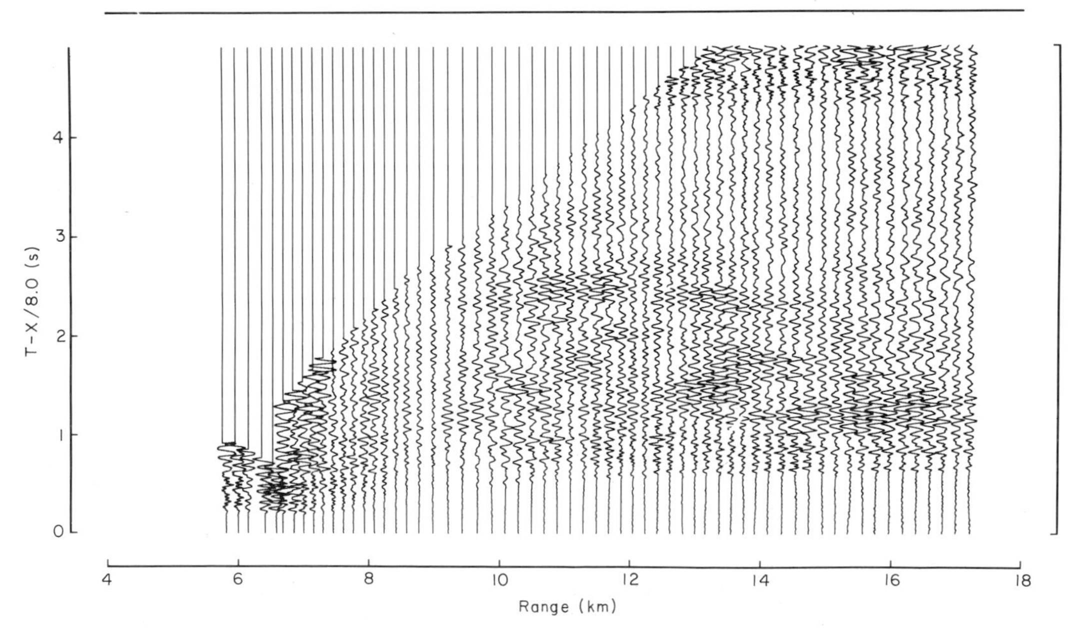

FIG. 8. Short refraction lines parallel to the rise axis on 0.11-Ma crust. Note the pronounced shadow zone appearing at a range of 8–11 km and the relatively impulsive arrivals at greater ranges.

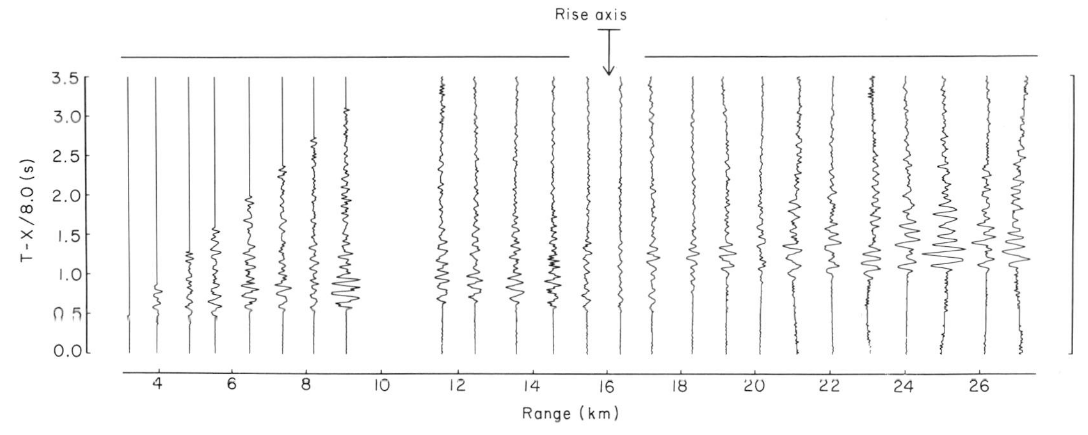

FIG. 9. Refraction line crossing rise axis as recorded on the hydrophone channel from Site I 16 km to the east. Note the time delay and change in wave form which occurs about two shots beyond the rise axis.

the solidification of the lower portion of the chamber as a function of age with a distal wing of melt remaining shallow in the crust.

Data on the cross-ridge profile, Fig. 9, indicate a much more subtle effect over the rise—i.e. a small time delay which appears to result from anomalous material in the lower crust. Examination of data from these various profiles indicates the low-velocity zone (magma chamber) is somewhat greater than 6 km in half width and morphologically resembles the model proposed by Pallister & Hopson (1981) based on observations of the Semail ophiolite in Oman.

Evolution of the oceanic crust

Several studies in the past have provided evidence that the oceanic crust evolves as a function of age. Le Pichon *et al.* (1965), Shor *et al.* (1971), Goslin *et al.* (1972), Christensen & Salisbury (1975) and Woollard (1975) conducted studies which indicated that the oceanic crust thickened regularly with age. Mechanisms proposed to explain the thickening included regular ongoing intrusion of the lower crust from below (Christensen & Salisbury 1975) and serpentinization of the uppermost mantle (Lewis 1978). The ophiolite model for the oceanic crust, on the other hand, presumes that the crust is formed at the rise axis and moves away with the underlying lithosphere. Although intraplate volcanism is widespread there is no good evidence for the regular migration of basaltic melts to the base of the crust. Substantial serpentinization is unlikely since little evidence exists for large volumes of hydrothermal circulation within the lower oceanic crust.

McClain (1981) has reviewed available evidence for crustal thickening and, after a careful winnowing of the data set employed by most of the authors cited above, concluded there is not compelling evidence for crustal thickening. He pointed out that the bulk of the thin crustal solutions at small ages were clustered north of the Mendocino fracture zone where the slow-spreading Gorda Rise has historically produced very thin oceanic crust. He eliminated these solutions from the data set asserting that any thickening pattern should persist when the remainder of the extensive data set was examined. He also eliminated stations which were directly over fracture zones, seamounts or the volcanoes of the Hawaiian Archipelago. Finally McClain pointed out that the non-tangency of the ground waves and the bottom-reflected, water-borne phases were usually corrected assuming the delay was caused by propagation through sediments. This was done (recall that many of the profiles were collected in the 1940s and 1950s before sea-floor spreading was understood) even when the stations were near the rise crests where little sediment can be found. McClain corrected his data set for this effect by recomputing crustal thicknesses for crust less than 6 m.y. of age where sedimentation rates are very low. Rather than use the assumed sediment velocity of 2.15 km s^{-1} he substituted a 'Layer-2A' velocity of 3.5 km s^{-1}. This, of course, increased the total crustal thickness. The compound effect of these very reasonable modifications was to cast considerable doubt on the hypothesis of crustal thickening beyond an age of a few million years.

Spudich & Orcutt (1980) pointed out that the compressional velocity of the sea floor acts as a notch filter for the transmission and conversion of shear waves. Figure 10 depicts the compres-

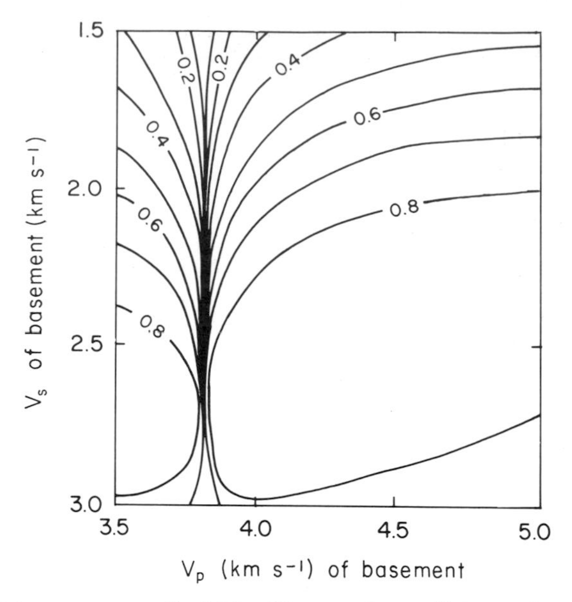

FIG. 10. Magnitude of the energy normalized PS or SP conversion coefficient at the sediment–basement interface. Phase velocity of the incident wave is 3.8 km s^{-1}, a typical crustal S-wave velocity. A basement velocity of 3.8 km s^{-1} will clearly filter out crustal shear waves with this phase velocity.

sional to shear conversion coefficient for an ocean overlying the basaltic crust where the phase velocity of the incident plane wave is taken as 3.8 km s^{-1}. The coefficient is quite large for basement velocities above and below 3.8 km s^{-1} and for high shear velocities. However, a notch appears near 3.8 km s^{-1} and broadens as the shear velocity drops. Clearly shear waves cannot exist in the crust for phase velocities near the compressional velocity of the sea floor. Spudich & Orcutt (1980) exploited this notch filter to constrain the compressional velocity of the shallow crust in this study. The compressional velocity was required to exceed 4.3 km s^{-1} and values in the range 4.4–4.6 km s^{-1} provided the best results.

We have subsequently applied this technique to the analysis of a set of Marine Seismic System (MSS) borehole data collected in the Atlantic during Leg 78B of the Deep Sea Drilling Project. These data were characterized by large amplitude shear waves and the compressional velocity at the site, on 9-Ma crust, was 4.4 km s^{-1}. Clowes & Au (1982) were also able to use the notch-filter effect to resolve high compressional velocities near 4.4–4.7 km s^{-1} at the sea floor for small crustal ages of 1–3 Ma.

Houtz & Ewing (1976) previously published an extensive treatment of sonobuoy data which outlined a systematic increase of shallow crustal velocity (Layer 2A) as a function of time. The crust near the rise axis is, statistically, typified by a velocity of near 3.3 km s^{-1}, while crust at an age of 35 Ma possesses a velocity of 4.5 km s^{-1}. The increase in velocity is attributed to a decrease in rock porosity by infilling with sediments and alteration products produced by hydrothermal circulation.

Figure 11 is modified from Fig. 8 of Houtz & Ewing (1976) where the triangles (Pacific) and squares (Atlantic) represent averaged values for crust of various ages. The straight line is a least-squares' regression of these data and indicates that the Layer-2A compressional velocity does, indeed, increase with age. However, the results from the studies of Spudich & Orcutt (1980) and Au (1981) and the MSS data, when plotted on the same figure, lie at much higher velocities. These results are inconsistent with the ageing hypothesis and serve as counter-examples to the interpretation of the sonobuoy data. Furthermore, recent ocean-crust, multichannel expanding-spread data collected during the Lamont Doherty, University of Texas, University of Rhode Island, Woods Hole Oceanographic Institution TRANSECT programme between the Mid-Atlantic Ridge and Blake-Bahama Plateau are typified by strong shear wave propagation from short ranges to ranges in excess of the mantle triplication (Peter Buhl, pers. comm. 1982). These observations will add considerably to this data set which is very sensitive to the upper-crustal velocity and will

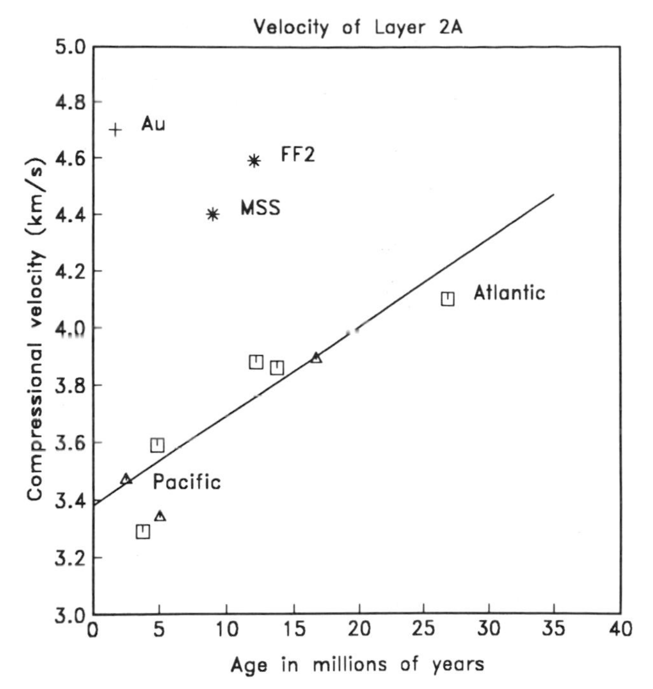

FIG. 11. Regressions of 'Layer-2A' velocities in the Pacific (triangles) and Atlantic (squares) as a function of age from Houtz & Ewing (1976). Asterisks and plus signs are taken from studies cited in the text. These latter velocities serve as well-constrained counter-examples to the velocity–age systematics proposed for 'Layer-2A'.

provide a clearer picture of the nature of the evolution of the shallow oceanic crust. Nevertheless, it has been possible to document sites in which the simple Layer-2A ageing phenomenon does not apply and the previously apparent variation of upper-crustal velocity with age may more aptly be described by regional or even local variations in crustal genesis.

ACKNOWLEDGMENTS: This research was sponsored by the National Science Foundation under grants OCE81-08971 and OCE79-26438. Analytical methods used in the analysis of these data were developed under the sponsorship of the Office of Naval Research. Participation in the Ophiolites and Oceanic Lithosphere meeting was made possible through the Office of Naval Research and the Geological Society of London. We thank the captain and crew of the R/V Melville for the splendid support given to the OBS group during both expeditions. Roger Hekinian made the recent refraction work possible by kindly providing us with R/V Charcot SEABEAM maps of the MAGMA Expedition area.

References

AU, D. 1981. *Crustal Structure from an Ocean Bottom Seismometer Survey of the Nootka Fault Zone*, Ph.D. Thesis, University of British Columbia, 131 pp.

BIBEE, L. D. 1979. *Crustal Structure in Areas of Active Crustal Accretion*. Ph.D. Thesis, University of California, San Diego, 155 pp.

CHRISTENSEN, N. I. & SALISBURY, M. H. 1975. Structure and constitution of the lower oceanic crust. *Rev. Geophys. and Space Phys.* **13**, 57–86.

CLOWES, R. M. & AU, D. 1982. *In-situ* evidence for a low degree of S-wave anisotropy in the oceanic upper mantle. *Geophys. Res. Lett.* **9**, 13–6.

FRANCIS, T. J. G., PORTER, I. T. & MCGRATH, J. R. 1977. Ocean bottom seismograph observations on the Mid-Atlantic Ridge near 37°N. *Bull. Geol. Soc. Am.* **88**, 664–77.

FORSYTH, D. W. 1975. Fault plane solutions and tectonics of the South Atlantic and Scotia Sea. *J. Geophys. Res.* **80**, 1429–43.

GOSLIN, J., BAUZART, P., FRANCHETEAU, J. & LE PICHON, X. 1972. Thickening of the oceanic layer in the Pacific Ocean. *Mar. Geophys. Res.* **1**, 418–27.

HALE, L. D., MORTON, C. J. & SLEEP, N. H. 1982. Reinterpretation of seismic reflection data over the East Pacific Rise. *J. Geophys. Res.* **87**, 7707–17.

HERRON, T. J., STOFFA, P. L. & BUHL, P. 1980. Magma chamber and mantle reflections—East Pacific Rise. *Geophys. Res. Lett.* **7**, 989–92.

HOUTZ, R. & EWING, J. 1976. Upper crustal structure as a function of plate age. *J. Geophys. Res.* **81**, 2490–98.

KEEN, C. E. & TRAMONTINI, C. 1970. A seismic refraction survey on the Mid-Atlantic Ridge. *Geophys. J. R. astr. Soc.* **20**, 473–91.

KLITGORD, K. D. & MAMMERICKX, J. 1982. Northern East Pacific rise: magnetic anomaly and bathymetric framework. *J. Geophys. Res.* **87**, 6725–50.

LE PICHON, X., HOUTZ, R. E., DRAKE, C. L. & NAFE, J. E. 1965. Crustal structure of the mid-ocean ridges. I. Seismic refraction measurements. *J. Geophys. Res.* **70**, 319–40.

LEWIS, B. T. R. 1978. Evolution of ocean crust seismic velocities. *Ann Rev. Earth Planet. Sci.* **6**, 377–404.

LILWALL, R. C., FRANCIS, T. J. G. & PORTER, I. T. 1981. A microearthquake survey at the junction of the East Pacific Rise and the Wilkes (9°S) fracture zone. *Geophys. J. R. astr. Soc.* **66**, 407–16.

MACDONALD, K. C. & FOX, P. J. 1983. Overlapping spreading centres: new accretion geometry on the East Pacific Rise. *Nature, Lond.* **302**, 55–8.

MCCLAIN, J. S. 1981. On long-term thickening of the oceanic crust. *Geophys. Res. Lett.* **8**, 1191–94.

—— & LEWIS, B. T. R. 1980. A seismic experiment at the axis of the East Pacific Rise. *Marine Geology*, **35**, 147–69.

MOORE, R. D., DORMAN, L. M. HUANG, C. Y. & BERLINER, D. L. 1981. An ocean bottom microprocessor based seismometer. *Mar. Geophys. Res.* **4**, 457–77.

ORCUTT, J. A., KENNETT, B. L. N., DORMAN, L. M. & PROTHERO, W. A. 1975. Evidence for a low-velocity zone underlying a fast-spreading rise crest. *Nature, Lond.* **256**, 475–76.

PALLISTER, J. S. & HOPSON, C. A. 1981. Samail ophiolite plutonic suite: field relations, phase variation, cryptic variation and layering, and a model of a spreading ridge magma chamber. *J. Geophys. Res.* **86**, 2593–2644.

POEHLS, K. 1974. Seismic refraction on the Mid-Atlantic Ridge at 37°N. *J. Geophys. Res.* **79**, 3370–73.

PROTHERO, W A. & REID, I. D. 1982. Microearthquake on the East Pacific Rise at 21°N and the Rivera fracture zone. *J. Geophys. Res.* **87**, 8509–18.

REID, I. D., ORCUTT, J. A. & PROTHERO, W. A. 1977. Seismic evidence for a narrow zone of partial melting underlying the East Pacific Rise at 21°N. *Bull. Geol. Soc. Am.* **88**, 678–82.

RIEDESEL, M., ORCUTT, J. A., MACDONALD, K. C. & MCCLAIN, J. S. 1982. Microearthquakes in the black smoker hydrothermal field, East Pacific Rise at 21°N. *J. Geophys. Res.* **87**, **10**, 613–23.

RISE PROJECT GROUP 1980. East Pacific Rise: hot springs and geophysical experiments. *Science*, **207**, 1421–33.

SHOR, G. G., JR, MENARD, H. W. & RAITT, R. W. 1971. Structure of the Pacific Basin. *In*: MAXWELL, A. E. (ed.) *The Sea, 4, Part II*, J. Wiley & Sons, pp. 3–27.

SPUDICH, P. & ORCUTT, J. A. 1980. Petrology and porosity of an oceanic crustal site: results from wave form modelling of seismic refraction data. *J. Geophys. Res.* **85**, 1409–1433.

WEIDNER, P. J. & AKI, K. 1973. Focal depth and mechanism of mid-ocean ridge earthquakes. *J. Geophys. Res.* **78**, 1818–81.

WHITMARSH, R. B. 1975. Axial intrusion zone beneath the median valley of the Mid-Atlantic Ridge at 37°N detected by explosion seismology. *Geophys. J. R. astr. Soc.* **42**, 189–215.

WOOLLARD, G. P. 1975. The interrelationships of crustal and upper mantle parameter values in the Pacific. *Rev. Geophys. and space Phys.* **13**, 87–137.

J. A. ORCUTT, Geological Research Division (A-015), Scripps Institution of Oceanography, La Jolla, CA 92093, USA.

J. S. MCCLAIN, Department of Geology, University of California, Davis, CA 95616, USA.

M. BURNETT, Geological Research Division (A-015), Scripps Institution of Oceanography, La Jolla, CA 92093, USA.

Depths and temperatures of mid-ocean-ridge magma chambers and the composition of their source magmas

M. R. Fisk

SUMMARY: Electron-microprobe analyses of submarine basalt glass and coexisting olivine from the Galapagos Rise and the Mid-Atlantic Ridge (60–63°N) reveal that olivine phenocrysts are virtually unzoned and olivine compositions are identical to those predicted by assuming that the olivine crystallized at the liquidus temperature of the host glass. The absence of chemical zoning of olivine indicates that their host magmas must have cooled by less than 15°C and the MgO/FeO of the magma must have been nearly constant (less than 3 wt% olivine removed) during the time that olivine phenocrysts existed in the magma. Olivine phenocrysts re-equilibrate with their host magma relatively quickly and any changes in temperature or composition that occurred in the 24 h previous to eruption would be recorded in the olivine composition. The settling rates of 1-mm olivine phenocrysts are too slow compared to magma ascent rates to allow early-formed olivine to settle out. There, many magmas that erupt at mid-ocean-ridge crests must ascend from their final reservoir nearly isothermally and isochemically, or alternatively they ascend at superliquidus temperatures and cool below the liquidus just prior to eruption. In either case the magmas identify the temperature (±15°C) and the composition of the magma reservoir immediately prior to eruption. The depth of these magma reservoirs may be estimated through experimental and empirical determination of the pressure–temperature phase diagrams of oceanic basalts. For Galapagos Rise and Mid-Atlantic Ridge basalts these reservoirs are at 3–25 km beneath the ridge crust and in some cases individual reservoirs may be located within ±3 km.

The mineral and liquid chemistry of magmas that erupt at mid-ocean ridges can be used to determine the temperatures and depths of mid-ocean-ridge magma chambers and to reveal the chemistry of the magmas contained in these magma chambers. This same mineral and liquid chemistry can, however, tell us little about the composition of the magmas that are fed into the magma chambers from the mantle below. The evidence presented here indicates that magmas that reach the sea floor are virtually identical to the magmas that emerge from the magma chambers, but are quite different from the magmas that are injected into the chambers from deeper in the mantle.

The basalts considered in this study are from the Mid-Atlantic Ridge from 60 to 63°N (Reykjanes Ridge) and from the Galapagos Spreading Center from 85 to 100°W. Fig. 1 shows the locations of the dredged basalts from both areas. The major and trace-element chemistry of these basalts and their phenocrysts have been examined by a number of authors (Schilling 1973; Hermes & Schilling 1976; Fisk *et al.* 1982; Schilling *et al.* 1982; Verma & Schilling 1982). The chemistry and abundance of the phenocrysts and host glass are used here to determine the liquidus temperatures, pressures of phenocryst formation and the eruption kinetics of basalts in these two areas.

Magma temperatures

The liquidus temperatures of sixteen Mid-Atlantic Ridge basalts were measured experimentally (Fisk *et al.* 1980) and the liquidus temperatures of the nine basalts from the Reykjanes Ridge are shown in Fig. 2. The olivine geothermometer of Roeder & Emslie (1970) accurately predicts the olivine liquidus temperatures of oceanic basalts as has been demonstrated by a number of melting experiments on MORBs (Bender *et al.* 1978; Dungan *et al.* 1978; Fisk *et al.* 1980; Fujii *et al.* 1978; Fukuyama & Hamuro 1978; Walker *et al.* 1979). For the basalts from the Reykjanes Ridge the calculated and measured liquidus temperatures are within 15°C. This same geothermometer was applied to Galapagos Spreading Center basalts (Fisk *et al.* 1982) and the results are shown in Fig. 3. These calculated liquidus temperatures are therefore probably within a few degrees of the actual liquidus temperatures. For these temperature calculations a ferric–ferrous ratio of 0.135 was assigned to the basalts based on the wet chemical analyses of ferric and ferrous iron in these same basalts (Schilling *et al.* 1982).

There are several interesting features about the calculated and measured temperatures of these ocean-ridge basalts. The maximum calculated or measured liquidus temperature from the Reykjanes Ridge or Galapagos Spreading Center is 1230°C and the minimum temperature is about 1150°C. The most primitive erupted MORBs have liquidus temperatures of no more than 1270°C (Bender *et al.* 1978). Erupted magmas are therefore limited to the small temperature range of 120°C. Another interesting feature of the temperatures is that they are highly variable near fracture zones but quite uniform over 'normal'

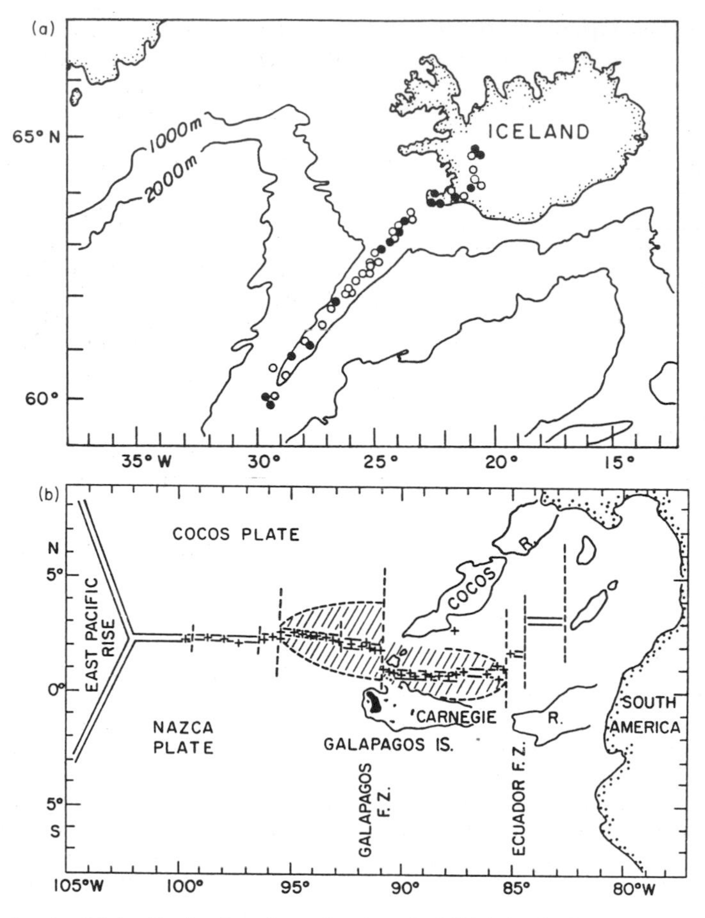

FIG. 1. Maps showing (a) the distribution of samples along the Mid-Atlantic Ridge and (b) the Galapagos Spreading Center. The major geographical features of the two areas are outlined on the maps. All samples used in this study are dredged pillow basalts from the ridge axes of these ridges. ○ ●, Reykjanes Ridge and +, Galapagos Spreading Center.

segments of the ridge, and there may be a tendency for magmas to erupt at lower temperatures nearer to Iceland and to the East Pacific Rise. Temperatures over the Galapagos Platform are highly variable, possibly reflecting the thickness of the crust and the complexity of the plumbing system. Over the 'normal' ridge, temperatures are much less variable suggesting that steady-state conditions exist there.

Depths to magma chambers

The depths of magma chambers can be estimated if first the observation is made that fractionation of most mid-ocean-ridge magmas appears to involve the removal of olivine, plagioclase and augite (e.g. Byerly 1980; Clague *et al.* 1981; Flower & Robinson 1981a, b; Ludden *et al.* 1980; O'Donnell and Presnall 1980; Thompson *et al.* 1980) and if it is therefore assumed that the magmas in most mid-ocean-ridge magma chambers are saturated with these three minerals even though augite is not a 1-atmosphere liquidus phase due to decompression. The pressure (depth) of three-phase saturation has been estimated by the method of Fisk *et al.* (1980). First, the pressure–temperature phase diagrams of the basalts were determined once the 1-atmosphere melting points of olivine, plagioclase and augite were measured by experiment (Fisk *et al.* 1980) or calculated by the computer program 'MAGMAS' (C. Ford pers. comm.) if experimental data were not available. The $\Delta T° \Delta P(\text{kb})^{-1}$ of olivine, plagioclase and augite that were used in the calculations were 2°C kb^{-1}, 6°C kb^{-1} and 10°C kb^{-1}, respectively. The depth at which a magma would be saturated with all three phases is taken as the depth of the magma chamber.

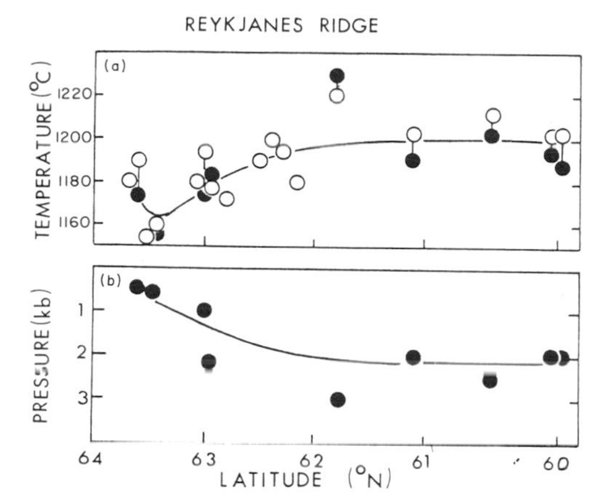

FIG. 2. (a) Experimentally measured ● (Fisk *et al.* 1980) and calculated ○ (Hermes & Schilling 1976) liquids temperatures along the Mid-Atlantic Ridge from 60–63.7°N (Reykjanes Peninsula). (b) Calculated depths of three-phase saturation of nine basalts for which the 1-atmosphere phase relations are known.

Basalts that contain olivine, plagioclase and augite must have come from at least this depth if the phenocrysts are cognate. The magma chambers could be deeper than this but small increases in pressure will greatly increase the crystal content of the magmas at a given temperature. Since these magmas have only a few percent phenocrysts (Schilling *et al.* 1982), the depth of the magma chambers is probably close to the depth of three-phase saturation. Basalts that have olivine alone or with plagioclase could have come from the depth of three-phase saturation but, if they did, augite would have to be resorbed before the magmas reached the surface.

The depths to magma chambers that were calculated using measured 1-atmosphere crystallization temperatures of the Reykjanes Ridge basalts are given in Fisk *et al.* (1980) and Fig. 2. The depth of three-phase saturation is shown for Galapagos Spreading Center in Fig. 3.

For basalts erupted along 'normal' ridge segments of the Reykjanes Ridge, the depth to three-phase saturation is ⩾6 km, but near Iceland the depth is 3–6 km. For the Galapagos Spreading Center most basalts from the 'normal' ridge segments have three-phase saturation depths greater than 6 km. Near the centre of the Galapagos Platform (91°W) the depths are highly variable but outside this central area the depths tend to be less than 6 km.

Kinetics of eruption

Figure 4b shows the composition of olivines in thirty-five basalts from the Galapagos Spreading Center. The interesting feature of these olivine phenocrysts and those from the Reykjanes Ridge (Hermes & Schilling 1976; Fisk 1978) is that most of them are virtually unzoned compared to fresh basalts from other oceanic volcanic environments such as ocean island or island arcs. (Compare the range of zoning in Reunion Island basalts (Fig. 4a) with those of the Galapagos Spreading Center.) The absence of zoning of olivine phenocrysts in submarine basalts appears to be an ubiquitous feature.

The cause of the difference in degree of zoning between ocean-floor and ocean-island basalts is probably one of kinetics because the amount of

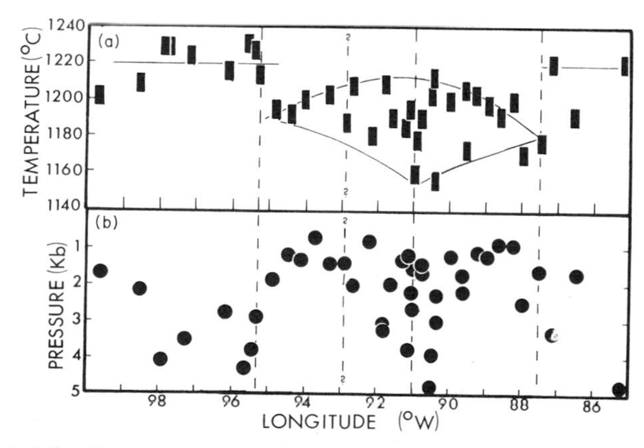

FIG. 3. (a) Calculated liquidus temperatures of glass exteriors of pillow basalts from the Galapagos Spreading Center. Vertical dashed lines represent fracture zones or the tips of propagating rifts. Horizontal lines extend over the regions of 'normal' oceanic crust. Envelope encloses temperatures of basalts from the Galapagos Platform high-magnetic intensity region. (b) Calculated pressure of three-phase saturation of the Galapagos Spreading Center pillow basalt glasses.

plagioclase crystallized. A lower limit of the CaO/Na_2O of the host liquid can be determined from the normative anorthite/albite of the liquid. For plagioclase of An_{90} to An_{95} the lower limit of CaO/Na_2O is 10.0. Fig. 6 shows that most choices of parental liquids of MORBs do not have CaO/Na_2O large enough to crystallize An_{90} plagioclase. The types of material that do have high enough CaO/Na_2O are mafic segregations in ophiolites and komatiites. Komatiites, however, have FeO/MgO that is too low to produce olivine of $Fo_{92.5}$ or less.

Conclusion

The absence of zoning in olivine phenocrysts in oceanic basalts indicates that during the time that magmas are carrying olivine they do not cool more than 15°C. The presence of phenocrysts of augite and plagioclase that are not stable at the 1-atmosphere liquidus temperature at which they erupt indicates that the magmas come from three-phase saturated magma chambers at depth. These facts allow us to calculate the depth, temperature and composition of mid-ocean-ridge magma chambers, and we find that they range in temperature from 1150 to 1270°C. The depths of the magma chambers appear to be variable but in general they are closer to the surface on oceanic platforms than on 'normal' ridge segments.

The magmas that erupt from the magma chambers are probably mixed, steady-state end-products (O'Hara & Mathews 1981) and can only tell us something of the parent magmas that were injected into the bottom of the magma chamber by the phenocrysts that they carry. The most primitive olivine and plagioclase phenocrysts indicate that the parent magmas had an FeO/MgO of 0.45–0.75 and a CaO/Na_2O greater than 10.0. The only ultramafic rocks that appear to fit these criteria are mafic segregates found in ophiolites and komatiites that have undergone olivine fractionation.

References

Basaltic Volcanism Study Project 1981. *Basaltic Volcanism in the Terrestrial Planets*. Pergamon Press, New York, 1286 pp.

Bender, J. F., Hodges, F. N. & Bence, A. E. 1978. Petrogenesis of basalts from the Project FAMOUS area: experimental study from 0 to 15 kbars. *Earth planet. Sci. Lett.* **41**, 277–302.

Bottinga, Y. & Weill, D. F. 1970. Densities of liquid silicate systems calculated from partial molar volumes of oxide components, *Am. J. Sci.* **269**, 169–82.

Bryan, W. B. 1979. Regional variation and petrogenesis of basalt glasses from the FAMOUS area, Mid-Atlantic Ridge, *J. Petrol.* **20**, 293–325.

Byerly, G. 1980. The nature of differentiation trends in some volcanic rocks from the Galapagos spreading centre, *J. Geophys. Res.* **85**, 3797–810.

Clague, D. A., Frey, F. A. Thompson, G. & Rindge, S. 1981. Minor and trace element geochemistry of volcanic rocks dredged from the Galapagos Spreading Center: role of crystal fractionation and mantle heterogeneity, *J. Geophys. Res.* **86**, 9469–82.

Donaldson, C. H. & Brown, R. W. 1977. Refractory megacrysts and magnesium-rich melt inclusions within spinel in oceanic tholeiites: indicators of magma mixing and parental magma composition. *Earth planet. Sci. Lett.* **37**, 81–9.

Drake, M. J. 1972. *The distribution of Major and Trace Elements between Plagioclase Feldspar and Magmatic Silicate Liquid: an Experimental Study*, Ph.D. Thesis, University of Oregon, 190 pp.

Dungan, M. A., Long, P. E. & Rhodes, J. M. 1978. The petrography, mineral chemistry, and one-atmosphere phase relations of basalts from site 395. Melson, W. G., Rabinowitz, P. D. *et al.* (eds). *Init. Rep. Deep Sea Drill. Proj.* **45**, 461–72.

Elthon, D. 1979. High magnesia liquids as the parental magma for ocean floor basalts, *Nature, Lond.* **278**, 514–8.

Fisk, M. R. 1978. *Melting Relations and Mineral Chemistry of Iceland and Reykjanes Ridge Tholeiites*. Ph.D. Thesis, University of Rhode Island, 254 pp.

——, Bence, A. E. & Schilling, J.-G. 1982. Major element chemistry of Galapagos Rift Zone magmas and their phenocrysts. *Earth Planet. Sci. Lett.* **61**, 171–89.

——, Schilling, J.-G. & Sigurdsson, H. 1980. An experimental investigation of Iceland and Reykjanes Ridge tholeiites. I. Phase relations. *Contrib. Mineral. Petrol.* **74**, 361–74.

Flower, M. F. J. & Robinson, P. T. 1981a. Basement drilling in the Western Atlantic Ocean. I. Magma fractionation and its relation to eruptive chronology, *J. Geophys. Res.* **86**, 6273–98.

—— & —— 1981b. Basement drilling in the Western Atlantic Ocean. II. Synthesis of construction processes at the Cretaceous ridge axis, *J. Geophys. Res.* **86**, 6299–309.

Frey, F. A., Bryan, W. A. & Thompson, G. 1974. Atlantic ocean floor: geochemistry and petrology of basalts from legs 2 and 3 of the Deep Sea Drilling Project, *J. Geophys. Res.* **79**, 5507–27.

Fujii, T., Kushiro, I. & Hamuro, K. 1978. Melting relations and viscosity of an abyssal olivine tholeiite. *Init. Rep. Deep Sea Drill. Proj.* Melson, W. G., Rabinowitz, P. D. *et al.* (eds). **45**, 513–7.

Fukuyama, H. & Hamuro, K. 1978. Melting relations of leg 46 basalts at atmospheric pressure. *Init. Rep. Deep Sea Drill. Proj.* Dmitriev, L., Heirtzler, J. *et al.* (eds). **45**, 235–9.

Gibb, F. G. F. 1970. Crystal-liquid relationships in some ultrabasic dykes and their petrographic significance. *Contrib. Mineral. Petrol.* **30**, 103–18.

Hermes, O. D. & Schilling, J.-G. 1976. Olivine from

Reykjanes Ridge and Iceland tholeiites, and its significance to the two mantle source model. *Earth planet. Sci. Lett.* **28**, 345–55.
LANGMUIR, C. H., BENDER, J. F., BENCE, A. E., HANSON, G. N. & TAYLOR, S. R. 1977. Petrogenesis of basalts from the FAMOUS area: Mid-Atlantic Ridge. *Earth planet. Sci. Lett.* **36**, 133–56.
—— & HANSON, G. N. 1981. Calculating mineral-melt equilibria with stoichiometry, mass balance and single-component distribution coefficients. NEWTON, R. C., NAVROTSKY, A. & WOOD, B. J. (eds.) *Thermodynamics of Minerals and Melts.* Springer-Verlag, New York, pp. 247–71.
LUDDEN, J. N., THOMPSON, G., BRYAN, W. B. & FREY, F. A. 1980. The origin of lavas from the ninety east ridge, Eastern Indian Ocean: an evaluation of fractional crystallization models. *J. Geophys. Res.* **85**, 4405–20.
MELSON, W. G., VALLIER, T. L., WRIGHT, T. L., BYERLY, G. & NELEN, J. 1976. Chemical diversity of abyssal volcanic glasses erupted along Pacific, Atlantic and Indian ocean sea-floor spreading centers. *In*: SUTTON, G. H., MANGHNANI, M. H. & MOBERLY, R. (eds). *The Geophysics of the Pacific Ocean Basin and its Margin. Geophysical Monograph 19.* Geological Society of America, Washington, 351–67.
O'DONNELL, T. H. & PRESNALL, D. C. 1980. Chemical variations of the glass and mineral phases in basalt, dredged from 25–30°N along the Mid-Atlantic Ridge. *Am. J. Sci.* **280-A**, 845–68.
O'HARA, M. J. 1965. Primary magmas and the origin of basalts, *Scott. J. Geol.* **1**, 19–40.
—— 1977. Geochemical evolution during fractional crystallization of a periodically refilled magma chamber, *Nature, Lond.* **266**, 503–7.
—— & MATHEWS, R. E. 1981. Geochemical evolution in an advancing, periodically replenished, periodically tapped, continuously fractionating magma chamber. *J. Geol. Soc., Lond.* **138**, 237–77.
ROEDER, P. L. & EMSLIE, R. F. 1970. Olivine-liquid equilibrium, *Contrib. Mineral. Petrol.* **29**, 275–89.
SCHILLING, J.-G. 1973. Icelandic mantle plume, geochemical evidence along the Reykjanes Ridge. *Nature, Lond.* **242**, 565–71.
——, KINGSLEY, R. H. & DEVINE, J. D. 1982. Galapagos hot spot-spreading center system. I. Spacial, petrological and geochemical variations (83°W–101°W). *J. Geophys. Res.* **87**, 5593–610.
SHAW, H. R. 1972. Viscosities of magmatic silicate liquids: an empirical method of prediction. *Amer. J. Sci.* **272**, 870–93.
SPARKS, R. S. J., MEYER, P. S. & SIGURDSSON, H. 1980. Density variation amongst mid-ocean ridge basalts: implications for magma mixing and the scarcity of primitive magmas. *Earth planet. Sci. Lett.* **46**, 419–30.
SPRAY, J. G. 1982. Mafic segregations in ophiolite mantle sequences. *Nature, Lond.* **299**, 524–8.
STOLPER, E. & WALKER, D. 1980. Melt density and the average composition of basalt. *Contrib. Mineral. Petrol.* **74**, 7–12.
THOMPSON, G., BRYAN, W. B. & MELSON, W. G. 1980. Geological and geophysical investigation of the Mid-Cayman Rise spreading center: geochemical variation and petrogenesis of basalt glasses. *J. Geol.* **88**, 41–55.
VERMA, S. P. & SCHILLING, J.-G. 1982. Galapagos hot spot-spreading center systems. II. 87Sr/86Sr and large ion lithophile element variations (85°W–101°W). *J. Geophys. Res.* **87**, 10838–56.
WALKER, D., SHIBATA, T. & DELONG, S. E. 1979. Abyssal tholeiites from the Oceanographer Fracture Zone. *Contrib. Mineral. Petrol.* **70**, 111–25.

M. R. FISK, Grant Institute of Geology, University of Edinburgh, Edinburgh EH9 3JW, Scotland.

Spreading-rate parameters in ocean crust: analogue for ophiolite?

M. F. J. Flower

SUMMARY: Geophysical and petrologic studies of ocean crust generated at different spreading rates reveal specific rate-dependent characteristics, e.g. relative thickness of eruptive and intrusive crustal layers, stability of axial magma reservoirs in time and space, and pressure–temperature regimes for magma fractionation. Petrogenetic interpretation of whole-rock and glass selvedge compositional variation in basalts from different spreading axes reflects these relations. Inverse correlation of plagioclase accumulation (separately from mafic phases) and increasing spreading rate is attributed to the predominance of polybaric fractionation system at slow-spreading ridge axes, in contrast to low pressure, isobaric, systems at fast-spreading axes. Evidence for 'cryptocumulates' (aphyric lavas showing mass-balance attributes of cumulates) at axes of intermediate spreading rate, appears to reflect a combination of slow-spreading character (e.g. plagioclase accumulation) with fast-spreading character (e.g. relatively 'open' magma supply). It is desirable to monitor these effects on low-K_D incompatible-elements variation, and to test the applicability of spreading-rate parameters on ophiolite lava associations of 'oceanic' affinity.

The global mid-ocean-ridge (MOR) system represents a locus of prolific partial melting at relatively shallow levels ($< c$ 30 km), under essentially anhydrous conditions (e.g. Bottinga & Allegre 1978; Green *et al.* 1979; Presnall *et al.* 1979.) Several lines of evidence suggest that mantle-derived magmas are processed through complex crustal reservoir systems, prior to their consolidation as solid crust. The source is presumed to be 'fertile' peridotite mantle, but remains elusive regarding its physical stratification and constituent chemical 'domains' in time and space. Structural and petrogenetic models for oceanic crust have been based in large part on information derived from ophiolite complexes (e.g. Gass 1968; Moores & Vine 1971; Cann 1974; Pallister & Hopson 1981). We now know that ophiolites were probably generated under a range of tectonic conditions, relatively few of which are confirmed as typically 'oceanic' in character. Since the inception of ocean-crust sampling, sufficient information has become available to model ocean-crust-building processes and, in particular, to constrain primary magma composition, petrogenesis and crustal structure at MORs as a function of spreading rate. Such data can then be usefully compared to ophiolite complexes of unknown or imprecisely known provenance.

There has been considerable success in the use of 'immobile' elements and radiogenic isotope contents, in ascribing ophiolite lava associations to a particular tectonic environment. Several discriminants have proved effective, e.g. Ti-Zr-Y, Ti-Cr (Pearce & Cann 1973; Pearce 1975), Ta-Hf-Th (Wood, *et al.* 1979), and Cr-Y (Pearce 1980). Sr, Nd and Pb isotopes are used in the interpretation of genetic environment magmatic source history and in monitoring interaction of ophiolite lithologies with flushing hydrothermal solutions. Attempts to find correlations between magma chemistry and paleo-spreading rate have been less successful, in view of the apparent lack of sensitivity of low-K_D element and isotope ratios to this factor, or more specifically, to the intrasystem fractionation processes. Despite some confusion about ophiolite genetic environments, it seems clear that most are formed at some type of dilating system, whether far removed from, or close to, major zones of crustal subduction.

From this perspective, a re-evaluation of ocean-crust models is required, in order to: (i) directly determine crustal structure and MOR petrogenetic systems, (ii) distinguish features unique to 'true' (MOR-generated) oceanic crust, and (iii) assess geodynamic factors (including spreading rate) and their influence on magma fractionation processes. Such evaluation should contribute to a sharper focus on the character of ophiolites, their genetic environments and corresponding magmatic character.

The global mid-ocean-ridge (MOR) system

The geometry and, by implication, longevity of MOR magma supply systems has been characterized variously as an 'infinite onion', 'infinite leek', 'string of pearls' and so on. Such models have been defined in terms of ophiolite analogues (e.g. Cann 1974; Dewey & Kidd 1977; Pallister & Hopson 1981), seismic refraction and reflection profiles (Fowler 1976; Reid *et al.* 1977; Nisbet & Fowler 1978; Lewis 1979; Orcutt, this volume), and petrologic/chemical information from erupted submarine lavas (e.g. Flower *et al.* 1977; Bryan, Thompson & Michael 1979; Rhodes & Dungan 1979). The axial magma supply system has the capacity to diversify magma chemistry from its primitive mantle-equilibrated character as a result of cooling, crystallization, and mixing processes, and is clearly the major source of compositional 'noise' in the ocean crust. The system is generally conceived as a complex of tubes and chambers, through which magma passes up to the sea floor or becomes trapped in lower crustal regions during dilation of the constructive plate margin. It has been further assumed that despite chemical fractionation resulting from processes in the system, parameters such as radiogenic/non-radiogenic isotope and low-K_D element ratios remain unmodified from their primitive mantle character and may be taken to be chemical 'signatures' of the mantle source (Schilling 1973, 1975; O'Nions *et al.* 1977, 1978). However, 'open-system' fractionation models (e.g. Stern 1979; O'Hara & Mathews 1981) indicate that incompatible and compatible element abundances may be decoupled, with significant variation of the inter-element ratios. Mantle signatures may thus be affected during magma evolution as a function of the type of magma supply system. It is thus important to understand the physical nature of MOR magma systems and to monitor constraints such as spreading rate (and corresponding thermal/kinematic effects) on the extent to which the system is open or closed during the ascent course of a specific batch of magma.

Rhodes & Dungan (1979) and Walker *et al.* (1979) described several effects resulting from magma mixing in open systems. They postulated the existence of compositionally 'steady-state' magmas, maintained in sub-axial reservoirs through periodic eruption, replenishment and fractionation, as crustal dilation proceeds. Sparks *et al.* (1980) and Stolper & Walker (1980) considered magmatic and solid crustal density distribution at MORs and concluded that (primitive) high-density picrite and (evolved) iron-enriched magmas are probably excluded from the eruptive crust due to their high densities. These and other considerations led to experimental investigations of fluid dynamic relationships, in particular, of processes involving the injection of hot, dense, primitive magma into reservoirs of lighter, more fractionated magma (Huppert & Sparks 1980a, b). Recognition of the control of spreading rate on the geometry and longevity of oceanic magma systems (Cann 1974; Sleep 1975, 1978; Kuznir & Bott 1976) permits further refinement of MOR fractionation models.

Mid-ocean-ridge (MOR) magma supply

'Onions' *v.* 'leeks'

Fig. 1 shows two current perceptions of open-system axial magma reservoirs, considered to be characteristic of fast-spreading axes (after Hale *et al.* 1972). The first, based on Pallister & Hopson's (1981) ophiolite model, indicates a substantial molten region, isotropic gabbro plating the chamber top, with layered gabbro forming the base, and a plagiogranite sandwich zone near the roof–floor intersection. The second scheme is derived from a thermal-conductivity model (Sleep 1975) and Dewey & Kidd's (1977) model for the Bay of Islands' ophiolite. In this scheme, the chamber is largely filled with cumulate mush, preserving a thin, intermittent melt zone only, isotropic gabbro forming as isolated cupolas in the roof zone. Hale *et al.* (1982) report that the seismic velocity profile at the East Pacific Rise (EPR), 21°N (spreading rate 6.2 cm yr^{-1}), corresponds to the second of these models, and cite evidence for an intermittent melt zone at depths of 2.5–3.5 km beneath the ridge axis. Orcutt (this volume) draws a similar conclusion on the basis of a recent seismic experiment conducted at 21°N, essentially confirming the Sleep/Dewey & Kidd model for this segment of the EPR axis (Lewis 1979). Either of these models is consistent with the persistence of open magma systems, although the internal dynamics may differ significantly.

Seismic studies of the Juan de Fuca Ridge (spreading rate: 5–6 cm yr^{-1}) (McClain & Lewis 1982) fail to reveal an axial magma chamber, possibly indicating the existence of 'leek-like' magma supply (Lilias & Rhodes 1982). The infinite leek model (Nisbet & Fowler 1978) was originally proposed for the slow-spreading Mid-Atlantic Ridge (MAR) (1–4 cm yr^{-1}) on the basis of seismic reflection and refraction profiles at 36 and 45°N. These observations are consistent with the predictions of Sleep (1975, 1978) and Kuznir & Bott (1976) for progressive restriction of the chamber size with decreasing spreading rate. It is therefore presumed that 'onion-like' reservoirs probably only characterize fast-spreading sections of the EPR (10–20 cm yr^{-1}). In following sections, differences in eruptive magma chemistry associated with spreading rate changes are reviewed, and an attempt is made to explain these in terms of corresponding changes in melting and magma supply systems. Primary constraints appear to be polybaric tholeiite phase equilibria and the relative extent of open- *v.* closed-system fractionation.

Correlation of magma chemistry and spreading rate

Early attempts to apply geochemical spreading-

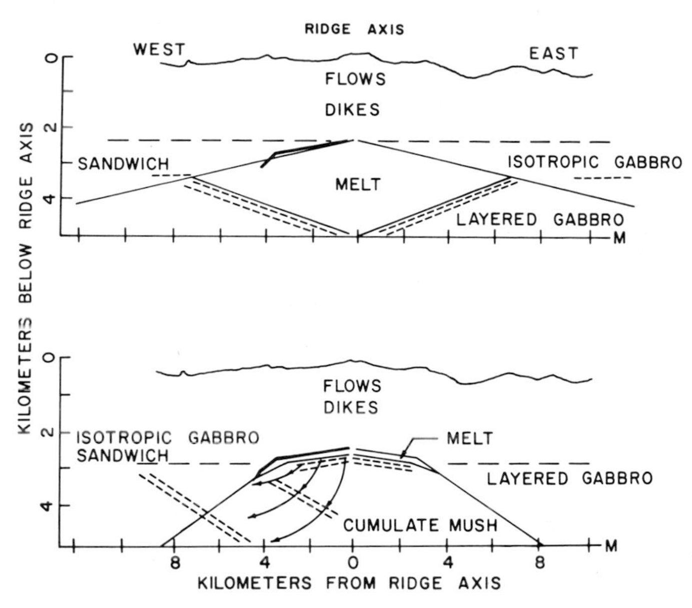

FIG. 1. 'Open-system' magma supply—two variants: based on (a) ophiolite (Pallister & Hopson 1981), (b) mid-ocean-ridge thermal and seismic models (Sleep 1975, 1978; Dewey & Kidd 1977).

rate discriminants (e.g. Bass 1971; Nisbet & Pearce, 1973; Scheidegger 1973) did not distinguish compositional variation of glass (representing magmatic liquid fractions) from whole rock (representing a crystal-liquid mixture). Flower (1980, 1981) made preliminary comparisons of MORB glass and whole-rock compositions from ocean ridges of different spreading rate and concluded that there was an increasing divergence of glass and whole-rock variation with decreases in spreading rate. He attributed this primarily to the tendency for plagioclase to accumulate in magma erupted at slow-spreading axes. Least-squares' analysis of the compositional variation in an expanded data base (representatives given in Fig. 2) essentially confirms these conclusions. Data for glass selvedges from drilled and dredged pillow basalt generated at different MORs have been compared in terms of spreading rate. No systematic correlation with spreading rate is observed for glass compositions (e.g. between MAR and EPR). However, 'noise', expressed by CaO/Al_2O_3, P_2O_5/TiO_2, for given MgO content, correlates with intersection of the MOR axis with transform fracture zones, 'hot spots', 'plumes', etc. (Shibata *et al.* 1979; Morel & Hekinian 1980; Eissen *et al.* 1981; Scheidegger & Corliss 1981), irrespective of spreading rate. Intra-site mass-balance solutions for glass variation (Table 1a) reflect similar compositional patterns for different ridge axes, suggesting the variation results from fractionation of ol + pl + cpx, in roughly similar proportions. While not unexpected, these fractionation solutions appear to correspond with lavas whose phenocryst (and microphenocryst) assemblages are dominated by ol + pl, a characteristic reinforced by experimental data for liquidus co-precipitation of these phases (Fisk *et al.* 1980, see also this volume). Where clinopyroxene does appear as phenocrysts, these are often resorbed and appear to be out of equilibrium with their host liquid. Comparative studies of phenocryst assemblages, coexisting glass and whole-rock compositions, and the experimentally determined phase equilibria (Bryan 1983) indicate a correlation of plagioclase with total phenocryst content, and also to incompatible element depletion.

The apparent conflict of mass balance solutions and experimental and petrographic indications has been attributed by Rhodes & Dungan (1979) to the steady-state character of magma in open MOR systems. Periodic infusions of picrite melt into two- or three-phase cotectic liquids, followed by further fractionation, results in recycling of fractionating magma through an interval of ol, followed by ol + pl (co-) precipitation before regaining the three-phase cotectic. Magma at the ol + pl stage will thus reflect mass-balance indications for a previous component of cpx fractionation. This explanation for the mass-balance solutions is not unique, and it may equally reflect a history of polybaric fractionation, involving a higher-pressure cpx-dominated stage (see O'Hara 1968; Bryan 1981; Casey *et al.* 1972; Elthon *et al.*, this volume). It is important to note

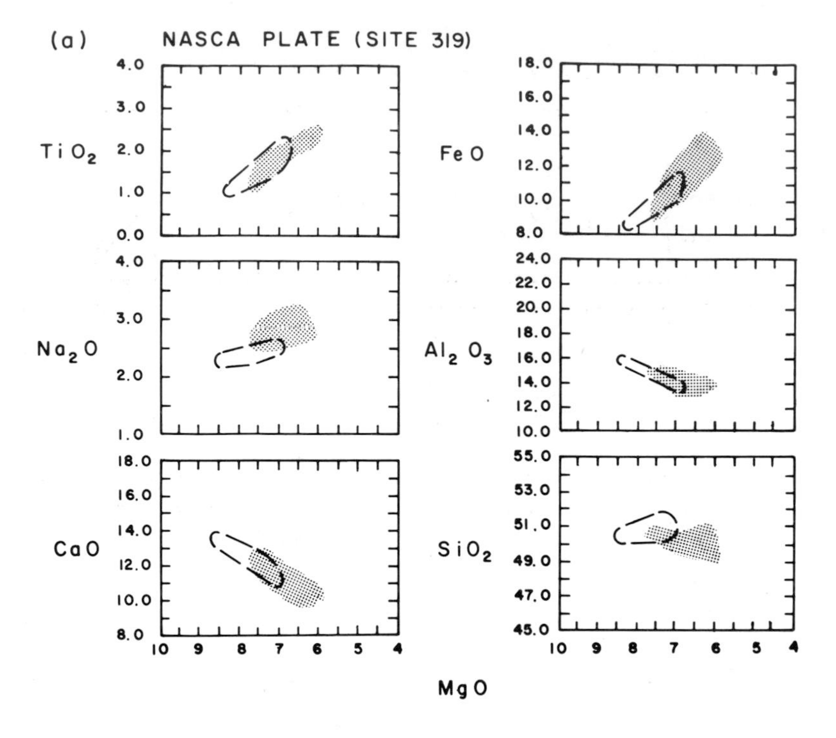

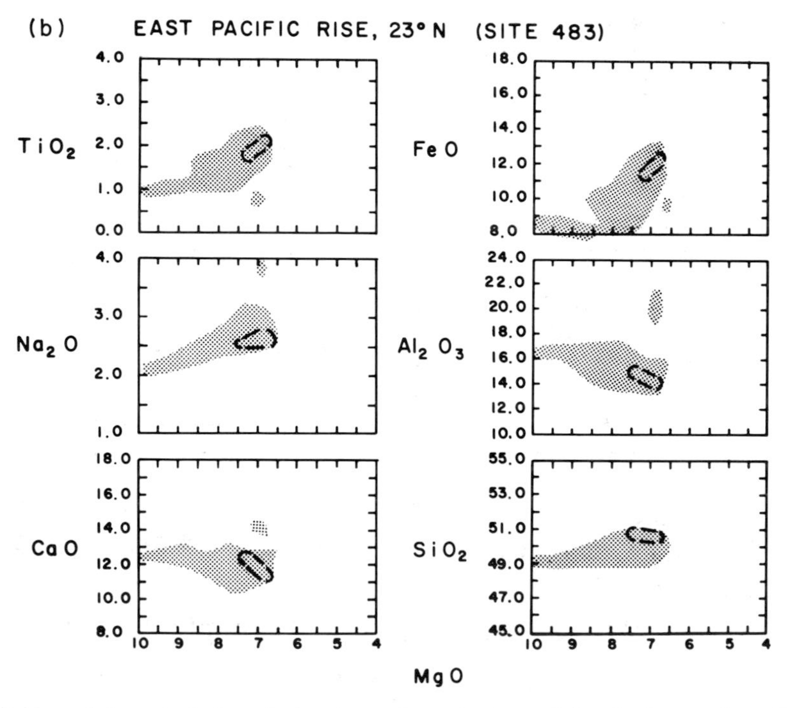

FIG. 2. Oxide variation of whole-rock chemistry; glass selvedge variation is shown schematically by dashed lines; data sources cited in text and in Table 1.

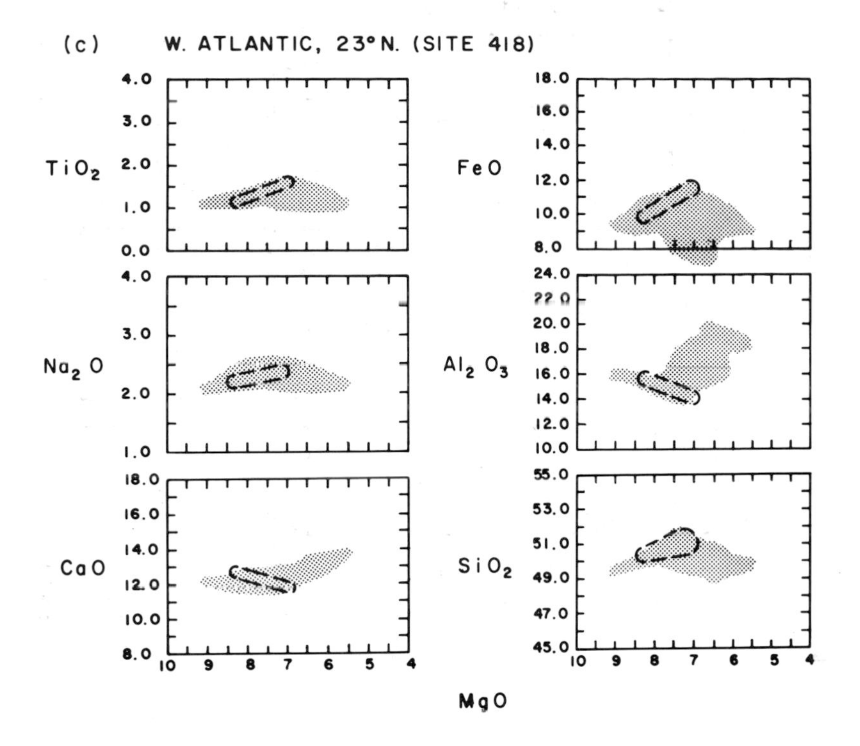
(c) W. ATLANTIC, 23°N. (SITE 418)
TiO2
FeO
Na2 O
Al2 O3
CaO
SiO2
4.0
3.0
2.0
1.0
0.0
18.0
16.0
14.0
12.0
10.0
8.0
24.0
22.0
20.0
18.0
16.0
14.0
12.0
10.0
55.0
53.0
51.0
49.0
47.0
45.0
10 9 8 7 6 5 4
MgO

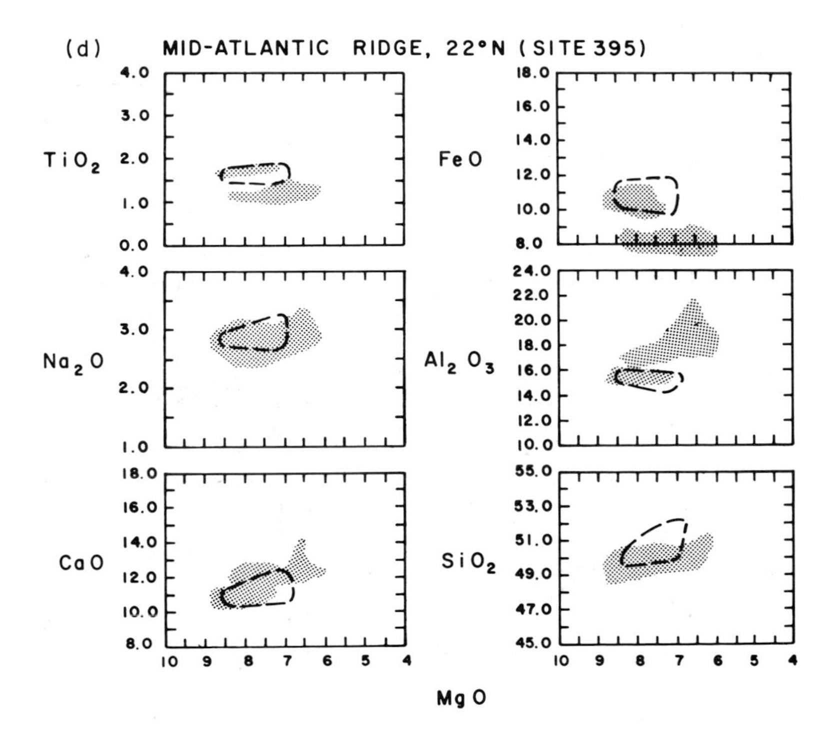
(d) MID-ATLANTIC RIDGE, 22°N (SITE 395)
TiO2
FeO
Na2 O
Al2 O3
CaO
SiO2
4.0
3.0
2.0
1.0
0.0
18.0
16.0
14.0
12.0
10.0
8.0
24.0
22.0
20.0
18.0
16.0
14.0
12.0
10.0
55.0
53.0
51.0
49.0
47.0
45.0
10 9 8 7 6 5 4
Mg O

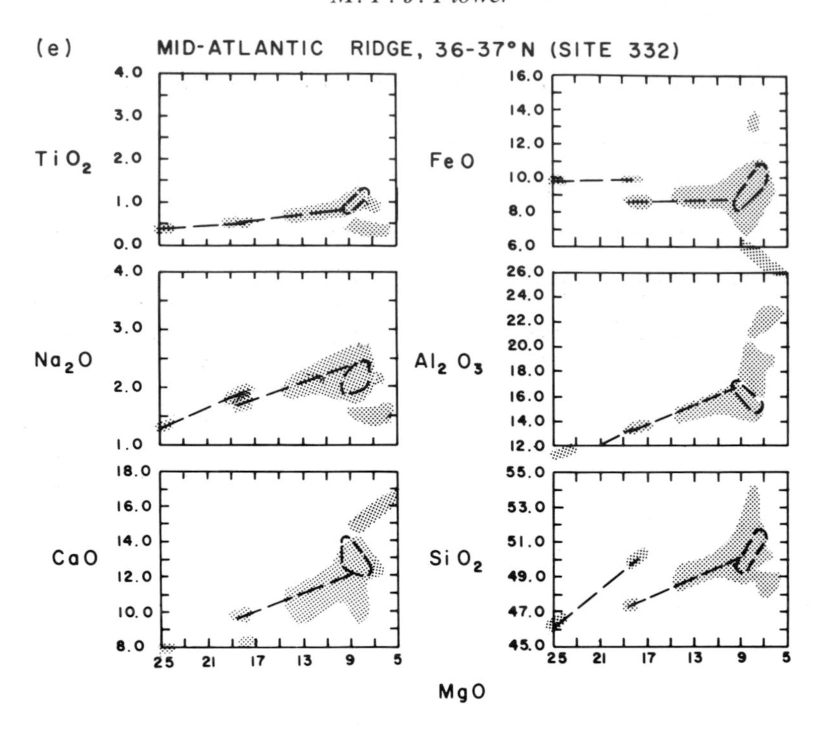

that the overall mass-balance solutions, and glass variation trends, characterize both fast- and slow-spreading ridge axes.

In contrast, whole-rock variation reflects a distinct correlation with spreading rate. Basaltic whole-rock compositions from the (fast-spreading) Nazca Plate (DSDP Sites 319–321) and EPR correspond closely with glass compositional trends, and their associated mass balances, as depicted in oxide variation diagrams (Fig. 2). This correspondence indicates that whole-rock compositional variation effectively represents that of the liquid fraction (reflecting a range of steady state), rather than a suspended emulsion of accumulated crystals and liquid. This is consistent with the exclusively aphyric and sparsely phyric character of basalts at fast-spreading MOR segments, except in the vicinity of major transform fracture zones (Eissen *et al.* 1981). Basalt lavas from the (intermediate-spreading) Gulf of California (EPR, 23°N) reflect similar relations, with many whole-rock compositions adhering to the glass-defined liquid line (Juteau *et al.* 1980; Flower & O'Hearn 1983; Flower *et al.* 1983; Saunders *et al.* 1983). Lichtman *et al.* (1982) report correlation of petrology and geomorphic expression at the EPR near 21°N, and show that phenocryst and microphenocryst contents decrease southwestwards from 3.9 to 0.2, and 6.7 to 2.3 vol. %, respectively. Incipient departure of whole-rock variation from the liquid line was observed in several 23°N lava groups (Flower *et al.* 1983) and attributed to accumulation, and subsequent resorption, of plagioclase (Goode 1977; Flower 1982) (see Table 1b).

Basaltic crust generated at the slow-spreading MAR axis (1–4 cm yr^{-1}) is considerably more diverse than that formed at the faster-spreading Pacific and Indian Ocean axes (Frey & Sung 1974; Kempe 1974). Whole-rock variation shows far greater divergence from the associated glass variation. This is best exemplified in basalts from DSDP Sites 417, 418 (Byerly & Sinton 1980; Flower & Robinson 1981), 395 (Melson 1979; Rhodes *et al.* 1979), 396 (Dmitriev 1979; Flower *et al.* 1979), and 332–335 (Flower *et al.* 1977; Byerly & Wright 1978) (Fig. 2). Most Atlantic basement sections consist of interlayered massive and pillowed lavas, ranging from aphryic, through sparsely to moderately phyric, to very coarsely phyric basalts. Phenocryst assemblages are commonly ol + pl + cpx, ol, and pl, the latter two reflecting cumulates of these phases. Least-squares' mass-balance solutions for variation between whole-rock compositions and between the latter and glass compositions suggest that plagioclase accumulation is a major process contributing to the overall whole-rock variation patterns (Table 1b). This is confirmed by the abundance of plagioclase phenocrysts in Atlantic

TABLE 1. *Least-squares' solutions for parent-derivative magma compositions from different spreading axes*

Parent-derivative file Nos (C.T.)	%ol	%pl	%cpx	ΣR^2	location (DSDP Site)	Ref.
(a) Glass fraction						
319A (a) 319A (hc)	-3.72	-23.58	-18.48	0.26	Nazca (319)	1
319A (b) 319A (hc)	-4.26	-20.70	-17.43	0.18	Nazca (319)	1
3227 3229	-3.71	-17.65	-12.25	0.20	EPR 12°N (421)	2
3228 3229	-3.41	-17.96	-12.98	0.19	EPR 12°N (421)	2
710 760	-5.56	-11.72	-5.40	0.45	EPR 23°N (474)	3
710 770	-6.29	-18.63	-3.57	0.05	EPR 23°N (474)	3
550 330	-5.01	-12.60	-8.35	0.06	EPR 23°N (483)	4
540 370	-4.34	-10.19	-6.21	0.08	EPR 23°N (483)	4
JF-69 (F) JF-75 (F)	-15.27	-36.38	-11.68	0.04	JFR 45°N	5
JF-114 (F) JF-76 (F)	-23.01	-43.73	-5.28	0.11	JFR 45°N	5
20 (F) 15 (C)	-4.0	-17.8	-11.6	0.04	W.Atl. 23°N (418)	6
29 (H) 21 (G)	-2.3	-12.9	-10.8	0.06	W.Atl. 23°N (418)	6
N68 N58	-6.32	-6.74	-1.54	0.14	MAR 22°N (395)	7
N65-G P11-G	-8.05	-7.87	-2.03	0.44	MAR 22°N (395)	7
1080 (X3) 450 (Y2)	-2.75	-13.46	-12.71	0.14	MAR 22°N (396)	8
1100 (X3) 450 Y2)	-3.24	-15.01	-7.20	0.10	MAR 22°N (396)	8
B-DG B-EG	-5.99	-17.36	-9.03	0.02	MAR 36–37°N (332)	9
B-BG1 B-BG2	-1.81	-6.21	-3.10	0.10	MAR 36–37°N (332)	9
(b) Wholerock fraction						
RH319 (A) RH319 (E)	-5.09	-23.45	-17.38	0.52	Nazca (319)	10
RH319 (B) RH319 (D)	-3.47	-24.58	-20.30	0.03	Nazca (319)	10
3030 (A) 3000 (A)	-8.50	-1.29	-1.84	0.17	EPR 23°N (474)	11
3250 (G) 3240 (G)	-1.53	+4.34	-0.82	0.50	EPR 23°N (474)	11
6200 (D) 6190 (D)	-1.45	+5.63	-0.62	0.12	EPR 23°N (483)	12
6610 (K) 6600 (K)	-0.99	+7.11	-1.35	0.16	EPR 23°N (483)	12
30 (I) 37 (J)	-1.6	+9.8	-2.7	0.14	W.Atl. 25°N (418)	6
9 (B) 14 (B)	-3.7	+12.4	-3.0	0.18	W.Atl. 25°N (418)	6
113 (A2) 127 (P2)	-5.10	+10.29	-3.11	0.21	MAR 22°N (395)	13
109 (A2) 129 (P2)	-2.82	+12.64	-2.02	0.20	MAR 22°N (395)	13
870 (X2) 1030 (B2)	-1.03	+16.02	-6.9	0.26	MAR 22°N (396)	8
750 (X2) 710 (B1)	-0.99	+3.73	-0.25	0.14	MAR 22°N (396)	8
B-CG 3F32B	-4.60	+42.98	-3.21	0.20	MAR 36–37°N (332)	9
B-FG 6F32B	-2.85	+13.14	-2.18	0.14	MAR 36–37°N (332)	9

Source references:

1. Hart (1976)
2. O'Hearn (unpubl.)
3. Fornari (1983)
4. Flower & O'Hearn (1983)
5. Melson (unpubl.)
6. Byerly & Sinton (1980)
7. Melson (1979)
8. Flower *et al.* (1977)
9. Byerly & Wright (1978)
10. Rhodes *et al.* (1976)
11. Saunders *et al.* (1983)
12. Flower *et al.* (1983)
13. Rhodes *et al.* (1979).

basalts, representing clear petrographic expression of the computed mass balances. In contrast, mass-balance solutions for Atlantic glass variation are invariably similar to those for Pacific glass (Table 1a). These relations may represent evidence for some degree of open-system fractionation at slower-spreading axes (Rhodes & Dungan 1979), although assuming different geometric forms from faster-spreading systems, in view of the whole-rock petrographic and chemical variation, and mass balances.

To summarize, the continuously expanding chemical data base for ocean-crust samples reflects an increasing divergence of whole-rock from glass compositional variation, with decreasing spreading rate. This signifies a tendency for massive accumulation of plagioclase at slow-spreading, but not fast-spreading, axes (Flower 1980, 1981). Supporting evidence is reported from the slow-spreading Red Sea axis (Juteau *et al.* 1983, pers. comm.), where plagioclase phenocryst abundances may be as high as those observed in MAR basalts.

Spreading rate and thermal distribution

Flower (1980, 1981) proposed that pressure–temperature characteristics of magma systems will vary in response to changes in spreading rate. Sleep (1975), Kuznir & Bott (1976) and others have computed ocean-ridge isotherm distributions in relation to different rates of spreading. The model of Sleep (1975) is reproduced in Fig. 3 with the temperature region exceeding 1185°C shaded to denote an approximate tholeiitic suprasolidus region. This region is clearly a function of both pressure and bulk composition and therefore not strictly isotherm-bounded. However, the 1185°C isotherm can be taken to approximate the schematic stabilities of magma reservoirs at contrasting spreading rates. Interpolation of the axial geotherm at different rates provides a qualitative picture of fractionation tendencies, without recourse to complex magmatic ascent models. This picture implies two or more pressure intervals of cooling and fractional crystallization for the 'slow' geotherm (consistent with 'infinite leek' models), while low-pressure isobaric fractionation is implied for fast-spreading axes (consistent with the 'infinite onion'). With this in mind, Flower (1980, 1981) cited the experimental results of Fujii & Kushiro (1977) indicating that plagioclase accumulation in cotectic magma is favoured by increasing pressure, due to the greater compressibility of tholeiite liquid, and suggested that polybaric fractionation regimes at slow-spreading axes may be largely responsible for enhancing this tendency.

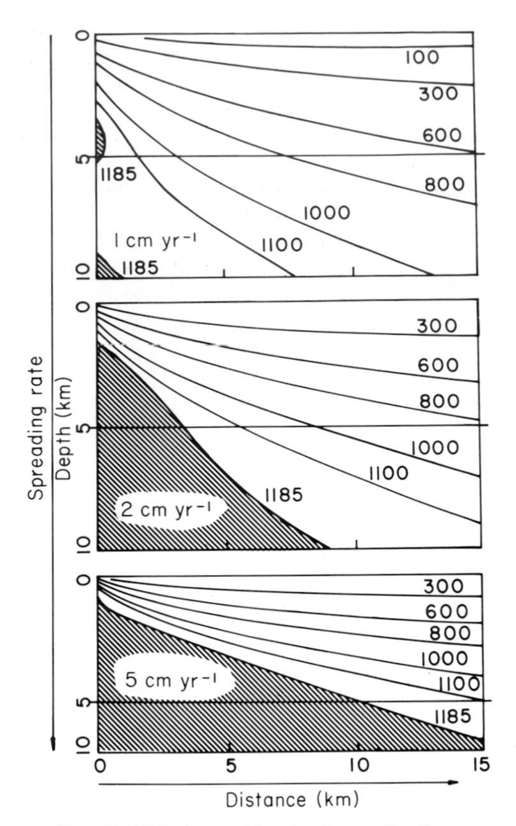

FIG. 3. Mid-ocean-ridge isotherm distribution for 'slow' (*c* 1 cm yr^{-1}) and 'fast' (> *c*. 5 cm/yr) spreading rates (after Sleep 1975).

Cumulates and cryptocumulates

The above model is admittedly simplistic, but the evidence of 'cryptocumulates' at intermediate-rate spreading axes (e.g. the Gulf of California), and cumulates at slow-spreading axes (e.g. MAR), provides support for its basic tenets. Cryptocumulate compositions are believed to reflect the chemistries and phase equilibria of plagioclase-cumulitic basalt, in which accumulated phenocrysts have been largely resorbed as a result of superheating by, and/or mixing with, later injections of picrite (Flower 1982). The relationship between cumulates and cryptocumulates can be illustrated with respect to two representative basement sites: DSDP Sites 483 and 332. Both are described more fully elsewhere (e.g. Flower *et al.* 1977, 1983). DSDP Site 483 is located on EPR basement at 23°N and reveals nearly 200 m of interlayered massive and pillowed lavas. Basalt 'chemical types' (determined from whole-rock major-element oxide variation) represent compositional/stratigraphic groupings, apparently comprising discrete eruptive packets. Overall variation of chemical types (i.e. between-type variation) corresponds to the fast-spreading, liquid-type, and is reflected by data for glass selvedges from pillow lavas (Flower & O'Hearn 1983). However, intra-type variation is distinct from, and superimposed upon, the between-type variation, especially for chemical types comprising massive lava units. The latter are for the most part aphyric, or very nearly so. Mass-balance solutions for intra-type variation reflect accumulation of plagioclase concomitant with olivine and clinopyroxene fractionation (Table 2), further supported by positive Eu anomalies (Cambon *et al.* 1983). Petrogenetic roles for plagioclase are clearly indicated by the variation of Sr (incompatible with respect to ol and cpx, but strongly compatible with respect to plagioclase) and Zr (strongly incompatible with respect to all phases). Plotted in the form: 1/Sr *versus* Zr/Sr, straight-line covariance is consistent with binary mixing (Langmuir *et al.* 1978). Variation of these parameters for Site 483 lavas, plotted according to chemical type (Fig. 4a), confirms the chemical effects of plagioclase accumulation in aphyric types C, D, H and K, but not in phyric types F, J, L and M. Site 332 includes up to 580 m of basement drilled near the MAR at 36–37°N. Plots of 1/Sr *v.* Zr/Sr for Site 332 chemical types show strikingly similar patterns to those for 483 (note difference in scale, however) (see Fig. 4b) for types Z, G and B, although these are spectacularly plagioclase-phyric (Flower *et al.* 1977; Byerly & Wright 1978). There appears to be no qualitative distinction in the major- and trace-element patterns between demonstrably cumulitic lavas (Site 332), and aphyric lavas interpreted to be cryptocumulates (Site 483). The correlation of +ve Eu anomalies with Sr in both lava sections, and the consistent major-element mass-balance constraints argue against the Eu anomalies being source-derived, as suggested by Saunders (1983). The latter interpretation is *ad hoc* and disregards both mass balances and *a priori* conditions of magma supply at intermediate rate (cf. slow-spreading) axes. Moreover, petrographic evidence for resorption in Site 483 basalts is convincing (Barker *et al.* 1983; Flower *et al.* 1983), and analogous to that described by Goode (1977) for the Kalka Intrusion.

Stratigraphic variation at both sites appears to reflect a repeated sequence of fractionation events acting upon magmas already exhibiting a distinct range of variation (*c.* 10–5 wt % MgO). It is suggested that the Site 483 cryptocumulates result from the accumulation of plagioclase alone (possibly in upper parts of the reservoir) during temporary stagnation of the supply system, followed by renewed (and repeated) injection of primitive magma (Flower 1982). This hypothesis is supported by two lines of evidence. First, the least phyric cryptocumulates show highest values for $Mg/(Mg + Fe^{2+})$, and Cr/Ni, suggesting maximum thermal input. Second, there is a strong indication of binary mixing (as opposed to fractional crystallization) from the intra-type major- and trace-element variation. Variation between compatible elements Cr, Ni and (in some cases) Sr, and incompatible elements Ti, Zr and Y is linear (Flower *et al.* 1983), in contrast to curved variation paths produced by fractional crystallization. The most convincing evidence for plagioclase accumulation, mixing and reheating occurs in chemical type 483-D, a thick massive unit in the upper part of the Site-483 section, reflecting high MgO content (7–10 wt %) and aphyric lithology. Analyses of cryptocumulates from this unit are compared with coarsely plagioclase-phyric lavas sampled from slow-spreading Atlantic crust (Table 3). Despite petrographic contrasts these samples reflect remarkably similar chemistries, especially in regard to

TABLE 2. *Least-squares' fractionation solutions for intra-chemical type major-element variation: DSDP Site 483 (with Zr, Ti and Sr enrichment factors)*

Sample file Nos* (C.T.)	%ol	%plag	%cpx	ΣR^2	[Zr]e	[Ti]e	[Sr]e
6200 (D) 6190 (D)	−1.45	+5.63	−0.62	0.12	0.97	0.96	1.31
5950 (C) 6180 (C)	−1.36	+1.63	−0.41	0.29	1.03	1.03	0.99
6140 (B) 5920 (B)	−0.91	+1.71	−0.01	0.15	0.99	0.99	1.20
6090 (H) 6100 (H)	−0.95	+0.72	−0.41	0.01	1.00	1.00	1.01
6472 (H) 6440 (H)	−0.51	+5.55	−0.20	0.24	0.96	0.95	1.02
6580 (K) 6540 (K)	−1.32	+4.91	−0.98	0.19	0.88	0.95	1.30
6660 (K) 6710 (K)	−0.75	+5.20	−0.83	0.21	0.96	0.92	1.28
6610 (K) 6600 (K)	−0.99	+7.11	−1.35	0.16	0.90	0.85	1.37

*Selected samples are MgO-rich and MgO-poor representatives of the chemical type indicated, or in the case of types H and K, individual massive units comprising these types.

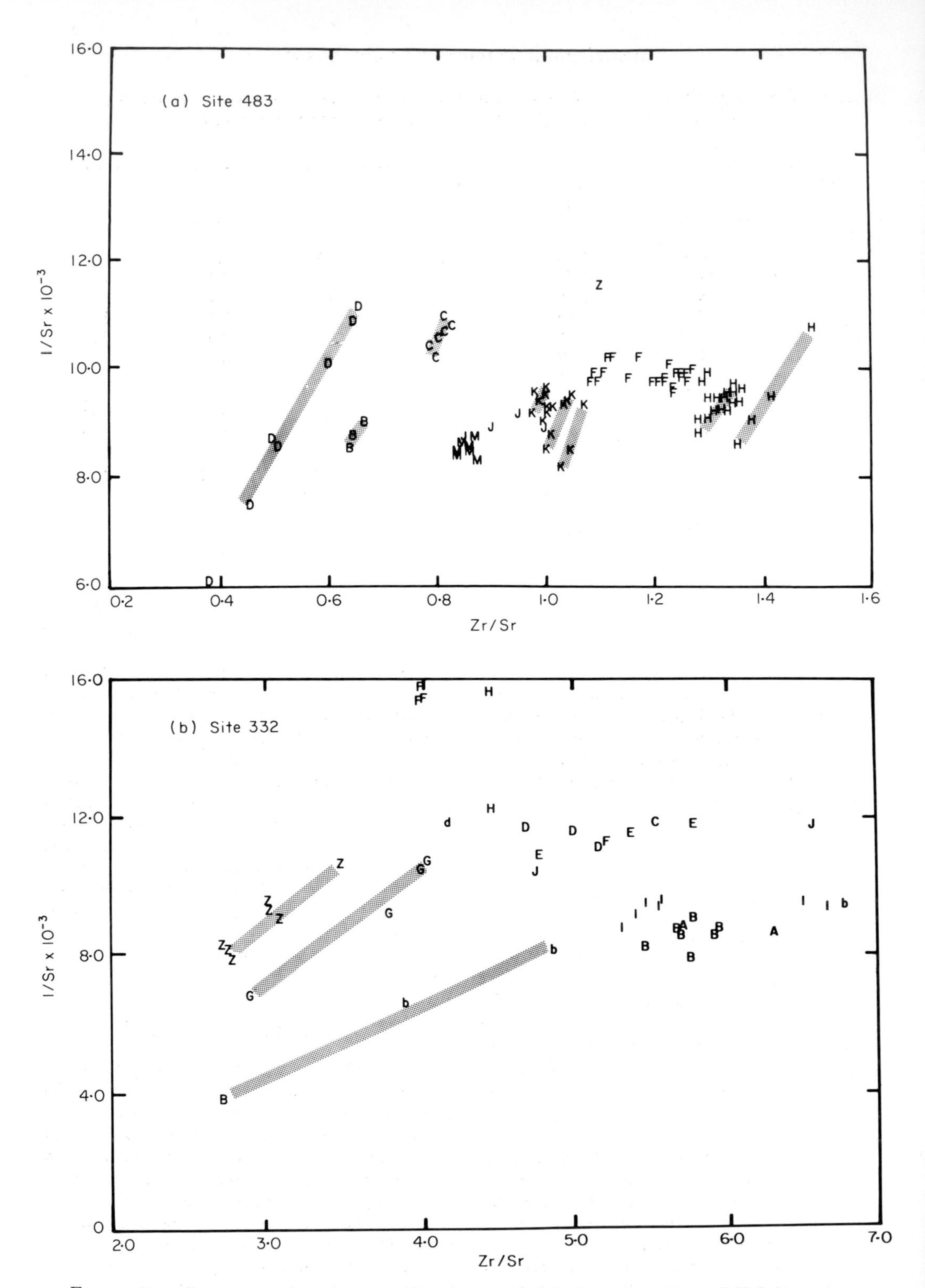

FIG. 4. 1/Sr *v*. Zr/Sr: (a) Gulf of California DSDP Site 483, (b) Mid-Atlantic Ridge DSDP Site 332; chemical types as indicated; data sources cited in text and Table 1. 1. 483B, 8–1, 29–35; 2. 483B, 8–2, 121–6; 3. 483, 17–2, 17–21; 4. 332A, 40–3, 35–7; 5. 396B, 16–5, 34–6; 6. 418A, 48–0, 70–2. References: (1) Flower (1982), (2) Byerly & Wright (1978), (3) Flower *et al.* (1977), and (4) Flower & Robinson (1981).

TABLE 3. *Chemical analyses of aphyric 'cryptocumulates' from DSDP Site 483, and plagioclase-phyric lavas from Atlantic Sites 332, 396 and 418 (normalized to dry weight)*

Sample No. Ref.	1. (1)	2. (1)	3. (1)	4. (2)	5. (3)	6. (4)
wt %						
SiO_2	50.26	50.08	49.56	49.69	50.67	49.62
Al_2O_3	17.42	16.51	16.86	16.88	17.36	17.16
FeO*	8.00	8.26	8.25	6.62	7.94	9.18
MgO	7.76	8.86	9.76	10.16	7.31	7.16
CaO	12.63	12.15	12.09	14.21	12.08	12.93
Na_2O	2.40	2.49	2.09	1.58	2.54	2.32
K_2O	0.04	0.05	0.05	0.11	0.24	0.04
TiO_2	1.01	1.10	0.90	0.56	1.19	1.26
P_2O_5	0.07	0.08	0.06	0.07	0.15	0.13
MnO	0.16	0.16	0.17	0.12	0.14	0.17
Total	99.75	99.74	99.79	100.00	99.62	99.97
ppm						
Zr	60	63	117	42	76	42
Y	25	26	37	12	26	21
Sr	118	165	174	124	131	105
Ni	109	87	96	145	137	102
Cr	320	301	285	411	294	182

their CaO, Al_2O_3 and Sr contents. As cryptocumulate magma appears to be absent from fast-spreading axes (Scheidegger & Corliss 1981) and there is clear petrographic expression of accumulated plagioclase in slow-spreading basalts (Flower 1981), these liquid compositions may uniquely characterize intermediate-rate spreading axes. From a phase-equilibrium standpoint, cryptocumulates will be characterized by their confinement at low pressures to the plagioclase crystallization field, in contrast to the cotectic (or near-cotectic) behaviour generally observed in non-cumulate basalts (e.g. Fisk *et al.* 1980).

Mixing is clearly not restricted to fast- and intermediate-rate MOR axes, however. The extent of variation of parameters such as Ti/Y versus Zr/Y (Fig. 5) is not compatible with simple phenocryst redistribution schemes (e.g. those modelled in Table 2) and variation of these and other major- and trace-element parameters is consistent (at least within chemical types) with simple binary mixing, both for Site 483 (intermediate-spreading) and Site 332 (slow-spreading) lavas. Such mixing is irrespective of whether Zr/Y and Ti/Y variation reflect distinct source signatures, derive from separate open fractionation systems of differing maturity, or the same system but with incomplete mixing (O'Hara & Mathews 1981). Mixing in such systems may of course be more complex, depending on the extent of layering due to double-diffusive convection (Huppert & Sparks, 1980b). In any case, the observed variation appears to confirm that mixing is an important process at most spreading axis types, although rarely taken to completion to result in truly steady-stage compositions. To summarize, cryptocumulates are believed to reflect (i) plagioclase accumulation in polybaric fractionation systems, (ii) resorption during picrite refilling in open phases of system evolution and (iii) an Al_2O_3-rich liquid of non-cotectic character. If this conclusion is correct and if cryptocumulate characterizes intermediate-rate spreading axes, it clearly reflects a combination of fast- and slow-spreading character, in terms of the fractionation regime and its spatial and temporal continuity.

Analogues for ophiolite?

Having established probable spreading-rate parameters for crust generated at MOR systems, the question arises as to whether these may be applied to ophiolites of unknown provenance. Valid application of such parameters depends on ophiolite magmas having developed in analogous magma systems, in respect to mantle source, thermal/kinematic regime and factors such as P_{H_2O}. Many studies of ophiolite lavas and plutonics assumed an analogy between ophiolite genetic environments and extensional MOR systems, based on interpretations of sheeted dykes, their chilling 'statistics' and the geometric relations of ophiolite lithologies. Recent research

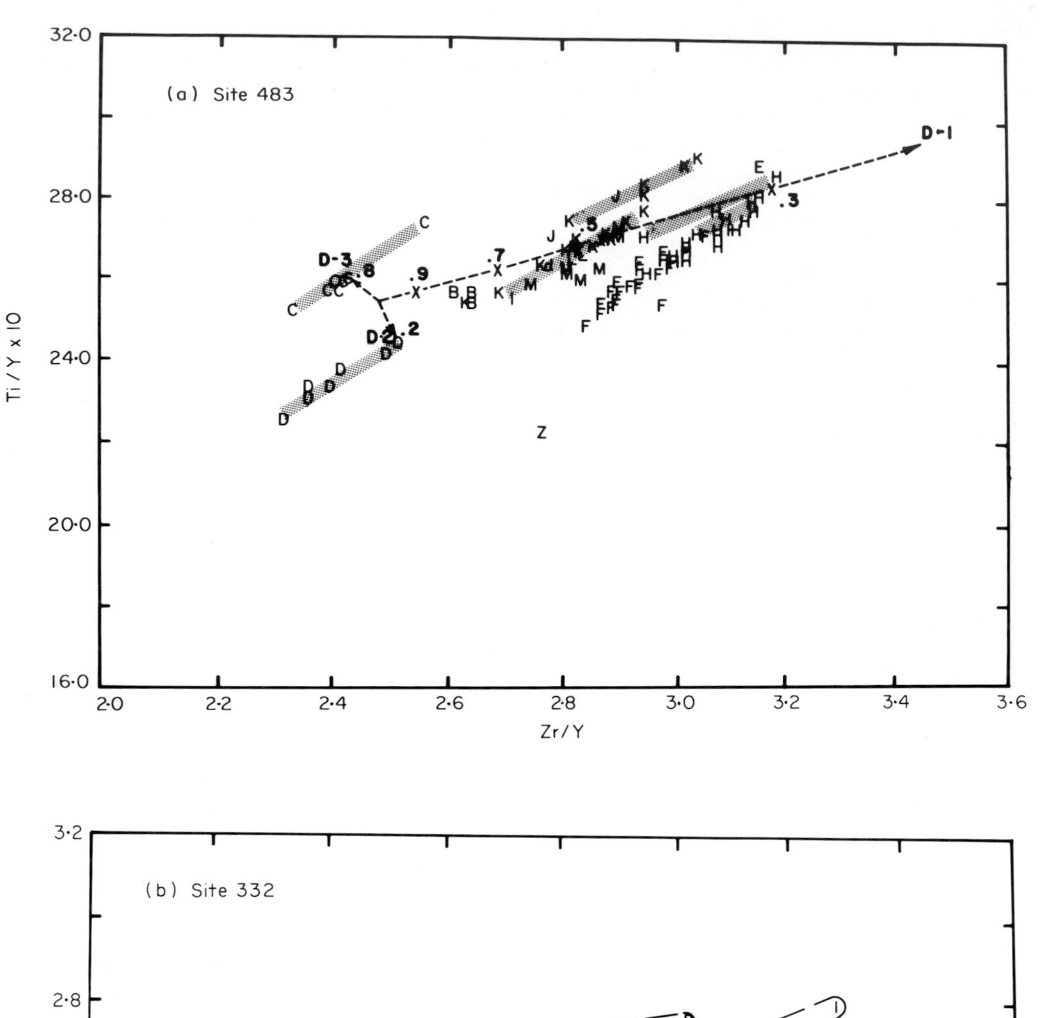
(a) Site 483
Ti / Y x 10
Zr/Y
D-1
D-3
D-2
32·0
28·0
24·0
20·0
16·0
2·0
2·2
2·4
2·6
2·8
3·0
3·2
3·4
3·6

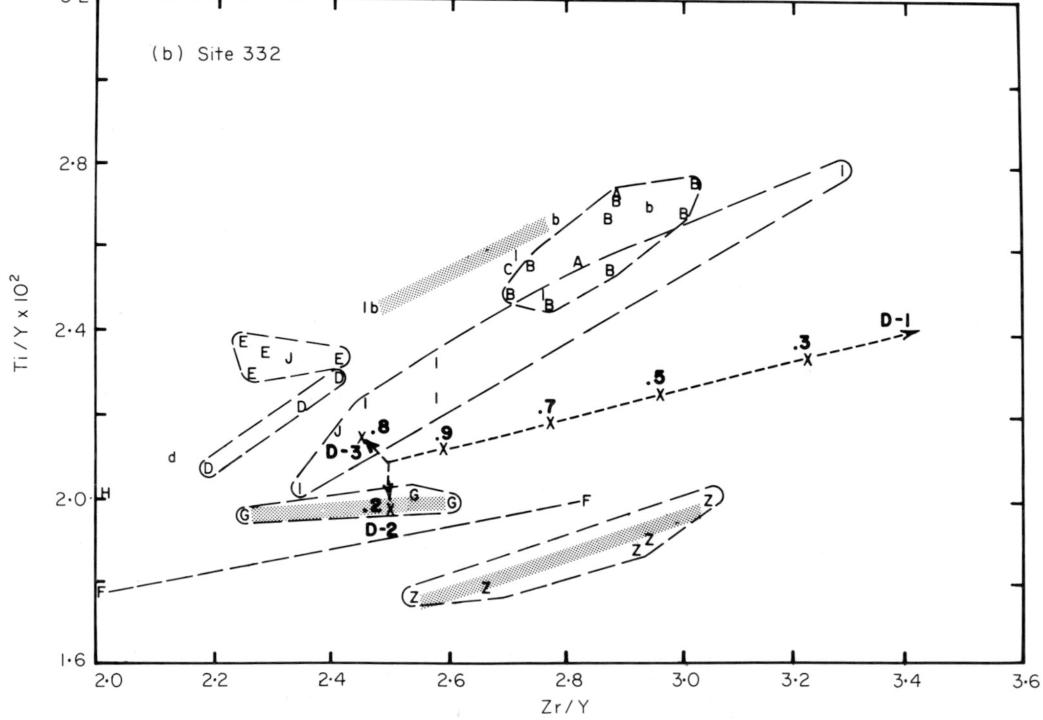
(b) Site 332
Ti / Y x 10²
Zr/Y
D-1
D-3
D-2
3·2
2·8
2·4
2·0
1·6
2·0
2·2
2·4
2·6
2·8
3·0
3·2
3·4
3·6

has revealed a number of discrepancies between the indications of petrography and chemical variation, and an extensional origin. Few ophiolite lava sequences show typical MORB-type phenocryst assemblages, and the apparent order of crystallization, indicated by ophiolite plutonics, often differs from that predicted for MORB (Cameron *et al.* 1980; Fisk *et al.* 1980). Several ophiolites (e.g. Troodos, Othris, Vourinos, Betts Cove) appear to include high-SiO_2 parental picrites, and orthopyroxene- and pigeonite-phyric basic lavas crystallizing very Cr-rich spinel but no plagioclase. Substantial sequences of andesite, dacite, and rhyodacite may also be present (e.g. Cameron *et al.* 1980; Robinson *et al.* 1983). These observations appear to reflect hydrous source material, a distinctive hierarchy of fractionation processes, and magma genesis at, or close to, subduction zones. These ophiolites also reveal refractory harzburgite mantle sequences, in contrast to the lherzolite commonly exposed by oceanic transform fracture zones (Hamlyn & Bonatti 1980). Plagioclase only appears on liquidi in andesitic or more evolved magmas, and is generally absent from basic magmas. The requisite analogy to the MOR environment does not exist therefore, even if these ophiolite associations were generated in a partly dilating environment.

It has been suggested (Cann, pers. comm. 1982) that MORB-like ophiolites are more fragmented, tectonized, and altered, due to their more extensive transport prior to obduction, and in consequence have been relatively neglected for stratigraphic and chemical analysis. Isotope studies (e.g. Allegre *et al.* 1982) and immobile element discriminants (Pearce 1980) appear to confirm that some ophiolite successions are indeed of oceanic or back-arc affinity. If these ophiolites reflect genesis in a relatively anhydrous environment, distinction between marginal back-arc and MOR spreading centres remains to be made. Saunders *et al.* (1980) note that back-arc magmatic diversity is probably as great as that of MORB, with the possibility of a volatile-rich component derived from the adjacent subduction zone. While this may have relatively marginal effects on magmatic phase equilibria, the resulting compositional discrepancy will be minimal compared to that between true ocean-ridge magmas and those generated at colliding plate boundaries (Green 1972; Hart *et al.* 1972; Hawkins 1976; Saunders & Tarney 1979; Weaver *et al.* 1979). Systematic chemical studies of 'oceanic' ophiolites are not yet available for rigorous application of petrologic spreading-rate parameters, although studies in progress on back-arc magma variation indicate significant variations in spreading rate (Flower, in prep.; P. Lonsdale, pers. comm.).

Conclusions

Dynamic factors such as spreading rate appear to determine the character of tholeiite magma fractionation systems beneath spreading MOR axes. The following scenario has been developed. Fast-spreading rates permit relatively simple, open systems, through which magma is processed and fractionated under low-pressure steady-state conditions. Slow-spreading rates appear to constrict the extent and continuity of the reservoir/transport system, such that fractionation is polybaric and characterized by phenomena such as plagioclase accumulation (at high pressure) and complex polybaric mineral–liquid disequilibria. Intermediate-rate magmas reflect both fast- and slow-spreading attributes: moderate plagioclase accumulation, whose effects are masked by superheating and mixing during replenishment of a partially open system. The appearance of slow-spreading petrologic character in magmas close to fast-spreading transform fracture zones (Eissen *et al.* 1981) may reflect localized polybaric fractionation in the transform environment.

Continuing studies of ophiolites reveal factors that constrain the analogies between their respective environments of genesis and MORs. Ophiolite magmas generated in the vicinity of subduction zones are likely to be hydrous, with derivative fractionation products consisting of calc-alkaline tholeiite, andesite, dacite and rhyolite. Their phase equilibria will be distinct from those of magma formed through anhydrous partial melting at MOR axes due to the shift of peridotite phase boundaries with P_{H_2O}, and corresponding increases in the degree of partial melting.

FIG. 5. Ti/Y *v*. Zr/Y: (a) Gulf of California DSDP Site 483, (b) Mid-Atlantic Ridge DSDP 332; chemical types as indicated data sources cited in text and Table 1. Fractionation trends D-1 to D-3 are calculated according to the Rayleigh equation:

$$\frac{C_1}{C_0} = F^{(D-1)}$$

where: C_1 = concentration of element in derivative liquid, C_0 = concentration of element in parent liquid, F = weight fraction of liquid remaining, and D = bulk distribution coefficient, for the following schemes: D-1 pl:cpx:ol (3:3:1), D-2 pl:ol (1:1), D-3 pl accumulation. Diffuse lines indicate sample associations whose major element mass balance reflects plagioclase accumulation (incl. cryptocumulates).

However, for ophiolite magmas indicating relatively anhydrous conditions of formation, ocean-crust spreading-rate parameters may be applicable and yield information on palaeo-spreading conditions.

ACKNOWLEDGMENTS: This work has been supported by grants from NSF (EAR-8212738) and UIC Research Board. I am grateful to Jean-Philippe Eissen (Strasbourg) and George Jenner (Mainz) for constructive reviews of the manuscript.

References

BARKER, S. E., KUDO, A. M. & KEIL, K. 1983. Mineral chemistry of basalts from Holes 483 and 483B. *Initial Reports of the DSDP*. **65**, 635–42.

BASS, M. M. 1971. Variable abyssal basalt populations and their relation to sea floor spreading rates. *Earth Planet. Sci. Lett.* **11**, 18–22.

BOTTINGA, Y. & ALLEGRE, C. J. 1978. Partial melting under spreading ridges. *Phil. Trans. R. Soc. Lond.* **A288**, 501–25.

BRYAN, W. B. 1981. The role and limitations of simple fraction and mixing processes in the genesis of basalts from the Mid-Atlantic Ridge. *Chapman Conference on 'Generation of Oceanic Lithosphere'*. Warrenton, VA, USA. April 1981, Programme with Abstracts.

—— 1983. Systematics of modal phenocryst assemblages in submarine basalts: petrologic implications. *Contr. Mineral. Petrol.* **83**, 62–74..

——, THOMPSON, G. & MICHAEL, P. J. 1979. Compositional variation in a steady-state zoned magma chamber, Mid-Atlantic Ridge at 36°50′N. *Tectonophys.* **55**, 63–85.

BYERLY, G. R. & WRIGHT, T. L. 1978. Origin of major element chemical trends in DSDP Leg 37 basalts, Mid-Atlantic Ridge. *J. Volcanol. Geotherm. Res.* **3**, 229–79.

—— & SINTON, J. A. 1980. Compositional trends in natural basaltic glasses from DSDP Holes 417D, 418A, B. *Initial Reports of the DSDP.* **51–53**, 229–79.

CAMBON, P., BOUGAULT, H., JORON, J. L. & TREUIL, M. 1983. Basalts from the East Pacific Rise: an example of typical oceanic crust depleted in hygromagmatophile elements. *Initial Reports of the DSDP.* **65**, 623–34.

CAMERON, W. E., NISBET, E. G. & DIETRICH, V. J. 1980. Petrographic dissimilarities between ophiolitic and ocean-floor basalts. *In*: PANAYIOTOU, A. (ed.) *Ophiolites: Proc. Int. Ophiolite Symposium, Cyprus*, Geological Survey Department, Nicosia, Cyprus, 182–92 pp.

CANN, J. R. 1974. A model for oceanic crustal structure developed. *Geophys. J. R. Astr. Soc.* **49**, 169–87.

CASEY, J. F., ELTHON, D. & KOMOR, S. 1982. A new model for crustal and upper mantle accretion at mid-ocean spreading centres involving high pressure polybaric crystallization. *Trans. Am. Geophys. Union, EOS.* **63**, 1136.

DEWEY, J. F. & KIDD, W. S. F. 1977. Geometry of plate accretion. *Geol. Soc. Am. Bull.* **88**, 960–8.

DMITRIEV, L. 1979. Petrology of the basalts, legs 45, 46. *Initial Reports of the DSDP.* **46**, 143–50.

——, SOBOLEV, A. V. & SUSCHEVSKAJA, N. M. 1979. The primary melt of the oceanic tholeiite and the upper mantle composition. *In*: TALWANI, M., HARRISON, C. G. & HAYES, D. (eds). *Deep Drilling Results in the Atlantic Crust*. Maurice Ewing Series, Vol. 2. Am. Geophys. Union. 302–13 pp.

EISSEN, J.-P., BIDEAU, D., JUTEAU, T. 1981. Présence de basaltes porphyriques dans les zones de fracture de la dorsale Est-Pacifique, *C. R. Acad. Sci. Paris* **293**, 61–6.

FISK, M. R., SCHILLING, J-G. & SIGURDSSON, H. 1980. An experimental investigation of Iceland and Reykjanes Ridge tholeiites. I. Phase relations, *Contr. Mineral. Petrol.* **74**, 361–74.

FLOWER, M. F. J. 1980. Accumulation of calcic plagioclase in ocean-ridge tholeiite: an indication of spreading rate? *Nature, Lond.* **287**, 530–2.

—— 1981. Thermal and kinematic control on ocean-ridge magma fractionation: contrasts between Atlantic and Pacific spreading axes, *J. Geol. Soc. Lond.* **138**, 695–712.

—— 1982. Cryptocumulate tholeiite as evidence for magma mixing at an intermediate-rate spreading axis, *Nature, Lond.* **299**, 542–5.

—— & ROBINSON, P. T. 1981. Basement drilling in the western Atlantic. I. Magma fractionation and its relation to eruptive chronology, *J. Geophys. Res.* **86**, 6273–93.

—— & O'HEARN, T. 1983. An electron microprobe study of basalt chemical variation at DSDP Sites 482, 483 and 485, Gulf of California. *Initial Reports of the DSDP.* **65**, 549–58.

——, ROBINSON, P. T., SCHMINCKE, H.-U. & OHNMACHT, W. 1977. Magma fractionation systems beneath the Mid-Atlantic Ridge near 36–37°N., *Contr. Mineral. Petrol.* **64**, 167–95.

——, PRITCHARD, R. G., SCHMINCKE, H.-U. & ROBINSON, P. T. 1983. Geochemistry of basalts: DSDP Sites 482, 483 and 485 near the Tamayo Fracture Zone, Gulf of California. *Initial Reports of the DSDP.* **65**, 559–78.

FORNARI, D. J., SAUNDERS, A. D. & PERFIT, M. R. 1983. Major-element chemistry of basaltic glasses recovered during DSDP Leg 64. *In*: CURRAY, J. R., MOORE, D. J. *et al.* (eds) *Initial Reports of the DSDP* **64, 2**, 643–8.

FOWLER, C. M. R. 1976. Crustal structure of the Mid-Atlantic Ridge crest at 37°N., *Geophys. J. R. Astr. Soc.* **47**, 459–591.

FREY, F. A. & CHIEN MIN SUNG 1974. Geochemical results for basalts from Sites 253 and 254. *Initial Reports of the DSDP.* **26**, 567–72.

FUJII, T. & KUSHIRO, I. 1977. Density, viscosity and compressibility of basaltic liquid at high pressure, *Ann. Report Dir. Geophys. Lab., Carnegie Inst. Wash. Yearb.* **76**, 419–24.

GASS, I. G. 1968. Is the Troodos Massif of Cyprus a fragment of Mesozoic ocean floor. *Nature, Lond.* **220**, 39–42.

GOODE, A. D. T. 1977. Flotation and remelting of plagioclase in the Kalka Intrusion, Central Australia: petrological implications for anorthosite genesis. *Earth Planet. Sci. Lett.* **34**, 379–97.

GREEN, T. H. 1972. Crystallization of calc-alkaline andesite under controlled high pressure hydrous conditions. *Contr. Mineral. Petrol.* **34**, 150–66.

GREEN, T. H., HIBBERSON, W. D. & JAQUES, A. L. 1979. Petrogenesis of mid-ocean ridge basalts. *In*: MCELHINNY, M. W. (ed.) *The Earth, its Origin, Structure and Evolution*. Academic Press, London.

HALE, L. D., MORTON, C. J. & SLEEP, N. H. 1972. Reinterpretation of seismic reflection data over the East Pacific Rise. *J. Geophys. Res.* **87**, 7707–17.

HAMLYN, P. R. & BONATTI, E. 1980. Petrology of mantle-derived ultramafics from the Owen Fracture Zone, NW Indian Ocean: implications for the nature of the oceanic upper mantle. *Earth Planet. Sci. Lett.* **48**, 65–79.

HART, S. R. 1976. LIL-element geochemistry, Leg 34 basalts. *Initial Reports of the DSDP.* **34**, 283–8.

——, GLASSLEY, W. E. & KARIG, D. E. 1972. Basalts and sea-floor spreading behind the Mariana Arc. *Earth Planet. Sci. Lett.* **15**, 12–8.

HAWKINS, J. W. 1976. Petrology and geochemistry of basaltic rocks of the Lau Basin. *Earth Planet. Sci. Lett.* **28**, 283–97.

HUPPERT, H. E. & SPARKS, R. S. J. 1980a. Restrictions on the compositions of mid-ocean ridge basalts: a fluid dynamical investigation, *Nature, Lond.* **286**, 46–8.

—— & —— 1980b. The fluid dynamics of a basaltic magma chamber replenished by influx of hot, dense ultrabasic magma. *Contr. Mineral. Petrol.* **75**, 279–89.

JUTEAU, T., EISSEN, J.-P. & FRANCHETEAU, J. *et al.* 1980. Homogeneous basalts from the East Pacific Rise at 21°N.: steady-state magma reservoirs at moderately-fast spreading centres. *Oceanol. Acta.* **3**, 487–503.

KEMPE, D. R. C. 1974. The petrology of the basalts, Leg 26. *Initial Reports of the DSDP.* **26**, 465–504.

KUZNIR, N. J. & BOTT, M. H. P. 1976. A thermal study of the formation of oceanic crust, *Geophys. J. R. Astr. Soc.* **47**, 83–95.

LANGMUIR, C. H., VOCKE, R. D., HANSON, G. N. & HART, S. R. 1978. A general mixing equation with application to Icelandic basalts. *Earth Planet. Sci. Lett.* **37**, 380–403.

LEWIS, B. T. R. 1979. Periodicities in volcanism and longitudinal magma flow on the East Pacific Rise at 23°N., *Geophys. Res. Lett.* **6**, 753–6.

LICHTMAN, G. & EISSEN, J.-P. 1983. Time and space constraints on the evolution of medium-rate spreading centres. *Geology* (in press).

——, ——, NORMARK, W. R. & JUTEAU, T. 1982. Concurrent variations in petrology and geomorphic expression along the axis of the EPR near lat. 21°N., *Trans. Am. Geophys. Union, EOS.* **63**, 1134.

LILIAS, R. A. & RHODES, J. M. 1982. Evolution of basaltic magma along the Juan de Fuca Ridge, *Trans. Am. Geophys. Union, EOS.* **63**, 1147.

MCCLAIN, K. J. & LEWIS, B. T. R. 1982. Geophysical evidence for the absence of a crustal magma chamber under the northern Juan de Fuca Ridge: a contrast with ROSE results, *J. Geophys. Res.* **87**, 8477–90.

MELSON, W. G. 1979. Chemical stratigraphy of Leg 45 basalts: electron probe analyses of glasses. *Initial Reports of the DSDP.* **45**, 507–12.

MOORES, E. M. & VINE, F. J. 1971. Troodos Massif, Cyprus and other ophiolites as oceanic crust: evaluation and implications. *Phil. Trans. R. Soc. Lond.* **A268**, 443–66.

MOREL, J. M. & HEKINIAN, R. 1980. Compositional variation of volcanics along segments of recent spreading ridges, *Contr. Mineral. Petrol.* **72**, 425–36.

NISBET, E. G. & FOWLER, C. M. R. 1978. The Mid-Atlantic Ridge at 37 and 45°N: some geophysical and petrological constaints. *Geophys. J. R. Astr. Soc.* **54**, 631–60.

—— & PEARCE, J. A. 1973. TiO_2 and a possible guide to past oceanic spreading rates. *Nature, Lond.* **246**, 468–9.

O'HARA, M. J. 1968. Are ocean floor basalts primary magma? *Nature, Lond.* **253**, 708–10.

—— & MATHEWS, R. E. 1981. Geochemical evolution in an advancing, periodically replenished, periodically tapped, continuously fractionated magma chamber. *J. Geol. Soc. Lond.* **138**, 237–77.

O'NIONS, R. K., HAMILTON, P. J. & EVENSEN, N. M. 1977. Variations in $^{143}Nd/^{144}Nd$ and $^{87}Sr/^{86}Sr$ ratios in oceanic basalts. *Earth Planet. Sci, Lett.* **34**, 13–22.

——, EVENSEN, N. M., HAMILTON, P. J. & CARTER, S. R. 1978. Melting of the mantle, past and present: isotope and trace element evidence. *Phil. Trans. R. Soc. Lond.* **A288**, 547–59.

PALLISTER, J. S. & HOPSON, C. A. 1981. Samail ophiolite suite: field relations, phase variation, cryptic variation and layering, and a model of a spreading ridge magma chamber. *J. Geophs. Res.* **86 (B4)**, 2593–644.

PEARCE, J. A. 1975. Basalt chemistry used to investigate past tectonic environment on Cyprus. *Tectonophys.* **25**, 41–68.

—— 1980. Geochemical evidence for the genesis and eruptive settings from Tethyan ophiolites. *In*: PANAYIOTOU, A. (ed.) *Proc. International Ophiolite Symposium, Cyprus*, 261–72.

—— & CANN, J. R. 1973. Tectonic setting of basic volcanic rocks determined using trace element analyses. *Earth Planet. Sci. Lett.* **19**, 290–300.

PRESNALL, D. C., DIXON, J. R., O'DONNELL, T. H. & DIXON, S. A. 1979. Generation of mid-ocean ridge tholeiite, *J. Petrol.* **20**, 3–35.

REID, I., ORCUTT, J. A. & PROTHERO, W. A. 1977. Seismic evidence for a narrow zone of partial melting underlying the East Pacific Rise at 21°N. *Bull. Geol. Soc. Am.* **88**, 678–82.

RHODES, J. M. & DUNGAN, M. A. 1979. The evolution of ocean-ridge basaltic magmas. *In*: TALWANI, M., HARRISON, C. G. & HAYES, D. E. (eds). *Deep Drilling Results in the Atlantic Crust.* Maurice Ewing Series, Vol. 2. Am. Geophys. Union, 262–72 pp.

——, Blanchard, D. P., Rodgers, K. V., Jacobs, J. W. & Brannon, J. C. 1976. Petrology and chemistry of basalts from the Nazca Plate. II. Major and trace element chemistry. *Initial Reports of the DSDP*. **34**, 239–44.
——, ——, Dungan, M. A., Rodgers, K. V. & Brannon, J. C. 1979. Chemistry of Leg 45 basalts. *Initial Reports of the DSDP*. **45**, 447–60.
Robinson, P. T., Melson, W. G., O'Hearn, T. & Schmincke, H.-U. 1983. Volcanic glass compositions of the Troodos ophiolite, Cyprus. *Geology* **VII**, 400–4.
Saunders, A. J. 1983. Geochemistry of basalts recovered from the Gulf of California during Leg 65 of the Deep Sea Drilling Project. *Initial Reports of the DSDP*. **65**, 591–622.
—— & Tarney, J. 1979. The geochemistry of basalts from a back-arc spreading centre in the East Scotia Sea, *Geochim. Cosmochim. Acta* **43**, 555–72.
——, ——, Marsh, N. G. & Wood, D. A. 1980. Ophiolites as oceanic crust or marginal basin crust: a geochemical approach, *In*: Panayiotou, A. (ed) *Proc. International Ophiolite Symposium*. Geol. Surv. Dept., Nicosia, Cyprus, 193–204 pp.
——, Fornari, D. J., Joron, J.-L., Tarney, J. & Treuil, M. 1983. Geochemistry of basic igneous rocks recovered from the Gulf of California, *Initial Reports of the DSDP*. **64**, 595–642.
Scheidegger, K. F. 1973. Temperatures and compositions of magmas ascending along mid-ocean ridges. *J. Geophys. Res.* **78**, 3340–55.
—— & Corliss, J. B. 1981. Petrogenesis and secondary alteration of upper layer 2 basalts of the Nazca Plate. *Mem. Geol. Soc. Am.* (Nazca Plate).
Schilling, J.-G. 1973. Iceland mantle plume: geochemical study of the Reykjanes Ridge, *Nature, Lond.* **242**, 565–71.
—— 1975. Azores mantle blob: rare earth evidence, *Earth Planet. Sci. Lett.* **25**, 103–15.
Shibata, T., Thompson, G & Frey, F. A. 1979. Tholeiite and alkali basalts from the Mid-Atlantic Ridge at 43°N. *Contr. Mineral. Petrol.* **70**, 127–41.
Sinton, J. M. & Christie, D. M. 1982. Geochemical anomalies of the Blanco and Cobb propagating rifts: an evolutionary model, *Trans. Am. Geophys. Union, EOS* **63**, 1147.
Sleep, N. H. 1975. Formation of ocean crust: some thermal constraints. *J. Geophys. Res.* **80**, 4037–42.
—— 1978. Thermal structure and kinematics of mid-ocean ridge axes: some implications to basaltic volcanism, *Geophys. Res. Lett.* **5**, 426–8.
Sparks, R. S. J., Meyer, P. & Sigurdsson, H. 1980. Density variation amongst mid-ocean ridge basalts: implications for magma mixing and the scarcity of primitive lavas. *Earth Planet. Sci. Lett.* **46**, 419–30.
Stern, C. 1979. Open and closed system igneous fractionation within two Chilean ophiolites and the tectonic implications. *Contr. Mineral. Petrol.* **68**, 243–58.
Stolper, E. & Walker, D. 1980. Melt density and the average composition of basalt. *Contr. Mineral. Petrol.* **74**, 7–12.
Walker, D., Shibata, T. & DeLong, S. E. 1979. Abyssal tholeiites from the Oceanographer Fracture Zone, *Contr. Mineral. Petrol.* **70**, 111–25.
Weaver, S. D., Saunders, A. D., Pankhurst, R. J. & Tarney, J. 1979. A geochemical study of magmatism associated with initial stages of back-arc spreading, *Contr. Mineral. Petrol.* **68**, 151–69.
Wood, D. A., Joron, J. L. & Treuil, M. 1979. A re-appraisal of the use of trace elements to classify and discriminate between magma series erupted in different tectonic settings, *Earth Planet. Sci. Lett.* **45**, 326–36.

M. F. J. Flower, Department of Geological Sciences, University of Illinois (Chicago), Chicago, IL 60680, USA.

The structure of the oceanic upper mantle and lower crust as deduced from the northern section of the Oman ophiolite

J. D. Smewing, N. I. Christensen, I. D. Bartholomew & P. Browning

SUMMARY: The lowermost 7 km of the Oman ophiolite consist of harzburgite, dunite and minor lherzolite of mantle origin overlain by layered gabbros and peridotites of the lower oceanic crust. Detailed structural measurements along three traverses through the ophiolite, complemented by olivine petrofabric determinations on oriented specimens from these traverses, reveal the following geometrical parameters of the spreading process:

(i) The ultramafic rocks of the mantle sequence have all suffered deformation by simple shear at temperatures exceeding 950°C beneath a spreading ridge.

(ii) The tectonite fabric which characterizes the mantle sequence passes upwards with no change in orientation or intensity into the layered gabbros and peridotites at the base of the crust.

(iii) The sense of shear and type of slip system vary at random with depth, reinforcing petrological arguments that the Oman mantle sequence results from the juxtaposition of various mantle 'packages' equilibrated at different depths beneath the ridge.

(iv) No unique kinematic model for the ophiolite can be formulated from the structural data of the three traverses; each traverse has a distinctive geometrical arrangement of the various structural elements relative to the reconstructed ridge directions. Dispersed mineral-layering planes indicate closing directions of magma chambers strung along the ridge.

The concept that ophiolites represent ancient oceanic crust, formed either at a major oceanic spreading centre, or more probably, one sited in a back-arc marginal basin, has been founded on a variety of geological and geophysical reasoning. Accepting this analogy, ophiolites can then be used to gain more insight into the processes of ocean-crust formation. Since the time when the ocean-crust analogy was originally formulated by such workers as Gass (1968) and Moores & Vine (1971) much of this work has concentrated on petrological and geochemical modelling (Refs in Coleman 1977). It is only relatively recently that attention has been paid to the pervasive tectonite fabrics that characterize the ultramafic rocks of the ophiolite suite and their possible relationship to the geometrical parameters of the spreading process (Juteau *et al.* 1977; George 1978; Nicolas *et al.* 1980; Prinzhofer *et al.* 1980; Girardeau & Nicolas 1981; Nicolas & Violette 1982).

Structural studies of ophiolites can augment modelling based on oceanic seismic studies in providing data on the kinematics of the upper mantle beneath spreading centres. Such studies in ophiolites are most meaningful, however, when a complete stratigraphy is preserved (so that a maximum number of structural parameters can be interrelated) and the ophiolite is extensive over a sufficiently wide area to display the extent of the structural pattern. Unfortunately, the structural studies on ophiolites to date suffer by either lacking the complete stratigraphy—e.g. Turkey (Juteau *et al.* 1977) and New Caledonia (Prinzhofer *et al.* 1980)—or, where the total stratigraphy is present, by not being laterally extensive—e.g. Troodos, Cyprus (George 1978) and Bay of Islands, western Newfoundland (Girardeau & Nicolas 1981). The Oman ophiolite has suffered relatively little dismemberment during emplacement, and throughout most of its 500 km length, the complete stratigraphy of the ophiolite is exposed. In this account, structural data, complemented by forty-eight olivine petrofabric diagrams are presented from three traverses through the northern section of the Oman ophiolite. Incorporating the data from a traverse described by Boudier & Coleman (1981) in the southern section of the Oman ophiolite allows an interpretation of the geometrical pattern of spreading from a complete ophiolite representing a ridge-parallel segment of Upper Cretaceous oceanic crust exceeding 300 km in length.

Geological setting

The field relations, petrology and geochemistry of the Oman ophiolite have been the subject of many recent papers (e.g. Hopson *et al.* 1981; Smewing 1980; Alabaster *et al.* 1983). This section is restricted to a brief account of the mafic and ultramafic plutonic rocks at the base of the ophiolite from which the structural measurements were taken. The Oman ophiolite is of Cenomanian age (95 Ma). (Tippit *et al.* 1981) and forms the major part of a 700 km arcuate mountain chain along the southern coast of the Gulf of Oman. It comprises the uppermost nappe of a series of Mesozoic thrust slices overlying the autochthonous Mesozoic carbonate platform of the Arabian shield.

The lowermost unit of the ophiolite consists of up to 7 km of ultramafic tectonites. The dominant

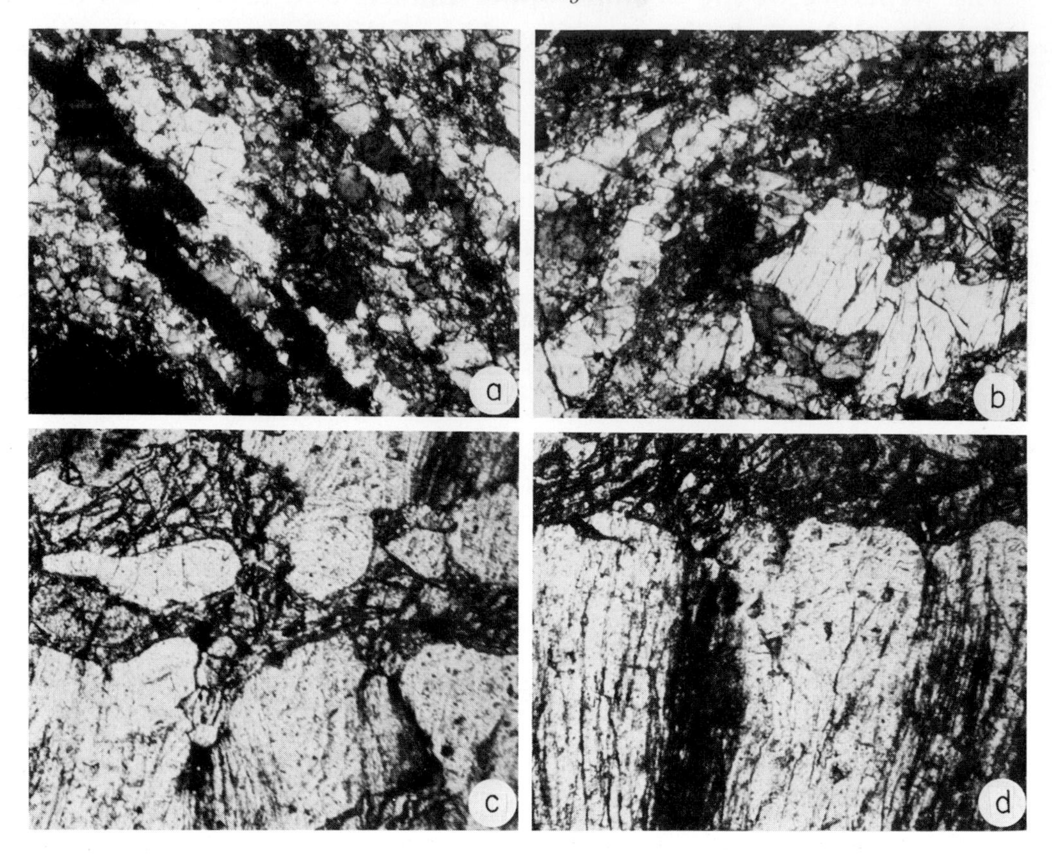

FIG. 1. (a) Harzburgite, Wadi Jizi. Tectonite foliation in harzburgite running from top left to bottom right. Elongate olivine crystal just left of centre with kink-band boundaries recording dextral shear sense with respect to orientation of photograph. Crossed nicols. (Width of photograph 3.5 mm). (b) Harzburgite, Wadi Jizi. Olivine fabric wrapping around enstatite porphyroclast. Crossed nicols. (Width of photograph 3.5 mm). (c & d) Base of layered sequence, Wadi Ragmi. Tectonized layered gabbro with foliation parallel to long dimension of photograph. Pull-apart fractures in plagioclase oriented perpendicular to lineation. PPL (Width of photograph 3.5 mm).

lithology (more than 80%) is harzburgite with a very uniform mineralogy of 80% olivine (Fo_{91}), 17–19% orthopyroxene (En_{91}), 1% chrome spinel and 0–2% clinopyroxene ($Wo_{45}En_{51}Fs_4$). On a centimetre scale, a ratio segregation into more dunitic and pyroxenitic layers is common. A tectonic foliation, defined by the planar orientation of orthopyroxenes is generally parallel to this layering. Chrome spinel lineations are also commonly visible within the foliation plane. A pronounced shearing fabric, associated with the emplacement of the ophiolite affects the lowermost 100–200 m of the harzburgite tectonite. The harzburgites unaffected by emplacement have a microscopic porphyroclastic to mylonitic texture; sub-boundaries in olivine (Fig. 1a) are oriented oblique to the foliation in sections cut perpendicular to the foliation and including the lineation (*xz* plane of Nicolas & Poirier (1976)), indicating deformation by simple shear, as typical of other ophiolite harzburgites (Prinzhofer *et al.* 1980; Girardeau & Nicolas 1981).

Throughout the harzburgite tectonite, sharply defined bodies of dunite (Fo_{91-92} + accessory chrome spinel) are encountered. For the most part, the dunites are concordant with the harzburgite foliation and possess a parallel tectonite fabric. They vary in size from layers 1–2 cm thick to pods with long axes up to 500 m long and thicknesses up to 50 m. In the upper part of the harzburgite tectonite, however, dunite generally appears as random, wispy stringers in harzburgite. In Oman chrome spinel concentrations are restricted to the dunite pods.

Apart from the dunite/chromite bodies in the harzburgite tectonite, a series of more mafic and ultramafic lenses, veinlets and dykes also appear. These bodies are most numerous in the uppermost 50 m of the harzburgite tectonite where they are represented by gabbro, wehrlite and pyroxenite. With increasing depth, these three petrological types persist but in decreasing abundance. At depths exceeding 2 km, olivine norites predominate and in the lowermost 1 km, websterites are most common. These bodies vary from being strictly concordant, with a tectonic fabric parallel to the enclosing harzburgite, to being sharply discordant and completely undeformed. Boudier & Coleman (1981) have proposed an explanation for this relationship by a mechanism where dykes are emplaced sequentially into the harzburgite and are then progressively rotated into the plane of the foliation.

Layered gabbros and peridotites, up to 3.5 km in total thickness, rest with a sharp contact on the harzburgite tectonite. The layers are organized in cyclic units up to 200 m thick, in which the most common sequence is dunite–wehrlite–olivine gabbro–pyroxene gabbro. The thickness of layers varies from several tens of metres to only a few millimetres. In lateral extent, layers rarely exceed 0.5 km and often pinch out after only a few metres. Cumulate* textures, generally adcumulate, and a strong igneous lamination, particularly in gabbroic types, characterize these layers.

The contact between the uniform harzburgite tectonite and the petrologically more diverse layered sequence is an easily mappable, planar boundary that can be traced throughout the length of the ophiolite. The petrology of the lowermost layers resting directly on the harzburgite tectonite varies considerably along the strike of the contact; in terms of the cyclic unit described above this may be anything from dunite to pyroxene gabbro. The mineralogy of the dykes cutting the uppermost harzburgite tectonite matches the phase assemblage developed at the base of the layered sequence. Regardless of whether the harzburgite tectonite is overlain by gabbro or peridotite, the cyclic pattern of layering described above characterizes the layered rocks up sequence.

Although this contact marks a sharp petrological break, the pervasive tectonite fabric of the harzburgite does not die out here but can be traced upwards into the layered rocks without any significant change in orientation or intensity in all the three areas studied. The tectonized layered rocks show a flattening of all constituent minerals and a lineation on the foliation plane defined by olivine and plagioclase elongation. In thin section, pyroxenes are severely fractured; olivine shows a similar wavy substructure to that noted in the harzburgite, with kink-band boundaries oriented at a high angle to the foliation; and plagioclase shows a distinctive pull-apart planar fabric perpendicular to its elongation (Fig. 1c & d). The tectonite fabric overprinting the base of the layered sequence becomes weaker upwards; tectonized cumulates alternate with non-tectonized rocks up sequence and eventually the tectonite fabric dies out completely, some 100–200 m above the base of the layered sequence.

The layered sequence passes upward with a loss of igneous lamination and the abrupt incoming of metabasaltic dykes into isotropic gabbros and plagiogranites, varying in thickness from 50–200 m. The ophiolite stratigraphy is completed by a sheeted dyke swarm, 1.5 km thick, and 1 km of largely pillowed volcanics.

*This term is used here in a purely descriptive, non-genetic sense.

The choice of a geometrical reference frame

In order to relate the structural measurements described later in this paper to processes at present-day oceanic spreading centres, a rigid geometrical reference frame for the Oman ophiolite has to be chosen. All measurements need to be referred to a palaeohorizontal plane, the direction of the ridge and the side of the ridge which the ophiolite represents.

There are several possible choices for a palaeohorizontal plane in the Oman ophiolite:

(i) depositional planes in sediments and volcanics;
(ii) a plane perpendicular to the orientation of the dyke swarm and intersecting the dyke swarm parallel to strike;
(iii) major lithological contacts within the ophiolite; and
(iv) the plane of magmatic layering.

All these choices depend upon certain assumptions being made with regard to the dynamics of the spreading process. It is the object of this discussion to decide which is geologically the most valid.

Depositional planes in pillowed volcanics can frequently be steeply dipping and this is borne out in the field where flow orientations may change dramatically over a short distance. The dip of pelagic sediments within and above the volcanic pile should be a reliable palaeohorizontal indicator; however, in practice, the occurrence of such outcrops is sparse in Oman and, as they occur near the top of the ophiolite, are often a considerable distance from the structures of interest in the gabbros and peridotites.

[100] direction and will produce the petrofabric diagrams shown in Fig. 2c & d. In the first case (Fig. 2c) the olivine petrofabric diagram is typified by clustering of [100] and girdles of [010] and [001]. In the second case, all crystallographic axes form clusters. In both the medium- and high-temperature ranges, the tectonite foliation, defined by mineral elongation, makes an angle θ with the slip planes. The smaller the angle θ the higher the amount of shear. The sense of asymmetry of the slip planes with respect to the tectonite foliation reveals the sense of shear. With increasing strain, recrystallization becomes more and more important.

In summary, therefore, olivine petrofabric diagrams have the potential of revealing:

(i) whether an olivine-bearing crystalline aggregate consists of randomly oriented olivine crystals (rare), olivine crystals that have undergone passive body rotation only in a viscous medium (e.g. layered gabbros and peridotites) or olivine crystals that have undergone both passive body rotation and intercrystalline gliding (tectonites). This is useful in distinguishing tectonized from non-tectonized rocks of the layered sequence;

(ii) the slip system, and hence estimates of the temperature of deformation in the tectonites;

(iii) the sense of shear in tectonites; and

(iv) in favourable conditions, the relative amount of shear in tectonites.

Structural measurements

The locations of the three traverses described in this study are shown in Fig. 3. The complete ophiolite stratigraphy is well exposed in each of these traverses with the exception of Wadi Bani Kharus where the pillowed volcanics are largely covered by recent deposits. The following structural measurements were recorded from the harzburgite tectonite and layered sequence of the three traverses:

S_{0t} : mineral layering in harzburgite tectonite;

S_{1t} : tectonite foliation in harzburgite tectonite;

L_{1t} : chrome spinel lineation in harzburgite tectonite;

S_{0c} : mineral layering in layered sequence;

S_{1c} : tectonite foliation in layered sequence; and

L_{1c} : tectonite lineation in layered sequence.

This notation is taken for the most part from Juteau *et al.* (1977). Several other structural measurements were recorded, notably dyke orientations in the harzburgite tectonite and mineral lineations in undeformed rocks of the layered sequence. However, these were never found in sufficient numbers to be quantitatively significant.

In each traverse, the structural measurements were related to the geometrical reference frame described above by rotating the petrological 'Moho' to horizontal and recording the strike of the dyke swarm (now sub-vertical after rotation). The results of this rotation on the means of the structural measurements for the three traverses are shown in Figs 4 & 5 and Table 1.

Mineral layering in the layered sequence in all three traverses is shallow dipping (<20°) and dips into the ridge axis in Wadi Ragmi and Ath Thuqbah and along the ridge in Wadi Bani Kharus. Mantle flow planes approximated by S_{1t} and S_{1c} are shallow dipping for Wadi Ath Thuqbah and Bani Kharus but moderately inclined (~45°) in Wadi Ragmi, and mantle flow directions given by L_{1c} and L_{1t} lie within or close to mantle flow planes. All mantle flow directions are oriented at moderate to high angles (>30°) to the palaeo-ridge.

In summary, the results of the structural measurements for the three traverses show that although there is an overall tendency for planar fabrics to be shallow dipping (<20°) and mantle flow direction to be oriented at high angles to the palaeo-ridge axis, the contrasting geometrical disposition of the various structural elements with respect to the palaeo-ridge direction in each traverse shows that no unique kinematic model can be formulated for the Oman ophiolite based on this data set.

Olivine petrofabric analysis

Forty-eight oriented specimens from the harzburgite tectonite and layered sequence of the three traverses were collected for olivine petrofabric analysis. Equal-area, lower-hemisphere projections of [100], [010] and [001] axes for most of these specimens are shown in Fig. 6.

The following points are significant:

(i) Olivine in undeformed layered gabbros and peridotites away from the base of the layered sequence shows an orientation due to passive body rotation alone (Specimens 36, 84). Some preferred orientation of olivine [100] and [001] within the layering plane is shown (specimen 84). This is consistent with the orientation of olivine by magmatic convection currents.

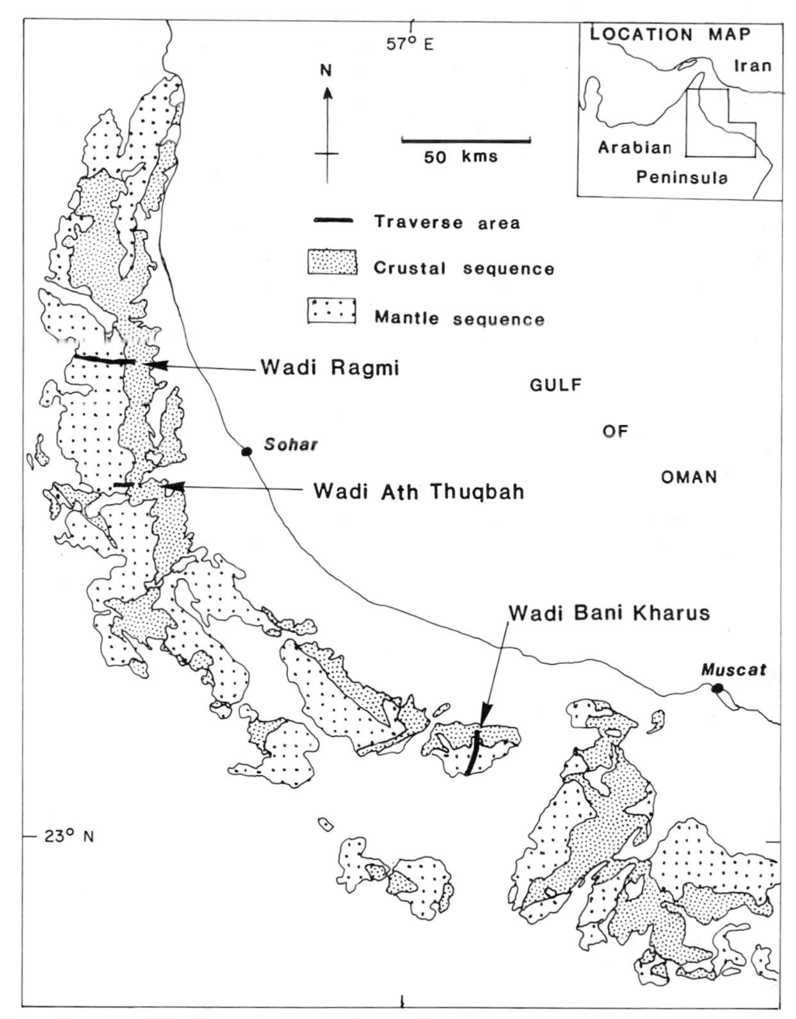

FIG. 3. Location map of the three traverses and outcrop area of the most northerly ophiolite blocks in the Sultanate of Oman.

(ii) The tectonized layered gabbros and peridotites at the base of the layered sequence have an olivine orientation consistent with deformation at both moderate (950–1300°C) and high (>1300°C) temperatures (specimens 33, 193, 2419).

(iii) Harzburgites immediately underlying the petrological 'Moho' have a fabric which is identical in terms of intensity, shear sense and the type of slip plane activated with the layered gabbros and peridotites immediately above the petrological 'Moho'.

(iv) All harzburgites show a tectonite olivine fabric.

(v) There is no obvious correlation between depth in the harzburgite tectonite and the olivine slip system activated. The [100] (0kl), moderate-temperature slip system is the most prevalent. Similarly shear senses are highly variable down section.

(vi) In a number of cases (specimens 60, 71, 2297) a [100] (001) slip system has been identified. This has not previously been described from either experimental or natural studies and cannot, at present, be explained.

Conclusions

Kinematic models of ocean-crust formation (Langseth *et al.* 1966; Sleep 1975; Lachenbruch 1976; Tapponnier & Francheteau 1978) portray mantle material upwelling beneath oceanic

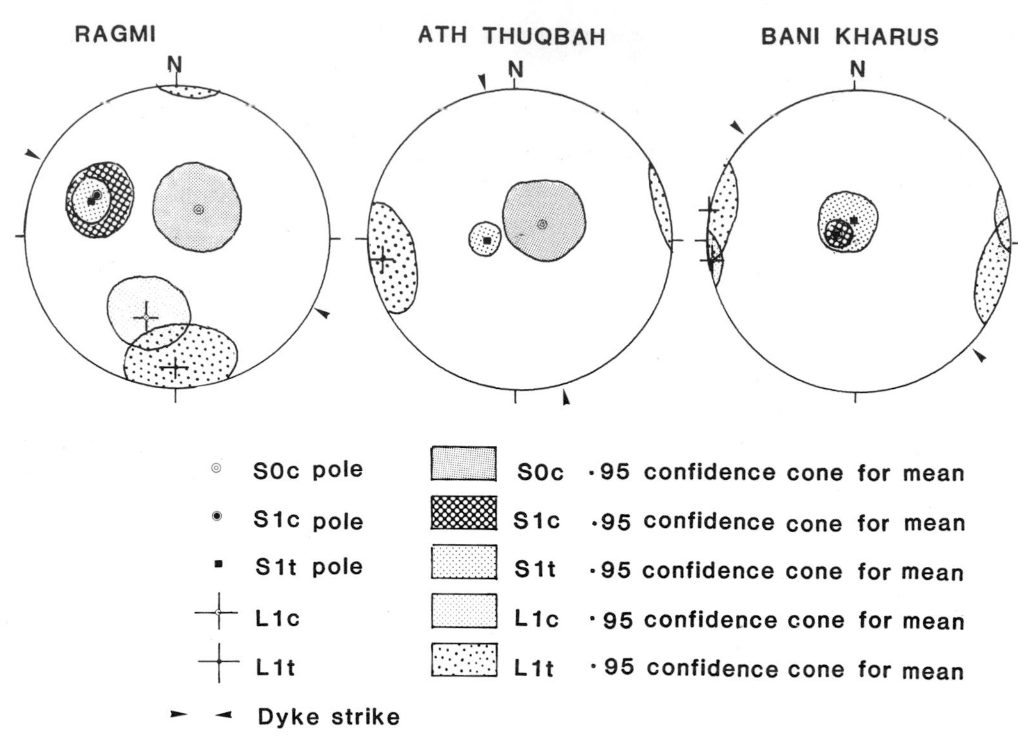

FIG. 4. Plotted mean structural measurements for each traverse rotated to a horizontal petrological 'Moho'. All projections are equal area, lower hemisphere. Each mean and confidence cone is the result of at least 100 measurements.

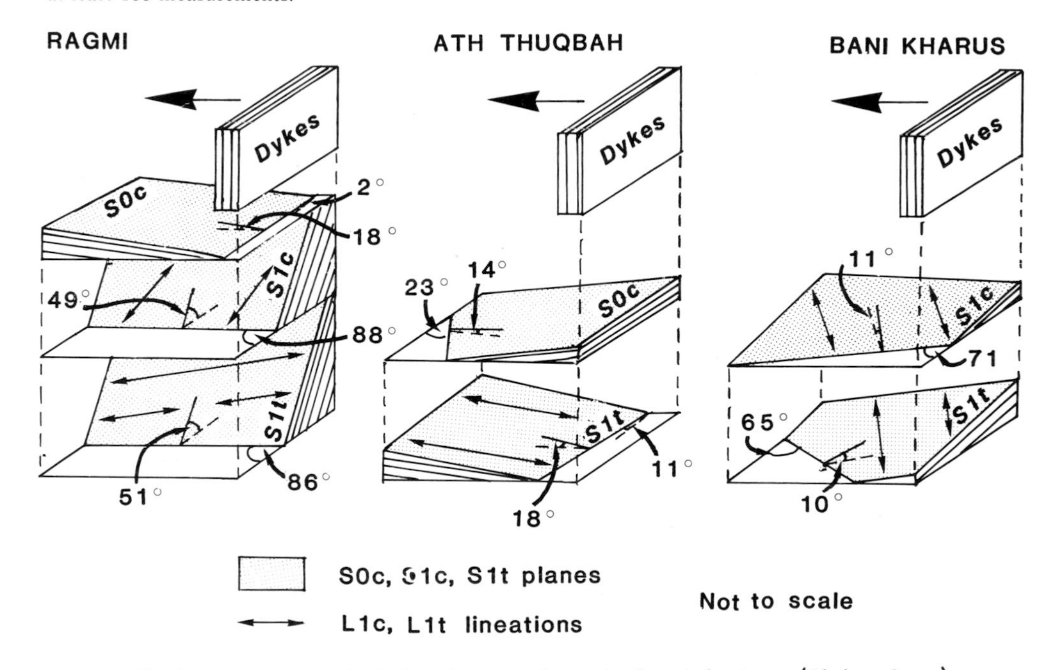

FIG. 5. Perspective views of the structural elements of the three traverses following rotation of the petrological 'Moho' to the horizontal.

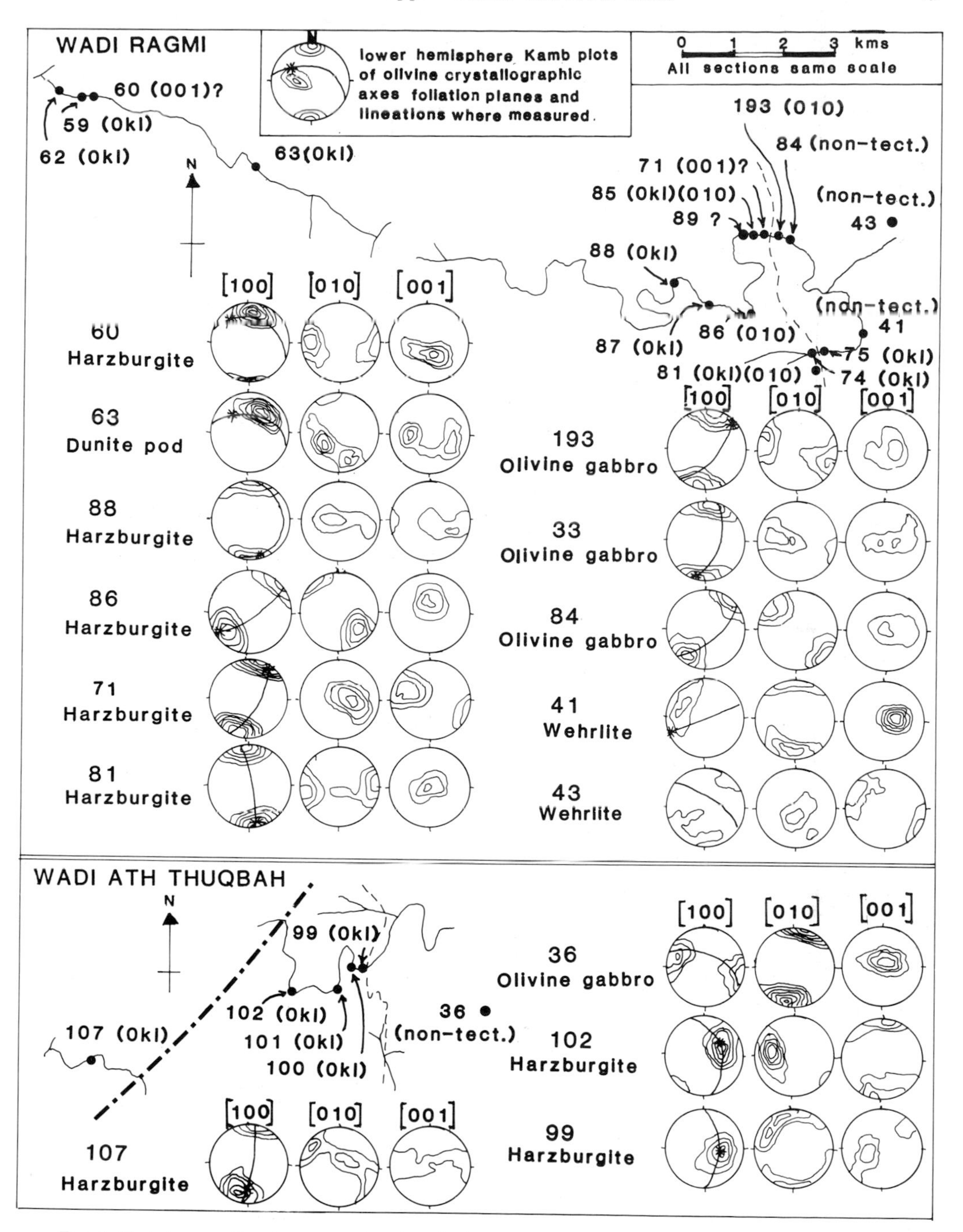

FIG. 6. Sketch map of each traverse showing the locations of specimens that were selected for olivine petrofabric analysis. Equal-area lower hemisphere projections of [100], [010] and [001] axes of at least sixty-five olivine grains, plotted as Kamb diagrams are shown for most of the specimens. The projections are contoured in 2σ intervals with the lowest contour equal to 4σ. Layering (S_0) or foliation (S_1) planes and lineation (L_1) are shown where discernible in outcrop. The slip direction is always [100] irrespective of which slip system is activated; the slip plane and the shear sense are discussed in the text.

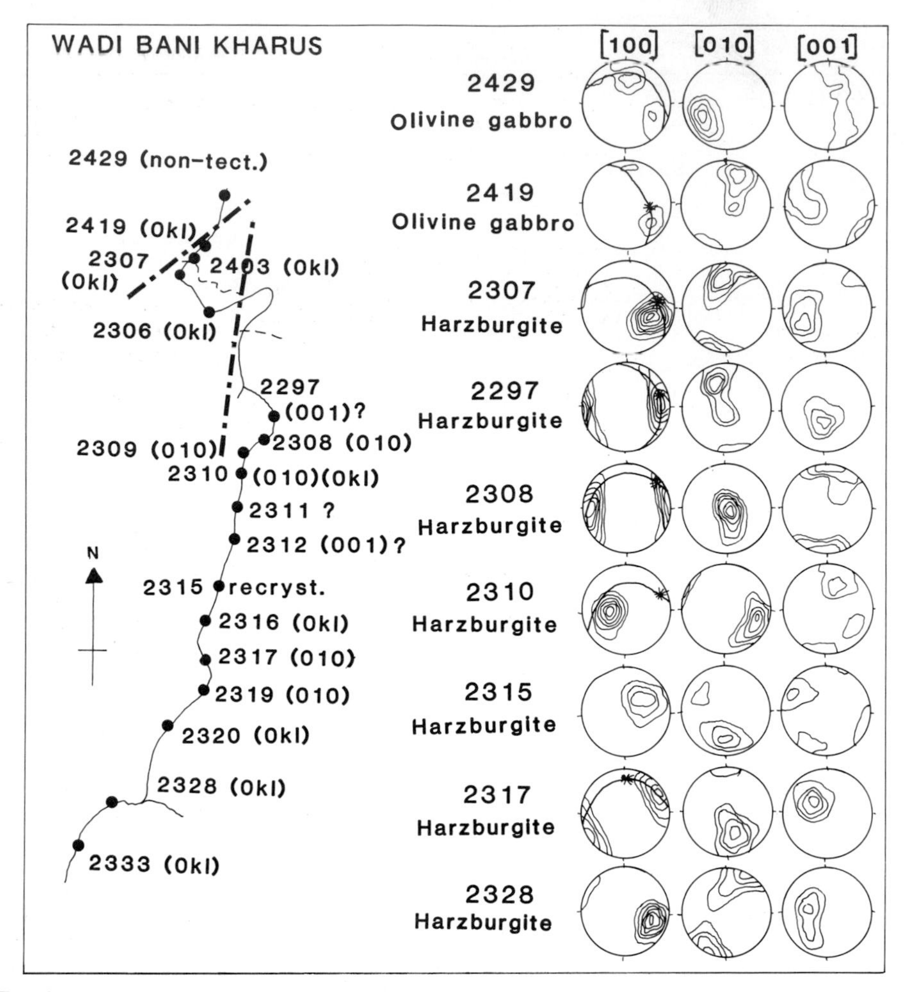

FIG. 6. (continued)

TABLE 1. *Mean structural measurements for each traverse following rotation of the petrological 'Moho' to the horizontal. Measurements shown as dip amount over dip direction; 95% confidence cones shown in brackets. Each mean is the result of at least 100 measurements.*

	Wadi Ragmi	Wadi Ath Thuqbah	Wadi Bani Kharus
Dyke strike	120°	165°	131°
S_{1t}	51/116° (10.2)°	18/086° (8.0)°	10/155° (16.4)°
L_{1t}	18/183° (24.5)°	09/260° (23.6)°	00/281° (20.6)°
S_{0c}	18/212° (24.0)°	14/232° (21.8)°	Sub-parallel to S_{1c}
S_{1c}	49/118° (19.0)°	Undeformed	11/112° (8.5)°
L_{1c}	46/213° (20.0)°	Undeformed	02/084° (14.0)°

*S_{0t} sub-parallel to S_{1t} in all cases.

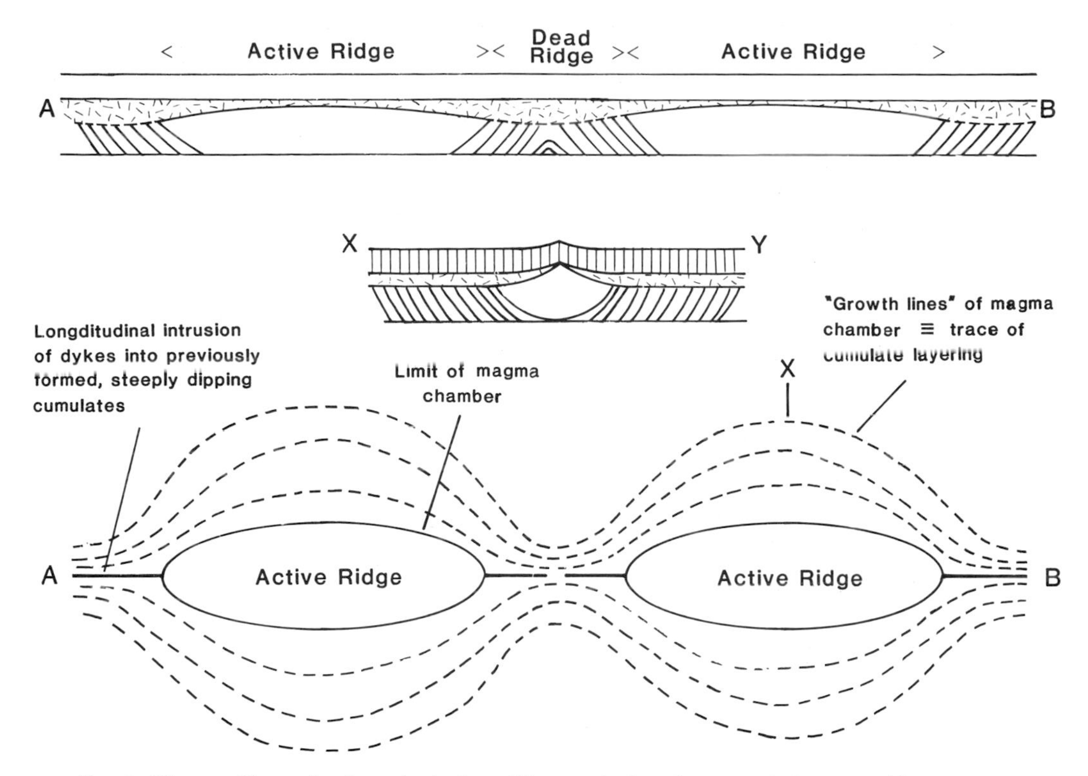

FIG. 7. Diagram illustrating hypothesis that oblique attitudes of structural elements with respect to orientation of palaeo-ridge may be due to discrete ellipsoidal magma chambers strung out along the ridge.

spreading centres and diverging along sub-horizontal flow planes beneath the newly created oceanic lithosphere. If ophiolites sample this oceanic lithosphere then they should show a small angle between mantle flow planes and the palaeohorizontal, a high angle between mantle flow directions, given by chrome spinel lineations and olivine [100] maxima, and the direction of the palaeo-spreading axis and a uniform shear sense. Nicolas & Violette (1982) have already shown that this idea may not be tenable based on a synthesis of structural measurements from eleven ophiolites. They recognize a 'Table Mountain type', approximating to the situation described above, and an 'Acoje type' which preserves vertical rather than horizontal fabrics in the harzburgite tectonite, deformed cumulates and large-scale folds which they ascribe to diapiric rather than laterally spreading asthenospheric flow.

Based on the structural measurements presented earlier, the Oman ophiolite would appear to be generally of the Table Mountain type. However the lack of correlation between the geometry of the structural elements in the three traverses suggests that no unique kinematic model for the Oman ophiolite can be formulated. Rather, a different kinematic interpretation for each traverse is valid based on their unique geometries. Furthermore, the lack of correlation with depth of shear sense and type of slip system activated suggests that the Oman harzburgite tectonite did not uniformly accrete against the base of the oceanic lithospheric plate. Petrologic data (Browning 1982) suggests rather that the Oman harzburgite tectonite results from the juxtaposition of various mantle areas just beneath the palaeo-spreading axis magma chamber during active spreading, these areas having equilibrated at different depths beneath the palaeo-spreading axis.

The angle between the strike of the mineral layering in the layered sequence and the strike of the dyke swarm varies from 0–70° in the three traverses studied. A spreading model with a continuous magma chamber beneath the ridge would predict that this angle should be zero. The data therefore suggest a model of discrete magma chambers beneath the ridge (Fig. 7); the closing of magma chambers parallel to the ridge axis occurring where the angles between the dyke swarm strike and the cumulate layering strike

Melt percolation beneath a spreading ridge: evidence from the Semail peridotite, Oman

R. T. Gregory

SUMMARY: Field mapping in the Semail ophiolite complex, southeastern Oman Mountains reveals a thick oceanic crustal section (on average >6 km thick) underlain by a 9–12 km thick upper-mantle sequence consisting of a basal harzburgite-dunite zone overlain by a foliated harzburgite containing concordant layers of dunite (cm–m scale) and olivine orthopyroxenite (cm scale). The harzburgite and the concordant layers are crosscut by discordant dunite bodies (cm–km scale) and by dykes of dunite, websterite, olivine clinopyroxenite, and gabbro (cm–m scale). The sequence and mineralogy of these features are all consistent with the interpretation that both the concordant and discordant layers represent crystallization or reaction products between ascending melts and host peridotite. Dunite is the only volumetrically significant discordant or concordant rock type crosscutting the peridotite representing ~50% of the basal 3–4 km of the peridotite section and <15% elsewhere. Addition of ~2 km of mantle dunite to the Semail average crustal composition suggests that the average melt percolating up through the Semail mantle had picritic affinities (PLAG 32, DIOP 16, OL 44, SI 08). Field evidence does not require that the harzburgite be cogenetic with the ascending melts that produced the overlying oceanic crustal section. The extremely depleted character of the Semail harzburgite results from interaction with transient melts in the regime of overall high melt/rock ratio (0.5–0.8) found beneath the fossil late-Cretaceous ridge. The melt/rock ratio is calculated assuming that the basal harzburgite-dunite represented the shallow-level, 'off-axis', boundary of the zone of upward magma migration beneath the ridge. The apparently thicker than normal crustal section and the depleted character of the Semail peridotites may suggest that the supply of melt to the Semail ridge system was greater than could be accommodated by spreading in the closing Hawasina ocean basin.

Field and petrologic studies of ophiolite complexes allow direct observation of the products of oceanic crustal genesis. Studies of the plumbing system in ophiolite complexes provide powerful constraints on the origin of mid-ocean-ridge basalts (MORB). Recent petrologic and geochemical studies in both ophiolite complexes (Stern 1979; Juteau & Whitechurch 1980; Hopson *et al.* 1981; McCulloch *et al.* 1981; Pallister & Hopson 1981; Pallister & Knight 1981; Smewing 1981) and on MORB (Rhodes *et al.* 1979; Walker *et al.* 1979) have confirmed that both have a low-pressure signature. This low-pressure signature is imparted to the basaltic liquids by processes occurring in crustal, open-system magma chambers. An unresolved question concerns the nature of the primary (mantle-derived) liquids capable of producing the MORB, and closely related ophiolite suites.

Recent high-pressure experimental studies have confirmed that picritic liquids in equilibrium with mantle mineralogy at 15–20 kb pressures are capable upon fractionation of producing the MORB suite (Green *et al.* 1979; Stolper 1980). A second group of hypotheses favours the production of primary MORB from a plagioclase-lherzolite assemblage at about 10 kb (Presnall *et al.* 1979). In the latter, no modification of the primary melt is required on the ascent to the surface, whereas in the former, a significant amount of olivine fractionation is required (O'Hara 1968). A critical issue is whether any primary melt can reach the crustal magma chamber unfractionated.

The purpose of this paper is twofold: (i) to review the crosscutting relationships in the Semail peridotite in the context of recent experimental work; and (ii) to suggest that, although subsequently deformed, the original character of the features within the peridotite is best explained by igneous processes. The Semail crustal section is dominated by layered gabbro composed of olivine, clinopyroxene, and plagioclase whereas the underlying harzburgite is an olivine, orthopyroxene and spinel assemblage. As pointed out by Malpas (1978) and Hopson *et al.* (1981), and confirmed experimentally (Stolper 1980), the magmas that produced the overlying crust are not in equilibrium with the harzburgite assemblage. Thus, by evaluating the products of the melt–harzburgite interaction, some constraints can be placed upon the composition of the transient melts that fed the crustal magma chambers.

Field relationships

The Semail peridotite is the predominant member of the Semail ophiolite sequence (exposed over 9000 km^2). Within the study area (Fig. 1), exposed

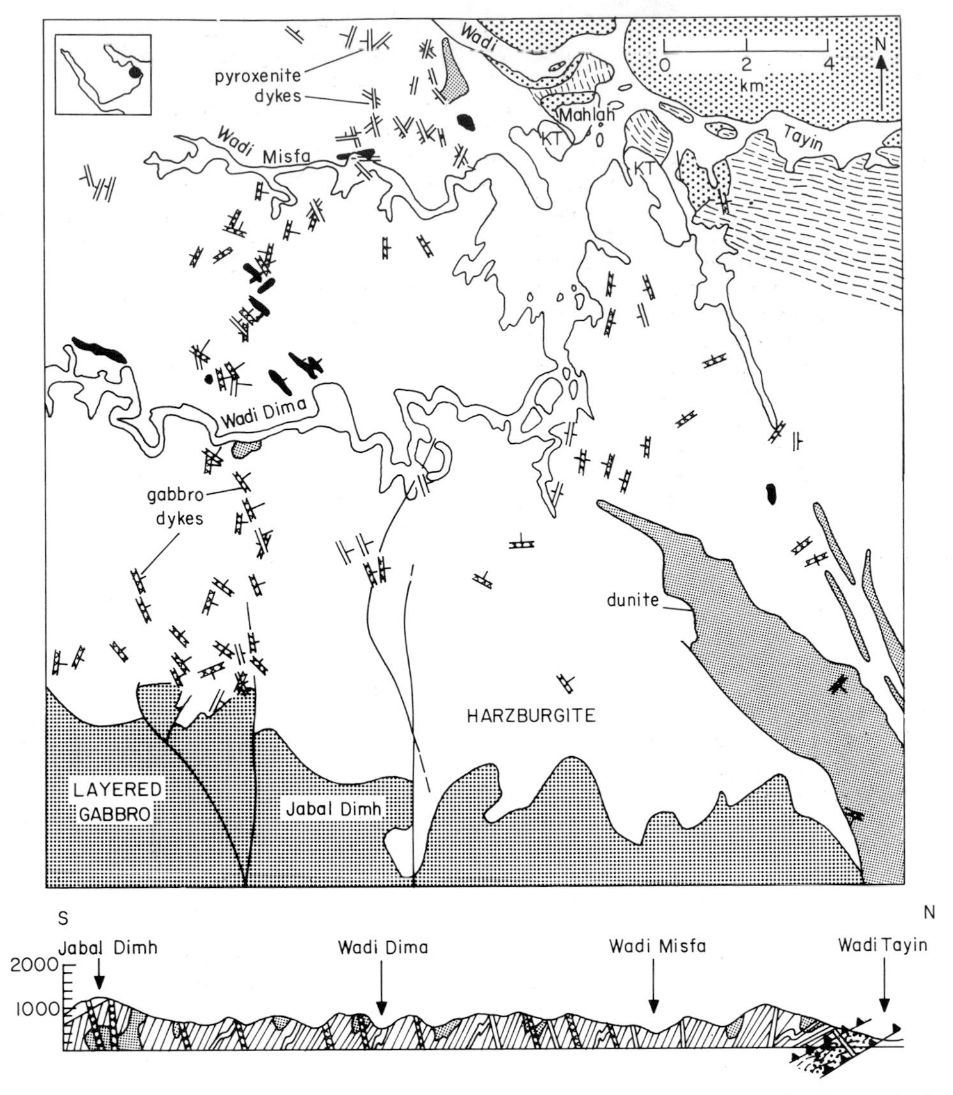

FIG. 1. Location map the study area showing the orientation of the largest dunite bodies (heavy stipple and black), and the attitudes of dykes within the Ibra peridotite section. The basal harzburgite-dunite zone is indicated by the dashed-line pattern. The dashed lines approximate the axial-plane foliation of the harzburgite-dunite zone. Figure modified after Boudier & Coleman (1981) original mapping by E. Bailey, F. Boudier, R. Coleman, R. Gregory, C. Hopson, and J. Pallister. Length of dip arrows on dyke attitudes corresponds inversely to the angle of dip (long: 5–30°, intermediate: 35–60°, short: 65–85°).

between Jabal Dim'h and Wadi Tayin, some 700 km^2 were examined as part of the Ibra transect (cf. Hopson *et al.* 1981; Boudier & Coleman 1981). Along the Ibra transect, the peridotite section underlies a 5–8 km thick section of oceanic crust. The Semail peridotite is predominantly a depleted harzburgite and consists of the following units: (i) a basal harzburgite-dunite zone, locally up to 4 km thick, (ii) a banded harzburgite crosscut by dunite and pyroxenite dykes, and (iii) banded harzburgite crosscut by dunite, pyroxenite and gabbro dykes. This pseudostratigraphy, which includes a plagioclase -in (gabbro) boundary, is crosscut by large (km-sized) dunite bodies indicating that the section is not repeated by folding and faulting. Thus, combining the oceanic crustal section thickness of the ophiolite (6–8 km) with the estimated 9–12 km thick peridotite section suggests that the 5 kb isobar is exceeded in most traverses.

The crosscutting features (Table 1) within the Semail peridotites can be classified into two groups: those that are discordant, and those that are concordant to the foliation of the harzburgite tectonite. The oldest features are concordant with the foliation developed in the tectonite peridotite. On centimeter scales the layering (or banding of Boudier & Coleman 1981) is dominated by isoclinally folded olivine orthopyroxenite. On metre scales, particularly in the basal harzburgite-dunite zone, the larger, concordant layers are dunite containing accessory chromite and diopside. The next oldest features within the peridotite are the concordant pyroxenite or websterite dykes. The early websterite dykes tend to exhibit fold limbs that lie parallel to foliation planes, and fold hinges that swell in size. Macroscopically, the pyroxene crystals in the fold hinges appear undeformed suggesting that the websterites crystallized during the deformation from melt pockets that formed at the hinges of the folds.

Crosscutting the tectonite peridotite and the concordant, isoclinally folded dunite, olivine orthopyroxenite and pyroxenite, are the discordant dykes of dunite, websterite, olivine clinopyroxenite and gabbro. The discordant features tend to occupy conjugate dyke sets trending NW and NNE (Fig. 1). Large dunite bodies (up to 2 km width) tend to develop along the NW fracture trend. The smaller (cm scale) websterite, clinopyroxenite and gabbro dykes tend to develop along both conjugate joint sets. Gabbro dykes occur only in the upper two-thirds of the peridotite section. The youngest discordant features are unaffected by the tectonite development. However, older discordant bodies may exhibit a mineral lineation subparallel to that of the harzburgite. The older, discordant dykes also appear to have rotated closer to the plane of the harzburgite foliation than the youngest, undeformed dykes (Boudier & Coleman 1981).

The youngest events within the Semail peridotite are: (i) the formation of the basal mylonite, (ii) intrusion of post-detachment diabase dykes that also crosscut the metamorphic contact aureole, and (iii) the serpentinization of the peridotite. The basal mylonite is of interest because it lies sandwiched between the contact aureole rocks underlying the early (high T) Semail detachment surface (Hopson *et al.* 1981) and the basal harzburgite-dunite zone. The basal mylonite affects only a portion of the harzburgite-dunite zone (F. Boudier, pers. comm. 1980), and thus appears to be a younger feature superimposed on an older feature within the Semail peridotite.

Volumetrically, the concordant and discordant dunite are the most significant crosscutting features within the Semail peridotite. Concordant dunites comprise up to 50% of the basal harzburgite-dunite zone. Elsewhere, the concordant and discordant dunite comprises less than 15 volume % of the peridotite section. The next most volumetrically significant feature in the peridotite is the concordant olivine orthopyroxenite. The development of the orthopyroxenite layering is not uniform within the section. Locally, the concordant orthopyroxenite comprises up to 50% of the exposure above the basal harzburgite-dunite zone. However, in most parts of the section, the olivine orthopyroxenite layers comprise less than 10% of the peridotite section. The discordant gabbro, websterite and clinopyroxenite dykes comprise less than 1% of the section, and are volumetrically insignificant compared to the dunite and olivine orthopyroxenite.

Crystallization histories of ascending melts

The crosscutting and discordant features discussed above, and summarized in Table 1 can be interpreted by consideration of two facts: (i) orthopyroxene is never an early cumulus phase in the overlying cumulate gabbro, and is virtually absent in the gabbro of the southeastern Oman mountains, and (ii) orthopyroxene is not a liquidus phase of MORB in experimental studies at pressures less than 10 kb (Stolper 1980; Green *et al.* 1979). Thus, the field observations, and corroborating experimental work require that all contacts between ascending melts and harzburgite are boundaries where magmatic reactions occur. The reaction relationship between liquid and wall-rock peridotite results in some combination of crystal accumulation in the conduits, and dissolution of a low melting fraction within the wall-rock peridotite. Abundant evidence for the origin of the crosscutting dunite as crystal cumulates exists from field and petrologic studies of peridotite. (For a summary, see Coleman 1977, and, more recently, Malpas 1978; Hopson *et al.* 1981; Quick 1981b; Boudier & Coleman 1981; Cassard *et al* 1981; Lago *et al.* 1982.)

Analysis of the distribution of crosscutting features within the Semail peridotite indicates that most melts resided in the olivine-stability field during the ascent through the Semail peridotite. Both the discordant and concordant dunite occupy conduits whose widths span metre to kilometre scales. Small scale (cm-sized) crosscutting features (olivine orthopyroxenite, websterite, olivine clinopyroxenite and gabbro) are always multiply saturated. Melts crystallizing along the walls of the smaller conduits appear to

TABLE 1. *Crosscutting relations in the Semail peridotite*

Rock type	Scale	Foliation	Remarks	Mineralogy	Relative Age*
Diabase	m	discordant	rare, also cuts metamorphic aureole underlying Samail thrust		5
Basal mylonite	m–?	concordant to foliation in contact aureole	dies out up section before top of hz-du zone		4
Gabbro	cm	discordant	conjugate dykes	ol, cpx, plag ± sp opx, cpx, plag ± sp	3c
Clinopyroxenite	cm	discordant	conjugate dykes	cpx, ol	3b
Websterite	cm–m	discordant	conjugate dykes	cpx, opx, sp, ol	3b
Dunite	cm–km	shape of bodies invariably discordant; internal foliation concordant to to discordant	occupy conjugate dyke sets; plag types near the top of the peridotite section	ol, sp; ol, sp ± cpx ± plag	3b
Websterite	cm	concordant with no tectonite fabric	often concentrated at the hinges of folds	cpx, opx, ol, sp	3a
Orthopyroxenite	cm	concordant	isoclinally folded	opx, ol, sp	2
Dunite	cm–10 m	concordant	isoclinally folded	ol, sp	2
Harzburgite	km	tectonite fabric	concordancy of all other features measured relative to harzburgite foliation	ol, opx, sp	1

*Rock types showing the same number designation are mutually crosscutting or virtually the same age. The lower-case letter indicates which rock type would be more likely found crosscutting the other rock types under the same number category.

lie on cotectics. The earliest dykes first intersected the olivine-orthopyroxene cotectic, whereas the latest dykes intersected first the olivine-clinopyroxene cotectic, and later the olivine-clinopyroxene-plagioclase cotectic.

By mass-balance arguments, the approximate major-element composition of the melts entering the Semail ridge system can be calculated if the composition of the crust and distribution of crystallization trails in the mantle section are known. Integrating the dunite volume occurring within the peridotite section requires at least 1–2 km of dunite to be added to the Semail average crustal composition. The results of this calculation are shown in Fig. 2. The Semail average crustal composition (Pallister & Gregory 1983) lies upon the 1-atmosphere cotectic for MORB determined by Walker *et al.* (1979). This composition lies squarely in the centre of the MORB cluster (Fig. 2). The addition of the upper-mantle dunite drives the Semail parental-melt composition (labelled 'Semail picrite' in Fig. 2) towards the 15 kb ol-cpx-opx saturation point of Stolper (1980) shown in Fig. 2. Shown for comparison is the field of Saudi Arabian picritic liquids associated with the opening of the Red Sea rift (Coleman *et al.* 1983).

The result of the calculation shown in Fig. 2 does not necessarily indicate that the parental or primary melts that generated the Semail oceanic crustal section were derived from the Semail peridotite at 15 kb pressure. Examination of the olivine corner (Fig. 3) of the system ol-di-si-pl projected from plagioclase onto the ol-di-si face illustrates this point. A primary-melt field corresponding to the 20 kb ol-cpx-opx point of Stolper (1980) can evolve along a variety of paths. In large conduits (m scale), or where the flow rate is high, a high melt/rock ratio is maintained. Magmas traversing such pathways, where the melt is effectively insulated from the peridotite wall-rocks, will remain in the olivine field during

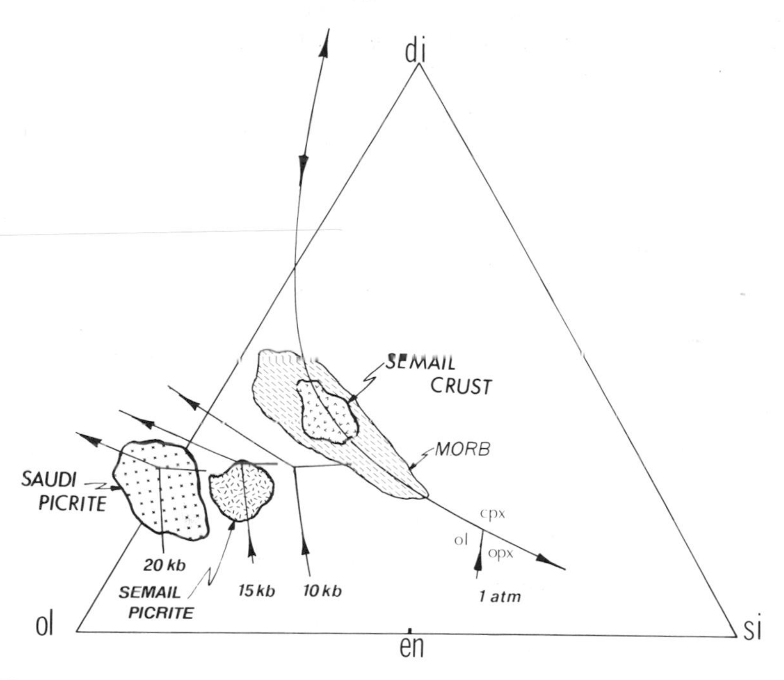

FIG. 2. The system di-ol-si projected from plagioclase using the method of Walker *et al.* (1979). Liquidus relationships are from Stolper (1980) for 20, 15 and 10 kb and from Walker *et al.* (1979), for 1 atmosphere. The fields for the Semail crust and Semail picrite are calculated from field relationships using mass-balance arguments. The field for Saudi picrite is from analyses of picritic liquids erupted during the early phase of Red Sea rifting (Coleman *et al.* 1983). The field for MORB is taken from Walker *et al.* (1979).

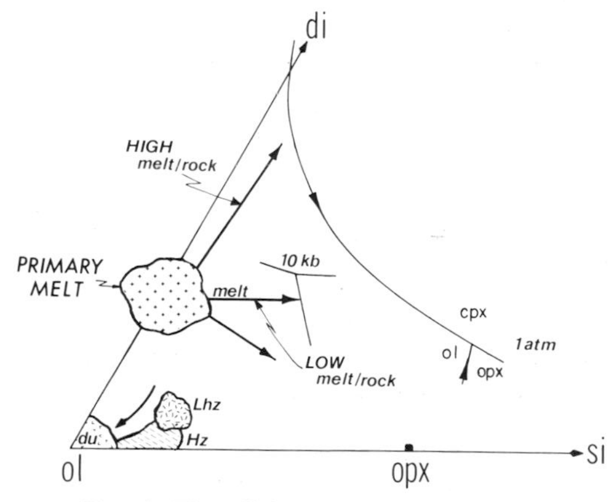

FIG. 3. The olivine corner of the di-ol-si system shown in Fig. 2, illustrating how primary picritic liquids might evolve due to fractionation during ascent through the upper mantle. In large conduits, melts follow a vector (marked high melt/rock) that approaches the olivine control vector. The magnitude and direction of the low melt/rock vector will be a function of the ratio of wall-rock assimilated to olivine crystallized. The trajectories suggest that melts in small (low m/r) conduits will evolve rapidly to cotectic assemblages.

the entire ascent to the crust (Fig. 3). In small-scale conduits, where the surface area of the dyke is large compared to the volume, the melts will react with the wall rocks more efficiently. The path along which the melt evolves will be some vector (marked 'low melt/rock' in Fig. 3) that is the superposition of the olivine control vector (marked 'high melt/rock') and the negative of an orthopyroxene control vector. In cases where lherzolite is the wall-rock, the melt will evolve along a vector that is the superposition of the olivine control vector and a vector that points towards the composition that is being extracted out of the wall-rocks through mineral-melt reactions. Initially, melts that had equilibrated with a mantle assemblage at depths greater than the level of exposure, would most likely follow a trajectory as indicated on Fig. 3, and intersect first the olivine-orthopyroxene cotectic, whereas at shallower depths, the melt vector would intersect the olivine-clinopyroxene cotectic first. These relationships are preserved in the Semail peridotite as the majority of the olivine-orthopyroxenite dykes are represented as older concordant features, whereas the olivine clinopyroxenites are preserved as late conjugate dykes. In addition, the topology of Fig. 3 does not rule out picritic compositions such as the Saudi picrites as possible primary melts for the Semail oceanic crust, nor picritic melts in equilibrium with garnet as proposed by O'Hara (1968). As Fig. 3 illustrates, the position of the Semail picrites may not indicate the composition of the primary melt, but instead indicates the *approach* to equilibrium with the surrounding wall-rocks during the ascent to the surface.

Relationship between harzburgite and crustal layers

The approach to equilibrium between the ascending melts and the wall-rock peridotite will be governed by parameters such as dyke density, conduit shape, temperature, pressure and the proportion of melt to mantle peridotite. The increasing abundance of originally crosscutting features towards the base of the peridotite is suggestive of increasing dyke density along the margins of the thermal disturbance associated with the Semail ridge system (see Fig. 4). The presence of lherzolite wedges in contact aureoles in Newfoundland (Malpas 1978) suggests that the depleted harzburgite layer underlying ophiolites may be a shallow-level feature that results from the interaction between ascending melts and mantle peridotite. In this interpretation, the Semail harzburgite and oceanic crustal sections are related only by their mutual involvement in the Semail spreading event. The harzburgite is viewed as the wall-rock that has been subjected to the 'magmathermal' event associated with the formation of the Semail oceanic crust. Thus, the last depletion event within the Semail peridotite may be a result of the melt/rock ratio associated with the formation of the Semail crust. The melt/rock can be crudely estimated by assuming that the basal harzburgite-dunite zone represents the boundary in the mantle between peridotite largely unaffected by the ridge event (such as the lherzolite wedges at the base ophiolitic thrust sheets), and the ridge magmatic system (Fig. 4). Given the estimated thicknesses for the Semail crust (5–8 km), and the thickness of the Semail peridotite (9–12 km), the bulk melt/rock ratio for the Semail peridotite is estimated to be 0.50–0.80 in weight units. Given the possible implications of a melt/rock ratio of this magnitude, the question of the possible cognetic relationship between ascending melts and wall-rock can be addressed.

Conflicting evidence (McCulloch *et al.* 1981; and Pallister & Knight, 1981) exists concerning whether isotopic and rare earth element (REE) data for the Semail peridotite point to a cogenetic relationship. The light REE-depleted patterns of the Semail crustal section are not considered to be complementary with the U-shaped patterns determined on harzburgite samples (Pallister & Knight 1981). On the basis of similar ϵ_{Nd} (harzburgite $\epsilon_{Nd} = 8.3$, and crustal rocks, $7.5 < \epsilon_{Nd} < 8.6$), McCulloch *et al.* (1981) argued that both the Semail mantle and crustal sections could be derived from similar isotopic reservoirs. Both groups suggested that the REE-abundance data may not rule out a cogenetic relationship because of uncertainties in the behaviour of the constituent phases within the peridotite (particularly olivine in a clinopyroxene-free harzburgite), and in the cumulate dunite. Consideration of the melt/rock calculation, above, suggests that the ϵ_{Nd} data also may not be a definitive test of a cogenetic relationship between peridotite and crustal rocks. Using the crustal and mantle thicknesses for the Semail ophiolite, and Nd abundance data (~ 5 ppm for the average crust, and ~ 0.04 ppm for the harzburgite (McCulloch *et al.* 1981)), the estimated Nd melt/rock ratio exceeds 50. A more realistic estimate for Nd melt/rock allows for the dissolution of clinopyroxene out of a lherzolite precursor (Nd ~ 3 ppm, ~ 15 modal % clinopyroxene). Allowing for a lherzolitic precursor reduces the Nd melt/rock ratio to a value on the order of 10. Assuming that the exchange is reasonably efficient, the ϵ_{Nd} of the wall-rock peridotite will be reset by melt migration of a magnitude comparable to the Semail spreading event, thus obliterating any original primary relationships.

In order to further constrain: (i) the relationship of the Semail peridotite to the overlying oceanic crustal section, and (ii) the efficiency of the reaction between ascending melts and mantle peridotite, it is instructive to compare the Semail ophiolite with the Trinity peridotite, northern California. In particular, Quick (1981a, b) has shown that the Trinity peridotite provides an example of a peridotite body that exhibits a wide range of compositions from plagioclase lherzolite to dunite. In contrast to the Semail peridotite, the Trinity peridotite has a thin (<3 km thick) section of oceanic crust. Thus, for each column of upper-mantle section, at least twice as much melt passed through the Semail section. The Trinity peridotite contains a suite of cross-cutting features similar to those of the Semail peridotite and is part of the Trinity ophiolite (Lindsley-Griffin 1977). Concentric around the crosscutting dunite, pyroxenite and gabbro dykes, zones of depleted wall-rocks are observed. Moving outward from the dunite bodies, harzburgite, lherzolite and plagioclase lherzolite are observed. These depletion halos are observed 5–10 dyke widths away from the contacts between wall-rock and dykes. Thus, from these relationships, the Trinity lherzolite is

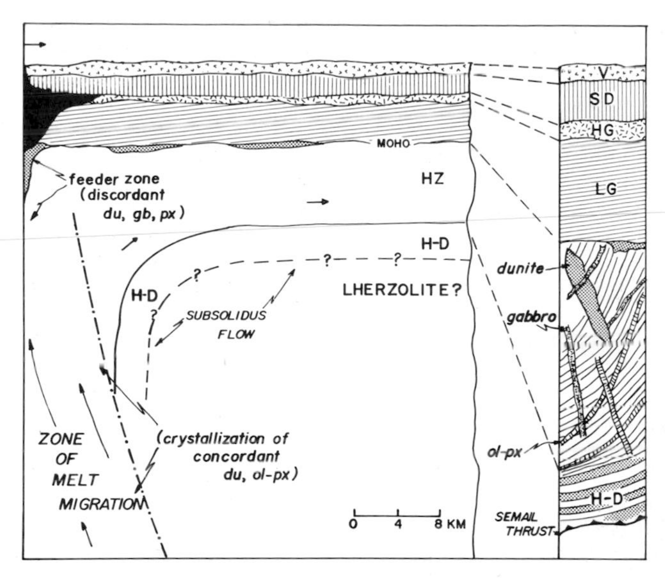

FIG. 4. Cartoon reconstruction showing the zone of melt migration beneath the Semail ridge. The harzburgite-dunite zone (H-D) is interpreted to be the fossil boundary between the zone of melt migration, and the uppermost mantle (lherzolite?) largely unaffected by the Semail magmatic event. The harzburgite-dunite zone is inferred to have formed when the upward-flowing mantle crossed the boundary of the melt-migration zone (a thermal barrier, schematically shown by the dash-dot line). Both the now concordant olivine-orthopyroxenite and dunite are interpreted to have formed at depths greater than the level of exposure, and subsequent to crystallization from a magma to have been rotated by the plastic flow of the upper mantle. The orientation of the harzburgite foliation and the discordant dunite (heavy stipple), and the vergence of the layered gabbro (LG, lined pattern), and the crosscutting gabbro and pyroxenite dykes are shown in the column on the right. The large inclined dunite bodies (Fig. 1 and stratigraphic column) are interpreted to have been initially vertical, and represent fossil feeder zones. Subsequent to their formation these discordant dunites are rotated by the local plastic flow beneath the ridge.

probably not cogenetic with the melts that produced the thin crustal section (Quick 1981a, b). The preservation of the lherzolite screens is interpreted as the result of less vigorous dyke injection in the Trinity peridotite, as compared to the Semail peridotite section, and the lower melt/rock ratio of the magmatic system that last affected the Trinity peridotite. The efficiency of the depletion of the Trinity lherzolite around melt conduits suggests that increasing the melt/rock ratio to a value similar to that of the Semail magmatic system would result in the conversion of the entire Trinity massif to a predominantly harzburgite assemblage. Thus, the Trinity peridotite may represent an example of a spreading system with a low melt/rock ratio, whereas the Semail system may represent a regime of high melt/rock ratio. This high melt/rock environment may result from the fact that the Semail ridge (Fig. 4) was operating in the terminal phase of the closing Hawasina ocean basin (Hopson *et al.* 1981). As an alternative to the fast-spreading hypothesis (e.g. Pallister & Hopson 1981), the Semail ophiolite crustal section may be relatively thick (~ 8 km) because of the 'piling up' of melt on the ridge system as the rate of extension decreased in the Hawasina ocean basin.

Conclusions

Studies of the Semail peridotite suggest that no primary melt reached the Semail crustal chamber unfractionated. Crosscutting features indicate that most melts remained in the olivine field during their ascent through the upper-mantle section. Mass-balance arguments that take into account crystal fractionation and partial melting in the Semail upper-mantle section indicate that although the Semail average crustal composition is similar to average MORB, the melts traversing the peridotite section were picritic. This agrees with a similar conclusion of Malpas (1978) for the Bay of Islands, Newfoundland and Smewing (1981) for the Semail ophiolite.

Comparison of the depleted Semail harzburgite with the less-depleted Trinity peridotite suggests that melt/wall-rock interactions in similar environments to that of the Semail peridotite may be sufficient to convert a Trinity-type lherzolite into a depleted harzburgite. The net effects of the upper-mantle peridotite–melt interactions at spreading ridges will be: (i) to progressively deplete the uppermost mantle layer resulting in a harzburgitic layer, and (ii) to modify the composition of the ascending melts in such a way as to preclude a unique determination of the mantle source-region characteristics in terms of simple isobaric melt-residue relationships.

ACKNOWLEDGMENTS: I would like to thank R. G. Coleman, J. E. Quick, J. S. Pallister, F. Boudier, C. A. Hopson, and H. P. Taylor, Jr for stimulating discussion of this work.

References

BOUDIER, F. & COLEMAN, R. G. 1981. Cross section through the peridotite in the Semail ophiolite, southeastern Oman mountains. *J. Geophys. Res.* **86**, 2573–93.

CASSARD, D., NICOLAS, A., RABINOWICZ, M., MOUTTE, J., LEBLANC, M. & PRINZHOFER, A. 1981. Structural classification of chromite pods in southern New Caledonia. *Econ. Geol.* **76**, 805–31.

COLEMAN, R. G. 1977. *Ophiolites: Ancient Oceanic Lithosphere?* Springer-Verlag, Berlin, 229 pp.

——, GREGORY, R. T. & BROWN, G. F. 1983. Cenozoic volcanic rocks of Saudi Arabia. *U.S. Geological Survey open-file report.* (In press)

GREEN, D. H., HIBBERSON, W. O. & JAQUES, A. L. 1979. Petrogenesis of mid-ocean ridge basalts. *In*: MCELHINNEY, M. W. (ed.) *The Earth: its Origin Structure and Evolution*, Academic Press, London, pp. 265–290.

HOPSON, C. A., COLEMAN, R. G., GREGORY, R. T., BAILEY, E. H. & PALLISTER, J. S. 1981. Geologic section through the Samail ophiolite and associated rocks along a Muscat–Ibra transect, southeastern Oman Mountains. *J. Geophys. Res.* **86**, 2527–44.

JUTEAU, T. & WHITECHURCH, H. 1980. The magmatic cumulates of Antalya (Turkey): evidence of multiple intrusions in an ophiolitic magma chamber, *In*: PANAYIOTOU, A. (ed.) *Ophiolites, Proceedings International Ophiolite Symposium, Cyprus*, Geol. Survey, Dept. Cyprus. pp. 377–91.

LAGO, B. L., RABINOWICZ, M. & NICOLAS, A. 1982. Podiform chromite ore bodies: a genetic model. *J. Petrol.* **23**, 103–25.

LINDSLEY-GRIFFIN, N. 1977.The Trinity ophiolite, Klamath Mountains, California. *Bull. Oreg. Dept. Geol. Mineral. Ind.* **95**, 107–20.

MALPAS, J. 1978. Magma generation in the upper mantle, field evidence from ophiolite suites, and application to the generation of oceanic lithosphere. *Phil. Trans. R. Soc. Lond.* **A 288**, 527–46.

MCCULLOCH, M. T., GREGORY, R. T., WASSERBURG, G. J., TAYLOR, H. P. Jr 1981. Sm-Nd, Rb-Sr, and $^{18}O/^{16}O$ Isotopic systematics in an oceanic crustal section: evidence from the Samail ophiolite. *J. Geophys. Res.* **86**, 2721–36.

O'HARA, M. J. 1968. Are any ocean floor basalts primary magma? *Nature, Lond.* **220**, 683–6.

PALLISTER, J. S. & GREGORY, R. T. in press. Composition of the Semail ocean crust. *Geology.*

—— & HOPSON, C. A. 1981. Samail ophiolite plutonic suite: field relations, phase variation, cryptic variation and layering, and a model of a spreading ridge magma chamber. *J. Geophys. Res.* **86**, 2593–644.

—— & KNIGHT, R. J. 1981. Rare-earth element geochemistry of the Samail ophiolite near Ibra, Oman. *J. Geophys. Res.* **86**, 2673–98.

PRESNALL, D. C., DIXON, J. R., O'DONNELL, T. H.& DIXON, S. A. 1979. Generation of midocean ridge tholeiite. *J. Petrol.* **20**, 3–35.

QUICK, J. 1981a. Petrology and petrogenesis of the Trinity peridotite, an upper mantle diapir in the eastern Klamath Mountains, Northern California. *J. Geophys. Res.* **86**, 11,837–63.

—— 1981b. The origin and significance of large, tabular dunite bodies in the Trinity peridotite, northern California. *Contrib. Mineral. Petrol.* **78**, 413–22.

RHODES, J. M., DUNGAN, M. A., BLANCHARD, D. P. & LONG, P. E. 1979. Magma mixing at mid-ocean ridges: evidence from basalts drilled near 22°N on the Mid-Atlantic Ridge. *Tectonophysics* **55**, 35–62.

SMEWING, J. D. 1981. Mixing characteristics and compositional differences in mantle-derived melts beneath spreading axes: evidence from cyclically layered rocks in the ophiolite of North Oman. *J. Geophys. Res.* **86**, 2645–60.

STERN, C. 1979. Open and closed system igneous fractionation within two Chilean ophiolites and tectonic implications. *Contrib. Mineral. Petrol.* **68**, 243–58.

STOLPER, E. 1980. A phase diagram for mid-ocean ridge basalts: preliminary results and implications for petrogenesis. *Contrib. Mineral. Petrol.* **74**, 13–27.

WALKER, D., SHIBATA, T., DELONG, S. E. 1979. Abyssal tholeiites from Oceanographer Fracture Zone II. Phase equilibria and mixing. *Contrib. Mineral. Petrol.* **70**, 111–25.

R. T. GREGORY, Department of Geology, Arizona State University, Tempe, AZ 85287, USA.

Parent magmas of the Semail ophiolite, Oman

J. S. Pallister

SUMMARY: The plutonic rocks of the Semail ophiolite in the southeastern Oman Mountains formed in a spreading-ridge magma chamber that was repeatedly replenished by primitive magma. An average composition of the parent magma is calculated by mass balance of the igneous stratigraphy. The average parent magma was in equilibrium with most early-formed cumulus minerals from the ophiolite and is very similar to primitive mid-ocean-ridge basalt (MORB). Comparison with experimentally determined liquidus relations for primitive MORB also suggests that most of the cumulus minerals could have been produced by fractionation of the average parent magma. The magma is basaltic—in contrast to the picritic melts proposed as parents for some ophiolites. Although the *average* parent magma composition is basaltic, individual batches of magma that fed the magma chamber varied in composition because of mantle processes. The primary mantle melts were probably picritic, but due to reaction during ascent through the upper mantle and mixing or density entrapment within the crustal-level magma chamber, these picritic melts were probably not erupted on the sea floor. A model with primary mantle melting at 40–60 km depth followed by reaction during ascent and possibly second-stage melting at $\leq$25 km depth is favoured for the ophiolite.

Constraints on Semail parent magma compositions

Crystallization sequence

The phase distribution in the cumulus suite of the Semail ophiolite in the Jabal Dimh area of the southeastern Oman mountains is described in detail by Pallister & Hopson (1981). Several points are reiterated here because they are important to the parent magma problem. Olivine and chromite, which were the first liquidus minerals, form basal cumulus dunite zones that are usually present immediately above the petrologic Moho. However, dunite is locally absent where olivine gabbro formed the floor of the magma chamber. Clinopyroxene and plagioclase joined olivine as liquidus phases nearly simultaneously, and very close to the chamber floor. Ultramafic layers recur interlayered with gabbro throughout the cumulus suite.

These relations apparently constrain parent magma compositions to lie along olivine control lines that intersect the four-phase (olivine-plagioclase-clinopyroxene-melt) cotectic of the tholeiitic basalt compositional tetrahedron (Pallister & Hopson, 1981, Fig. 20), or they were shifted onto such lines by magma mixing as explained in a later section.

The local presence of cumulus olivine gabbro instead of dunite at the floor of the chamber and sparse gabbro dykes in the mantle peridotite suggest that some batches of parent magma had already reached the four-phase cotectic when they entered the chamber. This indicates that significant fractionation of parent magma batches, dominated by olivine removal, took place within the oceanic upper mantle during ascent of mantle diapirs to the crustal-level magma chamber. Field evidence of this mantle fractionation is recorded in the mantle suite of the ophiolite (Hopson *et al.* 1979; Boudier & Coleman 1981; Gregory & Coleman 1981; Smewing 1981) and is discussed in a later section.

Mineral-melt equilibria

Minerals in the basal cumulates were probably in equilibrium with parent magmas that fed the magma chamber. As pointed out above, most parent-magma batches had olivine as a liquidus phase. Microprobe analyses of unzoned olivines from basal olivine adcumulates (dunites) in three different areas of Jabal Dimh yield $Fo_{89-90.5}$ compositions (Pallister & Hopson 1981). Mantle olivines from the underlying harzburgite and dunite tectonites are typically Fo_{90-91}.

Using a composition of Fo_{90} for olivine in the basal cumulus dunite and equilibrium distribution of FeO*/MgO between olivine and melt (Roeder & Emslie 1970), it is concluded that most parent magmas had FeO*/MgO ratios of about 0.7 (equivalent to Mg′ values (mol $Mg/Mg+Fe^{+2}$) of about 0.72). This constraint applies to estimates of parent magma composition made on the basis of mass balance.

Mass balance and calculation of SAVE

Measured stratigraphic sections through the ophiolite in Jabal Dimh yield the average ocean crustal column shown in Fig. 1. Mass balance of the column using representative chemical analyses of the various units, gives the Semail average parent magma (SAVE) composition listed in Table 1 and Fig. 1. Average whole-rock chemical analyses of the various igneous-rock units in Jabal Dimh, and the proportions used in the mass balance are shown in Fig. 1.

of the initial low-level cumulus clinopyroxenes in Jabal Dimh, which range from Mg_{84-93}. The total observed range in the core compositions of cumulus clinopyroxenes in Jabal Dimh fall in the range Mg_{74-93} (Pallister & Hopson 1981). Given that the clinopyroxenes formed in the experimental runs are as magnesian, or less, than those which would form at lower pressures (~2.5 kb—see below), primitive MORB such as DSDP3-18-7-1 could not be parental to those that are more magnesian than Mg_{88}, as pointed out by Green *et al.* (1979) and Duncan & Green (1980). As is the case with plagioclase, most, but not all, of the observed clinopyroxene compositions could be produced by crystallization and fractionation of primitive MORB.

The exceptional, highly calcic, plagioclase ($>An_{89}$) and the magnesian clinopyroxene ($>Mg_{88}$) from the ophiolite may be the products of batches of 'second-stage' melt (Duncan & Green 1980) as noted above.

The experimental work on primitive MORB indicates that plagioclase preceeds clinopyroxene as a liquidus phase at low pressure (1 atmosphere). However, the sequence of crystallization inferred from Jabal Dimh cumulates indicates the near-simultaneous appearance of clinopyroxene and plagioclase on the liquidus following olivine. This also occurs in experimental runs at higher pressure (10 kb; Green *et al.* 1979). However, the basal Jabal Dimh cumulates probably crystallized at only ~2.5 kb (based on a column consisting of 3 km H_2O + 0.5 km volcanics + 1.5 km dyke complex + 5 km plutonic rocks). This pressure may not be sufficient to explain the lack of olivine + plagioclase crystallization prior to cotectic olivine + plagioclase + clinopyroxene crystallization.

Another possibility is shown diagrammatically in the simplified tholeiitic basalt compositional tetrahedron of Presnall *et al.* (1978; 1979) (Fig. 2). SAVE and its picritic precursors may have evolved along an olivine-control line (such as α–β, Fig. 2) that intersects the olivine-plagioclase-melt divariant surface (a-b-c-d-p, Fig. 2), but they were shifted to olivine-control lines that intersect the four-phase cotectic by magma mixing. Upon entering the magma chamber, mixing of olivine-liquidus SAVE (such as at ϵ, Fig. 2) with a more evolved melt on the four-phase cotectic (such as 'z', Fig. 2) could shift the bulk composition onto an olivine control line (e.g. point σ, Fig. 2) that does intersect the cotectic. The resultant liquid compositional path is shown by the open arrow in Fig. 2. Various amounts of mixing are possible, but in any case, mixing of olivine-liquidus melts with cotectic chamber-resident melts would shift the bulk composition toward, if not onto , the surface composed of olivine-control lines that inersect the four-phase cotectic (surface Mg_2SiO_4-a-p, Fig. 2).

This effect of magma mixing is not apparent from the cumulus crystallization sequence alone, which suggests that SAVE lies on an olivine-control line that intersects the four-phase cotectic. The combination of the crystallization sequence inferred from the cumulates, experimentally determined liquidus relations for primitive MORB similar to SAVE, and abundant evidence for magma mixing derived from phase and cryptic variation studies of the cumulates (Pallister & Hopson, 1981), supports the mixing model proposed above. This mixing model also provides a partial explanation for the 'clinopyroxene problem' (Rhodes & Dungan 1979) with co-genetic suites of MORB.

The fact that primitive MORB samples such as ALV517-1-1 and DSDP3-18-7-1 (Table 1) are on the olivine liquidus is used as justification for olivine-addition experiments to determine mantle equilibration temperatures and pressures (Green *et al.* 1979). Even these melts could be the products of mixing of more primitive melts within the olivine liquidus volume with evolved melts on the four-phase cotectic. In this case, simple olivine addition would not lead back to the primary-melt composition. Experiments that determine the olivine-orthopyroxene cotectic for mantle melts using primitive MORB as starting material in harzburgite capsules (Stolper 1980) are also subject to the possible complication of magma mixing. Although in these experiments the melt composition is shifted to equilibrate with mantle-type olivine and orthopyroxene, the system is dependent upon the starting composition for components that are not major constituents in these two minerals, such as CaO and Al_2O_3.

Because SAVE is an *average* composition of oceanic crust (by the ophiolite analogy), magma mixing of co-genetic melts affects the crystallization sequence and cumulus stratigraphy but does not change the average composition. Therefore, the similarity of SAVE and primitive MORB indicates that these melts represent *average* parent melts. In this sense, these are reasonable starting compositions for olivine addition and harzburgite equilibration studies.

Evidence from the mantle suite

Boudier & Coleman (1981) and Gregory & Coleman (1981) cite field and chemical evidence from the mantle suite of the ophiolite that favours cumulus crystallization and wall-rock reaction of melt diapirs with host harzburgite as they

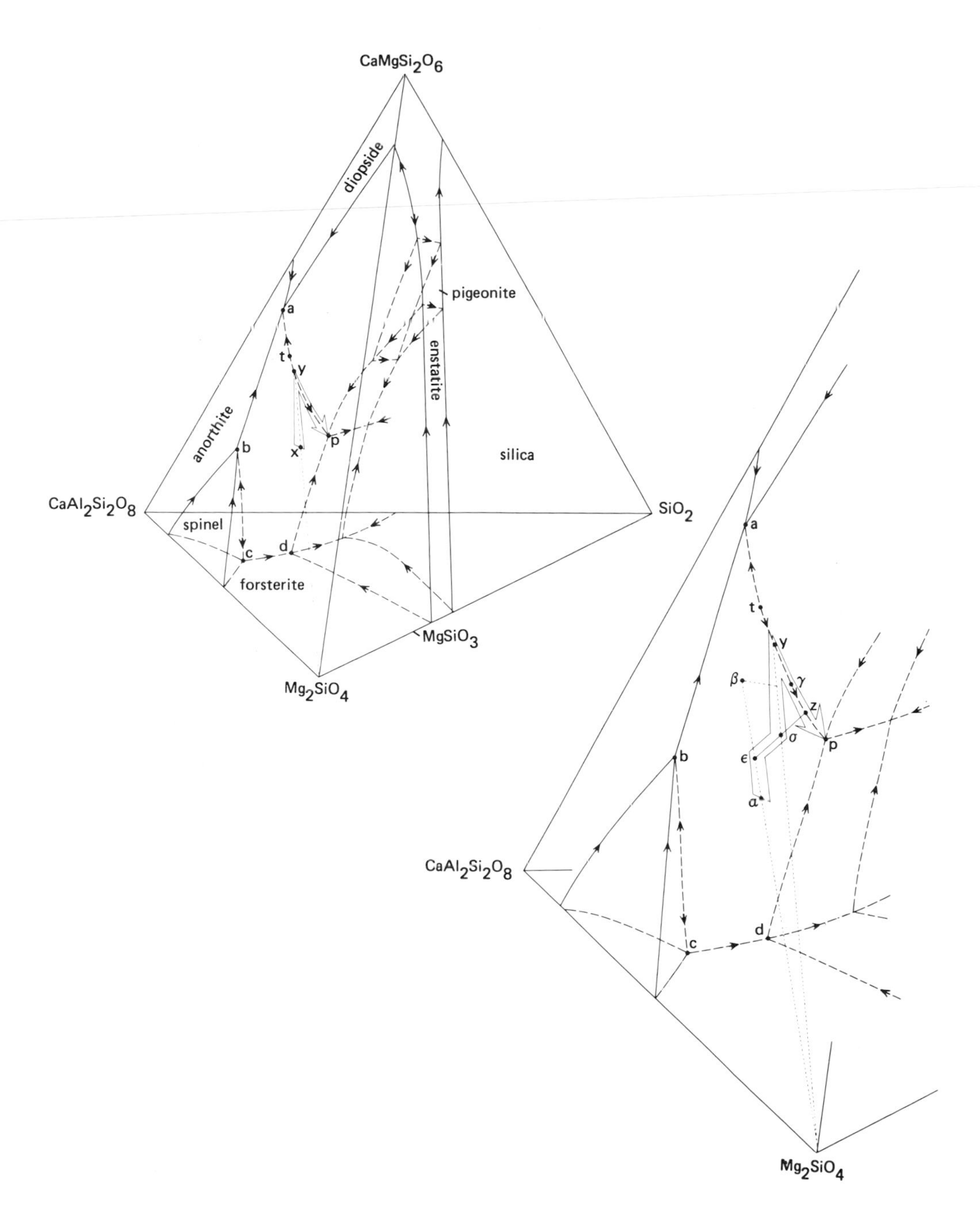

FIG. 2. Inferred phase relations of Jabal Dimh melt compositions in the simplified tholeiitic basalt tetrahedron of Presnall *et al.* (1978; 1979). Arrowheads indicate directions of decreasing temperature on liquidus univariant lines within the tetrahedron and on lines of intersection of quaternary divariant surfaces with faces of the tetrahedron. Phase relations on the back face, $CaMgSi_2O_6$-$CaAl_2Si_2O_8$-SiO_2 of the tetrahedron are omitted for clarity. The small tetrahedron is from Pallister & Hopson (1981). Hypothetical phase relations are shown diagramatically in the enlarged part of the tetrahedron (see text).

compositional differences in mantle-derived melts beneath spreading axes: evidence from cyclically layered rocks in the ophiolite of North Oman, *J. Geophys. Res.* **86**, 2645–60.
SPARKS, R. S. J., MEYER, P. & SIGURDSON, H. 1980. Density variation amongst mid-ocean ridge basalts: implications for magma mixing and the scarcity of primitive lavas. *Earth Planet. Sci. Lett.* **46**, 419–30.
STOLPER, E. 1980. A phase diagram for mid-ocean ridge basalts: preliminary results and implications for petrogenesis. *Contrib. Mineral. Petrol.* **74**, 13–27.
—— & WALKER, D. 1980. Melt density and the average composition of basalt. *Contrib. Mineral. Petrol.* **74**, 7–12.
WALKER, D., SHIBATA, T. & DELONG, S. E. 1979. Abyssal tholeiites from the Oceanographer Fracture Zone. II. Phase equilibria and mixing. *Contrib. Mineral. Petrol.* **70**, 111–25.

J. S. PALLISTER, US Geological Survey, Denver, Colorado 80225, USA.

Cryptic variation within the Cumulate Sequence of the Oman ophiolite: magma chamber depth and petrological implications

P. Browning

SUMMARY: The Cumulate Sequence of the Oman Ophiolite consists of layered mafic and ultramafic cumulates that represent the products of a dynamic, long-lived, crustal magma chamber that underlay a late-Cretaceous spreading ridge. Regional variations in Cumulate-Sequence stratigraphy reflect not only contrasting degrees of mantle fractionation but also contrasting primary magma compositions.

Microprobe data are presented from a 600 m section of modally layered olivine gabbros. Coherent cryptic cyclic variation is shown, with both 'normal' and 'reversed' geochemical gradients present; these are explained in terms of fractionation, mixing and eruption in an open-system magma chamber. The height of the liquid column required to precipitate the observed cyclic units is calculated to be about 100 m; the implications of this result are discussed.

A contrast between the Fo-An co-variation between olivine and plagioclase in the Oman Cumulate Sequence and that observed for MORB volcanics is noted; this may reflect compositional stratification in spreading-ridge magma chambers or significant differences between the petrogenesis of the Oman ophiolite and MORB.

Fresh insights into the physical properties of silicate liquids have led recently to a re-examination of the classic basaltic magma-chamber model of Wager & Brown (1968). Magma chambers in which convective circulation occurs on a large scale and in which the fractionation process is dominated by crystal settling are now in question. Double-diffusive convection may result in vigorous circulation within a magma chamber, but on limited vertical scales. Layering in cumulates has been accounted for by a mechanism of *in-situ* oscillatory crystallization (McBirney & Noyes 1979).

Better understanding of the genesis of mid-oceanic-ridge basalts (MORB) has been reached by the appreciation of the open-system nature of the sub-ridge magma chamber (O'Hara 1977; Walker *et al.* 1979) and the importance of mixing—or more importantly, the delay in mixing of primitive magma inputs (Sparks *et al.* 1979, Stolper & Walker 1980)—in restricting MORB to relatively evolved steady-state compositions as compared to their likely picritic primary magmas (O'Hara 1968; Green *et al.* 1979; Stolper 1980).

As yet deep-sea drilling has only sampled the extrusive basaltic carapace that forms the uppermost layer of the oceanic lithosphere; likely samples of deeper oceanic crust and mantle have only been recovered by dredge haul from fracture zones, and as a result their original geometric relationships are obscure. Direct testing of hypotheses regarding the nature and processes that operate within MOR magma chambers must for the present proceed by examination of the lower levels of ophiolite complexes, considered to be the landbound fossil analogues of the oceanic crust and upper mantle (Gass 1967; Coleman 1971; Dewey & Bird 1971; Moores & Vine 1971).

This paper presents the results of an electron-microprobe study of the layered cumulate mafic and ultramafic rocks which represent the direct crystallization products of a sub-spreading ridge magma chamber. The objective was to investigate magma-chamber processes such as fractionation, extrusion and mixing.

The geology of the Oman ophiolite

The Oman ophiolite is the largest and best preserved of a chain of ophiolite complexes that extends through the Alpine-Himalayan fold mountain belt. The 700 km long arcuate chain of the Oman Mountains (Fig. 1) is dominated by the outcrop of the Oman ophiolite; in areal extent (about 30,000 km^2), it must rank as one of the largest masses of mafic and ultramafic rock in the world.

The generalized stratigraphy of the Oman ophiolite is found to correspond in all respects to the idealized Penrose stratigraphy (Anon. 1972). The stratigraphic nomenclature adopted by Open University workers, and followed in this work, is shown in Fig. 1. A fundamental two-fold division of the ophiolite is drawn between the Mantle Sequence and the Crustal Sequence (which comprises the Cumulate Sequence, the High Level Intrusives, the Sheeted Dyke Complex and the Extrusive Sequence); the boundary between the Mantle and Crustal Sequence has been called the Petrological Moho (Malpas 1978).

In terms of the consensus ophiolite petrogenetic model (Gass & Smewing 1981), melts derived from the partial fusion of an aluminous peridotite upper-mantle rise to feed a sub-axial crustal magma chamber beneath a spreading oceanic

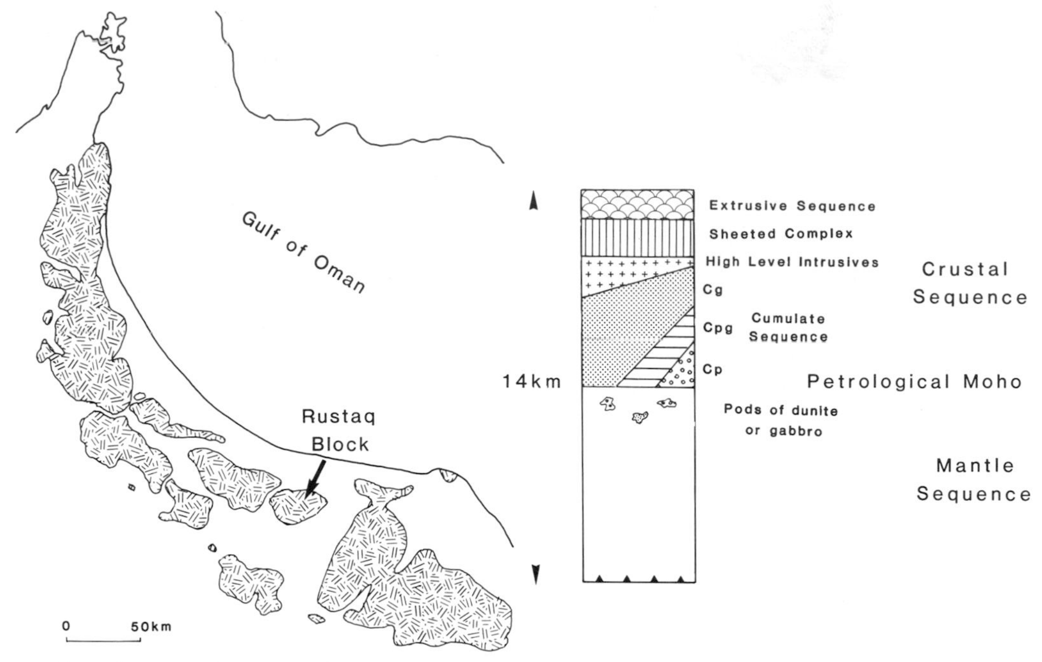

FIG. 1. Map showing the location and outcrop extent of the Oman ophiolite. The location of the Rustaq Block is indicated. Vertical column shows ophiolite stratigraphy adopted by Open University workers.

ridge. The tectonite harzburgite of the Mantle Sequence represents the residuum formed after partial melting; pods of dunite within the Mantle Sequence represent fractionates from the rising partial melts. In the sub-ridge magma chamber, layered mafic and ultramafic rocks of the Cumulate Sequence form on the walls and floor, whilst isotropic gabbros of the High-Level Intrusives underplate the roof. Episodic injection at the spreading-ridge crest results in the generation of the Sheeted Dyke Complex, and the eruption of the pillowed Extrusive Sequence.

The Cumulate Sequence is the subject of the present contribution. The rock types and the character of the layering within the Cumulate Sequence of the Oman ophiolite are described by reference to the Rustaq Block.

The Cumulate Sequence of the Rustaq Block

Regional stratigraphy

Significant regional contrasts are present within the Cumulate Sequence of the Rustaq Block (Fig. 2); four areas (bounded by major faults) have been subdivided on the basis of:

(i) the thickness of the Cumulate Sequence;
(ii) the nature of the cumulate assemblage above the Petrological Moho;
(iii) the character of the pod types in the underlying Mantle Sequence; and
(iv) the crystallization order shown by the rocks of the Cumulate Sequence.

In the area to the west of the Wadi Bani Suq Fault, the Cumulate Sequence attains a thickness of up to 4 km. It is characterized by ultramafic assemblages on the Petrological Moho and dunitic pod types in the underlying Mantle Sequence. The cumulates display the crystallization order olivine → clinopyroxene → plagioclase. Between the Wadi Bani Suq and Wukabah Faults only a limited (1 km) development of exclusively ultramafic cumulate is encountered. Again dunitic pods occur in the Mantle Sequence beneath. Within the area bounded by the Wukabah and Sulawah Faults, approximately 2 km of gabbroic cumulate assemblages rest directly on the Petrological Moho; gabbroic pods are seen in the Mantle Sequence below. A return to a stratigraphy reminiscent of the area to the west of the Wadi Bani Suq Fault is seen in the part of the Rustaq Block that lies to the east of the Sulawah Fault. A thick (>3 km) succession of mafic and ultramafic cumulates, with ultramafic assemblages occurring above and below (in pods) the Petrological Moho is observed. However, in contrast to the western areas of the Rustaq Block, the crystallization order displayed is olivine → plagioclase → clinopyroxene.

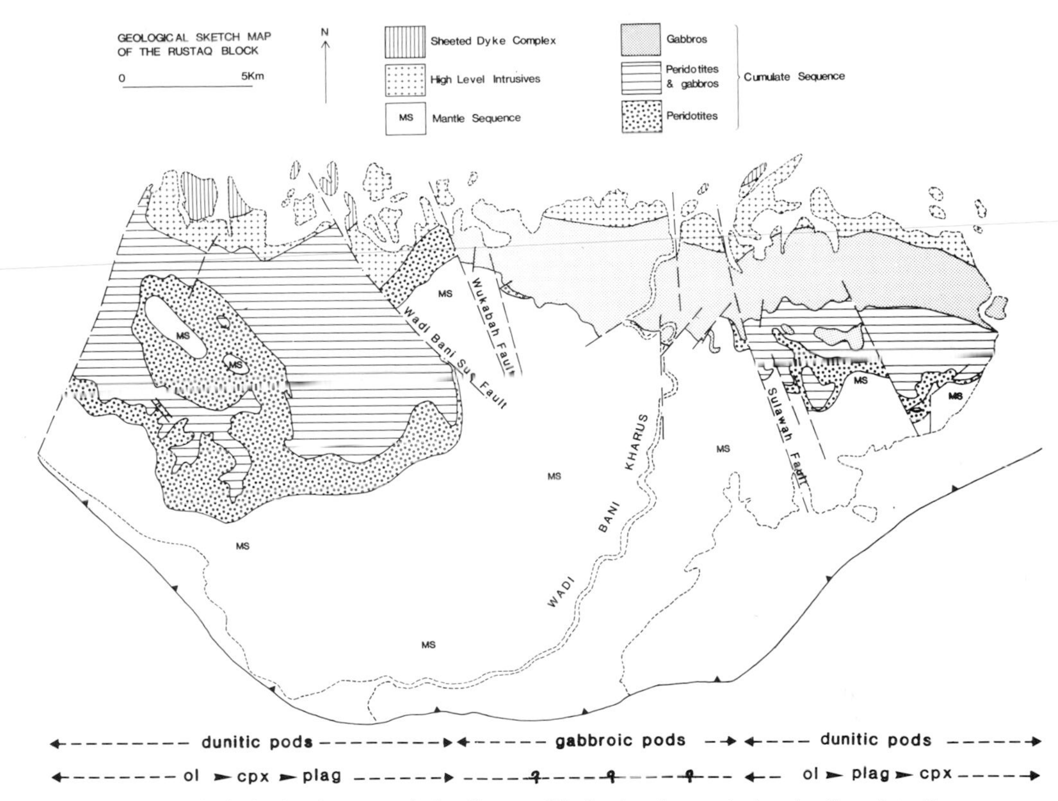

FIG. 2. Geological sketch map of the Rustaq Block showing variation in Cumulate Sequence stratigraphy. Also indicated are the distribution of pod types in the Mantle Sequence and the characteristic crystallization orders shown by the cumulates.

Such regional contrasts in Cumulate Sequence stratigraphy are interpreted in terms of differing degrees of mantle fractionation and, more significantly, differences in parental-magma compositions (Browning & Smewing 1981; Browning 1982). Areas within the Rustaq Block where gabbroic cumulate assemblages rest directly on the Petrological Moho and where gabbroic pods are present in the underlying Mantle Sequence are considered to reflect the supply of a relatively evolved basaltic parental magma to the sub-ridge magma chamber (a result of enhanced mantle fractionation). In those areas where ultramafic assemblages are present above and below the Petrological Moho, more picritic parental magmas were supplied. Differences in the CaO/Al_2O_3 ratio of the parental magmas are inferred to account for the contrasting crystallization orders between the western and eastern portions of the Rustaq Block; those magmas that underwent an olivine → clinopyroxene → plagioclase order had higher CaO/Al_2O_3 ratios than those that displayed an olivine → plagioclase → clinopyroxene sequence.

Layering in the Cumulate Sequence

Two types of layering are widespread in the Cumulate Sequence; modal (and sometimes also phase) layering on a metre scale and phase layering on a 5–100 m scale (the qualifier referring to the type of contact that defines the limits of the layer).

The modal layering (equivalent to the classic rhythmic layering of Wager & Brown 1968) is ubiquitous to the rocks of the Cumulate Sequence. The most common variety is defined by laminated, modally graded layers between 5 and 100 cm thick, which are often laterally discontinuous on an outcrop scale and are defined by sharp modal, sometimes phase contacts.

Such modal layering may be superimposed on a much larger-scale, laterally continuous phase layering. Phase layers, defined by phase contacts that reflect regionally significant cumulus appearances and terminations, combine to form cyclic units, within which consistent crystallization orders are displayed. (This contrasts with the often apparently random ordering of phase layering that may occur on the metre scale.)

The cryptic variation occuring within cyclic units composed of large-scale phase layers in the Wadi Ragmi area has been described by Smewing (1981). The following sections document the results of a microprobe study through a 600 m section of modally layered olivine gabbros.

Analytical procedures

All mineral chemistry data presented in this work were collected on a fully automated, wavelength dispersive, Cambridge Instruments Microscan 9 microprobe at the Open University. Analysis was performed at an accelerating potential of 20 kV, a specimen current of 30.5 nA and a typical spot size of 10–15 μm, with a ZAF correction being provided on-line.

Five analyses were completed per mineral phase per rock sample, checking for zoning on the core and rim on at least one grain. In the graphs that follow the mean, maximum and minimum values for any compositional parameter are shown. An error bar for one analysis, based on counting statistics at the 2σ level, is also indicated.

A traverse through modally layered olivine gabbros from Wadi Bani Kharus

Browning (1982) has shown that variation in phase chemistry is absent on the scale of individual modally graded layers. In an effort to establish at what scale any geochemical variation does occur within modally graded cumulates, the mineral chemistry of a 2000+ m thickness of the Cumulate Sequence in Wadi Bani Kharus has been studied (Fig. 2). In particular, the lowermost 600 m of modally layered gabbroic cumulates that form the base of the Cumulate Sequence were sampled at a 20 m spacing.

Petrography

The mineralogy of the lower, closely sampled 600 m section of the Cumulate Sequence in Wadi Bani Kharus is straightforward; plagioclase and clinopyroxene are always present, with or without olivine and very minor Fe-Ti oxides. Typical modes for this section are estimated as plagioclase 55%, clinopyroxene 35% and olivine 10%. Plagioclase usually shows clouding, olivine is serpentinized to varying degrees, and clinopyroxene may show retrograde breakdown to actinolite.

The primary igneous textural relationships are complicated by sub-solidus deformation (Christensen & Smewing 1981; Smewing *et al.*, this volume). The degree of recrystallization is seen to decrease with height (none has been identified above 600 m), though not uniformly. Original igneous features may be found interspersed between recrystallized zones; only meso- or adcumulate textures have been identified. Such specimens are usually coarser grained (maximum grain size 3–6 mm) and show a well-developed lamination of prismatic grains. The recrystallized equivalents are finer grained (less than 3 mm, often 0.25–0.5 mm mean grain size), and show the development of a foliation defined by lenticular rather than prismatic-shape fabrics in the same plane, or at a small angle to, the original igneous lamination.

Phase chemistry

Only data for the closely sampled 600 m section are presented here. Over the entire 2000+ m succession of Cumulate Sequence and High Level Intrusives in Wadi Bani Kharus, for all elements and compositional parameters, no overall trend of fractionation is seen, and the range in abundances shown is similar, irrespective of the height above the Petrological Moho (also noted by Pallister & Hopson 1981).

Plots of the variation in phase chemistry with height over the 600 m closely sampled section are shown in Fig. 3. Zoning is only well developed with respect to TiO_2 and Na_2O in clinopyroxene, and An content of plagioclase. The section has been divided into a number of cycles; three major cycles I, II and III, based on the Cr_2O_3 content of clinopyroxene, and a number of smaller divisions (a, b, c, etc.) based on the overall geochemical coherence, have been identified (Fig. 3). The data for the divisions Id, IIc, and IIId is insufficient to allow unambiguous interpretation of the geochemical trend.

The dominant features, as would be anticipated are 'normal' negative geochemical gradients with height for compatible elements (or ratios) (i.e. Fo and NiO in olivine, Mg*, Cr*, and Cr_2O_3 in clinopyroxene, and An in plagioclase), and positive gradient trends for incompatible elements (i.e. MnO in olivine, Na_2O, TiO_2 and MnO in clinopyroxene). No overall trend to more evolved compositions with height is observed. Although the overall geochemical pattern is dominated by 'normal' trends, 'reversed' trends where compatible elements show sustained enrichment with height and incompatible elements show a progressive depletion, occur in cycles IIe and IIf. Also, apparently incoherent geochemical behaviour ('decoupling') is shown in cycles IId and IIf for Na_2O, in cycles Ib and IIf for TiO_2, and in cycle IIIb by Cr_2O_3.

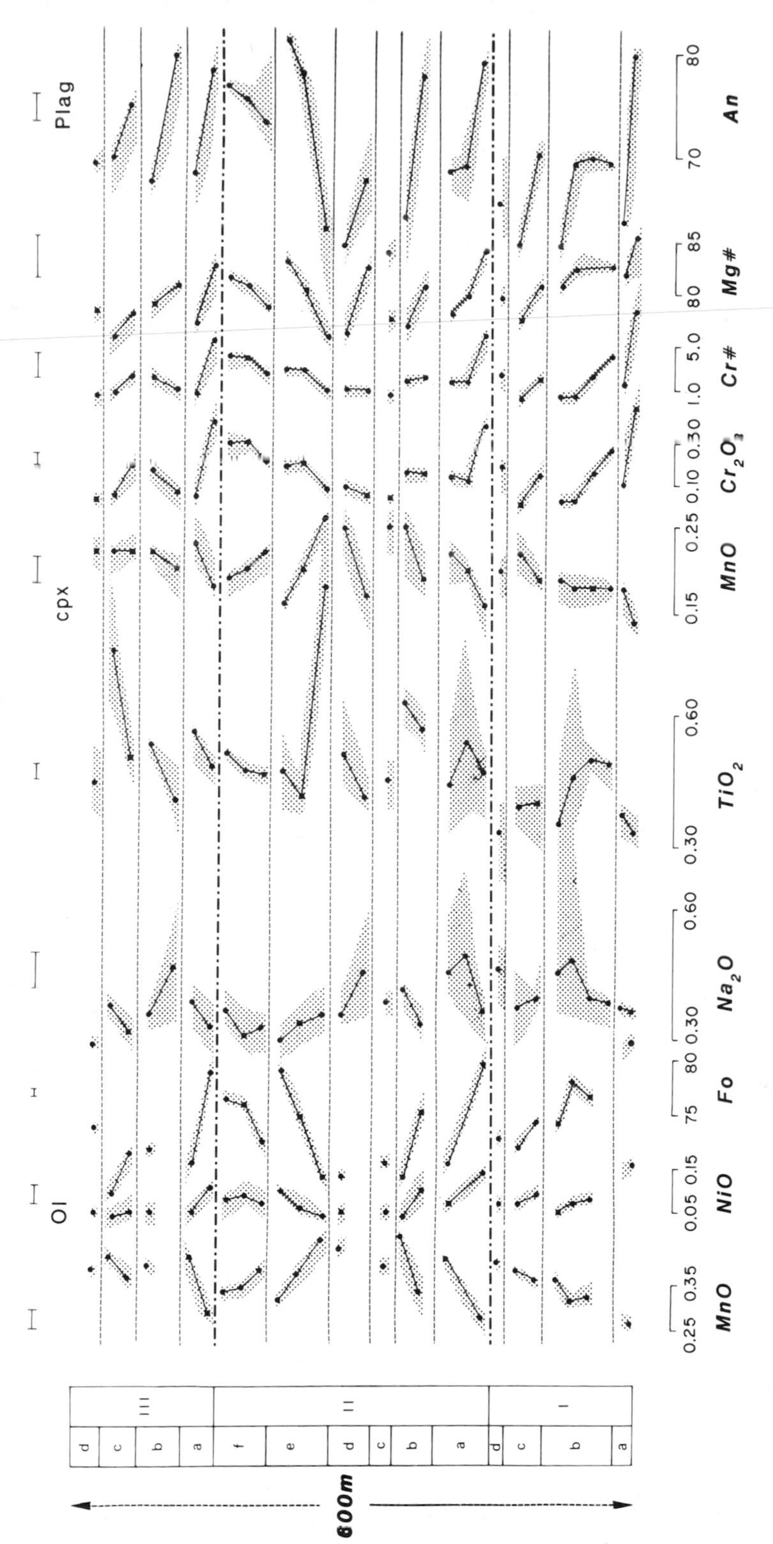

FIG. 3. Variation of phase chemistry with height over a 600 m section of modally layered olivine gabbros in Wadi Bani Kharus. Data from Browning (1982). Data points are means of five analyses; absolute range shown by shading. Error bars for an individual determination are shown. Cycles erected on basis of cryptic variation are indicated.

Wadi Bani Kharus cycles in the light of open-system fractionation

The interpretation of the data in Fig. 3 indicates that cyclic compositional variation is present on a vertical scale of about 50 m. Within each of the larger cycles I, II and III (ignoring, for the moment, the 'reversed' cycles IIe and f) it is noted that for compatible elements or ratios, as well as a progressive depletion in terms of the abundances within each of the smaller constituent cycles (a, b, c, etc.), a progressive depletion in terms of the successive primitive 'resets' that start each constituent cycle is also seen. This is characteristic of an open system tending towards a steady-state composition.

In an attempt to model the cumulate compositional profiles produced during open-system fractionation Browning (1982) identified four mechanisms by which sustained 'reversed' geochemical gradients may be produced.

(i) eruption/isolation of magma from the main body of the chamber;
(ii) a change of the input composition to still more primitive compositions;
(iii) change in bulk distribution coefficient; and
(iv) mixing in of overlying, evolved magma.

In the case of cycles IIe and f we may reject the third possibility; no change in the assemblage of ol + cpx + plag is seen at any point in the section. No discontinuity is observed at the junction between IId and e, merely a reversal in gradient; the possibility of mixing in an overlying, evolved magma is excluded, though it remains an option to account for the discontinuity between IIe and f. It is difficult to differentiate between the end results of mechanisms (i) and (ii); this is to be expected as eruption of magma temporarily reduces the mass of the chamber allowing the effects of primitive magma from below to momentarily reset the mixed liquid compositions to more primitive compositions—exactly the same effect that is achieved by making the input composition more primitive itself. Both these mechanisms are considered viable alternatives in the case of Wadi Bani Kharus.

The 'decoupling', notably by TiO_2 and Na_2O in clinopyroxene, does not have a satisfactory explanation. Precipitation of magnetite would account for sudden depletion rather than enrichment in TiO_2 in clinopyroxene, but no petrographic evidence supports this. The fact that it is two relatively highly incompatible elements, Na and Ti, that show decoupling, and that it is also these elements that show the strongest development of zoning, suggests that some later-stage intercumulate process may be the mechanism responsible.

The petrogenesis of the gabbroic cumulates in Wadi Bani Kharus is to be contrasted with the bimodal mafic and ultramafic cumulates from Wadi Ragmi (Smewing 1981). The cyclic units delineated in Wadi Bani Kharus on the basis of cryptic variation in phase chemistry are not matched by any parallel phase layering. Furthermore, the relatively evolved nature of the magmas—at the 'density minimum' of Sparks *et al.* (1979) and Stolper & Walker (1980)—reaching the Petrological Moho in Wadi Bani Kharus resulted in immediate mixing of the melt input with the 'perched' basaltic liquids of the open-system magma chamber. In the Wadi Ragmi area however, picritic melts were supplied to the Petrological Moho, and these ponded at the base of the magma chamber, fractionating olivine until their density became equal to the overlying basaltic melt, whereupon mixing occurred.

Magma-chamber height

Consider a magma chamber, tabular in form, of height l, width a, length b, and composition C_o (Fig. 4). For perfect Rayleigh fractionation, after crystallizing a mass fraction X, giving a thickness of cumulate c, the liquid composition will evolve to:

$$C_1 = C_o(1-x)^{(D-1)}.$$

In terms of the dimensions of the magma chamber, the mass fraction X that has fractionated is:

$$X = [c.a.b]/[l.a.b] = c/l.$$

Given that C_{xo}/C_o and C_{xl}/C_l are equal to D, we may combine these equations to give:

$$C_{xl} = C_{xo}\ (1-c/l)^{\ (D-1)}$$

where C_{xo} and C_{xl} are the compositions of the first and last formed cumulates in a given cycle. Rearranging gives:

$$l = c/[1-(C_{xl}/C_{xo})^{\ 1/(D-1)}]$$

which in principle allows calculation of the maximum magma-chamber height, as c, C_{xo}, and C_{xl} are known, provided the liquid remained homogeneous in composition at all stages of the fractionation cycle.

Of the 'normal' cycles Ib is the best defined. Compositional data for NiO in olivine and Cr_2O_3 in clinopyroxene are used to determine the maximum height of the column of liquid that fractionated to form the observed thickness of cumulate (see Fig. 4).

Determination of suitable bulk distribution coefficients is not straightforward; the rocks at all levels show evidence of crystal sorting; no

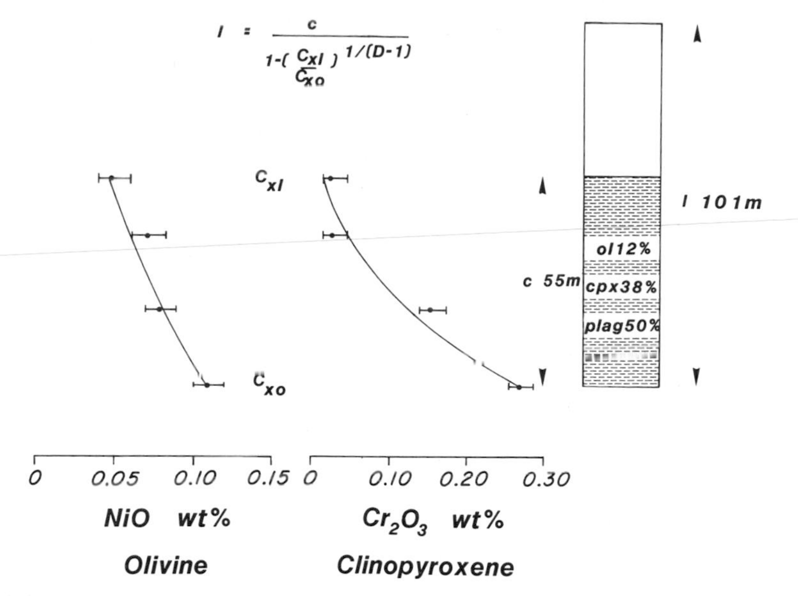

FIG. 4. Calculation of liquid-column height. Data for cycle Ib are shown. For derivation of expression see text.

'average rocks' are available which might represent unsorted crystal extracts. Mean modal proportions of the olivine-bearing assemblages in cycle Ib are olivine 10%, clinopyroxene 35% and plagioclase 55%. Using suitable density values these convert to olivine 12%, clinopyroxene 38%, and plagioclase 50% by weight.

The following values for mineral-liquid distribution coefficients are used (Cox *et al.* 1979):

	ol	cpx	plag
Cr	0.2	10	0.01
Ni	10	2	0.01

and the mean abundances and heights of the samples at the top and bottom of cycle Ib:

	NiO_{ol}	Cr_2O_{3cpx}	Height (m)
2393	0.11	0.27	26.60
2396	0.05	0.03	82.00

Bulk distribution coefficients for the observed assemblages are:

$$D_{Cr} = 3.83$$

$$D_{Ni} = 1.97$$

Values for Ni in clinopyroxene and plagioclase, and Cr in olivine and plagioclase, are determined by assuming equilibrium between olivine, clinopyroxene, plagioclase and liquid giving the following values for whole rock Cr_2O_3 and NiO:

	NiO	Cr_2O_3	
2393	0.022	0.103	(C_{xo})
2396	0.010	0.011	(C_{xl})

The equation above allows calculation of the maximum height of the column of liquid that cycle Ib could have fractionated from in a tabular shaped magma chamber. For Cr_2O_3

$$l = [82.00-26.60]/[1-(0.011/0.103)^{1/(3.83-1)}] = 101.4 \text{ m},$$

and for NiO

$$l = [82.00-26.60]/[1-(0.010/0.022)^{1/(1.97-1)}] = 99.6 \text{ m}$$

which are in good agreement. The ratio of the observed cumulate thickness to the calculated magma-chamber height indicates that a mass fraction (X) of about 0.55 had crystallized during cycle Ib.

Similar expressions can be derived for other magma-chamber shapes; for a magma chamber triangular in cross-section, the walls of which dip inwards at 30°, a liquid-column height of about 190 m is calculated. For a semi-circular magma chamber, a liquid-column height of about 160 m is obtained.

These values for the thickness of the column of liquid that fractionated to form the rocks of cycle Ib are an order of magnitude less than the vertical extent of the Cumulate Sequence in Wadi Bani Kharus. If it is argued that the thickness of the Cumulate Sequence should reflect the depth of the magma chamber in which it formed, then the values obtained above can only be explained in terms of a stratified magma chamber (Turner & Chen 1974; McBirney & Noyes 1979; Huppert &

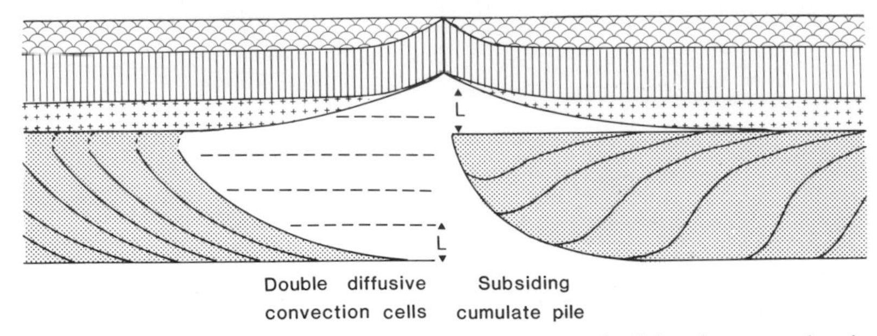

FIG. 5. Alternative magma-chamber models for spreading ridge in light of magma-chamber height calculation. Left: Limited magma-chamber height is allowed for by compositional stratification caused by double-diffusive convection. Right: extensive thickness of cumulates is achieved by subsidence.

Sparks 1980a, b) in which bodies of liquid of limited vertical extent behaved temporarily as closed fractionating systems (Fig. 5). If, however, the thickness of the Cumulate Sequence is the result of substantial subsidence (as in the model proposed by Dewey & Kidd (1977)), it may be that a value of 100–200 m for the height of the liquid column reflects the true depth of the magma chamber (Fig. 5).

Olivine–plagioclase co-variance: a contrast with MORB

A plot between the An content of plagioclase and Fo content of olivine for the gabbro traverse in Wadi Bani Kharus is shown in Fig. 6a; at first sight the correlation is much less well defined than for similar trends from continental layered intrusions (see for example Wager 1967). If however, the graph is considered in the light of the cycles erected on the basis of the variation of phase chemistry with height (Fig. 3), several important features emerge (Fig. 6b):

(i) For those cycles that contain three or more data points, good correlations are present; if more data points were available for the remaining cycles it is suggested that the same would be true for these.

(ii) The observed correlations appear to radiate from the composition point Fo_{85},An_{85}.

(iii) The gradients of the observed correlation for any one cycle decrease systematically with the height of the cycle from the Petrological Moho.

The contrasting gradients between different cycles provide independent support of the cyclic units erected on the basis of the variation of phase chemistry with height.

Prediction of the major-element evolution of magmas and their coexisting phases is still in its infancy (e.g. Nathan & van Kirk 1978); only a qualitative treatment of the major-element co-variation between coexisting phases of the Wadi Bani Kharus Cumulate Sequence is attempted here.

Consider a magma of a given composition fractionating an assemblage of olivine, clinopyroxene and plagioclase under conditions of fixed total and volatile pressure. The line of liquid descent will be fixed; the locus of compositions will not in general be a straight line (major-element distribution coefficients change with both temperature and composition, and the proportions of the phases fractionating may change down temperature), nor will the locus of coexisting mineral compositions (see the hypothetical graph of successive coexisting olivine and plagioclase compositions in Fig. 6c). A family of similar loci will exist for all other starting liquid compositions.

Browning (1982) has considered three mechanisms for producing the dispersion in the Fo-An correlation seen in the Wadi Bani Kharus cyclic units (Fig. 6d):

(i) If the proportions of olivine, clinopyroxene, and plagioclase fractionating were to change in different ways in different cycles then contrasting loci of coexisting mineral compositions will result to reflect the change in lines of liquid descent.

(ii) An increase in the partial pressure of water not only depresses but also 'steepens' the solidus of the plagioclase feldspars (Yoder *et al.* 1957, Yoder 1965). For a given decrease in temperature this results in a smaller change in An content of plagioclase relative to the change that occurs in Fo content of olivine under water-free conditions.

(iii) McBirney & Noyes (1979) have interpreted the systematic change in the rates of evolution of coexisting olivine and plagioclase compositions within the upper part of the Layered Series of the Skaergaard intrusion (Naslund 1976), as being due to

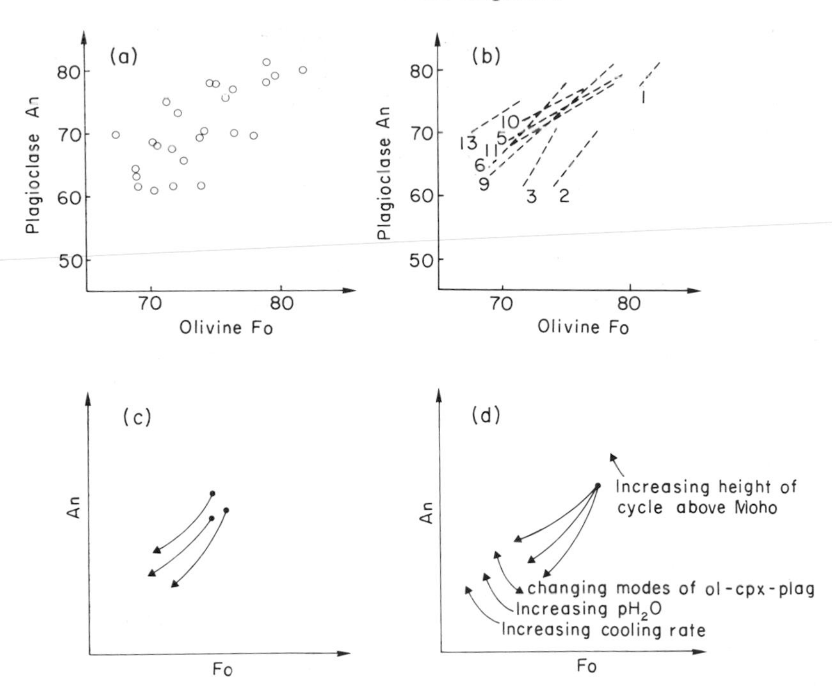

FIG. 6. (a) Fo-An co-variation in coexisting olivine and plagioclase from Wadi Bani Kharus. (b) Correlations between Fo-An in light of cryptic cycles erected on basis of phase chemistry. (Cycles are numbered consecutively from 1 to 13 from bottom to top.) (c) Hypothetical loci for coexisting olivine and plagioclase fractionating from different parent compositions. (d) Mechanisms for changing the gradients of the Fo-An correlation; for discussion see text.

the effects of more rapid cooling rates at the margins of the intrusion in decreasing the rates of diffusion of some chemical components more than others. Where cooling rates were more rapid, the diffusion rates of complex silica and alumina ions were lower than those of the simple ions of iron and magnesium; the rate of fractionation of plagioclase was lower where the rate of crystallization was too rapid for equilibration to be approached as closely as it was in olivine and pyroxene.

A change in the mode of the fractionating assemblage may be rejected in the case of Wadi Bani Kharus; the average cpx/plag ratio is not seen to vary significantly with height in the traverse. Neither higher partial pressures of water, or higher rates of cooling with concomitant lower diffusion rates, can be rejected as mechanisms to explain the observed features of the Fo-An co-variation; both are plausible in that upward migration of volatiles and faster rates of cooling are likely to occur at higher levels in any magma chamber. However, it may be noted that if the partial pressure of water was elevated during the formation of the higher levels of the gabbro traverse, then it was never high enough to stabilize amphibole; also no textural differences have been discerned within the section that could be attributed to a variation in cooling rates.

In the previous section it was indicated that the cumulates of the Wadi Bani Kharus section may have formed in a stratified magma chamber in which layers of liquid of limited vertical extent (between 100–200 m) acted temporarily as closed systems to precipitate the individual cyclic units. If such liquid stratification was the result of double-diffusive convection (Turner & Chen 1974; McBirney & Noyes 1979), it may be that the diffusive processes in such a system combined to produce contrasting compositional gradients within the magma body, which in turn led to the contrasting correlations with height seen between Fo-An contents of the coexisting olivine and plagioclase.

A comparison of the liquid major-element compositions of the Oman ophiolite with those of MORB cannot be attempted due to the highly altered nature of the ophiolite basalts. However, one approach is to compare the phenocryst compositions of MORB with the cumulus-phase compositions of the Oman ophiolite. The compositions of coexisting olivine and plagioclase from the Cumulate Sequence and those from

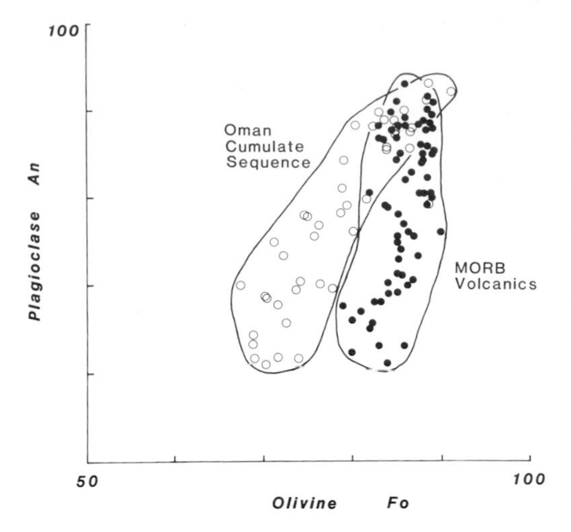

FIG. 7. Coexisting olivine and plagioclase variation for the Cumulate Sequence of the Oman ophiolite (Browning 1982) and MORB volcanics (for data sources see text).

MORB are shown in Fig. 7. Data have been taken from studies by Kirkpatrick (1978), Hodges (1978), Mevel *et al.* (1978), Sato *et al.* (1978), Dungan *et al.* (1978), Donaldson *et al.* (1976), Bryan *et al.* (1977), Flower *et al.* (1977; 1978), Clarke & Loubat (1977), Frey *et al.* (1974), Hekinian *et al.* (1976), Shibata *et al.* (1978) and Bryan (1979). In basalts in which two or more generations of phenocrysts or microphenocryst were present, the most primitive olivine, or plagioclase composition was selected. Also included on Fig. 7 are data on oceanic gabbros and peridotites from Symes *et al.* (1977).

The two suites are clearly distinct; for a change in plagioclase composition from An_{90} to An_{60}, the MORB field shows only a limited range in olivine composition from Fo_{90} to Fo_{80}, whilst the Oman field shows a more substantial variation from Fo_{90} to under Fo_{70}. Both fields apparently have a common origin.

If it is argued that eruptions at the spreading ridge sample only the top of the magma chamber (the area where heat loss is most rapid), it may be that the differences in Fo-An co-variation between phenocrysts in MORB lavas and cumulus phases from the Oman Cumulate Sequence reflect the contrasting cooling rates between their respective sites of crystallization and fractionation. However such an explanation cannot account for the observed contrasts; for more rapid cooling rates and inhibited plagioclase fractionation, a shallower not steeper gradient to the Fo-An co-variation should be seen for the MORB volcanic olivine-plagioclase pairs compared to the Oman cumulate pairs.

Several authors (e.g. Pearce 1979, Pearce *et al.* 1981) have advocated a marginal or back-arc basin setting for the site of formation of the Oman ophiolite. They identify higher abundances of K, Rb, Sr, Ba and Th, and a lower content of Cr with respect to MORB for the Extrusive Sequence. Such geochemical features may be attributed to selective enrichment of mantle source regions by mobile elements in aqueous fluids driven off subducted oceanic crust. The shallower trend in the Fo-An co-variation of the Oman ophiolite with respect to MORB could plausibly be assigned to the effects of fractionation under conditions of elevated P_{H_2O}, a result of subduction-zone influence in the source region of the magmas.

If temporarily isolated magma cells of limited vertical extent developed within a regime of double-diffusive convection (McBirney & Noyes 1979), and led to gross compositional differences between the top and bottom of the magma chamber, it may be that *in-situ* bottom crystallization produced cumulates that have contrasting Fo-An co-variation to those of the liquids erupted from the upper half of the magma chamber. Testing of such a hypothesis is difficult for the Oman ophiolite since olivine and plagioclase phenocrysts are rarely preserved in the Sheeted Dyke Complex and the Extrusive Sequence. As such it must remain a plausible explanation for the observed contrasts in Fo-An co-variation between the Oman plutonic and the MORB volcanic rocks.

Finally, the possibility that the primary magmas of the Oman and MORB had significantly different major-element compositions must be considered. A relatively low CaO/Al_2O_3 ratio for MORB primary magmas is indicated by the experimental work of Bender *et al.* (1978) who showed that a primitive basalt glass from the FAMOUS area (considered to have fractionated about 10% olivine) crystallized plagioclase and not clinopyroxene after olivine at low pressure. As stated earlier, the ol → cpx → plag → crystallization sequence displayed by most of the Cumulate Sequence indicates a comparatively higher CaO/Al_2O_3 ratio for the Oman primary magma.

Conclusions

The Cumulate Sequence of the Oman ophiolite represents the products of a sub-spreading ridge magma chamber which underwent open-system fractionation. A height of 100–200 m for the liquid column that formed the cyclic units indicates either that the magma chamber was of limited vertical extent, or that it was compositionally stratified. Support for the latter derives from the contrasts in Fo-An co-variation observed between successive cycles.

Differences between the Fo-An co-variation in the Cumulate Sequence of the Oman ophiolite and MORB may reflect such magma-chamber heterogenieties, or it may point to fundamentally different petrogenetic histories between ophiolites and MORB.

ACKNOWLEDGMENTS: This work was completed whilst in receipt of a NERC studentship. My supervisors Ian Gass and John Smewing are thanked for critical reviews, as is Liz McLean, who drafted all the diagrams.

References

ANONYMOUS 1972. Penrose field conference on ophiolites. *Geotimes* **17**, 24–5.

BENDER, J. F., HODGES, F. N. & BENCE, A. E. 1978. Petrogenesis of basalts from the project FAMOUS area: Experimental petrology study from 0–15 kbars. *Earth Plan. Sci. Lett.* **41**, 277–302.

BROWNING, P. 1982. *The petrology, geochemistry, and structure of the plutonic rocks of the Oman Ophiolite*. Unpublished Open University Thesis, 404 pp.

—— & SMEWING, J. D. 1981. Processes in magma chambers beneath spreading axes: evidence from magmatic associations in the Oman Ophiolite. *J. geol. Soc. London*; **138**, 279–80.

BRYAN, W. B. 1979. Regional variation and petrogenesis of basalt glasses from the FAMOUS area, Mid-Atlantic Ridge. *J. Petrol.* **20**, 293–325.

——, THOMPSON, G., FREY, F. A., DICKEY, J. S. & ROY, S. 1977. Petrology and geochemistry of basement rocks recovered on Leg 37. *DSDP, Initial Reports of the Deep Sea Drilling Project*, **37**, 695–704.

CHRISTENSEN, N. I. & SMEWING, J. D. 1981. Geology and seismic structure of the northern section of the Oman ophiolite. *J. Geophys. Res.* **86**, 2545–555.

CLARKE, D. B. & LOUBAT, H. 1977. Mineral analysis from the peridotite-gabbro-basalt complex at Site 334. *DSDP Leg 37, Initial Reports of the Deep Sea Drilling Project*, **37**, 847–56.

COLEMAN, R. G. 1971. Plate tectonic emplacement of upper mantle peridotites along continental edges. *J. Geophys. Res.* **76**, 1212–22.

COX, G., BELL, J. D. & PANKHURST, R. J. 1979. *The Interpretation of Igneous Rocks*. Allen & Unwin, London, 450 pp.

DEWEY, J. F. & BIRD, J. M. 1971. Origin and emplacement of the ophiolite suite: Appalachian ophiolites in Newfoundland. *J. Geophys. Res.* **76**, 3179–206.

—— & KIDD, S. F. 1977. Geometry of plate accretion. *Bull. geol. Soc. Am.* **88**, 960–8.

DONALDSON, C. H., BROWN, R. W. & REID, A. M. 1976. Petrology and chemistry of basalts from the Nazca Plate. I. Petrography and mineral chemistry. *DSDP, Initial Reports of the Deep Sea Drilling Project*, **34**, 227–38.

DUNGAN, M. A., LONG, P. E. & RHODES, J. M. 1978. The petrology, mineral chemistry and one-atmosphere phase relations of basalts from Site 395. *Initial Reports of the Deep Sea Drilling Project*, **45**, 461–78.

FLOWER, M. F. J., ROBINSON, P. T., SCHMINCKE, H. H. & OHNMACHT, W. 1977. Petrology and geochemistry of igneous rocks. *DSDP Leg 37, Initial Reports of the Deep Sea Drilling Project*, **37**, 653–80.

——, OHNMACHT, W., SCHMINCKE, H. U., GIBSON, I. L., ROBINSON, P. T. & PARKER, R. 1978. Petrology and geochemistry of basalts from Hole 396B, Leg 46. *Initial Reports of the Deep Sea Drilling Project*, **46**, 179–213.

FREY, F. A., BRYAN, W. B. & THOMPSON, G. 1974. Atlantic ocean floor: Geochemistry and petrology of basalt from Legs 2 & 3 of the Deep Sea Drilling Project, *J. Geophys. Res.* **79**, 5507–5527.

GASS, I. 1967. The ultrabasic volcanic assemblage of the Troodos massif, Cyprus, *In*: WYLLIE, P. J. (ed.) *Ultramafic and related rocks*. Wiley, New York, 121–134.

—— & SMEWING, J. D. 1981. Ophiolites: Obducted oceanic lithosphere. *In*: EMILIANI, C. (ed.) *The Sea, Vol. 7: The Oceanic Lithosphere*. Wiley, New York, 339–361.

GREEN, D. H., HIBBERSON, W. O. & JAQUES, A. L. 1979. Petrogenesis of mid-ocean ridge basalts. *In*: MCELHINNEY, M. W. (ed.) *The Earth, Its Origin, Structure and Evolution*. Academic Press, London, pp. 265–97.

HEKINIAN, J. G., MOORES, J. G. & BRYAN, M. B. 1976. Volcanic rocks of the Mid-Atlantic rift valley near 36°49′N. *Contrib. Min. Pet.* **58**, 83–110.

HODGES, F. N. 1978. Petrology and chemistry of basalts from DSDP Leg 46. *Initial Reports of the Deep Sea Drilling Project* **46**, 227–33.

HUPPERT, H. E. & SPARKS, R. S. J. 1980a. Restrictions on the composition of mid-ocean ridge basalts: a fluid dynamical investigation. *Nature, Lond.* **286**, 46–8.

—— & SPARKS, R. S. J. 1980b. The fluid dynamics of a basaltic magma chamber replenished by influx of hot, dense, ultrabasic magma. *Contrib. Min. Pet.* **75**, 279–89.

KIRKPATRICK, R. J. 1978. Petrology of basalts: Hole 396B DSDP Leg 46. *Initial Reports of the Deep Sea Drilling Project* **46**, 165–78.

MALPAS, J. 1978. Magma generation in the upper mantle, field evidence from ophiolite suites and application to the generation of oceanic lithosphere. *Phil. Trans. Roy. Soc. Lond.* **288**, 527–46.

MCBIRNEY, A. R. & NOYES, R. M. 1979. Crystallisation and layering of the Skaergaard intrusion. *J. Petrol.* **20**, 487–554.

MEVEL, C., OHNENSTETTER, D. & OHNENSTETTER, M. 1978. Mineralogy and petrography of Leg 46 basalts. *Initial Reports of the Deep Sea Drilling Project*. **46**, 151–64.

MOORES, E. M. & VINE, F. J. 1971. Troodos Massif, Cyprus and other ophiolites as oceanic crust: evaluation and implications. *Phil. Trans. Roy. Soc. Lond.* **A268**, 443–66.

NASLUND, H. R. 1976. Mineralogical variations in the upper part of the Skaergaard intrusion, East Greenland. *Yb. Carn. Inst. Wash.* **75**, 640–4.

NATHAN, H. D. & VAN KIRK, H. D. 1978. A model of magmatic crystallization. *J. Petrol.* **19**, 66–94.

O'HARA, M. J. 1968. The bearing of phase equilibria studies in synthetic and natural systems on the origin and evolution of basic and ultrabasic rocks. *Earth Sci. Rev.* **4**, 69–133.

—— 1977. Geochemical evolution during fractional crystallization of a periodically refilled magma chamber. *Nature, Lond.* **266**, 503–7.

PALLISTER, J. S. & HOPSON, C. A. 1981. Semail ophiolite plutonic suite: Field relations, phase variation, cryptic variation and layering and a model of a spreading ridge magma chamber. *J. Geophys. Res.* **86**, 2593–644.

PEARCE, J. A. 1979. Geochemical evidence for the genesis and eruptive setting of lavas from Tethyan ophiolites. *In*: PANAYIOTOU, A. (ed.) *Ophiolites, Proceedings of the International Ophiolite Symposium, Cyprus 1979*, pp. 261–72.

——, ALABASTER, T., SHELTON, A. W. & SEARLE, M. P. 1981. The Oman ophiolite as a Cretaceous arc-basin complex: evidence and implications. *Phil. Trans. Roy. Soc. Lond.* **A300** 299–317.

SATO, H., AOKI, K., OKAMOTO, K. & FUJITA, B. 1978. Petrology and chemistry of basaltic rocks from Hole 396B, IPOD/DSDP Leg 46. *Initial Reports of the Deep Sea Drilling Project* **46**, 115–42.

SHIBATA, T. DE LONG, S. E. & WALKER, D. 1978. Abyssal tholeiites from the Oceanographer Fracture Zone. I. Petrology and fractionation. *Contrib. Min. Pet.* **70**, 89–102.

SMEWING, J. D. 1981. Mixing characteristics and compositional differences in mantle-derived melts beneath spreading axes: Evidence from cyclically layered rocks in the ophiolite of North Oman. *J. Geophys. Res.* **86**, 2645–60.

SPARKS, R. S. J., SIGURDSSON, H. & MEYER, P. 1979. Density variation amongst mid-ocean ridge basalts: implications for magma mixing and the scarcity of primitive lavas. EOS **60**, 971 (abstract).

STOLPER, E. 1980. A phase diagram for mid-ocean ridge basalts: preliminary results and implications for petrogenesis. *Contrib. Min. Pet.*, **74**, 13–27.

—— & WALKER, D. 1980. Melt density and the average composition of basalt, *Contrib. Min. Pet.* **74**, 7–12.

SYMES, R. F., BEVAN, J. C. & HUTCHINSON, R. 1977. Phase chemistry studies on gabbro and peridotite rocks from Site 334, DSDP Leg 37. *Initial Reports of the Deep Sea Drilling Project* **37**, 841–6.

TURNER, J. S. & CHEN, C. F. 1974. Two-dimensional effects in double-diffusive convection. *J. Fluid Mech.* **63**, 577–92.

WAGER, L. R. 1967. Rhythmic and cryptic layering in mafic and ultramafic plutons. *In*: HESS, H. H. & POLDERVAART, A. (eds) *Basalts, Vol. 2*, Wiley, New York, pp. 573–622.

—— & BROWN, G. M. 1968. *Layered Igneous Rocks*, Oliver and Boyd, Edinburgh, 588 pp.

WALKER, D., SHIBATA, T. & DELONG, S. E. 1979. Abyssal tholeiites from the Oceanographer Fracture Zone. II. Phase equilibria and mixing. *Contrib. Min. Pet.* **70**, 111–25.

YODER, H. S. 1965. Diopside-anorthite-water at five and ten kilobars and its bearing on explosive volcanism. *Yb. Carn. Inst. Wash.* **64**, 82–89.

——, STEWART, D. B. & SMITH, J. R. 1957. Ternary feldspar. Yb. Carn. Inst. Wash. **55**, 206–14.

P. BROWNING, Department of Earth Sciences, The Open University, Walton Hall, Milton Keynes MK7 6AA. Present address: Department of Geology, University of Western Australia, Nedlands 6009, Western Australia.

Cryptic mineral-chemistry variations in a detailed traverse through the cumulate ultramafic rocks of the North Arm Mountain massif of the Bay of Islands ophiolite, Newfoundland

D. Elthon, J. F. Casey & S. Komor

SUMMARY: The compositions of cumulus olivine, orthopyroxene, and clinopyroxene from a detailed traverse through the upper portion of the cumulate ultramafic rocks from the North Arm Mountain massif of the Bay of Islands ophiolite complex have been determined. The compositions of olivine ($Fo_{91.5-83.6}$), clinopyroxene ($100 \times Mg/(Mg + Fe) = 94.2–87.5$) and orthopyroxene ($100 \times Mg/(Mg + Fe) = 90.8–85.1$) within this traverse, when compared to the compositions of minerals that crystallize from oceanic basalts at various pressures, suggest that these cumulate ultramafic rocks crystallized from basaltic liquids at high pressures ($>$10 kb) in a subaxial magma conduit system. The mineral compositions, when plotted against stratigraphic height within the complex, define eleven different cryptic units over a 'stratigraphic' thickness of 136 m. The boundary of each of these cryptic units is marked by an abrupt discontinuity in mineral chemistry from somewhat fractionated to very primitive compositions, suggesting that batches of magma passing through the conduit system do not intermix with other batches of magma undergoing crystallization in the system.

Cumulate gabbroic and ultramafic rocks within ophiolite complexes and those recovered from contemporary mid-ocean spreading centres indicate that magma fractionation commonly occurs in sub-axial magma chambers. The shape and size of these sub-axial magma chambers, as well as the processes operative in them, have been the subject of much discussion and controversy (e.g. Cann 1970, 1974; Greenbaum 1972; Dewey & Kidd 1977; Nisbet & Fowler 1978; Sleep 1978; Kuznir 1980; Pallister & Hopson 1981; Smewing 1981; Casey & Karson 1981; Elthon *et al.* 1982; Hale *et al.* 1982). Magma-chamber models based on the geology and petrology of plutonic rocks from the Bay of Islands ophiolite have undergone a series of refinements over the last several years as new data have become available (Church & Riccio 1977; Dewey & Kidd 1977; Casey & Karson 1981; Elthon *et al.* 1982). Our preferred model (Elthon *et al.* 1982), which has emerged from these refinements, suggests that the plutonic portion of the ophiolite crystallized from a composite magma chamber that consisted of a large, continuously evolving, steady-state magma chamber at shallow crustal levels and a long, narrow dyke-like feeder conduit extending from the base of the shallow-level magma chamber to depths in excess of 30 km (see Fig. 4). This conduit presumably represents the channel through which magmas derived by melting of the mantle at greater depths are delivered to the crustal-level magma chamber. Mineral compositions from rocks within the cumulate ultramafic unit and the transition zone indicate that these rocks have been produced by high-pressure crystal fractionation along the walls of a conduit, whereas upper-level gabbroic rocks in the plutonic section have undergone low-pressure fractionation in a crustal-level magma chamber (Fig. 4 and Elthon *et al.* 1982). These ultramafic cumulates precipitated along the walls of the conduit system ultimately suffered high-temperature ductile deformation (Casey 1979, 1980) and apparently were tectonically transported to shallow levels near the base of the crust as a consequence of continuous mantle upwelling in the zone of mantle divergence beneath a mid-ocean ridge. The upper parts of the oceanic crust (gabbroic cumulates, sheeted dykes, and extrusives) are accreted on to the deformed ultramafic and transition-zone cumulates at shallow levels in a low-pressure environment.

Recorded within these cumulate rocks is evidence bearing on the complex crystallization history that basaltic liquids undergo within magma chambers. Constraining these crystallization processes requires knowledge of the manner in which mineral compositions vary with height in the plutonic section. Cryptic chemical variations, which are changes in the composition of cumulus minerals as a function of stratigraphic height (Wager & Deer 1939), have been documented in a number of basic layered intrusions and ophiolites (e.g. Wager & Deer 1939; Wager & Brown 1968; Jackson 1970, 1971; Moores 1969; Irvine 1974; Church & Riccio 1977; McBirney & Noyes 1979; Morse 1979; Pallister & Hopson 1981; Smewing 1981; Mussallam *et al.* 1981; Jaques 1981). Typical petrogenetic indices of crystal fractionation that have been used for defining the nature of these cryptic variations include the atomic $Mg/(Mg + Fe)$ (the Mg number) of olivine, clinopyroxene, or orthopyroxene and the mole % An in plagioclase.

In previous studies of cryptic compositional variations in the plutonic sections of ophiolite complexes (e.g. Church & Riccio 1977; Jaques 1981; Pallister & Hopson 1981; Smewing 1981), the stratigraphic interval between each sample has been characteristically quite large, generally on the order of one sample per 150–400 m of stratigraphic thickness. Because of the very large sampling interval used in these previous reconnaissance studies, it has not been possible to define either the thicknesses of individual cryptic units, the range of mineral compositions or the extent of crystallization that occurs within any individual cryptic unit. In this paper, a cryptic unit—cryptic layer of Wager & Deer (1939) or chemical-graded layer of Jackson (1971)—is defined as a cumulate unit in which the compositions of cumulus minerals change as a function of stratigraphic height in a manner that is consistent with the crystallization of a liquid. Detailed information on these cryptic units is extremely important in order to establish a more complete understanding of magma-chamber dynamics at mid-ocean-ridge spreading centres.

In the process of formulating this investigation, it seemed that the thousands of individual fine-scale layers found within the plutonic section indicated that the stratigraphic distances between samples in previous studies might be much too large to detect those changes in magma composition that would result from either crystallization or episodic replenishment of the chamber. In an effort, therefore, to further improve on models for both magma-chamber geometry and magma fractionation processes, we have undertaken a mineral-chemistry investigation of several detailed sampling traverses through the plutonic section of the North Arm Mountain massif of the Bay of Islands ophiolite. Samples in these traverses were collected at closely spaced intervals so that these cryptic compositional variations and their petrogenetic implications could be more adequately evaluated. The results obtained from a part of one of these traverses are reported here.

Field relationships

The cumulate ultramafic section investigated in this study is part of the plutonic section of the North Arm Mountain massif (Fig. 1) of the Bay of Islands ophiolite complex (Smith 1958; Williams 1973). The Bay of Islands complex is interpreted as an obducted remnant of early Paleozoic oceanic crust and upper mantle created at an accreting plate boundary (Church & Stevens 1971; Williams 1973; Church & Riccio 1977; Karson & Dewey 1977; Malpas 1978; Salisbury & Christensen 1978; Casey & Karson 1981; Casey *et al.* 1981; Elthon *et al.* 1982).

Detailed mapping (1:15,000) of the North Arm Mountain massif (Casey 1980) has demonstrated that the eastern portion of the massif north of the Gregory River Fault (Figs 1 & 2) is largely devoid of obduction-related faulting. This area has been chosen for our detailed study because it is structurally intact and exposure is excellent. Although this area is not complicated by obduction-related faulting, major lithologic contacts have been rotated or tilted as a result of obduction-related, large-scale folding about a NE–SW trending synclinal axis lying to the W of the study area along the line of pillow-lava exposures (Fig. 1). One manifestation of this large-scale folding is that the residual harzburgite/cumulate ultramafic contact strikes ~NE–SW and is nearly vertical in the study area (Figs 1 and 2).

The ophiolite stratigraphy may be divided into two petrogenetically distinct components. The lower component in the ophiolite stratigraphy consists of a thick (>5 km) section of harzburgite with minor dunite and orthopyroxenite. This component is interpreted to represent the residual mantle produced by partial fusion and the extraction of basaltic liquids (Irvine & Findlay 1972). The upper part of the ophiolite consists of the 'magmatic component' (~5–7 km thick), which is interpreted to be produced by the crystallization and differentiation of those basaltic liquids extracted from the underlying residual mantle (Irvine & Findlay 1972; Church & Riccio 1977; Casey *et al.* 1981). The magmatic component can be further subdivided into: (i) an upper intrusive/extrusive carapace composed of an uppermost unit of basaltic pillow lavas and other extrusive rocks, and an underlying unit of sheeted diabase dykes, and (ii) a plutonic section composed of four mappable and laterally continuous units. These plutonic units, from base to top, are: *Unit 1*—interlayered cumulate ultramafics composed dominantly of dunite, wehrlite, and clinopyroxenite with lesser amounts of websterite, lherzolite, and harzburgite, and rare chromitite; *Unit 2*—interlayered mafic and ultramafic cumulates (the 'transition zone') composed of complexly interlayered troctolite, olivine gabbro, gabbro, anorthosite, dunite, wehrlite, clinopyroxenite, websterite, and lherzolite; *Unit 3*—layered gabbroic cumulates composed of olivine gabbro, troctolite, gabbro, and anorthosite; *Unit 4*—non-layered ('isotropic') gabbroic rocks composed of gabbro, olivine gabbro, and metagabbro (locally includes small quartz diorite or trondhjemite intrusions). In addition to the above units, a laterally discontinuous unit composed of gabbros and metagabbros intruded by >10% diabase dykes is also locally present at

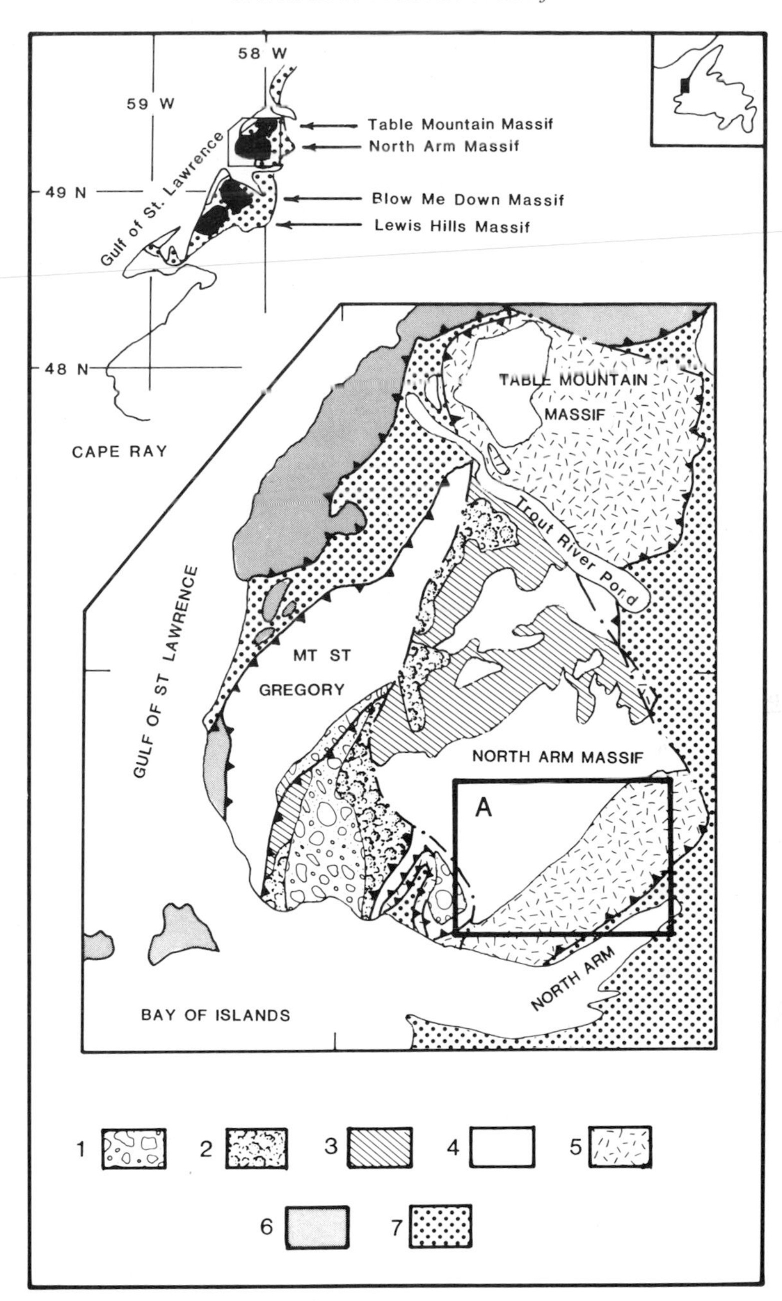

FIG. 1. Generalized geologic map of the North Arm Mountain massif in south-western Newfoundland. 1, parallochthonous sedimentary rocks; 2, mafic pillow lavas; 3, sheeted diabase dykes; 4, mafic and ultramafic plutonic rocks; 5, residual harzburgite tectonites; 6, the coastal complex (Karson & Dewey 1978); 7, Humber Arm sedimentary rocks. Dot-dash lines, major steep faults; lines with solid triangles, thrust faults. Area covered by Fig. 2 is shown by box labelled A.

the top of the plutonic sequence adjacent to the contact with the base of the sheeted dyke unit.

We have collected closely spaced samples from several traverses (Fig. 2) through the plutonic section of the North Arm Mountain massif in an effort to document the nature and scale of mineral-chemistry variations in these rocks as a function of stratigraphic height. Along each of

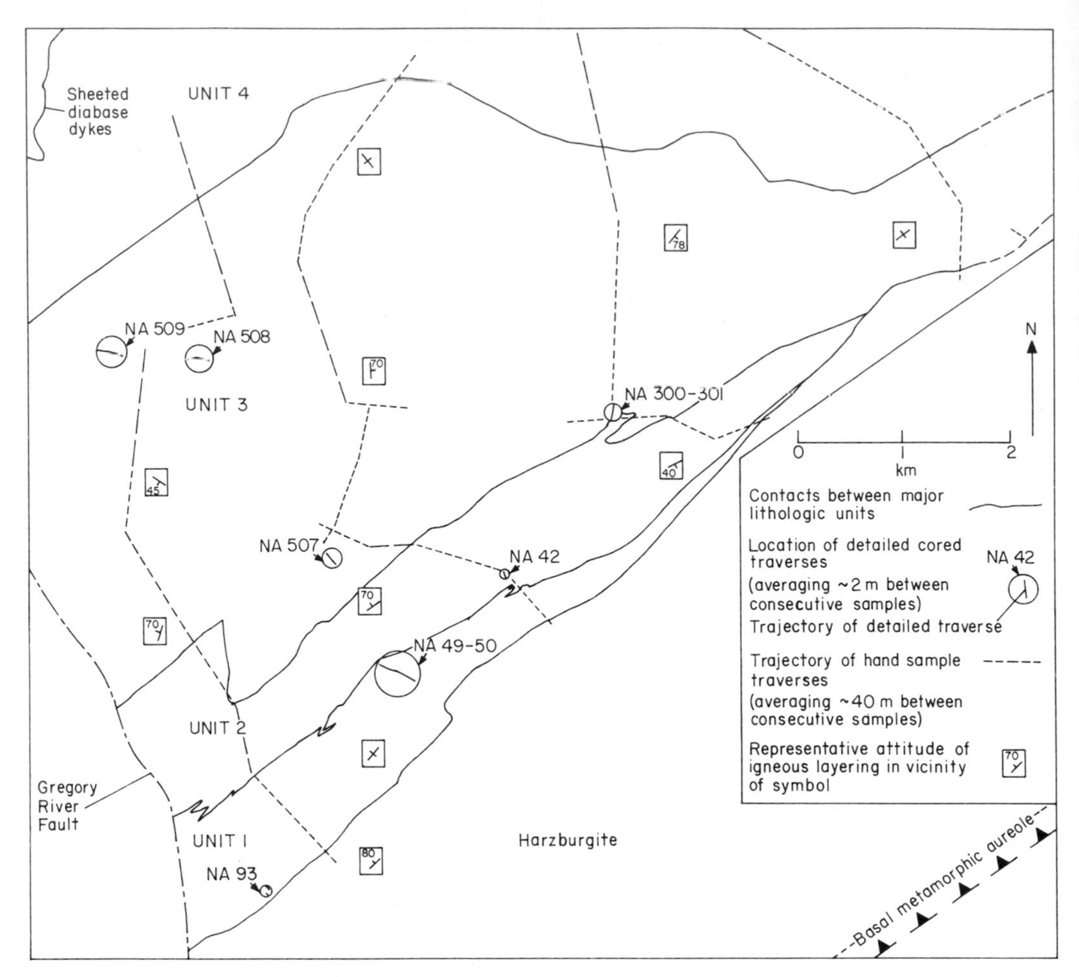

FIG. 2. Geological map indicating the locations of samples from the NA49–50 traverse and other traverses that have been collected. The mappable units indicated on the map are described in the text. The location of this map is shown by the box that is labelled A in Fig. 1.

these traverses, we have chosen limited segments for very detailed studies. These segments usually correspond to areas were exposure is complete for hundreds of metres perpendicular to layering. Detailed sampling was accomplished with the aid of a rock-coring device.

In this paper, we present the results from one such segment, hereafter referred to as the NA 49–50 traverse. The study involved a detailed mineral-chemistry study of fifty-seven samples along a traverse through the basal cumulate ultramafic unit of the plutonic section that begins approximately 700 m upsection of the cumulate ultramafic/residual harzburgite contact, spans 136 m of stratigraphic section and terminates near the cumulate ultramafic/transition-zone contact (Fig. 2). The different lithologies present and the thicknesses of the minor lithologic units (measured perpendicular to layering) are listed in Table 1 (minor lithologic units, as used here, are a layer or combination of layers that are laterally discontinuous and non-mappable at scales of <1:15,000, but are grouped together according to the distinctive lithologies present). These relationships are also shown in the columnar section along the left-hand side of Fig. 3.

Layering characteristics

The most pervasive and characteristic structure within the cumulate ultramafic and transition-zone rocks of the North Arm Mountain massif is fine-scale layering that is interpreted to have an igneous origin. This pervasive fine-scale layering is well exhibited in the ultramafic plutonic rocks along the NA 49–50 traverse. Individual layers

TABLE 1. *Lithological variations in the NA 49–50 traverse*

Lithologic unit no.	Lithology	Height of top of unit above base of section (m)
1	Interlayered dunite and wehrlite (opx in some layers)	0
2	Wehrlite	2.4
3	Interlayered dunite and wehrlite	2.6
4	Clinopyroxenite	3.4
5	Interlayered dunite and wehrlite	3.6
6	Interlayered wehrlite, dunite and clinopyroxenite	19.1
7	Interlayered dunite and wehrlite	25.1
8	Wehrlite with minor interlayered dunite (some layers contain minor plag)	58.3
9	Clinopyroxenite with minor olivine	60.2
10	Dunite	60.7
11	Clinopyroxenite with minor olivine	61.1
12	Interlayered dunite and wehrlite (opx in some layers)	61.4
13	Interlayered olivine gabbro and olivine-rich wehrlite (opx in some layers)	91.8
14	Dunite with minor clinopyroxene	94.4
15	Interlayered anorthositic gabbro and wehrlite	97.6
16	Dunite with minor wehrlite	101.4
17	Olivine gabbro (opx in some layers)	115.9
18	Interlayered wehrlite and clinopyroxenite	118.7
19	Feldspathic wehrlite	122.9
20	Clinopyroxenite (olivine and/or opx in some layers)	135.3
21	Feldspathic wehrlite	135.6

range in thickness from less than a centimetre to several metres. The thicker layers are often laterally continuous for several tens of metres and occasionally for up to 100 m. The thinner layers are generally laterally continuous for only very short distances.

It should be noted that the boundaries between lithologic units in the traverse studied (Fig. 3 and Table 1) may not correspond to the boundaries of individual layers. Especially in the thicker lithologic units, these units may include numerous individual layers and are distinguished from adjacent units by distinctive lithologies. Layering is too fine scale, in general, to represent each successive layer in a stratigraphic columnar section of reasonable size.

Two mutually exclusive types of fine-scale layering have been identified in undeformed plutonic sections of the North Arm Mountain massif (Casey 1980). These layer types are uniform layers and stratified layers. Uniform layers have both uniform mineral proportions and physical properties throughout the layer. Stratified layers are those in which a gradual change in the size of cumulus minerals and/or the proportions of these minerals occur as a function of stratigraphic height within the layer.

The dominant type of fine-scale layering in the Bay of Islands ophiolite and in the particular section under discussion here is uniform layering. Stratified layers are extremely scarce. Layer sequences are characterized by uniform layer after uniform layer without intervening statified layers. This type of sequence is not consistent with the model of Wager *et al.* (1960) for layer formation from periodic, crystal-laden density currents. If density currents were responsible for layer formation, intervening stratified layers between these uniform layers would be required (Wager *et al.* 1960). These characteristics of the layering, in conjunction with other observations—i.e. the relationship of the fine-scale layering attitudes to assumed palinspastic horizontals defined by large-scale lithologic unit contacts

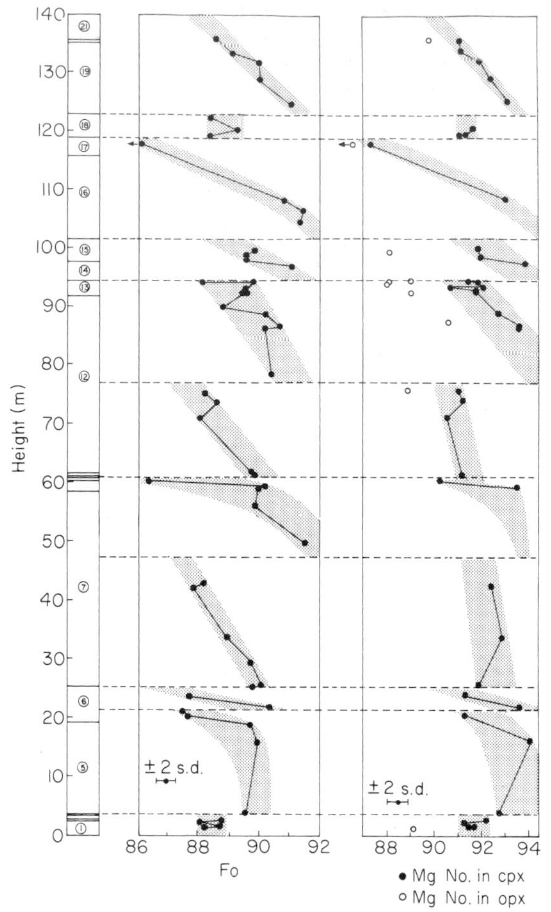

FIG. 3. Compositional variations of cumulus olivine and pyroxenes as a function of stratigraphic height within the traverse. The circled numerals in the left-hand column correspond to the lithologic units in Table 1. Also shown are representative standard deviations for the mineral-chemistry variations. The shaded regions for the cpx mineral compositions are drawn to correspond to the olivine shaded regions for rocks that do not contain cpx. The arrow for some of the data points indicate that the correct position lies to the left, at lower mg numbers. Mg No. = 100 × Mg/(Mg + Fe).

within the ophiolite (Casey 1980; Casey & Karson, 1981)—strongly suggest that layer formation is the result of *in situ* crystallization processes (cf. Campbell 1978; McBirney & Noyes 1979) along the boundary surface of a sub-axial magma chamber. Phase, ratio, and form contacts (Jackson 1967) are all observed along the traverse, although phase and ratio contacts are the predominant types in the basal cumulates of the plutonic section.

Geochemical evidence, e.g. the extremely low abundances of incompatible trace elements (Zr, Y and Ti; our unpublished data) and the very high Mg numbers of mafic minerals, indicate that these plutonic rocks originated as adcumulates or heteradcumulates that were essentially devoid of trapped intercumulus liquid (Wager *et al.* 1960). The low abundances of these incompatible trace elements enable one to estimate the percentage of trapped intercumulus liquid in most of these ultramafic plutonic rocks to be <5%. The occurrence of abundant monomineralic cumulate layers in these ultramafic rocks is additional evidence that many of these cumulates originally crystallized as adcumulates, devoid of any intercumulus liquid (Jackson 1971).

All of the rocks from the NA 49–50 traverse have undergone some degree of subsolidus deformation and recrystallization. We infer that this high-temperature ductile deformation took place beneath a mid-ocean ridge as these ultramafic cumulates were tectonically transported to the base of the oceanic crust from depths of formation that probably exceeded 30 km (Elthon *et al.* 1982). Petrofabric and textural studies of these deformed cumulates (Christensen & Salisbury 1979; Casey *et al.*, in prep.) suggest that this solid-state, ductile deformation and re-equilibrium occurred at very high temperatures (1000–1200°C: e.g. Nicolas *et al.* 1980). Locally, this high-temperature deformation has produced complex structural relationships within the plutonic section.

The complications resulting from the complex structures within the ultramafic cumulate unit could, under certain circumstances, present problems when assuming that the samples collected across strike represent successively younger rocks in the upsection direction. For example, whereas no mesoscopic isoclinal folds have been identified along the NA 49–50 traverse, such folds have been identified in similar rocks along strike. In these areas with isoclinal folds, traverses of the type that we have undertaken might cross fold axes, thereby crossing inverted sections of these layered ultramafic rocks. We believe that this potential problem is minimized for the NA 49–50 traverse because folds that have been identified are generally small-scale features whose amplitudes and wavelengths are generally less than a metre and larger-scale folding is not observed in the study area of this traverse nor is it apparent along contacts between the major lithologic units (residual mantle/cumulate ultramafic or cumulate ultramafic/transition-zone contacts). If large-scale folds were crossed during the sample collecting and some repetition of lithologic units occurred, the mineral-chemistry trends observed in these traverses would not be expected to show only normal trends with the sharply-defined stepped resets (or cryptic contacts) shown in Fig. 3, but would include inverse trends where the Mg numbers of the mafic minerals would be symmetrically disributed about the axes of these folds. On the contrary, mineral-chemistry trends observed along the traverse are all normal trends that have sharply defined resets in the compositions of minerals rather than symmetrical trends (Fig. 3). Whereas minor, local perturbations in the mineral-chemistry trends may be caused by the repetition of units by small-scale folding, these minor effects do not appear to have significantly complicated the mineral composition trends in Fig. 3, perhaps because the scale of folding is generally smaller than typical sampling intervals and reset intervals.

Certain chemical characteristics of the mineral compositions of these rocks probably changed during the subsolidus re-equilibration. It is unlikely, however, that the Mg numbers, which are sensitive indices of crystal fractionation in cumulate ultramafic rocks, have been significantly changed during this equilibration. Fe-Mg cation exchange during this subsolidus equilibration will produce only minor changes in the mineral Mg numbers because co-crystallizing olivine, clinopyroxene and orthopyroxene have very similar Mg numbers (Matsui & Nishazawa 1974; Mori 1977; Elthon & Scarfe 1983). Further evidence indicating that any changes of the mineral Mg numbers during subsolidus re-equilibration have been minimal is provided by the observation that, aside from the major resets (cryptic contacts) in mineral compositions, any change in the composition of a particular phase from one layer to the next is almost always small even if the modal proportions of minerals are very different in the adjacent layers.

Analytical methods

Within the NA 49–50 traverse, the mineral chemistry of fifty-seven samples has been investigated. The olivine, clinopyroxene and orthopyroxene analyses, listed according to stratigraphic position in the traverse, are given in Table 2. The compositions of pyroxenes were determined on the ARL-SEMQ electron microprobe at the Department of Mineral Sciences of the Smithsonian Institution. These analyses were made with a defocused beam (30 μm) in order to include the fine-scale exsolution lamellae that commonly occur. These averages generally incorporate twelve to fifteen individual analyses with a 10 s counting interval per analysis. These compositions of the olivines in the NA 50 samples were determined in the same manner except that the beam diameter was 3–5 μm. The compositions of olivines in the NA 49 samples were determined on the ARL electron microprobe at the Johnson Space Centre (complete olivine analyses are available upon request). For all of the above analyses, natural or synthetic mineral standards were used and all results are corrected for matrix effects (Bence & Albee 1968; Albee & Ray 1970) and instrumental drift.

Results

Our major objective in this mineral-chemistry investigation is to determine how the compositions of cumulus minerals vary as a function of stratigraphic position within this traverse through the cumulate ultramafic rocks. We interpret these compositional variations to have probably developed due to small changes in the composition of the liquid (or possibly the pressure of crystallization) during magma ascent within magma conduit systems (see Elthon *et al.* 1982). The compositions of cumulus minerals would thus reflect changes in the liquid composition as a consequence of crystallization processes operating within this magma conduit system. By documenting the manner in which mineral compositions change as a function of height within these layered rocks, additional constraints on these fractionation processes can be established.

The vertical variations of the most sensitive petrogenic indicators (Fo in olivine and the Mg numbers of clinopyroxene and orthopyroxene) are shown in Fig. 3. Shown along the left-hand side of Fig. 3 is the corresponding columnar section in which the lithologic units, numbered 1 through 21, are indicated by the circled numerals (compare with Table 1). The NiO content of olivine (Table 2; 1 s.d. × 0.015 wt % NiO) may be a sensitive indicator of crystallization processes as well. The interpretation of these NiO contents, however, is complicated by the presence of Ni-bearing sulphides in many of these rocks. Consequently, we do not believe that these NiO data can be quantitatively interpreted until the sulphide mineral assemblages and compositions are documented.

An examination of the variations in the olivine composition as a function of height in Fig. 3 indicates that there are eleven major compositional resets (or cryptic contacts) in this traverse. (There may be small, more subtle compositional resets that are obscured by the major resets, but the present data do not permit us to identify these.) At each of these resets, shown as horizontal dotted lines in Fig. 3, there is a

TABLE 2. *Mineral chemistry*

Sample no.	49B		49A	49C	49D
Height above base of section (in metres)	1.5		1.7	2.5	2.8
Olivine					
Fo	88.3		88.8	88.1	88.8
NiO	0.19		0.19	0.30	0.33
Pyroxenes	cpx	opx	cpx	cpx	cpx
SiO_2	52.63	55.55	52.62	52.12	53.05
TiO_2	0.18	0.06	0.13	0.14	0.12
Al_2O_3	3.63	2.33	3.34	3.52	3.04
Cr_2O_3	0.97	0.49	1.11	0.97	0.98
FeO	2.90	7.29	2.85	2.90	2.57
MgO	17.48	33.42	17.58	17.17	17.12
CaO	23.19	1.13	22.74	22.33	22.73
Na_2O	0.21	0.03	0.23	0.20	0.21
Total	101.19	100.30	100.60	99.35	99.82
100×Mg/(Mg+Fe)	91.5	89.1	91.7	91.3	92.2

Sample no.	49E	49H	49G	49F	49L
Height above base of section (in metres)	3.5	3.9	16.1	19.0	20.7
Olivine					
Fo		89.5	90.0	89.7	87.7
NiO		0.23	0.28	0.27	0.26
Pyroxenes	cpx		cpx		cpx
SiO_2	53.09		53.59		52.93
TiO_2	0.11		0.09		0.12
Al_2O_3	3.21		1.46		3.44
Cr_2O_3	1.04		0.72		0.97
FeO	2.37		1.94		2.87
MgO	16.94		17.64		16.95
CaO	23.09		24.87		22.83
Na_2O	0.21		0.26		0.28
Total	100.06		100.57		100.39
100×Mg/(Mg+Fe)	92.7		94.2		92.3

Sample no.	49K	49J	49I	49U	49T
Height above base of section (in metres)	21.4	22.0	23.8	25.4	25.7
Olivine					
Fo	87.5	90.4	87.8	89.8	90.1
NiO	0.26	0.28	0.22	0.32	0.28
Pyroxenes	cpx	cpx	cpx		
SiO_2	Not Analysed	53.32	51.94		
TiO_2		0.09	0.17		
Al_2O_3		2.01	3.69		
Cr_2O_3		0.95	1.02		
FeO		2.03	2.88		
MgO		16.88	17.13		
CaO		24.36	22.51		
Na_2O		0.24	0.21		
Total		99.88	99.55		
100×Mg/(Mg+Fe)		93.7	91.4		

Sample no.	49S	49R	49Q	49P	49O
Height above base of section (in metres)	29.7	33.6	37.0	42.2	43.1
Olivine					
Fo	89.7	89.0	88.1	87.9	88.2
NiO	0.25	0.24	0.24	0.25	0.24
Pyroxenes		cpx	cpx	cpx	
SiO_2		53.54	51.95	52.53	
TiO_2		0.20	0.01	0.20	
Al_2O_3		2.67	3.16	3.19	
Cr_2O_3		0.86	0.97	0.98	
FeO		2.28	2.29	2.35	
MgO		16.91	16.32	16.14	
CaO		24.5	23.77	24.63	
Na_2O		0.25	0.26	0.34	
Total		101.21	98.73	100.36	
100×Mg/(Mg+Fe)		93.0	92.7	92.4	

Sample no.	49N	49M	49W	49V	49X
Height above base of section (in metres)	49.8	55.9	58.9	59.6	60.2
Olivine					
Fo	91.5	89.7	90.0	90.1	86.4
NiO	0.27	0.24	0.35	0.31	0.23
Pyroxenes			cpx		cpx
SiO_2			53.06		52.14
TiO_2			0.34		0.15
Al_2O_3			3.31		3.40
Cr_2O_3			1.01		0.95
FeO			2.11		3.34
MgO			17.30		17.43
CaO			23.03		21.73
Na_2O			0.49		0.23
Total			100.65		99.37
100×Mg/(Mg+Fe)			93.6		90.3

Sample no.	49Y	49Z	50I	50H	50G
Height above base of section (in metres)	61.0	61.3	71.1	73.6	75.2
Olivine					
Fo	89.9	89.7	88.1	88.7	88.2
NiO	0.33	0.25	0.17	0.18	0.17
Pyroxenes	cpx	cpx	cpx	cpx	opx
SiO_2	52.99	52.91	52.53	51.96	57.12
TiO_2	0.15	0.17	0.16	0.16	0.20
Al_2O_3	3.19	3.48	3.49	3.37	2.52
Cr_2O_3	0.90	1.05	1.04	1.05	0.40
FeO	2.88	3.22	2.91	2.91	7.30
MgO	16.91	17.43	17.16	16.73	32.32
CaO	22.86	22.95	23.18	22.08	0.57
Na_2O	0.23	0.25	0.24	0.24	0.05
Total	100.11	101.46	100.71	98.50	100.48
100×Mg/(Mg+Fe)	91.3	90.6	91.3	91.1	88.8

Sample no.	50F	50E	50D	
Height above base of section (in metres)	78.4	86.2	86.7	
Olivine				
Fo	90.4	90.2	90.7	
NiO	0.31	0.30	0.34	
Pyroxenes		cpx	cpx	opx
SiO_2		53.11	52.16	56.63
TiO_2		0.33	0.28	0.10
Al_2O_3		3.29	3.12	2.33
Cr_2O_3		1.09	1.08	0.59
FeO		2.01	2.02	6.14
MgO		16.66	17.04	34.00
CaO		23.89	22.70	0.84
Na_2O		0.36	0.42	0.02
Total		100.74	98.82	100.65
100×Mg/(Mg+Fe)		93.7	93.8	90.8

Sample no.	50L	50K		50R	50U
Height above base of section (in metres)	94.2	94.3		97.1	97.9
Olivine					
Fo	89.8	88.1		91.1	89.6
NiO	0.20	0.20		0.20	0.17
Pyroxenes		cpx	opx	cpx	cpx
SiO_2		52.11	57.09	52.59	52.89
TiO_2		0.21	0.21	0.14	0.21
Al_2O_3		3.68	2.26	3.24	3.47
Cr_2O_3		1.10	0.47	1.00	1.09
FeO		2.71	6.97	1.95	2.65
MgO		16.46	32.95	16.99	17.25
CaO		22.04	0.92	24.32	22.45
Na_2O		0.36	0.02	0.28	0.38
Total		98.67	100.89	100.51	100.39
100×Mg/(Mg+Fe)		91.5	89.4	94.0	92.1

Sample no.	50C	50B	50Q		50P
Height above base of section (in metres)	88.6	89.7	92.3		92.5
Olivine					
Fo	90.2	88.8	89.5		89.6
NiO	0.29	0.24	0.15		0.18
Pyroxenes	cpx		cpx	opx	cpx
SiO_2	53.48		53.03	57.08	53.52
TiO_2	0.32		0.20	0.22	0.24
Al_2O_3	3.30		3.09	1.87	3.50
Cr_2O_3	1.07		0.95	0.41	1.12
FeO	2.44		2.70	6.95	2.68
MgO	17.65		17.19	32.65	16.98
CaO	22.32		22.92	0.60	22.43
Na_2O	0.37		0.37	0.03	0.38
Total	100.95		100.45	99.81	100.85
100×Mg/(Mg+Fe)	92.8		91.9	89.3	91.8

Sample no.	50T	50S	50Y	50X	50W
Height above base of section (in metres)	98.1	99.6	104.6	106.4	108.0
Olivine					
Fo	89.6	89.8	91.4	91.5	90.8
NiO	0.18	0.17	0.25	0.19	0.22
Pyroxenes	cpx	cpx			cpx
SiO_2	52.89	52.29			52.35
TiO_2	0.24	0.24			0.14
Al_2O_3	3.45	3.91			3.02
Cr_2O_3	1.07	1.25			0.97
FeO	2.69	2.69			2.12
MgO	17.25	17.06			16.34
CaO	22.18	21.77			24.32
Na_2O	0.39	0.33			0.25
Total	100.13	99.52			99.51
100×Mg/(Mg+Fe)	92.0	91.9			93.2

Sample no.	50O		50N	50M	
Height above base of section (in metres)	92.9		93.2	94.0	
Olivine					
Fo			89.6	88.9	
NiO			0.21	0.23	
Pyroxenes	cpx	opx	cpx	cpx	opx
SiO_2	52.78	57.11	52.67	52.67	56.82
TiO_2	0.19	0.18	0.23	0.19	0.19
Al_2O_3	3.20	2.13	3.35	3.34	2.29
Cr_2O_3	0.90	0.51	1.06	0.79	0.63
FeO	3.18	7.78	2.54	2.75	7.69
MgO	17.47	31.87	16.81	17.69	32.06
CaO	22.56	0.84	22.12	22.82	0.87
Na_2O	0.33	0.04	0.30	0.20	0.02
Total	101.61	100.46	99.08	100.45	100.57
100×Mg/(Mg+Fe)	90.7	88.0	92.1	92.0	88.1

Sample no.	50Z		50DD	50CC	50BB
Height above base of section (in metres)	117.6		119.0	119.3	120.2
Olivine					
Fo	83.6			88.4	89.3
NiO	0.16			0.23	0.34
Pyroxenes	cpx	opx	cpx	cpx	cpx
SiO_2	52.73	57.34	53.44	53.11	52.77
TiO_2	0.27	0.17	0.13	0.11	0.21
Al_2O_3	3.39	1.96	3.11	3.07	3.36
Cr_2O_3	0.78	0.30	0.88	0.80	1.01
FeO	4.24	9.67	2.97	2.88	2.70
MgO	16.54	30.95	17.06	17.32	16.82
CaO	21.67	0.63	23.20	22.46	22.49
Na_2O	0.37	0.05	0.21	0.19	0.33
Total	100.13	101.07	101.00	99.92	99.69
100×Mg/(Mg+Fe)	87.5	85.1	91.6	91.5	91.7

TABLE 2 (cont.)

Sample no.	50AA	50HH	50GG	50FF	50EE
Height above base of section (in metres)	122.5	124.7	129.0	132.1	133.4
Olivine					
Fo	88.4	91.1	90.0	90.0	89.2
NiO	0.19	0.37	0.28	0.30	0.29
Pyroxenes	cpx	cpx	cpx	cpx	cpx
SiO_2	52.51	53.21	52.15	52.71	53.14
TiO_2	0.12	0.37	0.31	0.37	0.52
Al_2O_3	3.28	3.03	3.36	3.56	3.68
Cr_2O_3	0.63	1.04	1.13	1.11	1.16
FeO	3.18	2.24	2.41	2.59	2.89
MgO	17.35	17.43	16.51	16.93	16.94
CaO	22.27	22.61	22.39	22.32	22.81
Na_2O	0.20	0.42	0.43	0.42	0.53
Total	99.54	100.32	98.69	100.01	101.67
100×Mg/(Mg+Fe)	92.6	93.3	92.4	92.1	91.3

Sample no.	50II	
Height above base of section (in metres)	135.5	
Olivine		
Fo	88.4	
NiO	0.27	
Pyroxenes	opx	cpx
SiO_2	56.12	53.51
TiO_2	0.21	0.12
Al_2O_3	2.26	3.27
Cr_2O_3	0.54	0.81
FeO	6.73	3.01
MgO	33.87	17.61
CaO	1.10	21.98
Na_2O	0.03	0.22
Total	100.86	100.54
100×Mg/(Mg+Fe)	90.0	91.2

—·— indicates reset.

distinctive increase in the Fo contents of the olivines. In each cryptic unit, which is the stratigraphic interval located between these resets, the Fo contents of olivines and the Mg numbers of clinopyroxenes systematically decrease with increasing height in the stratigraphic column (see Fig. 3). This decrease in the Fo content of olivine and the Mg number of clinopyroxene is interpreted to develop by crystal fractionation (see Irvine 1980). The episodic resets to higher Fo contents in olivine and higher Mg numbers in clinopyroxene are interpreted to indicate that the crystal fractionation processes have been interrupted by the influx of a more 'primitive' basaltic liquid.

It is also important to note that, following these resets, the initial olivines to crystallize generally have Fo contents ranging between $Fo_{90.5}$ and $Fo_{91.5}$. This range in olivine compositions is almost exactly the same as the olivine compositions of residual harzburgites ($Fo_{90.5-91.6}$; Malpas 1978) from the Bay of Islands, suggesting that the liquids from which the basal layers of these cryptic units crystallized were primary magmas or had undergone only very small amounts of crystallization. The olivine compositions typically decrease during crystal fractionation to approximately $Fo_{88.5-87}$ before the next reset occurs. The average stratigraphic distance between resets is 13 m and the average difference between the Fo contents of olivines of rocks just above and just below these resets is 3.2% Fo.

The origin of these ultramafic rocks

We have previously outlined the reasons for our interpretation that the cumulate ultramafic rocks from the North Arm Mountain massif, of which this section is a part, originally crystallized from a basaltic liquid at high pressures ($\gtrsim$10 kb) and were subsequently tectonically transported to a near-crustal environment in the zone of mantle upwelling and divergence beneath a mid-ocean ridge (Elthon *et al.* 1982). We believe that the data reported in this present study further support our proposal that these cumulate ultramafics originally crystallized from a basaltic liquid at high pressures.

Specifically, Elthon *et al.* (1982) noted that the mineral associations and mineral chemistry of these ultramafic cumulates are not consistent with the low-pressure crystallization of mid-ocean-ridge-basaltic (MORB) liquids. One-atmosphere experimental studies of MORBs (e.g. Dungan *et al.* 1978; Bender *et al.* 1978; Walker *et al.* 1979) show that olivine and plagioclase are the earliest silicate minerals to crystallize, clinopyroxene crystallizes only from moderately fractionated liquids, and orthopyroxene crystallizes from extremely fractionated liquids. As a consequence of the late crystallization of both clinopyroxene and (especially) orthopyroxene at 1 atmosphere, these pyroxenes have Mg numbers that are $\lesssim$84 for clinopyroxene and $\lesssim$75 for orthopyroxene.

Elthon *et al.* (1982) noted, and we re-emphasize on the basis of data in Table 2, that the Mg numbers of coexisting olivine and clinopyroxene or coexisting olivine, clinopyroxene, and orthopyroxene are much too high to have crystallized from a MORB at 1 atm. The contraction of the olivine-phase volume with increasing pressure and the expansion of the phase volumes of both orthopyroxene and clinopyroxene with increasing pressure results in the crystallization of pyroxenes with high Mg numbers at high pressures (Bender *et al.* 1978; Elthon *et al.* 1982). A special appeal to some exotic parental-basalt composition that might crystallize these pyroxenes with high Mg numbers in a low-pressure environment seems unjustified at present because of the strong chemical similarity between basalts from the Bay of Islands ophiolite (Suen *et al.* 1979; F. Siroky, unpublished data) and MORB.

The observed mineral associations in this NA 49–50 traverse very closely match the types of mineral associations that are anticipated to develop by the crystallization of MORBs at high pressures. Elthon *et al.* (1982) have proposed that the types of cumulates expected during the early stages of the crystallization of MORB liquids at 10–25 kb are (spinel-bearing) dunites, wehrlites or olivine-bearing clinopyroxenites, and finally either websterites or lherzolites as the basaltic liquid reaches the isobaric pseudo-invariant point that corresponds to the pressure of crystallization. In the NA 49–50 section, (spinel-bearing) dunites and wehrlites predominate with lesser amounts of olivine-bearing clinopyroxenite, orthopyroxene-bearing wehrlite, or lherzolite.

The basal layers in most of these crystallization intervals are dunites (some of these dunites may contain a trace amount of clinopyroxene as well). This observation is consistent with the crystallization of olivine as a result of the expansion of the olivine primary-phase volume with decreasing pressure (O'Hara 1968) during polybaric crystallization. As the extent of crystallization increases, the olivine + clinopyroxene + spinel + liquid isobaric pseudo-univariant curve is reached (actually, a series of these curves needs to be considered because of the likelihood that this crystallization took place over a range of pressures). The abundance of olivine, clinopyroxene, and spinel in this cumulate ultramafic sequence indicates that most of the crystallization of the liquid occurs along this isobaric pseudo-univariant curve or series of curves.

The data listed in Table 2 and shown in Fig. 3 demonstrates that, when orthopyroxene is found in these cumulate rocks, it is typically found at the top of each cryptic unit. Thus, the crystallization of orthopyroxene in the upper layers of many of these cryptic units indicates that the basaltic liquids undergoing crystal fractionation commonly equilibrate to the isobaric pseudo-invariant points during magma ascent as a consequence of high-pressure crystal fractionation.

The uppermost cryptic unit in this traverse is anomalous in the respect that the basal layer contains olivine, clinopyroxene, plagioclase, and spinel. It is apparent from the composition of olivine in this rock ($Fo_{91.1}$) that the liquid from which these minerals crystallized had not undergone significant crystallization, even though the liquid was apparently crystallizing olivine, clinopyroxene, plagioclase, and spinel. The forsterite-rich olivine composition and the absence of a basal layer of dunite, which would crystallize as a result of a decrease in pressure in the system, suggest that the pressure at which this cryptic unit crystallized from a basaltic liquid was similar to the pressure of origin of the basaltic liquid by melting of the mantle. We suggest that this crystallization interval did not result from crystallization of a basaltic liquid that had been undergoing polybaric crystallization while rising through the conduit system. Instead, we suggest that, in this case, a 'primitive' basaltic liquid that had undergone only minor previous crystallization leaked into the magma conduit system from the surrounding mantle at a shallow level. Evidence supporting the rare occurrence of this leaking process is provided by small dykelets and irregular intrusions that cross cut the ultramafic layers (Casey 1980). These rare dykelets and intrusions might represent the solidified traces where basaltic liquids leaked into the conduit system. This interpretation would suggest that the entry of magma into the magma conduit may, in some circumstances, occur at various pressures and that the compositions of magmas delivered to the magma conduit would, consequently, be chemically different.

Implications for magma-conduit systems

The mineral compositions, mineral assemblages and the thicknesses of cryptic units observed within these cumulate ultramafic rocks from the North Arm Mountain massif indirectly record the physico-chemical processes operating during the earliest stages of the crystallization of the basaltic liquids within the subaxial magma-conduit system (see Fig. 4). Although unravelling the details of the operation of these magma-conduit systems requires further study, we present at this stage our first order observations on the nature of these conduits.

The magnitude of the change in the Fo contents of olivines or the Mg number of clinopyroxene from the base to the top of any individual cryptic unit is chiefly a function of the amount of crystallization that occurs within the magma-conduit system. As the amount of crystallization that occurs in the magma conduit increases, the Mg number of clinopyroxene and the Fo content of olivine that crystallize from the liquid both

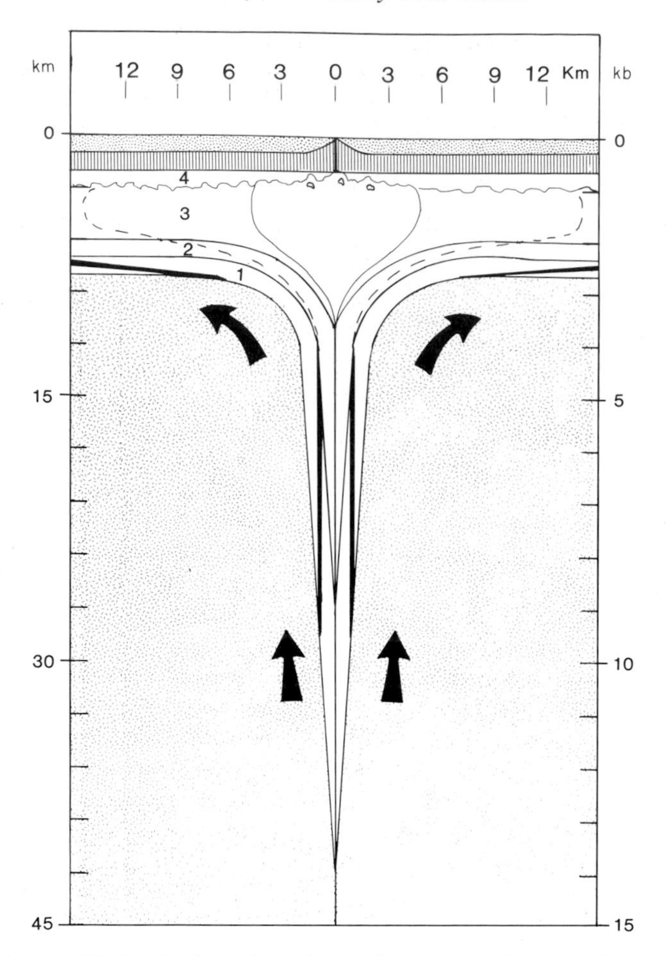

FIG. 4. Geological model for the formation of oceanic crust at a fast-spreading centre, based on geological and petrological evidence from the Bay of Islands ophiolite (Elthon *et al.* 1982). See Elthon *et al.* (1982) for details of the model and for the corresponding model for slow-spreading centres. This cross-sectional view is oriented perpendicular to the axis of spreading. Units 1, 2, 3, and 4 are described in the text. The lowermost stippled area is residual upper mantle. The large arrows indicate the motion of the mantle during upwelling.

decrease. The proportions in which olivine, clinopyroxene, orthopyroxene, or spinel crystallize from the liquid will, in detail, determine the precise magnitude of these changes in mineral chemistry per percent crystallization.

By considering several crystal fractionation models, we estimate that the individual cryptic units depicted in Fig. 3 are cumulates formed by approximately 5–10% crystallization of the liquid in the magma conduit, indicating that the mass of the liquid undergoing crystallization in the magma conduit is approximately ten to twenty times the mass of the cumulate rocks precipitated along the conduit. The widths of these dyke-like magma conduits, at the pressure (depth) of the formation of these rocks, are thereby estimated by assuming bilateral symmetry in crystallization and divergence within the dyke to be at least 26 m across (twice the 13 m average thickness of a cryptic unit), which assumes continuous flow through an elongate crack during crystallization, and up to a maximum of 250–500 m, which is calculated assuming a cylindrical shape for these magma conduits and no flow during crystallization.

Our observation that the mineral chemistry of these cumulate ultramafic rocks indicates only 5–10% crystallization of the basaltic liquid in the magma conduit requires that the remainder of the liquid in the conduit after this crystallization has occurred is physically isolated from these cumulates before any further crystallization can occur.

If separation of the remaining liquid and the cumulates did not occur, more Fe-rich mineral compositions would develop.

A related observation bearing on the transport of liquids within these magma conduits is that, judging from the very high Fo contents in olivines in the basal layers of the crystallization intervals, the incoming basaltic liquids that produce the episodic resets shown in Fig. 3 have undergone only minor amounts of crystallization and they are not intermixed with a significant amount of evolved basaltic liquid that remains in the magma conduit from a prior crystallization interval. If a significant amount of intermixing between the evolved, residual basaltic liquid, which crystallizes olivine of Fo $\lesssim$87, and the incoming basaltic liquid, which crystallizes olivine of Fo $\gtrsim$90.5, occurred, then the Fo contents of the first olivines to crystallize following this mixing would not be as high as they are (see Fig. 3).

On the basis of the limited data that we present here, two models are considered capable of explaining the evolution of the conduit system (Fig. 4). It is suggested that these conduits either close episodically, only to reopen along the axis of closure with the next incoming pulse of mantle-derived liquid, or that they seldom solidify completely, but instead remain unobstructed, permitting the episodic flow of magma through them. The principle behind this interpretation is similar to that developed in the analysis of chilled margins in sheeted dyke complexes (Kidd 1977; Rosencrantz 1980). In the cooling and solidification of sheeted dykes, a pair of inwardly facing chilled margins are produced. Similarly, in the formation of the cumulates along the margins of the magma conduit, mineral compositions in opposing sides of the magma-conduit walls will have trends indicating a decrease in Fo concentration (increasing fractionation) towards the centre of the conduit. Therefore, the complete closure of a conduit would result in mineral-chemistry trends that are symmetrical to the axis of closure of the magma conduit. Unless the magma conduit was reopened along the same axis as its axis of closure, these symmetrical mineral-chemistry trends would be preserved in the cumulate ultramafic section. The lack of such symmetrical trends in the NA 49–50 traverse (see Fig. 3) or in the other traverses that we are evaluating (our unpublished data), suggests to us that these magma conduits either must reopen along the axis of closure or that they seldom close completely. If they do not close completely, however, the amount of previously fractionated magma residing in the conduit column must be very small compared to the volume of the incoming batch or this fractionated magma must be forced upward, ahead of the new pulse of magma, to prevent significant intermixing and to explain the 'primitive' nature of the resets that we observe.

The observations listed above lead us to suggest that these subaxial magma conduits are generally narrow, continuously or semi-continuously operating channels through which basaltic liquids are transported to crustal-level magma chambers. Pulses of magma that pass through this conduit system apparently do not significantly intermix with other batches of magma undergoing crystallization in other parts of the system.

Because of this limited degree of intermixing and the apparent chemical 'primitiveness' of most of these magmas, we suggest that detailed petrological and geochemical investigations of carefully documented sections of cumulate ultramafic rocks in ophiolites will provide critical information for evaluating the true compositional variations of mantle-derived melts.

ACKNOWLEDGMENTS: We are grateful to W. G. Melson, E. Jarosewich, and J. Collins of the Smithsonian Institution, G. MacKay of NASA/JSC, and D. Deuring of SMU for access to and assistance with their electron microprobes. This work has been supported by National Science Foundation grants EAR80-26445 and OCE81-21232.

References

ALBEE, A. L. & RAY, L. 1970. Correction factors for electron probe microanalysis of silicates, oxides, carbonates, phosphates, and sulfates. *Anal. Chem.* **42**, 1408–14.

BENCE, A. E. & ALBEE, A. L. 1968. Empirical correction factors for the electron microanalysis of silicates and oxides. *Geology* **76**, 382–403.

BENDER, J. F., HODGES, F. N. & BENCE, A. E. 1978. Petrogenesis of basalts from the project FAMOUS area: experimental study from 0 to 15 Kbars. *Earth Planet. Sci. Lett.* **41**, 277–302.

CAMPBELL, I. H. 1978. Some problems with cumulus theory. *Lithos* **11**, 311–23.

CANN, J. R. 1970. New model for the structure of oceanic crust. *Nature, Lond.* **226**, 928–30.

—— 1974. A model for oceanic crustal structure developed. *Geophys. J. Royal Astr. Soc.* **39**, 169–87.

CASEY, J. F. 1979. Structure within the plutonic section, North Arm Mountain massif, Bay of Islands Ophiolite Complex: its bearing on tectonic processes at accreting plate margins. *In*: *Abstracts Volume, International Ophiolite Symposium, Cyprus.* Geological Survey Department, Nicosia, pp. 88–9.

—— 1980. *The geology of the southern part of the North Arm Mountain massif, Bay of Islands Ophiolite*

Complex, western Newfoundland with application to ophiolite obduction and the genesis of the plutonic portions of oceanic crust and upper mantle. Ph.D. Dissertation, SUNY, Albany, 594 pp.

—— & KARSON, J. A. 1981. Magma chamber profiles from the Bay of Islands Ophiolite Complex: implications for crustal-level magma chambers at mid-ocean ridges. *Nature, Lond.* **292**, 295–301.

——, DEWEY, J. F., FOX, P. J., KARSON, J.A. & ROSENCRANTZ, E. 1981. Heterogeneous nature of oceanic crust and upper mantle: a perspective from the Bay of Islands Ophiolite. *In*: EMILIANI, C. (ed.) *The Sea, Vol. VII, The Oceanic Lithosphere.* John Wiley, New York, pp. 305–38.

CHRISTENSEN, N. I. & SALISBURY, M. H. 1979. Seismic anisotropy in the oceanic upper mantle: evidence from the Bay of Islands Ophiolite Complex. *J. Geophys. Res.* **84**, 4601–10.

CHURCH, W. R. & RICCIO, L. 1977. Fractionation trends in the Bay of Islands ophiolite of Newfoundland: polycyclic cumulate sequences in ophiolites and their classification. *Can. J. Earth Sci.* **14**, 1156–65.

—— & STEVENS, R. K. 1971. Early Paleozoic ophiolite complexes of the Newfoundland Appalachians as mantle-oceanic crust sequences. *J. Geophys. Res.* **76**, 1460–6.

DEWEY, J. F. & KIDD, W. S. F. 1977. Geometry of plate accretion. *Bull. Geol. Soc. Amer.* **88**, 960–8.

DUNGAN, M. A., RHODES, J. M., LONG, P. E., BLANCHARD, D. P., BRANNON, J. C. & RODGERS, K. V. 1978. The petrology and geochemistry of basalts from Site 396, Legs 45 and 46 of the Deep Sea Drilling Project. *Initial Reports of the Deep Sea Drilling Project* **46**, 89–113.

ELTHON, D., CASEY, J. F. & KOMOR, S. 1982. Mineral chemistry of ultramafic cumulates from the North Arm Mountain massif of the Bay of Islands Ophiolite: evidence for high-pressure crystal fractionation of oceanic basalts. *J. Geophys. Res.* **87**, 8717–34.

—— & SCARFE, C. M. 1983. High-pressure phase equilibria of a high-magnesia basalt and the genesis of primary oceanic basalts. *Am. Mineral.* (in press.)

GREENBAUM, D. 1972. Magmatic processes at oceanic ridges, evidence from the Troodos Massif, Cyprus. *Nature Phys. Sci.* **238**, 18–21.

HALE, L. D., MORTON, C. J. & SLEEP, N. H. 1982. Reinterpretation of seismic reflection data over the East Pacific Rise. *J. Geophys. Res.* **87**, 7707–17.

IRVINE, T. N. 1974. Petrology of the Duke Island ultramafic complex, southeastern Alaska. *Geol. Soc. Amer. Mem.* **138**, 240 pp.

—— 1980. Magmatic infiltration metasomatism, double-diffusive fractional crystallization, and adcumulus growth in the Muskox Intrusion and other layered intrusions. *In*: HARGRAVES, R. B. (ed.) *Physics of Magmatic Processes.* Princeton University Press, pp. 325–83.

—— & FINDLAY, T. C. 1972. Alpine type peridotite with particular reference to the Bay of Islands igneous complex. *Univ. Ottawa Earth Phys. Branch Publ.* **42**, 97–128.

JACKSON, E. D. 1967. Ultramafic cumulates in the Stillwater, Great Dyke, and Bushveld Intrusions. *In*: WYLLIE, P. J. (ed.) *Ultramafic and Related Rocks.* Wiley & Sons, New York, pp. 20–38.

—— 1970. The cyclic unit in layered intrusions—a comparison of the repetitive stratigraphy in the ultramafic parts of the Stillwater, Muskox, Great Dyke, and Bushveld Complexes. *Spec. Publ. Geol. Soc. & S. Afr.* **1**, 391–424.

—— 1971. The origin of ultramafic rocks by cumulus processes. *Fortschr. Miner*, **48**, 128–74.

JAQUES, A. L. 1981. Petrology and petrogenesis of cumulate peridotites and gabbros from the Marum Ophiolite Complex, northern Papua New Guinea, *J. Petrology* **22**, 1–40.

KARSON, J. A. & DEWEY, J. F. 1977. Coastal Complex, western Newfoundland: an early Ordovician oceanic fracture zone. *Bull. Geol. Soc. Amer.* **89**, 1037–49.

KIDD, R. G. W. 1977. A model for the process of formation of the upper oceanic crust, *Geophys. J. Royal Astr. Soc.* **50**, 149–83.

KUZNIR, N. J. 1980. Thermal evolution of the oceanic crust: its dependence on spreading rate and effect on crustal structure. *Geophys. J. Royal Astr. Soc.* **61**, 167–81.

MCBIRNEY, A. R. & NOYES, R. M. 1979. Crystallization and layering of the Skaergaard Intrusion. *J. Petrology* **20**, 487–554.

MALPAS, J. 1978. Magma generation in the upper mantle, field evidence from ophiolite suites, and application to the generation of oceanic lithosphere. *Phil. Trans. Royal Soc. Lond.* **A288**, 527–46.

MATSUI, Y. & NISHAZAWA, O. 1974. Iron (II), magnesium exchange equilibrium between olivine and calcium-free pyroxene in the temperature range 800° to 1300°C. *Bull. Soc. Mineral. Crystallogr.* **97**, 122–30.

MOORES, E. M. 1969. Petrology and structure of the Vourinos ophiolitic complex, northern Greece. *Geol. Soc. Am. Spec. Paper 118*, 74 pp.

MORI, T. 1977. Geothermometry of spinel lherzolites. *Contrib. Mineral. Petrol.* **59**, 261–279.

MORSE, S. A. 1979. Kiglapait geochemistry. I. Systematics, sampling, and density. *J. Petrology* **20**, 555–90.

MUSSALLAM, K., JUNG, D. & BURGATH, K. 1981. Textural features and chemical characteristics of chromites in ultramafic rocks. Chalkidiki Complex (north-eastern Greece). *Tschermaks Min. Petr. Mitt.* **29**, 75–101.

NICOLAS, A., BOUDIER, F. & BOUCHEZ, J. L. 1980. Interpretation of peridotite structures from ophiolitic and oceanic environments. *Am. J. Sci.* **280–A(1)**, 192–210.

NISBET, E. G. & FOWLER, C. M. R. 1978. The Mid-Atlantic Ridge at 37° and 45°N: some geophysical and petrological constraints. *Geophys. J. R. Astr. Soc.* **54**, 631–60.

O'HARA, M. J. 1968. The bearing of phase equilibria studies on the origin and evolution of basic and ultrabasic rocks. *Earth Sci. Rev.* **4**, 69–133.

PALLISTER, J. S. & HOPSON, C. A. 1981. Samail ophiolite suite: Field relations, phase variation,

cryptic variation and layering, and a model of a spreading ridge magma chamber. *J. Geophys. Res* **86**, 2593–644.
ROSENCRANTZ, E. J. 1980. *The geology of the northern half of the North Arm Mountain Massif, Bay of Islands Ophiolite Complex with application to the upper ocean crust lithology, structure, and genesis.* Ph.D. Dissertation, SUNY, Albany, 250 pp.
SALISBURY, M. H. & CHRISTENSEN, N. I. 1978. The seismic velocity structure of a traverse through the Bay of Islands Ophiolite Complex, Newfoundland, an exposure of oceanic crust and upper mantle. *J. Geophys. Res.* **83**, 805–17.
SLEEP, N. H. 1978. Thermal structure and kinematics of mid-ocean ridge axis, some implications to basaltic volcanism. *Geophys. Res. Lett.* **5**, 426–8.
SMEWING, J. D. 1981. Mixing characteristics and compositional differences in mantle-derived melts beneath spreading axes: evidence from cyclically layered rocks in the ophiolite of north Oman. *J. Geophys. Res.* **86**, 2645–59.
SMITH, C. H. 1958. Bay of Islands Igneous Complex, western Newfoundland. *Mem. Geol. Soc. Can.* **290**, 132 pp.
SUEN, C. J., FREY, F. A. & MALPAS, J. 1979. Bay of Islands Ophiolite Suite, Newfoundland: petrologic and geochemical characteristics with emphasis on rare earth element geochemistry. *Earth Planet. Sci. Lett.* **45**, 337–48.
WAGER, L. R. & BROWN, G. M. 1968. *Layered Igneous Rocks.* Oliver and Boyd, Edinburgh.
—— & DEER, W. A. 1939. Geological investigations in east Greenland. III. The petrology of the Skaergaard Intrusion, Kangerdlugssuaq, east Greenland. *Meddr. Grönland* **105**, 1–352.
——, BROWN, G. M. & WADSWORTH, W. J. 1960. Types of igneous cumulates. *J. Petrology* **1**, 73–85.
WALKER, D., SHIBATA, T. & DELONG, S. E. 1979. Abyssal tholeiites from the Oceanographer Fracture Zone. II. phase equilibria and mixing. *Contrib. Mineral. Petrol.* **70**, 111–26.
WILLIAMS, H. 1973. Bay of Islands map area, Newfoundland. *Geol. Surv. Can. Paper 72–74*, and map.

D. ELTHON, J. F. CASEY & S. KOMOR, Department of Geosciences, University of Houston, University Park, Houston, TX 77004, USA.

FRACTURE ZONES

the historic data set can be explained by systematic changes associated with (i) ageing of the crust, (ii) accretionary and tectonic processes along the spreading centre, or (iii) proximity to transform faults is discussed. This paper concentrates on results from the North Atlantic, a mature, slow-spreading ridge (1–2 cm yr^{-1} half rate), though for completeness the differences between Atlantic crust and that formed at higher (Pacific) or lower (Arctic) spreading rates are reviewed briefly. The discussion is restricted to material accreted at the mid-ocean spreading centre. In the early stages of continental rifting, anomalous crust is likely to form, now to be found at the continent–ocean transition. Aseismic ridges, often formed off-axis, also typically exhibit abnormal, generally thickened crustal structure, (e.g. Azores-Biscay Rise, profile 14 in Fig. 2). These atypical regions are not considered further in this paper.

Variation of seismic structure with spreading rate

The crust of the very slow-spreading Arctic Ridge (0.5 cm yr^{-1}, Jackson *et al.* 1982) is much thinner (typically 2–3 km thick) than Atlantic crust (typically 6–8 km) and exhibits marked lateral thickness changes, presumably as a result of the episodic nature of the accretion and the low magma input to the crust. Fast-spreading ridges, such as the Pacific, produce generally more homogeneous seismic structure than the Atlantic, with subdued basement relief and widely spaced transform faults. The high magma supply at a fast-spreading centre maintains a steady-state crustal magma chamber detected beneath the East Pacific Rise by seismic reflection (Herron *et al.* 1978, 1980), by conventional refraction (Orcutt *et al.* 1976; Rosendahl *et al.* 1976; Burnett *et al.* 1982) and by fan shooting techniques. A major eruption is produced by a fast-spreading ridge typically once per 50–100 y (MacDonald 1982). Beneath the North Atlantic ridge crest there is no permanent high-level magma chamber and extrusions occur perhaps only once per 10,000 yr (Atwater 1979; MacDonald 1982).

Nevertheless, the variation of seismic velocity with depth on fast-spreading ridges is much the same as that of the Atlantic, with similar overall thickness (Reid & Jackson 1981), suggesting broadly similar igneous and tectonic processes, albeit operating at different rates.

Velocity–depth structure of normal crust

In Fig. 2a are illustrated all the velocity depth inversions from the North Atlantic that are constrained by synthetic seismogram modelling. Locations of the profiles are shown in Fig. 1 and the sources of the published data are listed in Table 1. Before discussing details of the profiles and their variability it is worth commenting that the maximum resolution that can be expected from a seismic wavelet is limited by the wavelength, which is typically about 500–1000 m. It will not be possible to resolve structure much smaller than the wavelength except in special circumstances. Synthetic seismograms generated from a model with a 'staircase' of layers of one-quarter wavelength or less in thickness are generally indistinguishable from those produced by models with much finer velocity steps, except where significant compressional to shear-wave-mode conversion occurs (White & Stephen 1980). Thus there is no physical reality to most of the velocity steps in Fig. 2a, and the profiles could equally well be represented by velocity gradients as in profile 9. When the profiles are stacked (Fig. 2b), the characteristics common to them all are emphasized: velocity gradients in the upper crust ('layer 2' of Raitt) are typically 1–2 s^{-1} with much smaller gradients of around 0.1 s^{-1} in the lower crust ('layer 3') and a more-or-less sharp crust–mantle transition at the base.

The overall change of seismic velocity with depth is governed primarily by the igneous petrology and the porosity of the rocks, modified by metamorphism, weathering and the effects of hydrothermal alteration. In the upper crust the steep velocity gradient is controlled mainly by the decrease in porosity with depth, and it is not until the lower crust is reached that the porosity is small enough that mineralogical changes become important in defining the seismic velocity. Porosities near the top of the basement are often very high (e.g. 12% in DSDP hole 504B, Becker *et al.* 1982) as a result of voids, caused by spaces between pillow lavas and rubble, ranging in size from microcracks, through cracks and joints up to fissures several metres in width. It is not known how deep into the crust the cracking extends at the spreading centre though on mature crust the seismic velocities near the base of layer 2 are similar to those expected from uncracked basalt at the appropriate pressures and temperatures (Whitmarsh 1978; Stephen *et al.* 1980; Spudich & Orcutt 1980). Earthquakes on the Mid-Atlantic Ridge extend as deep as 8 km in the vicinity of the median valley (Lilwall 1980; Toomey *et al.* 1982), showing that the crust is brittle at those depths and suggesting that cracking and water penetrates deeply into the crust along the spreading axis. However, hydrothermal circulation through the cracks will rapidly deposit and eventually fill them with secondary minerals, raising the overall

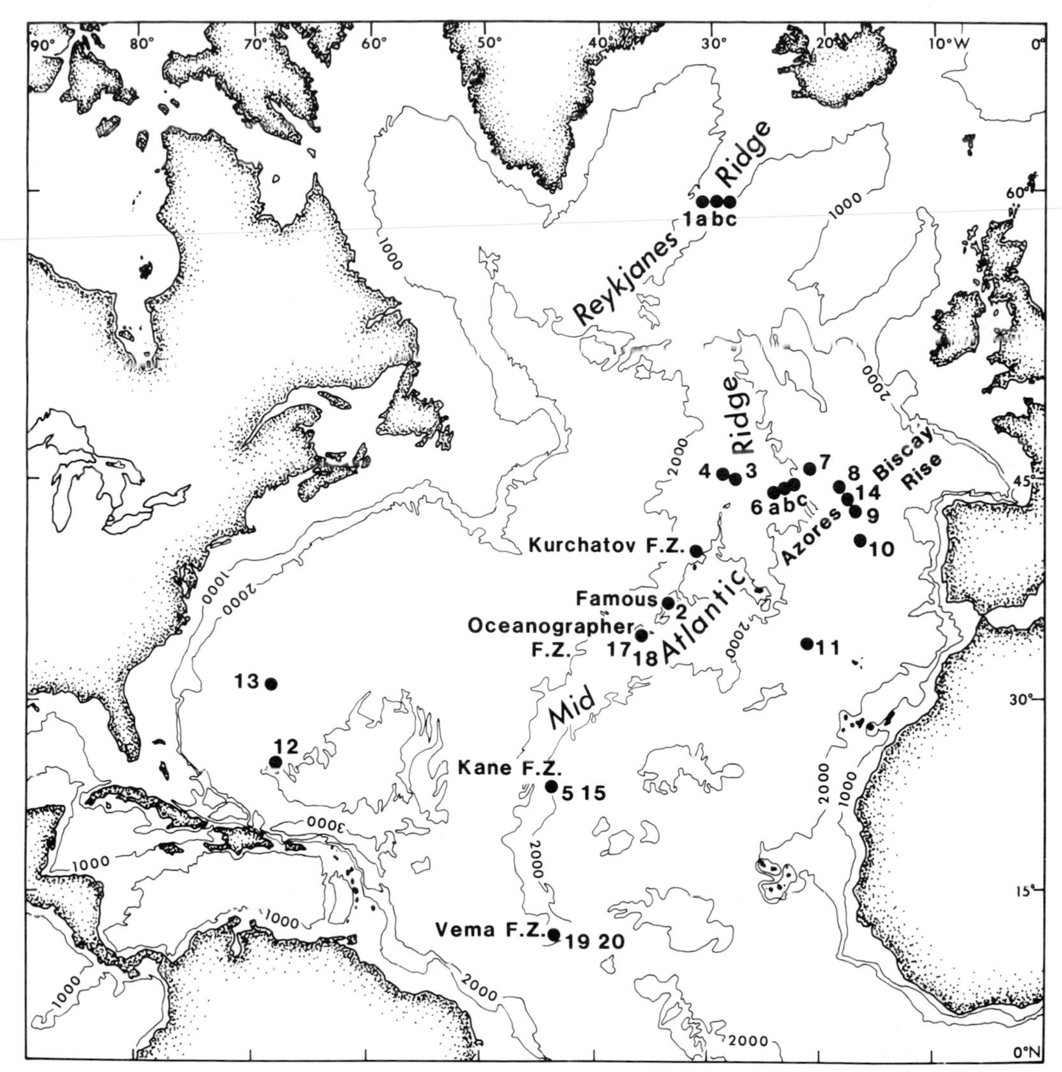

FIG. 1. Location of North Atlantic seismic profiles illustrated in Figs 2 and 3. See Table 1 for references to original publications. Named fracture zones are those from which seismic refraction information is available (see Fig. 3). Contour interval 1000 fathoms.

seismic velocity and reducing further water circulation. Retrograde metamorphism will occur as the crust is cooled by water circulation but this may subsequently be overprinted by renewed metamorphic changes as the cracks become sealed, and the cessation of hydrothermal circulation allows the crust to heat up again.

Where shear (S) wave measurements are available in addition to compressional (P) wave velocities the composition of the crust can be constrained much more tightly than from P-wave information alone (Spudich & Orcutt 1980). Unfortunately there are far fewer converted shear waves observed in the Atlantic than in the Pacific, because Atlantic crust is topographically much rougher and mode conversion highly variable spatially, so this review is restricted mainly to P-wave results.

High-velocity basal crustal material (7.2–7.7 km s^{-1}, 3B in layered solutions) is found at some sites (e.g. profiles 3b, 4), but is not widespread. Evidence for minor low-velocity layers in the lower crust comes only from profiles 1 and 12.

The transition from the crust to the mantle occurs over a transition zone of variable thickness, ranging from a step which is sharp compared to a seismic wavelength (e.g. profiles 2b, 3a, b, 4, 6a, 6b, 6c, 8, 9, 11) through 0.5 km (profile 13), 1 km

TABLE 1. *Sources of North Atlantic oceanic seismic structures illustrated in Figs 2 and 3*

Profile No.	Authors and year	Location	Profile name
1	Bunch & Kennett 1980	Reykjanes Ridge	a. line X b. line Y (solid outward: dotted inward) *c.* line Z
2	Fowler 1976	MAR crest in FAMOUS area	a. line C along median valley b. lines A and B perpendicular to m.v. (solid not crossing m.v.: dotted crossing m.v.)
3	Fowler 1978	MAR crest at 45°N	a. line E, oblique to spreading axis b. line F, perpendicular to spreading axis
4	Fowler & Keen 1979	MAR at 45°N	line 4 at s/b 1 oblique to spreading axis
5	Detrick & Purdy 1980	near Kane F.Z.*	OBH 1, parallel to spreading axis
6	Searle & Whitmarsh 1978	near King's Trough	a. lines 7284/7288 b. line 8895 c. line 7292
7	White & Matthews 1980	NE Atlantic	Aquaflex-PUBS line
8	Whitmarsh *et al.* 1982	Azores–Biscay Rise	NW flank line 9809
9	Whitmarsh *et al.* 1982	Azores–Biscay Rise	SE flank line 9807
10	Whitmarsh 1978	NE Atlantic	line 7756
11	White 1979	Madeira abyssal plain	Aquaflex line
12	Stephen *et al.* 1980	NW Atlantic	Oblique seismic expt. in DSDP hole 417D
13	Purdy 1983	NW Atlantic	composite from 5 OBH
14	Whitmarsh *et al.* 1982	Azores–Biscay Rise	crest lines 9802–9803
15	Detrick & Purdy 1980	Kane F.Z.	eastern fossil trace (soild OBH 4: broken OBH 5)
16(=7)	White & Matthews 1980	N.E. Atlantic F.Z.	reversed sonobuoy line
17	Sinha & Louden 1983	Oceanographer F.Z.	active transform line A (solid average structure: broken thinnest portion)
18	Fox *et al.* 1976	Oceanographer F.Z.	Western fossil trace
19	Ludwig & Rabinowitz, 1980	Vema F.Z.	active transform (solid SLF 5 & 6: broken SHF48)
20	Detrick *et al.* 1982a	Vema F.Z.	Western intersection (solid, FUKUI; broken FUKUI-E or FUKUI-W: dotted JM2-W)
21	Steinmetz *et al.* 1977	Kurchatov F.Z.	Western fossil trace

*F.Z., fracture zone.

(profile 5), and up to 1.5 km (profile 1c). Similar variability is found in the depth over which the change occurs from mafic to ultramafic rocks in ophiolite suites of presumed oceanic origin, the transition zone comprising an interfingering region of peridotitic and gabbroic pods with the ultramafic component decreasing upwards.

Crustal thicknesses in the compilation of profiles from the North Atlantic shown in Fig. 2 vary by 2–3 km. However, detailed surveys with

FIG. 2. (a) Velocity *v*. depth profiles from synthetic seismogram analyses of seismic refraction lines in the North Atlantic (profile numbers correspond to those on location map in Fig. 1 and in list of sources, Table 1). The profile numbers are shown above tick marks at the horizontal position corresponding to 4 km s^{-1} and each successive curve is displaced to the right. The horizontal scale of velocity shown in the figure applies to each profile. Dashed lines indicate relative uncertainty, dotted lines show solutions from different receivers on the same profile. Arrows mark position of Moho. Synthetic seismograms were not specifically constructed for profiles 9 and 10, though the solutions rely on general results from synthetic modelling. Profile 14 lies over anomalous crust of the Azores-Biscay Rise. (b) Compilation of profiles in (a) stacked by age groups.

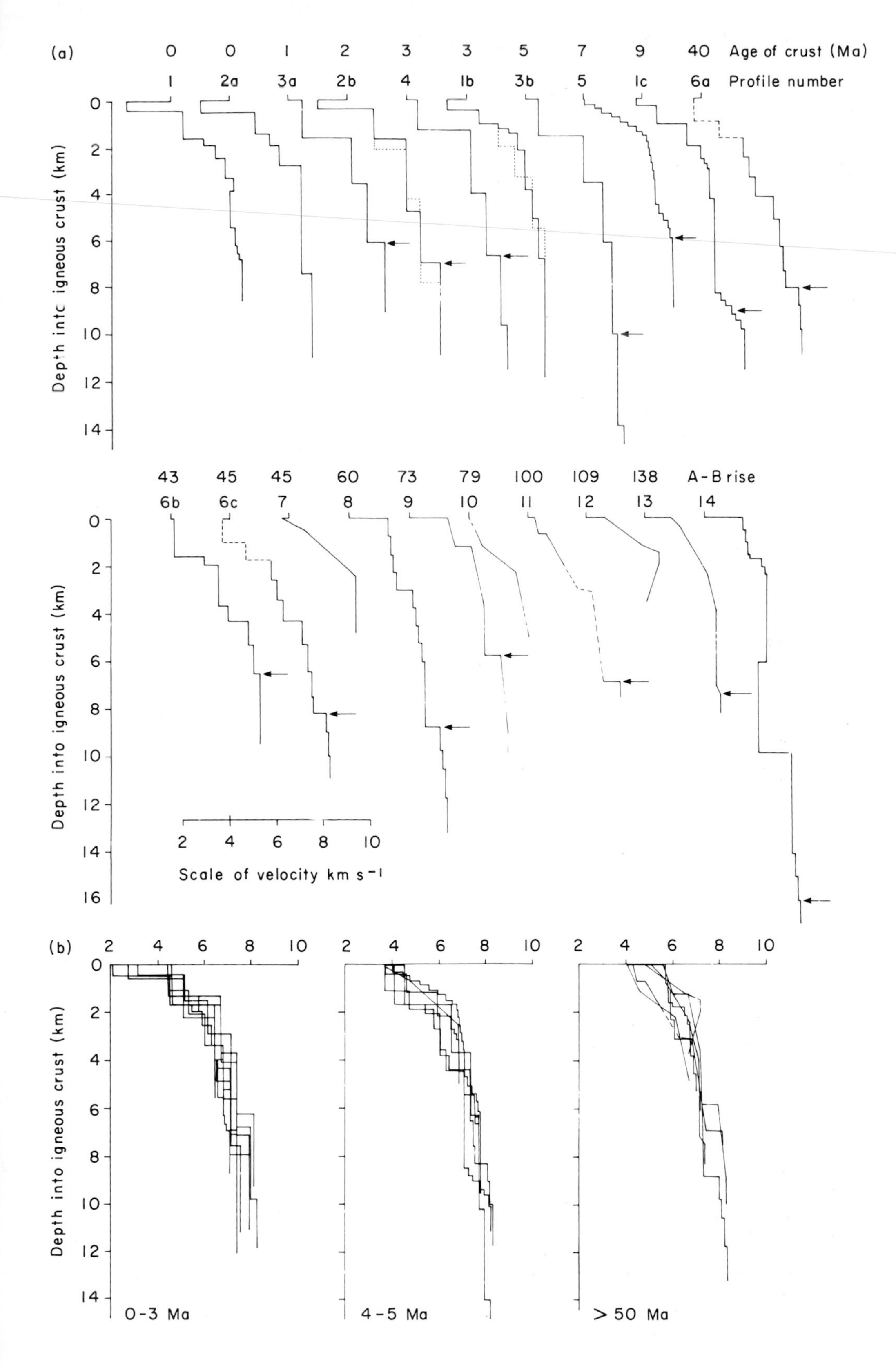

(a)
0 0 1 2 3 3 5 7 9 40 Age of crust (Ma)
1 2a 3a 2b 4 1b 3b 5 1c 6a Profile number
Depth into igneous crust (km)
43 45 45 60 73 79 100 109 138 A-B rise
6b 6c 7 8 9 10 11 12 13 14
Scale of velocity km s⁻¹
(b)
0-3 Ma
4-5 Ma
> 50 Ma

ocean-bottom seismometers in areas of normal crust away from fracture zones show little variation over distances in excess of 40 km (White & Purdy 1982, 1983; Purdy 1983), so it may be that much of the variation in thicknesses arises from sampling crust accreted in different segments of the spreading centre. The next two sections review the causes of lateral variability in the structure of normal crust formed away from transform faults.

Systematic variations in velocity of normal oceanic crust

Systematic variations are those that have predictable azimuthal, temporal or spatial control, such as anisotropic effects, ageing of the crust and changes in thickness related to faulting in the median valley.

Compressional wave anisotropy in the upper mantle is well documented (e.g. Christensen & Salisbury 1975), causing variations of up to ± 0.3 km s^{-1} propagation velocity. The maximum velocity is aligned parallel to the spreading direction and is thought to result from the crystallographic orientation of olivine crystals. Much of the scatter in the mantle velocities seen in compilations of layer solutions can be explained by random azimuthal orientation of the lines and the consequent measurement of a range of mantle velocities. Anisotropy in the lower crust (layer 3), if present is lower than can be detected (say less than ± 0.1 km s^{-1}). In the upper basaltic crust (layer 2), it is likely that the extensive faulting and fracturing which is well lineated perpendicular to the spreading direction will cause anisotropic propagation with the maximum P-wave velocity perpendicular to that of the mantle anisotropy. Such anisotropy is difficult to discern in P-wave arrivals because it tends to be obscured by the level of noise caused by the rugged topography and by lateral inhomogeneity in layer 2. However, Stephen (1981), using a three component borehole seismometer in DSDP hole 417D (profile 10), has reported S-wave splitting that suggests anisotropy in layer 2 caused by preferred crack orientation.

Marked changes occur in the velocity structure of the crust as it moves away from the spreading centre. Deep crustal velocities beneath the youngest crust of the Mid-Atlantic Ridge are rather low (7.2 km s^{-1}), with no Moho present (Fowler 1976, 1978; Bunch & Kennett 1980). Within 2 Ma a normal velocity profile with a well-developed Moho has developed. On a larger time scale, of about 40 Ma, the average velocity of the uppermost 0.5–1 km of the crust increases as the cracks and fractures of the basaltic layer become filled with minerals such as smectite and calcite (Houtz & Ewing 1976). Weathering of the topmost few tens of metres of the basement continues whilst it is exposed to the seawater, but practically ceases once it becomes buried by sediment.

The third type of systematic variations are thickness changes associated with the extensive faulting that lifts crustal blocks out of the median valley into the rift mountains. This generates the characteristic rugged basement topography of the Atlantic. Delay-time mapping of arrivals from a detailed seismic refraction survey over 9 Ma crust (White & Purdy, 1983) suggests that the isovelocity contours corresponding to the base of layer 2 are a subdued copy of the seafloor relief (= top of layer 2 since there is very little sediment present). Such a structure could be produced by growth faulting within the median valley. This faulting occurs within a few kilometres of the active spreading axis (Laughton & Searle 1979; MacDonald 1982) and penetrates well below the base of layer 2.

Non-systematic lateral variability of normal crust

This section reviews the magnitudes and possible causes of apparently random lateral inhomogeneity in normal crust on the scale of kilometres that is resolved by the seismic refraction method. There is clearly also considerable local, small-scale variability such as that between DSDP holes 417A and 417D which, although only 450 m apart, contained very different basaltic sections with virtually fresh glass in one and a highly weathered section in the other; such variations will be averaged out on the seismic scale unless they persist over distances of several kilometres.

An example of lateral variations constrained by synthetic seismogram modelling is shown by Line Y of Bunch & Kennett (1980), illustrated in profile 1b of Fig. 2. Lateral velocity gradients of up to 0.03 s^{-1} are found for velocities of around 6 km s^{-1}, an order of magnitude less than the vertical velocity gradients of about 0.5 s^{-1}. White & Purdy (1982, 1983) found lateral delay-time variations of typically 0.05–0.10 s over a 70×35 km area of 9 Ma crust, little more than the resolving power of the arrivals and Purdy (1983) found that arrivals at each of five ocean-bottom hydrophones (OBH) spaced over 20 km on 140 Ma crust could be combined within the uncertainty of the travel-time picks. Thus it seems that lateral variations are relatively small on the scale of kilometres within regions of crust away from fracture zones. A possible reason for the limited inhomogeneity is that the episodic intrusions at the spreading centre allow crustal accretion over a relatively broad (say 8 km) zone which may lead to lateral thickness changes over similar distances, while constructional features such as the volcanoes in the rift valley may be bodily transported out of the valley if subsequent rifts jump to one side. The degree and extent of hydrothermal circulation, crack filling and possible serpentinization at depth from water moving down fault zones (Francis 1981) will also vary spatially.

Lateral variations in the degree of cracking or weathering in the uppermost few tens of metres of the basement can have marked effects on the efficiency of P–S mode conversion, and are probably responsible for the large variability in the occurrence and amplitude of doubly converted S-wave returns (White & Stephen 1980).

Crustal structure in fracture zones

By far the largest cause of lateral inhomogeneity in Atlantic crust is the occurrence every 50–80 km along the ridge of a fracture zone with anomalous crust. This is also probably the single most important factor in generating scatter in compilations such as those by Raitt. For example, more than half the existing seismic refraction lines in the western central Atlantic are located across, or within a few kilometres of, a fracture zone (Purdy 1983).

In Fig. 3, the currently available refraction results from North Atlantic fracture zones are shown. There are two main features apparent on the profiles. In comparison to normal oceanic seismic structure, the fracture zones exhibit either very thin crust (profiles 15, 16, 17) or crust approaching normal thickness but exhibiting unusually low overall velocities (profiles 18, 19, 20). The crust may also be extremely variable along the length of a fracture zone, varying for example from 2–6 km thick over less than 10 km distance along the Oceanographer fracture zone (Sinha & Louden 1983). On a long profile parallel to, and on the western flank of, the Mid-Atlantic Ridge, Steinmetz *et al.* (1977) report that mantle arrivals from shots fired over the Kurchatov fracture zone are later by as much as 0.4 s again consistent with locally abnormal low-velocity crust beneath the fracture zone.

The thin crust in fracture zones suggests that they suffered a restricted magma supply. There are two reasons why this may have occurred. Firstly, where there was a large transform offset one segment of the spreading axis would be juxtaposed against old, cold lithosphere, which could lead to cooling and thus a reduction in the percentage of melt available (Fox & Gallo, in press). Secondly, it is observed that adjacent spreading-centre segments appear to accrete crust independently of one another, maintaining their separate identities for long periods even when the transform offsets are very small (Schouten & White 1980; Schouten & Klitgord 1982; Le Douran *et al.* 1982). Abnormally thin crust has been found even in a small offset fracture zone (White & Matthews 1980). In Iceland, magma has been observed to flow up to 70 km laterally from a single intrusive event (Sigurdsson & Sparks 1978), so it may be that the typical spacing between fracture zones of 50–80 km in the Atlantic is the distance over which magma from a single sub-crustal supply can feed a spreading-centre segment. Support for this idea comes from Sinha & Louden's (1983) observations which show the crustal thickness decreasing gradually over a 20 km distance on either side of both a major and a minor offset fracture zone (Fig. 2b), and from White & Matthews' (1980) determination of abnormal crust in a fracture zone with only a small offset.

The low crustal velocities typical of fracture zones are probably caused by the extensive faulting the crust has undergone whilst in the active transform domain. Although the surface trace of transform faults is generally very narrow, deep-towed reflection profiles across the Vema transform show that the sub-surface trace of the fault has moved laterally (Bowen, pers. comm.), so that faulting may eventually pervade the entire transform crust. In large transforms such as the Vema, there may also be rubble layers caused by talus derived from the fault scarps on the sides of the transform valley which will also have low velocities (Detrick *et al.* 1982a).

The highly fractured and faulted rocks of the transform zone will readily allow water to penetrate to the base of crust, particularly if the Moho is shallow. Serpentinization of upper-mantle peridotites could then readily occur. The low-velocity material in the lower part of the crust in those fracture zones exhibiting approximately normal crustal thicknesses (e.g. Vema, parts of Oceanographer) may represent original upper-mantle material that has been hydrated *in situ*.

Discussion: the formation of oceanic crust

The definition of the seismic structure of oceanic crust derived from carefully positioned and analysed refraction experiments has progressed to the stage where we can study not only the average variation of velocity with depth, but also the influence of different processes which cause lateral variations in structure. The broad picture of a slow-spreading ridge such as that in the North Atlantic is of the accretion of relatively homogeneous oceanic crust in independent spreading-centre segments typically 50–80 km long bounded by long-lived transform faults

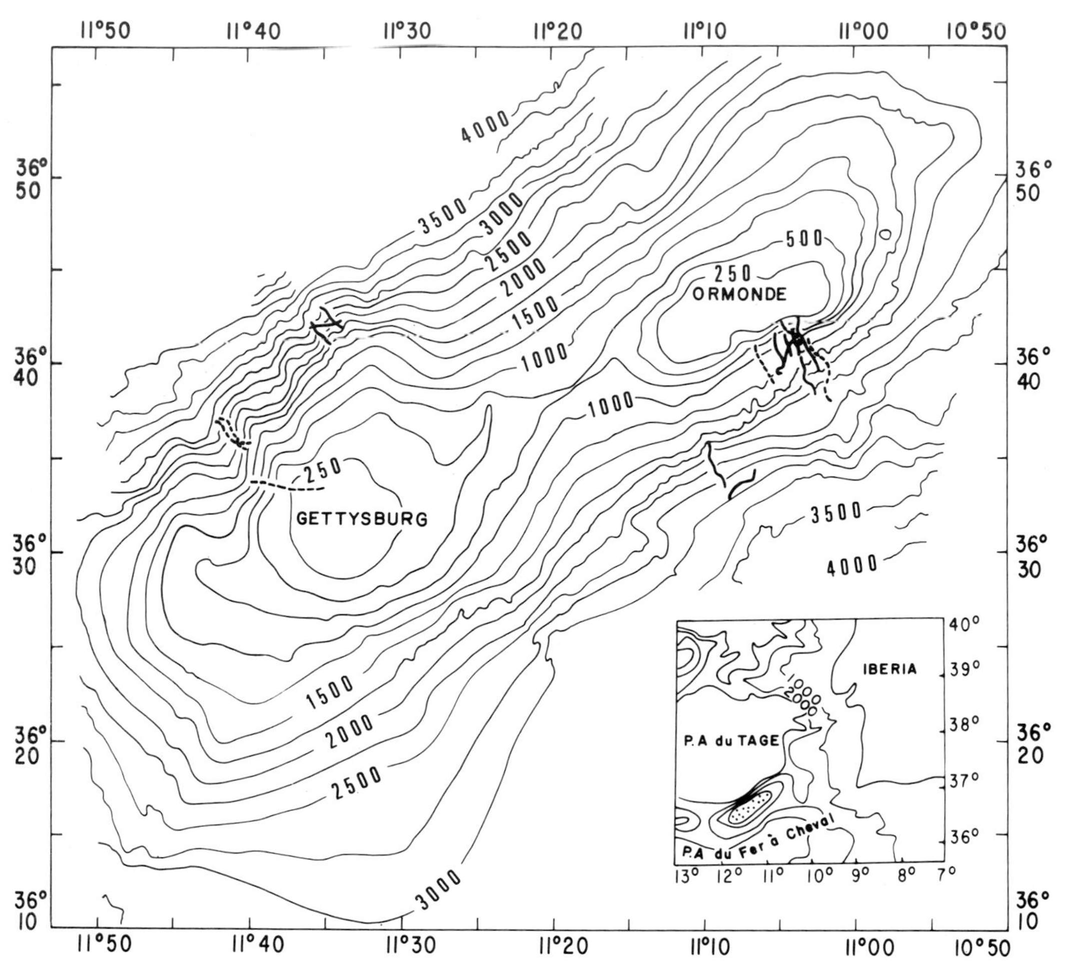

FIG. 1. Bathymetric map of the Gorringe Bank (after Beuzart *et al.* 1979) showing the tracks of the dives conducted during CYAGOR I (dotted lines) and CYAGOR II (full lines).

is capped by a thin formation of pillow basalts and dolerite breccias. Between 500 m and 100 m the top of the bank is completely covered by alkaline volcanics occurring as flows, breccias or conglomerates.

The SE flank of Mount Ormonde is cut by two major fault sets oriented N–S and N60°E. From the submersible, it was only possible to observe their normal component.

NW flank of Mount Gettysburg

In 1977, the main objective of the dives was the sampling of the serpentinite sequence. During CYAGOR II, we wanted to observe the contact between the serpentinites and the oceanic crust (gabbros); however, the three dives conducted give only a fragmented picture of its exact structure. Fig. 2b points out its complexity, evidenced by alternating exposures of gabbros, which correspond to a relatively high level of the magma chamber (Cpx-gabbros and Fe-Ti gabbros), and serpentinites. This type of relation between gabbro and serpentinites suggests that the contact is tectonic and that layered gabbros, supposed to form the lower part of the magma chamber have, if they existed, been tectonically removed.

The alkaline volcanics capping the bank have already been extensively studied (Féraud *et al.* 1982; Cornen 1982). We will now discuss the new information produced by a detailed study of the ultramafics and gabbros.

Ultramafic rocks

Most of the serpentinite samples were collected with the submersible on Mount Gettysburg, although a few were dredged on Mount Ormonde.

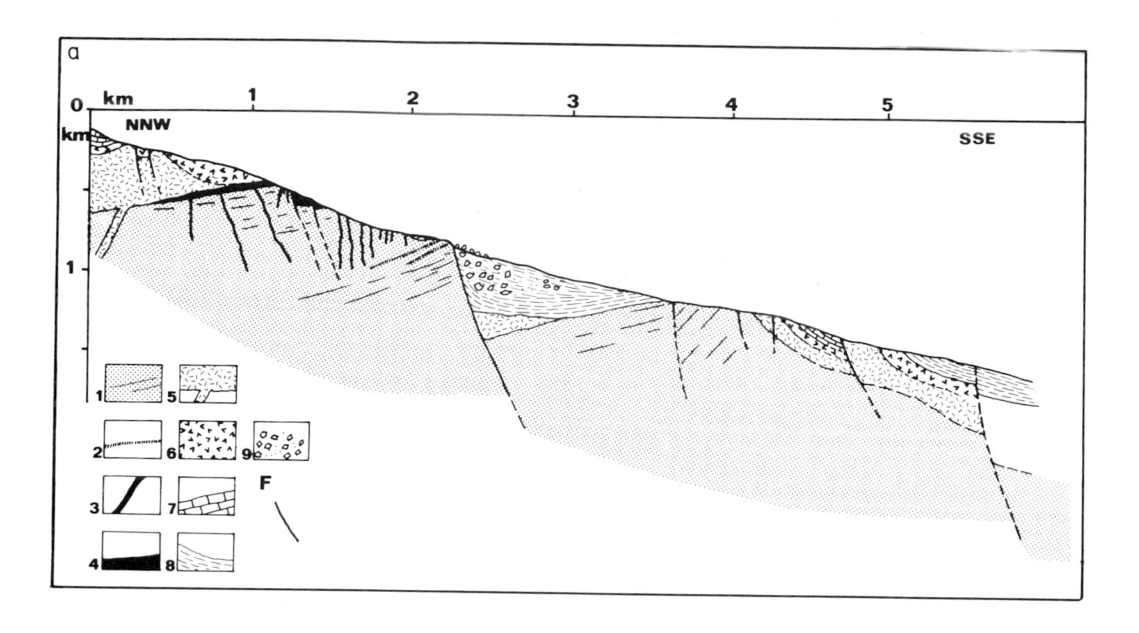

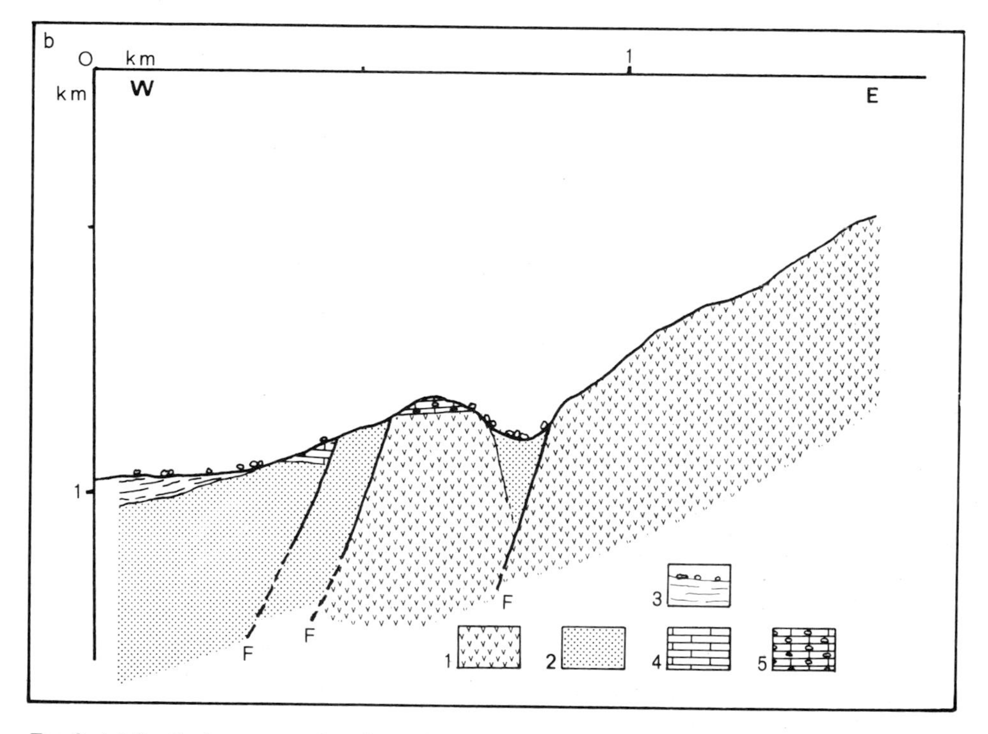

FIG. 2. (a) Synthetic cross-section through the SE flank of Ormonde Seamount, Gorringe Bank. 1: gabbros (lines as foliation); 2: flazer gabbros; 3: dolerite dykes; 4: pillow basalt and dolerite breccias; 5: alkaline volcanics; 6: alkaline breccias; 7: Tertiary chalks; 8: recent ooze; 9: recent scree deposits; F: faults. (b) Synthetic cross-section through the NW flank of Gettysburg Seamount, Gorringe Bank. 1: serpentinites; 2: gabbros; 3: recent ooze; 4: Tertiary chalks; 5: Barremian-Aptian sediments.

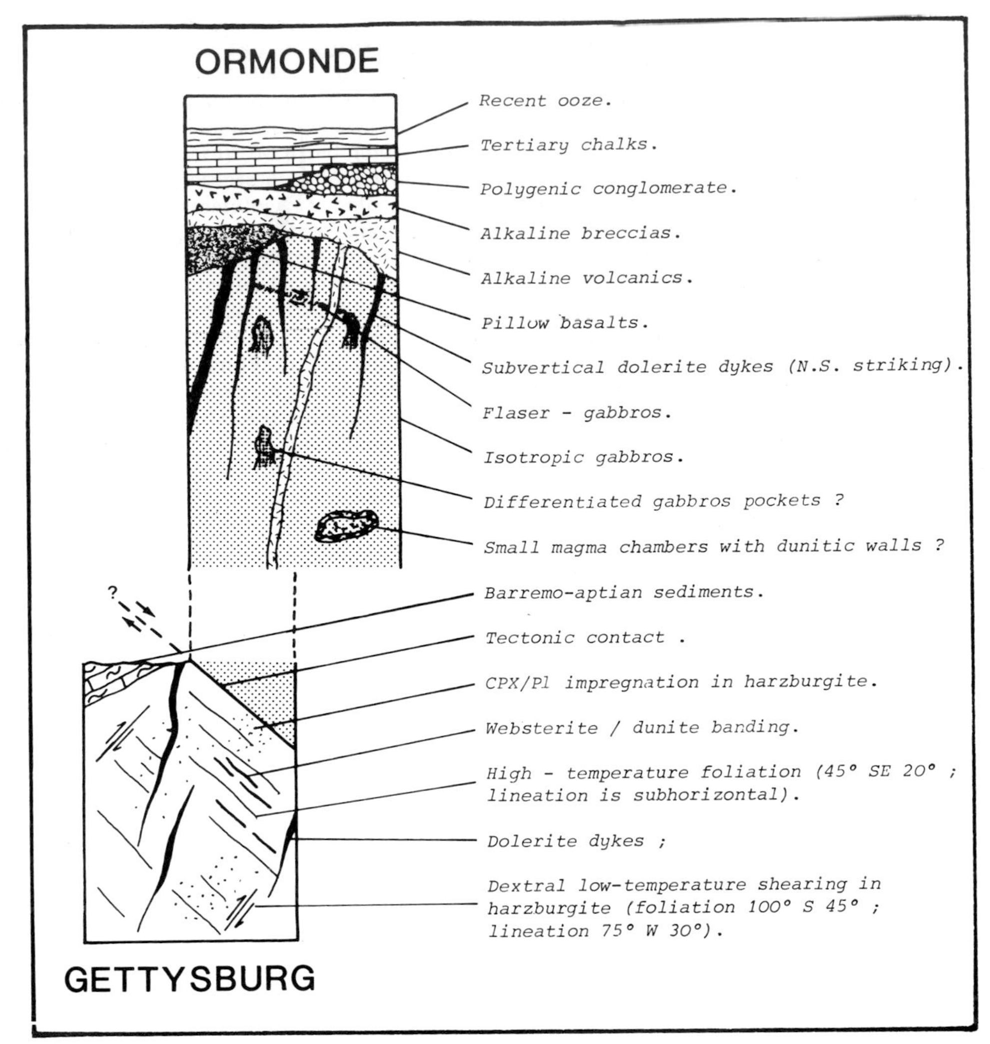

FIG. 5. Synthetic log of the Gorringe Bank.

Cayman Rise (Caytrough 1979; Malcolm 1981) suggest that deformation may occur in gabbros very soon after their crystallization. The fact that all the dolerite dykes observed and sampled were undeformed supports the hypothesis of an early deformation. K/Ar dating on Gettysburg gabbros suggests the influence of two main thermal events on secondary processes, one around 110 Ma and one around 82 Ma (Prichard & Mitchell, 1979). Further dating by $^{39}Ar/^{40}Ar$ methods is in progress.

Conclusions

The dives conducted during CYAGOR I and II cruises showed that the Gorringe massif is a section of serpentinites, gabbros and basalts analogous to ophiolite sequences capped by alkaline volcanics. The lack of layered gabbros, commonly observed between the isotropic gabbros and serpentinized peridotites may be explained either by an incomplete sampling (the area surveyed with CYANA is very small compared to the size of the massif), by the sedimentary cover deposited on the saddle between Gettysburg and Ormonde, by tectonic movements or by the type of magmatic differentiation.

The polyphase tectonic evolution of Gorringe bank is inferred from kinematic methods, and *in-situ* geological observations and is confirmed by detailed study of serpentinite and gabbro samples. A synthetic log of the present structure of the bank is shown in Fig. 5.

Gettysburg serpentinites are derived from spinel harzburgites, locally banded and more or less impregnated by a basaltic melt. We have demonstrated that the peridotites were plastically deformed by simple shear under mantle conditions. This high-temperature deformation (*c.* 1200°C) could have developed in a spreading-ridge environment. Low-temperature shearing (greenschist-facies conditions) locally overprints the high-temperature deformation. The orientations known from the few oriented samples collected by diving are shown in Fig. 3. Ormonde serpentinites are derived from dunites and their origin is still enigmatic: they could correspond to the walls of small magma chambers emplaced in isotropic gabbros.

The absence of systematic zonation in the distribution of the different gabbro types suggests the influence of tectonics. Detailed study of gabbro samples shows that they are highly deformed and that the deformation occurred at high and medium temperatures. It is likely linked to the location of the bank at the Eurasia–Africa plate boundary which acted as a transform fault. Hydrothermal circulation of seawater was concomitant and phases of static recrystallization (in amphibolite-facies conditions) occurred between shearing episodes. The brittle deformation observed in a few samples could be related to late tectonic movements responsible for the present morphology of the bank.

We would also like to point out that the study of the Gorringe Bank is important for the study of ophiolites on land. It should be considered as an example of intraoceanic massif, tectonized during successive deformation episodes. Recent and old sediments are deposited on the various rock types (serpentinites, gabbro, basalts) in a sequence similar to an ophiolite complex. Gorringe Bank observations have been compared to those made in several ophiolite massifs in Piemont and Queyras (Western Alps) where a pelagic, Upper Jurassic sedimentary cover, indiscriminately overlies gabbros, serpentinites, basalts and polygenic ophiolitic breccias (Lagabrielle 1982; Lagabrielle & Auzende 1982).

ACKNOWLEDGMENTS: We thank the officers and crew of N.O. Suroît, the submersible SP 3000 CYANA team and K. Garland for helping with the translation of the manuscript.

COB contribution No. 805.

References

ALLEGRE, C. J., MONTIGNY, R. & BOTTINGA, Y. 1973. Cortège ophiolitique et cortège océanique, géochimie comparée et mode de genèse. *Bull. Soc. géol. France* **15**, 461–77.

BEUZART, P., LE LANN, A., MONTI, S., AUZENDE, J. M. & OLIVET, J. L. 1979. Nouvelle carte bathymétrique au 1/100,000 du blanc de Gorringe (SW Portugal). *Bull. Soc. géol. France*, **XXI**, 557–62.

BONATTI, E. & HAMLYN, P. R. 1981. Oceanic ultramafic rocks. *In*: EMILIANI, C. (ed.) *The Sea, Vol. VII.* Wiley Interscience, New York, pp. 242–83.

CAYTROUGH 1979. Geological and geophysical investigation of the Mid-Cayman Rise spreading centre: initial results and observations. In: Maurice Ewing Series, 2, TALWANI, M., HARRISON, C. E., HAYES, D. E. (eds) *Am. Geophys. Union*, Washington, D.C., pp. 66–95.

CEULENEER, G. 1982. Structures des serpentinites et des gabbros du Banc de Gorringe (Océan Atlantique, SW Portugal), Campagnes CYAGOR II, 1981.

CORNEN, G. 1982. Petrology of the alkaline volcanism of Gorringe bank (Southwest Portugal). *Marine Geol.* **47**, 101–30.

CYAGOR 1977. Le Banc de Gorringe (Sud-Ouest du Portugal), un fragment de manteau et de croûte océanique reconnu par submersible. *C.R. Acad. Sci. Paris* **285**, série D, 1403–6.

—— 1979. Le banc de Gorringe: résultats de la campagne CYAGOR (août 1977). *Bull. Soc. géol. France* **XXI**, 545–56.

ENGEL, C. G. & FISHER, R. L. 1975. Granitic to ultramafic rock complexes of the Indian Ocean Ridge system, western Indian Ocean. *Bull. Geol. Soc. Am.* **86**, 1553–78.

FÉRAUD, G., GASTAUD, J., AUZENDE, J. M., OLIVET, J. L. & CORNEN, G. 1982. $^{39}Ar/^{40}Ar$ ages for the alkaline volcanism and the basement of Gorringe bank, North Atlantic Ocean. *Earth Planet. Sci. Lett.* **57**, 211–26.

FOX, P. J. & STROUP, J. P. 1981. The plutonic foundation of the oceanic crust. *In*: EMILIANI, C. (ed.) *The Sea, Vol. VII.* Wiley Interscience, New York, pp. 119–218.

LAGABRIELLE, Y. 1982. *Ophiolite et croûte océanique. Tectonique et environnement sédimentaire. Apports des données océaniques à l'interprétation géologique des séries ophiolitiques du Queyras (Alpes franco-italiennes).* Thèse de 3ème cycle, Université de Bretagne Occidentale, Brest, France.

—— & AUZENDE, J. M. 1982. Active *insitu* disaggregation of oceanic crust and mantle on Gorringe Bank: analogy with ophiolitic massifs. *Nature, Lond.* **297**, 490–493.

LE LANN, A., AUZENDE, J. M. OLIVET, J. L. 1979. Campagne CYAGOR: résultats des campagnes à la mer n° 17, pub. du CNEXO, 75 pp.

MALCOLM, F. L. 1981. Microstructures of the Cayman trough gabbros. *J. Geol.* **89**, 675–88.

MIYASHIRO, A. & SHIDO, F. 1980 Differentiation of gabbros in the mid-Atlantic Ridge near 24°N. *Geochemical J.* **14**, 145–54.

NICOLAS, A. & POIRIER, J. P. 1976. *Crystalline plasticity and solid state flow in metamorphic rocks.* Wiley & Sons, New York, 444 pp.

——, BOUDIER, F. & BOUCHEZ, J. L. 1980. Interpretation of peridotite structures from ophiolitic and oceanic environments. *Am. J. Sci.* **208-A**, 192–210.
——, GIRARDEAU, J., MARCOUX, J., DUPRE, B., WANG XIBIN, CAO YOUGONG, ZHENG HAIXIANG & XIAO XUCHAND 1981. The Xigaze ophiolite (Tibet): a peculiar oceanic lithosphere. *Nature, Lond.* **294**, 414–7.
OLIVET, J. L., BONNIN, J., BEUZART, P. & AUZENDE, J. M. 1981. Notice des cartes cinématiques de l'Atlantique Nord. (In press).
PRICHARD, H. M. 1979. A petrographic study of the process of serpentinization in ophiolites and the ocean crust. *Contrib. Mineral. Petrol.* **68**, 231–41.
—— & MITCHELL, J. G. 1979. K-Ar data for the age and evolution of Gettysburg bank, North Atlantic Ocean. *Earth Planet. Sci. Lett.* **44**, 261–8.
—— & CANN, J. R. 1982. Petrology and mineralogy of dredged gabbro from Gettysburg bank, Eastern Atlantic. *Contrib. mineral. Petrol.* **79**, 46–55.
PRINZ, M., KEIL, K. GREEN, J. A., REID, A. M. BONATTI, E. & HONNOREZ, J. 1976. Ultramafic and mafic dredge samples from the equatorial mid-Atlantic Ridge and fracture zones. *J. Geophys. Res.* **81**, 4087–103.
RUELLAN, E. 1982. Géologie du banc de Gorringe. Cartographie au 1/10,000 de deux secteurs: interprétation structurale. Unpublished, *D.E.A.*, Université de Bretagne Occidentale, Brest.
RYAN, W. B. F. & HSU, J. 1972. Gorringe bank—site 120. *Initial Reports of the D.S.D.P., Vol. XIII.* Washington (U.S. Government Printing Office), pp. 19–41.
SECHER, D. 1981. *Les lherzolites ophiolitiques de Nouvelle-Calédonie et leurs gisements de chromite.* Thèse de 3ème cycle, Nantes, France.

*J. M. AUZENDE, Chief Scientist, CNEXO-COB, BP 337, 29200 Brest, France.
G. CEULENEER, University of Nantes, France.
G. CORNEN, University of Nantes, France.
T. JUTEAU, University of Strasbourg, France.
Y. LAGABRIELLE, Université de Bretagne Occidentale, France.
G. LENSCH, University of Sarrebrücken, R.F.A.
C. MEVEL, University of Paris VI, France.
A. NICOLAS, University of Nantes, France.
H. PRICHARD, Open University, U.K.
A. RIBEIRO, Geological Survey of Portugal.
E. RUELLAN, Université de Bretagne Occidentale, France.
J. R. VANNEY, University of Paris VI, France.

Occurrence and significance of gneissic amphibolites in the Vema fracture zone, equatorial Mid-Atlantic Ridge

J. Honnorez, C. Mével & R. Montigny

SUMMARY: Numerous gneissic amphibolites were recovered together with variably deformed and metamorphosed gabbros and serpentinites in the Vema fracture zone during two separate cruises. Petrological studies and K/Ar dating show that these rocks result from recrystallization of gabbros under stress, at temperatures consistent with amphibolite facies, probably in the vicinity of the ridge axis. We suggest that gneissic amphibolites were formed in a large subvertical shear zone related to transform fault movement which favoured penetration of seawater in oceanic layer 3. Less deformed metagabbros recrystallized away from the fault zone under static conditions from interaction with evolved seawater-derived fluids.

Ocean floor metamorphism is mainly recorded as static, and models for oceanic metamorphism mostly put forward hydrothermal circulation of seawater beneath mid-ocean ridges by way of explanation (Miyashiro 1973; Spooner & Fyfe 1973; Cann 1979; Stern & Elthon 1979; Elthon 1981). The resulting rocks are characterized by preserved igneous textures and abundant relics and are quite different from continental metamorphic rocks. However, gneissic rocks have been known for some time in the oceanic environment (Honnorez *et al.* 1977; and see Honnorez *et al.* in press for a review). Recent studies of new sets of rocks suggest that gabbros of oceanic layer 3 are often deformed, sometimes so strongly that igneous textures are completely obliterated (Caytrough 1979; Malcolm 1981; Prichard & Cann 1982; CYAGOR II, this volume). This study deals with gneissic amphibolites, dredged together with gabbros and serpentinites during two separate cruises on the Vema fracture zone. These banded rocks, displaying alternating thin layers of plagioclase and amphibole, show no relic igneous texture. The purpose of this paper is to determine their condition of formation and the age of metamorphism by petrological studies and K/Ar dating. The significance of these rocks is discussed with respect to their geotectonic environment of formation in a major transform fault.

Sample location and description

The Vema fracture zone offsets the Mid-Atlantic Ridge at about 11°N in a left lateral direction by 320 km (Fig. 1). It is marked by a narrow

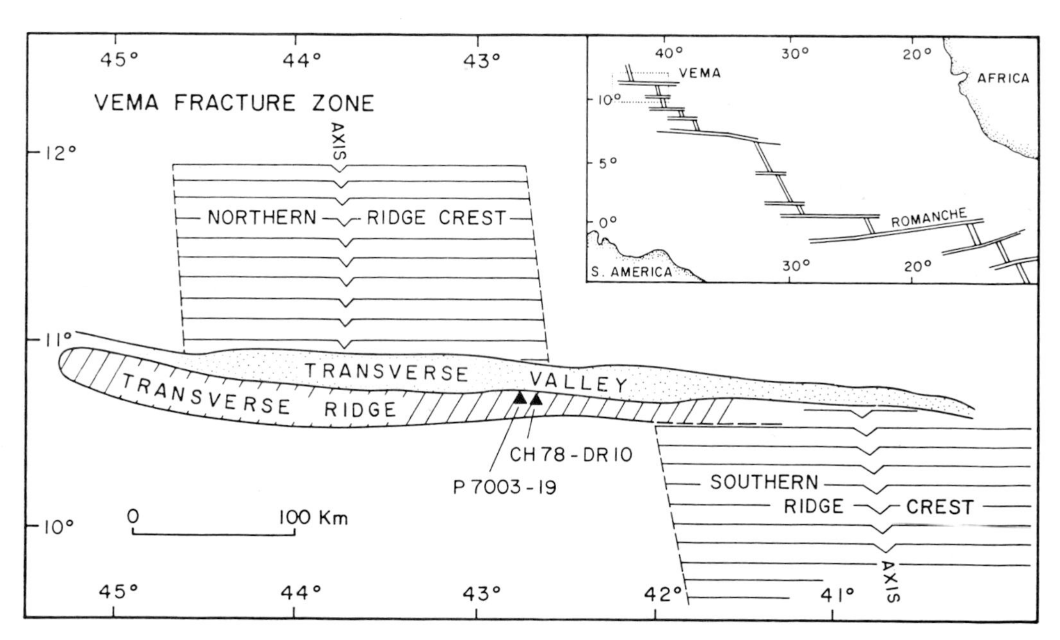

FIG. 1. Location of the two dredge hauls in the Vema fracture zone.

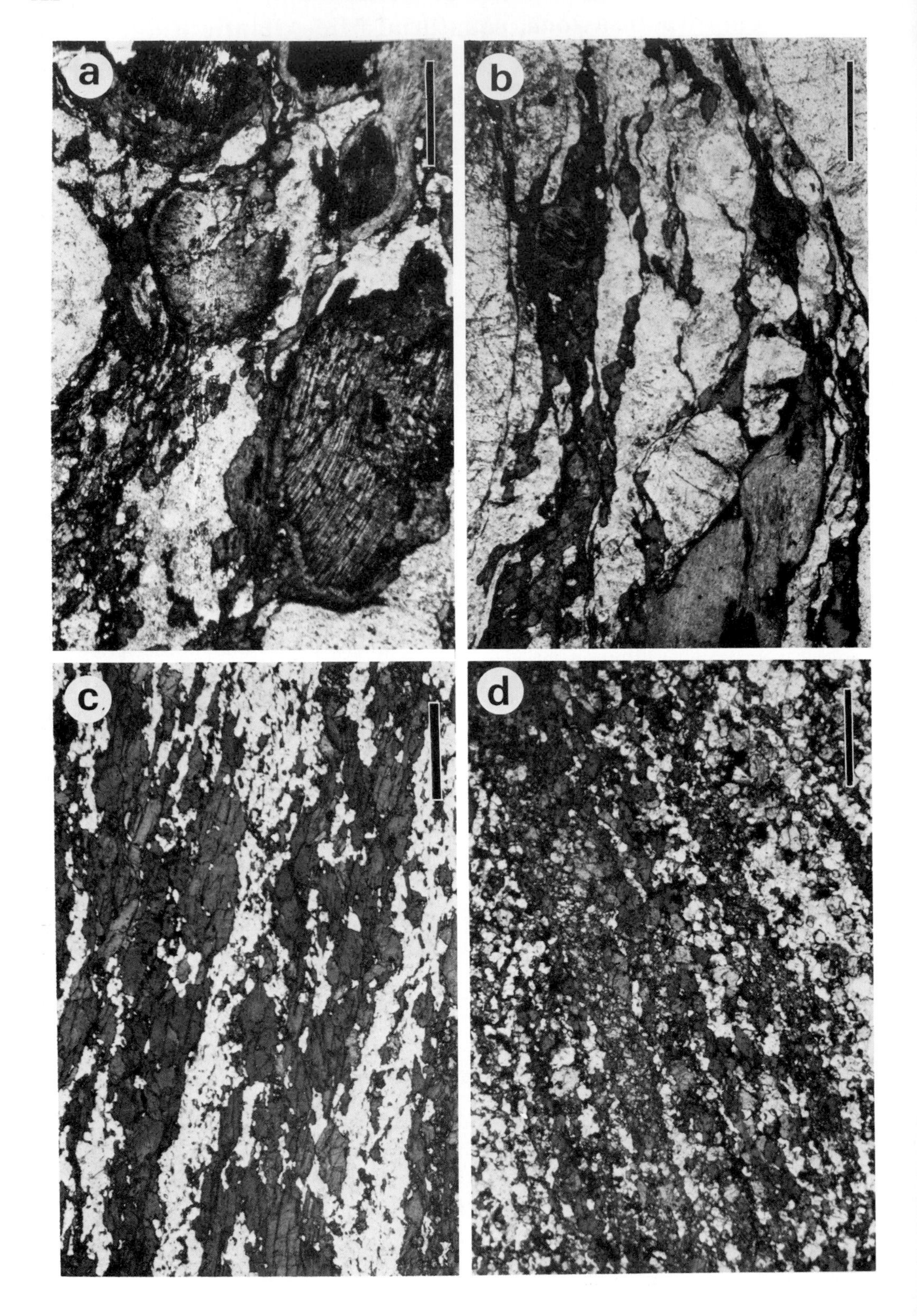
a
b
c
d

TABLE 1. *Location and composition of the dredge hauls*

Dredge haul No.	Geographic location	Tectonic location	Water depth range (m)	'Sub-bottom' depth range (m)	Rock type	Wt (%)	Total wt (kg)
P7003–19	10°46.0′N 42°48.5′W	Vema F.Z. North side of south transverse ridge	4000–4800	2050–2850	serpentinites	52	
					gabbros and	40	
					metagabbros		152
					amphibolites	4	
					basalt	4	
CH78–Dr10	10°41.5′N 42°40.8′W	Vema F.Z. - id -	4820	2870	serpentinites	60	
					amphibolites	35	
					gabbros + rodingites	5	232

(15–20 km wide) and almost straight E–W trough, 5 km deep. The fracture valley is bound by two steep walls whose bathymetries are strikingly contrasted. The south wall corresponds to a narrow (10–25 km) transverse ridge running parallel to the valley and reaching 550 m beneath sea level. Interpretations of this structure are summarized in Honnorez *et al.* (in press). High-resolution studies with multibeam side-scan sonars ('seabeam') during leg 78 of the R/V Jean Charcot (D. Needham *et al.*, unpublished data) indicate that the southern face of the transverse ridge is made up of numerous fault scarps with throws of a few hundred metres and narrow ledges, whereas the northern face appears to be formed by two or three fault scarps separated by two terraces.

The rocks studied were collected in two dredge-hauls from the base of the north-facing wall of the transverse ridge (Fig. 1). One dredge-haul, made by the University of Miami's R/V Pillsbury in 1970, recovered mainly mylonitized serpentinites and cataclastic gabbros with very few amphibolites and only two basalt fragments from depths ranging from 2050 to 2850 m below the crest of the transverse ridge. In 1978, a dredge-haul of the CNEXO's R/V Jean Charcot recovered numerous amphibolites and deformed serpentinites with very few gabbros at about 15 km east of the preceeding site and about 2900 m below the ridge crest. The locations and compositions of the two different dredge hauls are summarized in Table 1.

The gabbroic samples are more or less deformed and can be divided into three types depending on the intensity of the deformation. In *undeformed metagabbros*, the igneous texture is preserved and magmatic relicts are abundant, secondary minerals occur as pseudomorphs of igneous crystals or vein fillings. In *flaser-gabbros*, the igneous texture is still visible despite some deformation; in some rocks, igneous minerals (mostly plagioclase and pyroxene) form porphyroclasts in an oriented recrystallized matrix (Fig. 2a); but more commonly, igneous minerals are replaced by secondary phases (Fig. 2b). In gneissic amphibolites, igneous textures and minerals are completely obliterated and the resulting rocks display alternating streaks of hornblende and plagioclase (Figs 2c & d) and are very similar to amphibolites of continental metamorphic terranes. The transition between the three types is progressive (Figs 2a, b & c).

The gneissic amphibolite mineral assemblage is simple and constant: Ca-plagioclase + hornblende + ilmenite ± clinopyroxene, typical of the amphibolite facies. All the minerals have grown in the foliation and appear metamorphic, including clinopyroxene. Only occasional apatite and zircon, irregularly distributed in the rocks and concentrated in lenses aligned in the foliation, may be interpreted as magmatic crystals.

The gneissic amphibolites are commonly medium-grained, but a few of them are mylonitic: rounded hornblende and plagioclase porphyro-

FIG. 2. (a) Slightly deformed metagabbro. Porphyroclasts of clinopyroxene (with cleavages) rimmed by secondary amphiboles are preserved in an oriented recrystallized matrix made up of amphibole + plagioclase. Parallel nicols, scale bar = 1 mm. (b) Typical flaser-gabbro. Plagioclase and amphibole pseudomorphing magmatic pyroxene are elongated in the foliation. Despite strong deformation and recrystallization, the parental rock is still identifiable. Parallel nicols, scale bar = 1 mm. (c) Gneissic amphibolite with alternating layers of plagioclase, (white), hornblende (dark grey) and scarce ilmenite (black). Parallel nicols, scale bar = 1 mm. (d) Cpx-bearing gneissic amphibolite. The clinopyroxene occurs as rounded crystals (light grey) lining up in the foliation defined by alternating layers of plagioclase and hornblende. Parallel nicols, scale bar = 1 mm.

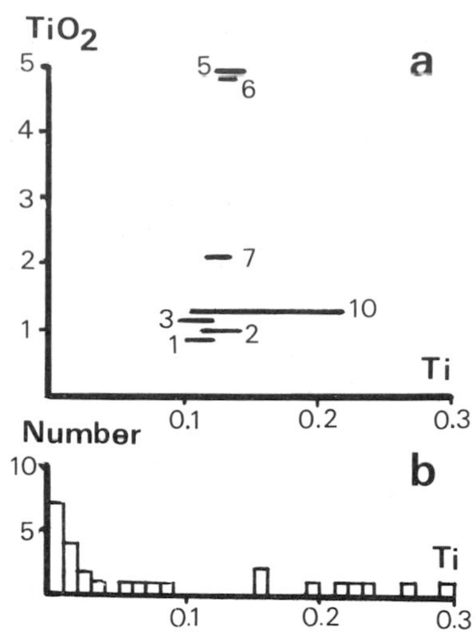

FIG. 5. (a) Ti range in amphiboles from gneissic amphibolites versus TiO_2 content of host rock (see Fig. 3 for keys of the numbers). Amphiboles from sample P19E are not figured in this diagram (see text for explanation). (b) Range of Ti content of amphiboles in one single sample of undeformed metagabbro (P19KKK). Number = number of analyses.

of hornblendes varies from sample to sample and is linked with the bulk-rock compositions (Fig. 6). Amphiboles grade from magnesio-hornblende to ferro-hornblende (Leake 1978). In all the amphiboles analysed, chlorine was below detection.

There is no significant compositional difference between the amphibole porphyroclasts and small grains in the matrix of the mylonite (CH81). This observation suggests that metamorphic conditions did not vary during mylonitization.

Amphiboles from the undeformed metagabbro sample have been analysed for comparison. They are much more heterogeneous. Most of them are strongly zoned from ferroan pargasitic hornblende to ferro-edenitic hornblende to edenite, but actinolite is also occasionally present. The Ti content of the amphiboles from this single sample is shown in Fig. 5b: their large scatter (from zero to 0.3 Ti) is due to the lack of homogeneization during static metamorphism. In the Al^{IV} *v.* alkalies diagram (Fig. 4), these amphiboles are strongly shifted toward more alkali-rich compositions with respect to hornblendes from gneissic amphibolites. It should be also pointed out that chlorine is present in most of them and reaches 4% in some crystals.

Clinopyroxene

The clinopyroxene composition is very homogeneous in a given sample but varies somewhat amongst the specimens (Fig. 6). All of the pyroxenes belong to the diopside-hedenbergite series and are always Al^{IV} alkali and Ti-poor, as expected from metamorphic pyroxenes. Like the amphiboles, their Mg/Fe ratio is related to the bulk-rock composition.

Plagioclase

Plagioclase in equilibrium with hornblende in the gneissic amphibolites is always calcic, but its composition varies from one specimen to another, ranging from An_{49} to An_{98}. The Fe and K contents of all plagioclases are always low (<0.4 and <0.05% respectively). Unzoned highly calcic plagioclase occurs in specimen CH77 which is characterized by a very high CaO/Na_2O ratio. In the other amphibolites, plagioclase crystals are commonly zoned from a calcic core to a more sodic rim. In the mylonites, the small plagioclase grains in the matrix are similar in composition to the more sodic rims of zoned porphyroclasts.

Age of metamorphism

Dating oceanic metamorphic rocks is difficult: the secondary minerals they contain are not favourable to geologic dating methods because they are always K-poor. K/Ar dating has nevertheless been attempted on isolated plagioclases and amphiboles, and on bulk-rock samples. The analytical precision of the results is rather poor because K concentrations are very low in

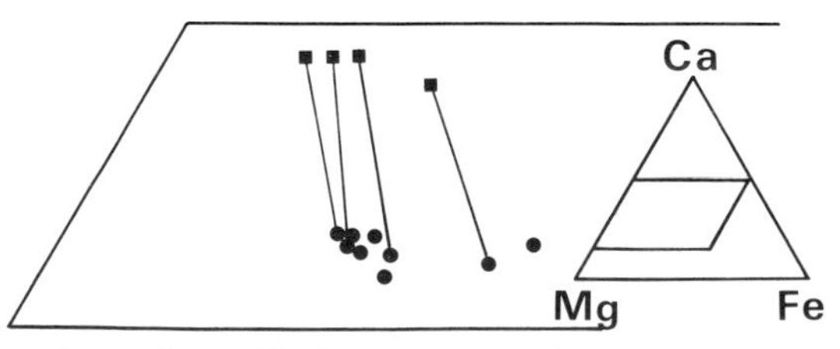

FIG. 6. Representative analyses of amphiboles (dots) and clinopyroxenes (squares) in a Ca:Fe:Mg diagram. The tie-lines join amphibole and clinopyroxene when they coexist in a sample (five samples do not contain pyroxene).

both amphiboles and plagioclases and hence in whole rocks.

Plagioclase and hornblende were separated by heavy liquids and magnetic separation.

The analytical procedure is explained in detail elsewhere (Westphal *et al.* 1979). Ages were computed with the K-decay constants proposed by Steiger & Jäger (1977). Plus-minus figures were estimates of analytical precision at one standard deviation (s.d.). They were calculated following the procedure given by Cox & Dalrymple (1967). The results of the K/Ar age analyses are given in Table 4.

The hornblendes of CH81 and CH20 display evidence of slight alteration (chloritization ?) that may explain both their high contamination in atmospheric argon and their relatively low age values, 7.8 ± 3.7 and 7.1 ± 2.3 Ma. The other hornblendes can be regarded as fresh. Two of them, i.e. CH77 and CH16 yield ages of the same order of magnitude, 8.6 ± 1.4 and 9.6 ± 1.5 Ma respectively. The amphiboles of CH83, on the other hand, display a higher value: 13.1 ± 1.8 Ma. Is there any real difference in the ages of CH83 and CH20, the extreme values found? The Critical Value Test (Dalrymple & Lanphere 1969) is indicated to claim at 95% confidence whether there is a real difference between the calculated ages of two minerals. In this case, it is not quite conclusive. If one assumes that the five rocks have been metamorphosed contemporaneously, the weighted average of the five amphiboles would indicate a 9.6 ± 1.6 (2σ) Ma age for the event. If the two altered amphiboles are eliminated, the event would be a little older: 10.1 ± 1.6 Ma.

One plagioclase, CH16, gives a K/Ar age, 10.9 ± 2.5 Ma, close to that of the coexisting hornblende. On the other hand, the other plagioclase, CH83, yields an age of 19.6 ± 4.1 Ma. This discrepancy might be due to an occurrence of excess argon in the feldspar. The analytical uncertainties, however, preclude any definite conclusion.

In spite of the high analytical errors, the bulk-rock sample CH83 yields a calculated age that is compatible with its amphibole separate. On the other hand, the bulk-rock sample CH23 indicates 17.8 ± 1.6 Ma. The lack of data on the mineral phases, however, prevents giving any sound and definite interpretation of that age.

Two observations lead us to conclude that a metamorphic event has occurred about 10 Ma: the minerals from sample CH16 yield almost concordant ages and the five dates on hornblendes also fall in this range. A 10 ± 2 Ma age has been determined by U-He method on sample P19E originating from the same area (D. E. Fisher, pers. comm.). But we cannot rule out the possibility that all the rocks were not metamorphosed during the same event.

Discussion

The gneissic amphibolites are completely devoid of relic minerals or textures but the nature of their parental rocks can be inferred from several observations: (i) they were collected in dredged

TABLE 4. *K/Ar ages obtained on the gneissic amphibolites*

Sample	Mineral	K_2O (wt %)	^{40}Ar rad (10^{-11} mol g^{-1})	^{40}Ar rad × 100 / ^{40}Ar total	Age (Ma)±2σ
CH78 Dr10-81	amphibole (80–160 μ)	0.089	0.09986	2.27	7.8 ± 3.7
CH78 Dr10-83	amphibole (80–160 μ)	0.095	0.1796	6.52	13.1 ± 1.8
	plagioclase (80–160 μ)	0.184	0.5229	2.01	19.6 ± 4.1
	whole rock (250–350 μ)	0.126	0.2703	5.71	14.8 ± 1.5
CH78 Dr10-16	amphibole (80–160 μ)	0.080	0.111	6.78	9.6 ± 1.5
	plagioclase (80–160 μ)	0.157	0.2470	4.92	10.9 ± 2.5
CH78 Dr10-20	amphibole (50–71 μ)	0.91	0.09281	2.19	7.1 ± 2.3
CH78 Dr10-77	amphibole (90–112 μ)	0.093	0.1148	5.17	8.6 ± 1.4
CH78 Dr10-23	whole rock (not crushed)	0.080	0.2056	7.58	17.8 ± 1.6

amphibolites from the Chenaillet ophiolite massif (Hautes Alpes, France). *Earth Planet. Sci. Lett.* **39**, 98–108.

MINSTER, J. B. & JORDAN, T. H. 1978. Present day plate motion. *J. Geophys. Res.* **83**, 5331–54.

MIYASHIRO, A. 1973. *Metamorphism and Metamorphic Belts*. J. Wiley and Sons, New York, 492 pp.

PRICHARD, H. M. & CANN, J. R. 1982. Petrology and mineralogy of dredged gabbros from Gettysburg bank, Eastern Atlantic. *Contrib. Mineral. Petrol.* **79**, 46–55.

PRINZ, M., KEIL, K., GREEN, J. A., REID, A. M., BONATTI, E. & HONNOREZ, J. 1976. Ultramafic and mafic dredged samples from the equatorial mid-Atlantic ridge and fracture zones. *J. Geophys. Res.* **81**, 4087–103.

RAASE, P. 1974. Al and Ti contents of hornblendes, indicators of pressure and temperature of regional metamorphism. *Contrib. Mineral. Petrol.* **45**, 231–6.

REED, S. J. B. & WARE, N. G. 1975. Quantitative electron microprobe analysis of silicates using energy-dispersive X-ray spectrometry. *J. Petrol.* **16**, 499–519.

RUCKLIDGE, J. & GASPARRINI, E. L. 1969. Electron microprobe analytical reduction EMPADR VII. Department of Geology, University of Toronto.

SMEWING, J. D. 1980. An upper Cretaceous ridge-transform intersection in the Oman ophiolite. *In*: PANAYIOTOU, A. (ed.) *Ophiolites–Proceedings International Ophiolite Symposium*, Cyprus, pp. 407–13.

SPEAR, F. S. 1981. An experimental study of hornblende stability and compositional variability in amphibolite. *Amer. J. Sci.* **281**, 697–734.

SPRAY, J. G. & RODDICK, J. C. 1981. Evidence for upper Cretaceous transform fault metamorphism in West Cyprus. *Earth Planet. Sci. Lett.* **55**, 273–91.

SPOONER, E. T. C. & FYFE, W. S. 1973. Sub-seafloor metamorphism, heat and mass transfer. *Contrib. Mineral. Petrol.* **42**, 286–304.

STEIGER, R. H. & JÄGER, E. 1977. Subcomission on geochronology: convention on the use of decay constants in geo- and cosmochronology. *Earth Planet. Sci. Lett.* **36**, 359–62.

STERN, C. & ELTHON, D. 1979. Vertical variations in the effects of hydrothermal metamorphism in Chilean ophiolites: their implications for ocean floor metamorphism. *Tectonophysics*, **55**, 179–213.

STROUP, J. B. & FOX, P. J. 1981. Geologic investigations in the Cayman trough: evidence for thin oceanic crust along the mid-Cayman rise. *J. Geol.* **89**, 395–420.

WESTPHAL, M., MONTIGNY, R., THUIZAT, R., BARDON, C., BOSSERT, A., HAMZEH, R. & ROLLEY, J. P. 1979. Paléomagnétisme et datation du volcanisme permien, triasique et crétacé du Maroc. *Can. J. Earth Sci.* **16**, 2150–64.

J. HONNOREZ, R.S.M.A.S., University of Miami, 4600 Rickenbacker Causeway, Miami, FA 33149, USA.

C. MEVEL, Laboratoire de Pétrologie, Université Pierre et Marie Curie, 4 place Jussieu, 75230 Paris Cedex 05, France.

R. MONTIGNY, I.P.G., Laboratoire de Paléomagnétisme, 5 rue René Descartes, 67084 Strasbourg, France.

Variations in structure and petrology in the Coastal Complex, Newfoundland: anatomy of an oceanic fracture zone

J. A. Karson

SUMMARY: The Coastal Complex of western Newfoundland is perhaps the world's best example of an ancient oceanic fracture zone preserved within an ophiolite complex. The structural style, metamorphic zonation, and magmatic history of the Coastal Complex contrasts sharply with that of the contiguous Bay of Islands Complex. The exposures of the Coastal Complex provide cross sections of the fracture zone at various vertical and lateral positions. A reconstruction of these sections reveals structural and petrological variations that are a function of depth as well as position along strike. For example, deep crustal sections show a history of strike-slip deformation that is fairly homogeneously distributed across a 4 km wide high-strain zone. High-temperature metamorphic rocks, partial melting relationships, and peridotite intrusive bodies are also present. Diabase dykes cut the entire assemblage. At shallow crustal levels, deformation is concentrated along relatively narrow, well-defined faults and shear zones that overall have an anastomosing aspect. High-grade metamorphic rocks occur only locally and have in most places suffered retrograde effects. Felsic intrusive bodies occur locally and serpentinite mélange is widespread. Volcanic rocks non-conformably overlie a tectonically brecciated horizon that transects the entire assemblage. The most striking lateral variations are the discontinuous nature of lithologic units and the apparent bifurcation of major high-strain zones along strike.

Oceanic fracture zones and their ophiolite analogues probably display a wide range of geological relationships that vary with depth, with position along strike, and in scale. Thus, the Coastal Complex should probably be viewed only as a single example of how exceedingly complex and variable the oceanic lithosphere may be near transform faults and their non-transform extensions. Clearly, similar ophiolitic assemblages should not be considered when attempting to model processes along purely extensional plate boundaries.

Bathymetric and physiographic maps of the sea floor show that ridge–ridge transform faults and their non-transform fracture-zone extensions affect a very significant portion of the contemporary oceanic lithosphere. Therefore, in order to fully understand the plate accretion process, the perturbations of the accretion process that occur near oceanic fracture zones must also be understood. Attempts to carry out direct observational studies of fracture zones (e.g. Bonatti & Honnorez 1976; Fox *et al.* 1976; Choukroune *et al.* 1978; Macdonald *et al.* 1979; Searle 1979; Karson & Dick 1983: OTTER 1983) have been hindered by deep water and the limited vertical section of the crust exposed there (Francheteau *et al.* 1976). While these studies are extremely important in describing the broad-scale surface morphology and first-order lithologic associations at fracture zones, structural studies are limited to essentially two-dimensional analyses. The lateral and vertical extent of structural and lithological associations at depth cannot be determined at present.

One of the most promising means of circumventing these problems is the study of fracture zones preserved within ophiolites. Several ophiolite fracture zones have been proposed in recent years (Moores & Vine 1971; de Wit *et al.* 1977; Saleeby 1977; Karson & Dewey 1978; Simonian & Gass 1978; Smewing 1980; Prinzhofer & Nicolas 1980), few of them, however, preserve the necessary geological relationships to unambiguously mark them as the result of ridge–ridge transform faulting. The Coastal Complex of western Newfoundland (Karson 1977; Karson & Dewey 1978; Karson 1982; Karson *et al.* 1983) is perhaps the least ambiguous of any of these.

Following Wilson's (1965) concept of ridge–ridge transform faults, Karson & Dewey (1978) presented a model outlining some of the geological relationships expected to occur along oceanic fracture zones. The Coastal Complex was cited as an example of an ophiolitic assemblage that contained many of the small- and large-scale relationships expected along the inactive fracture-zone extensions of transform faults. In this paper, these interpretations are extended to recently mapped areas where the local geology helps to bind together the variable array of structural and petrologic relationships that occur through this terrane.

In contrast to other potential ophiolite fracture zones, the Coastal Complex has clear-cut relationships with an undeformed, stratiform ophiolite, the Bay of Islands Complex. This permits the interpretation of structures in the Coastal Complex in terms of sea-floor spreading processes constrained by the geology of the Bay of Islands Complex. Exposures in the Coastal Complex provide several structural cross-sections of the fracture-zone assemblage at different structural levels of the crust and upper mantle and at different positions along strike. These exposures probably a relatively late, lower-temperature fault zone. The peridotites are probably related to similar intrusive bodies observed in the Lewis Hills, whereas the felsic bodies may be segregations of partial melt derived from a region similar to the migmatite zone exposed in the northern part of the Lewis Hills. NW-striking diabase dykes and the volcanics that they feed, on the E and W flanks of the area, transect and cover respectively, the previously faulted terrain (Williams & Malpas 1972; Williams 1973). This magmatic event probably occurred at a ridge-transform intersection as older, deformed crust moved past a spreading axis. The juxtaposition of amphibolites, serpentinites, and volcanics suggests considerable vertical displacement between crustal blocks in this region. The individual NW-trending dykes in the area usually cannot be traced across the faults that bound the major lithologic units. Tectonic offsets of the latest dykes might be expected outside the active transform domain due to vertical tectonic adjustments (DeLong *et al.* 1979; Karson & Dewey 1978).

Unfortunately, the contact with the adjacent BOIC is not preserved here. It is possible that a substantial section of deformed CC rocks, just to the east of the exposure described above, has been removed by erosion. Comparison with the Lewis Hills suggests that preferential removal of serpentinites may have taken place.

Look-out Hills Massif

The most northerly exposure of the CC is the Look-out Hills Massif located just south of Bonne Bay (Fig. 1). This area provides exposures of lithologies that formed at intermediate to very shallow structural levels (Fig. 5). Instead of a single high-strain zone as in the other localities described, two nearly parallel deformed belts (2–3 km wide) are separated by a much less deformed layered sequence (8 km wide).

The layered sequence consists of a basal unit of layered gabbroic rocks successively overlain by isotropic gabbroic rocks with isolated diabase dykes, a sheeted dyke complex, and basaltic pillow lavas. This sequence is very similar to those forming the upper members of the BOIC. Interlayered ultramafic cumulates in the basal unit suggest that a relatively thin (about 3 km) total crustal section may be present here. Within the middle units, diabase dykes are steeply dipping and trend predominately WNW. WNW-striking shear belts cut the gabbroic rocks and dykes.

To the E, the pillow-lava unit has been removed by erosion, but the other underlying units grade into a progressively more deformed and metamorphosed terrain similar to that described in the Lewis Hills Massif. The diabase dykes are tectonically rotated through almost 70° from strikes of WNW-ESE, where undeformed, clockwise to N–S where they reach amphibolite facies. The deformed dykes and surrounding metagabbros have a gently N or S plunging stretching lineation and steeply dipping, foliation. Variably deformed trondhjemite veins are present in many exposures. Along the eastern edge of this exposure, variably deformed pyroxenite and wehrlite dykes, sills, and at least one larger body up to several tens of metres in width, cut the metamorphic rocks. Lithologies further to the E have been removed by faulting and erosion. The maximum exposed width of the deformed zone in the eastern Look-out Hills is 3 km.

The layered sequence is truncated to the W by a felsic pluton, the largest one in the CC. This mass is extremely heterogeneous and includes mainly quartz-diorite and tonalite, but also substantial amounts of diorite. Texturally they vary from layered igneous cumulates to streaky gneisses. Intrusive relations are preserved around the margins of this body except along its western edge. Numerous trondhjemite veins invade the layered sequence and xenoliths of gabbro and diabase, in various stages of assimilation, occur throughout the quartz-diorite. Ductile shear belts and anastomosing fractures that cut the felsic material and the xenoliths strike NW. Along the southern margin of this felsic pluton several small fractured and veined sheeted dyke exposures occur and appear to be roof pendants. Undeformed WNW trending dykes with chilled margins cut the pluton.

The western edge of the felsic pluton is steeply dipping and faulted. This contact lies along an old fault zone; however displacement across this faulted contact that postdates the intrusion of the felsic pluton is believed to be small due to the present of abundant trondhjemite net-veins and dykes in lithologies to the W. These lithologies comprise perhaps the most complex segment of the CC. Overall it appears to be a phacoidal fault zone, approximately 2 km wide, made up of large (few hundred metres wide by few kilometres long) lensoid bodies of variably deformed and metamorphosed mafic material enveloped in a matrix of highly sheared serpentinite. A cross-section of this region produced from coastal exposures was presented in Karson & Dewey (1978); however, subsequent mapping has shown that significant along-strike variations occur inland. These variations and the more general aspects of the geology of this belt, as traversed from E to W, are stressed in the following discussion.

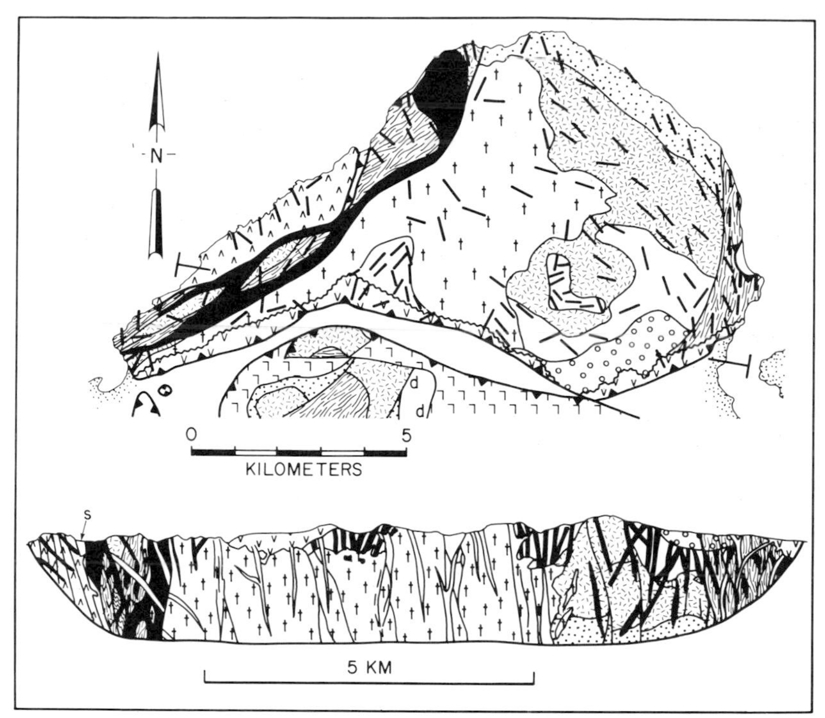

FIG. 5. Generalized geological map and cross-section of the Look-out Hills Massif (see text for discussion). Geology by Karson (1981–82, unpublished data). Cross-section projection plane dips 40° to the NE. Symbols are the same as those in Figs 3 and 4.

Immediately to the W of the felsic pluton an anastomosing belt of schistose to phacoidal serpentinites crops out. These were derived mainly from wehrlites with cumulate textures, but minor amounts of serpentinized harzburgites are also present. At the northern end of this belt serpentinites grade westward into a very weakly deformed, relatively fresh, body of cumulate wehrlite with intrusive contacts to the N and S. This relationship and the apparent clockwise rotation of the eastern edge of the wehrlite body suggest dextral displacement across this serpentinized fault zone. Massive diabase dykes with chilled margins cut the wehrlite body as well as the sheared serpentinites and strike NW.

In the southern and central parts of this belt several phacoidal slices of metagabbros occur. Their margins are, in most places, faulted against highly deformed serpentinites. Within the slices, somewhat different lithologies and structures are preserved including highly fractured diorite cut by trondhjemite dykes, mylonitic amphibolites and lineated, coarse-grained, layered gabbroic rocks. In general, compositional layering in the metagabbros dips gently to the S and the lineation is subhorizontal, trending SW–NE. In the gabbroic rocks throughgoing faults and shear zones separate intervals with closely spaced (few centimetres) parallel fractures. NE–SW striking diabase dykes are offset and rotated along the high-strain zones in these lensoid bodies, but a younger generation of undeformed dykes cut all the internal deformation features. The dykes are usually truncated at the faulted serpentinite contacts.

To the W, medium-grained gabbro cut by widely spaced shear belts grades rapidly (over about 50 m) into banded mylonites with anhydrous mineralogies. Locally, the mylonitic fabric is tightly to isoclinally folded about steep axes. Undeformed mafic dykes cut these mylonitic rocks and strike NW. The mylonites are faulted against a phacoidal assemblage of complexly interleaved serpentinites, greenschists, calc-silicate schists, and argillites. The westernmost slice in this assemblage is a lens of polydeformed greenschists with thin marble layers that are cut by diabase and trondhjemite dykes. Within this assemblage the least-deformed domains reveal a volcanic and coarse-volcaniclastic protolith. This slice is faulted against black to red slaty argillites with a few exotic carbonate, mafic volcanic, greywacke and greenschist blocks.

The westernmost unit in this assemblage consists of a steeply SE dipping, SE-facing sequence of mafic volcaniclastic breccias, tuffs,

Distribution of strain

At deep structural levels, strain is distributed in a fairly homogeneous, continuous fashion. A map-scale version of a longitudinally sliced ductile shear belt is present. Linear deformation fabrics dominate the deformed area. At shallower levels a different structural style occurs. Parts of large-scale ductile shear zones are still evident in a few places and planar as well as linear deformation fabrics are present. In general, however, areas with ductile deformation structures have been sliced and overprinted by an array of anastomosing fault zones that separate variable fractured lensoid masses of rock up to several kilometres in length. Fractures, faults and veins overprint earlier ductile deformation structures. In a given region these lensoid masses may display a wide range of lithologies and deformation features. Where protoliths may be determined, it is clear that a considerable amount of vertical shuffling has occurred, if a layered structure similar to that of the BOIC was initially present. Alternatively, it is possible that the lithologic units of the CC were assembled in a different vertical sucession from that of the BOIC. For example, volcanics and sediments may be found faulted against amphibolites and serpentinites with diabase dykes cutting all of the units. Vertical displacements are not so obvious at deeper levels and may have been obscured by igneous intrusions or high strains distributed over wider intervals.

Ultramafic rocks

Variably deformed ultramafic rocks occur throughout the CC. At deep levels these are clearly parts of crystal-mush intrusions (Karson *et al.* 1983) derived from accumulating magma chambers. Intrusive relations are obvious at deep levels, but are progressively destroyed at shallower levels where the ultramafics are intensely serpentinized and sheared. Serpentinite bodies derived from layered ultramafics, crystal-mush bodies, and residual upper mantle are faulted against sheeted dykes, volcanics, and argillites indicating substantial vertical penetration upward through the crust. Dykes cutting the serpentinites indicate that this faulting occurred, at least in part, near a spreading centre. These are probably analogous to the serpentinite bodies commonly reported from oceanic fracture zones (Miyashiro *et al.* 1969; Bonatti & Honnorez 1976; Fox *et al.* 1976).

Felsic rocks

At relatively deep crustal levels, felsic rocks occur mainly in variably deformed net-vein assemblages and migmatites with outcrop structures suggestive of partial melting from an amphibolite host. Abundant mafic gneisses in the vicinity may in part represent a residue left behind by partial melt and liquid extraction.

At shallow structural levels, large bodies of quartz-diorite to tonalite have intrusive contacts and may be the result of several processes, including segregation of partial melts derived from a deeper crustal level and/or segregation of highly fractionated magmas derived from a spreading-centre source, but modified by contamination by amphibolite and serpentinite xenoliths. The variable extent to which deformation has affected these bodies may be a result of their position relative to major fault zones in the complex and/or their greater mechanical strength relative to surrounding mafic to ultramafic lithologies. Their variably deformed nature indicates formation within the transform fault zone.

Undeformed mafic rocks

The undeformed mafic rocks in the CC are almost entirely dykes. At deep levels, the dykes are vertical and consistently oriented at a high angle to the overall trend of the CC. Only a few minor bodies of related gabbroic rocks have been identified. At shallow structural levels, the dykes have much less consistent orientations, probably reflecting the more complex and variable state of stress in the shallow crust. These dykes feed mafic to silicic-volcanics which may be analogous to those recently discovered at some Pacific ridge-transform intersections (Byerly *et al.* 1976; Fornari *et al.* 1983).

Contacts with stratiform ophiolite sequences

At the deepest structural levels of the CC, the contact with the BOIC reflects the evolution of a mid-ocean-ridge magma-chamber termination at a transform-fault offset (Karson 1977; Casey & Karson 1981). This interaction is recorded by relatively thick ultramafic cumulates; large, lensoid, mafic cumulate bodies; and localized fragmental cumulates. Compositional layering is lapped against the steep transform wall. This contact was also the locus of intrusion for ultramafic crystal-mush bodies.

At shallow levels the CC/BOIC contact is not represented; however, there is a contact between the western high-strain zone and the central layered assemblage of the Look-out Hills. Here the contact is marked by a large felsic intrusive body with a faulted western edge.

Lateral variations

Lateral variations through the CC are especially evident when considering different exposures of similar crustal levels—e.g. the Little Port Area, the exposure W of North Arm Mountain, and the northern part of the Look-out Hills. From these areas, the most important observation is that in most places a single highly deformed fault zone occurs, but elsewhere (Look-out Hills) there are two strands. This is most likely a result of an anastomosing fault geometry at mid-crustal levels that trapped a nearly undeformed crustal segment between two fault strands in the Look-out Hills.

Comparison of the cross-sections presented here as well as those in Karson and Dewey (1978) emphasizes the lateral heterogeneity of the CC. Particularly at middle to shallow crustal levels, cross-sections constructed for areas only a few hundred metres apart may show drastic changes in lithologies present and structural style. This is primarily due to the lensoid character of major lithologic units and the disruption along relatively late fault zones.

Summary and conclusions

When viewed on a broad scale and compared with the BOIC, some important features of the CC become evident. First, the CC is a relatively narrow linear belt of intense deformation and metamorphism that cuts the much less deformed, stratiform Bay of Islands ophiolite allochthon. Lithologic contacts, metamorphic belts, etc., are steeply dipping compared to the relatively gently dipping structure of the adjacent BOIC. Obduction-related folding has resulted in steep dips along the eastern edge of the BOIC. Significantly, hydrous metamorphic assemblages are nearly pervasive in the plutonic levels of the CC, whereas at comparable levels of the BOIC hydrous metamorphic assemblages are very discontinuous and sometimes completely absent even where local deformation has occurred. The CC appears to be a region where both felsic and ultramafic plutonic rocks are concentrated. The ultramafics are, in part, serpentinite diapirs localized on fault zones and rooted mainly in deeper-level crystal-mush intrusions. The felsic rocks appear to be more volumetrically significant relative to mafic rocks than they really are, due to preferential exposure of middle to shallow crustal levels where these bodies tend to be localized.

Many of the world's ophiolite complexes have a relatively simple internal structure and may be reasonably interpreted as obducted oceanic lithosphere with a simple plate-accretion history. Other ophiolite complexes, such as the CC, have internal structures that are difficult or impossible to reconcile with any steady-state plate-accretion process. These may represent oceanic lithosphere that was created or modified in any of several other tectonic environments. To include structural, petrological or geochemical data of such ophiolites in models of sea-floor spreading may serve to confuse our understanding of the plate-accretion process(es). Furthermore, studies of contemporary oceanic fracture zones indicating that the surficial geology and geomorphology of these areas are relatively simple, is probably only due to the limited view possible there.

ACKNOWLEDGMENTS: I would like to thank J. Casey, J. Dewey, D. Elthon, B. Idleman and W. Kidd for their helpful discussions over the past several years. J. Grippi, S. O'Connell, J. Sulanowski and K. Waldron provided assistance in the field. P. Barrows typed the manuscript and L. Raymond draughted the figures. This work was supported by NSF Grants EAR76 14459 and EAR80 26445. WHOI Contribution No. 5350.

References

BAKER, D. F. 1978. Geology-geochemistry of an alkali volcanic suit (Skinner Cove Formation) in the Humber Arm Allochthon, Newfoundland. M.Sc. thesis, Memorial University Newfoundland, St. John's Newfoundland, 202 pp.

BONATTI, E. & HONNOREZ, J. 1976. Sections of the Earth's crust in the Equatorial Atlantic. *J. geophys. Res.* **81**, 4104–16.

BYERLY, G. R., MELSON, W. G. & VOGT, P. R. 1976. Rhyodacites, andesites, ferrobasalts and ocean tholeiites from the Galapagos Spreading Centre, *Earth planet. Sci. Lett.* **30**, 215–21.

CASEY, J. F. & KARSON, J. A. 1981. Magma chamber profiles from the Bay of Islands ophiolite complex. *Nature, Lond.* **292**, 295–391.

——, DEWEY, J. F., FOX, P. J., KARSON, J. A. & ROSENCRANTZ, E. 1981. Heterogeneous nature of oceanic crust and upper mantle: a perspective from the Bay of Islands Ophiolite Complex. *In*: EMILIANI, C. (ed.) *The Oceanic Lithosphere The Sea, Vol. VII.* John Wiley and Sons, New York, pp. 305–38.

CHOUKROUNE, P., FRANCHETEAU, J. & LE PICHON, X. 1978. *In situ* structural observations along Transform Fault A in the FAMOUS area, Mid-Atlantic Ridge. *Bull. geol. Soc. Am.* **89**, 1013–29.

COMEAU, R. L. 1972. Transported slices of the Coastal Complex, Bay of Islands, western Newfoundland. M.Sc. thesis, Memorial University Newfoundland, St. John's, Newfoundland, 105 pp.

DALLMEYER, R. D. & WILLIAMS, H. 1975. $^{40}Ar/^{39}Ar$ ages from the Bay of Islands metamorphic aureole: their bearing on the timing of Ordovician ophiolite obduction. *Can. J. Earth Sci.* **12**, 1685–90.

DELONG, S. E., DEWEY, J. F. & FOX, P. J. 1979. Topographic and geologic evolution of fracture zones. *J. geol. Soc. London.* **136**, 303–10.

FORNARI, D., PERFIT, M. R., MALAHOFF, A. & EMBLEY, R. 1983. Geochemical studies of abyssal lavas

Mantle flow pattern at oceanic spreading centres: relation with ophiolitic and oceanic structures

A. Nicolas & M. Rabinowicz

SUMMARY: The rolling-mill effect concerning asthenosphere circulation beneath a fast-spreading centre (Rabinowicz *et al.* in press) is recalled and discussed in terms of its geological assessment. The model predicts a fast and narrow jet of asthenosphere which carries 70% of the total melt produced beneath the ridge. This explains both the structural data in most ophiolites and the formation of the crust within a few kilometres at fast-spreading ridges. It is proposed that the more diffuse activity at slow-spreading ridges from mid-oceans or marginal basins, along with the more diffuse magmatism recorded in a few ophiolites is related to the disappearance of this narrow stream as a consequence of an increase of viscosity due to lower temperatures beneath the ridge.

In a preceding paper (Nicolas & Violette 1982) a general model of asthenospheric flow beneath an oceanic spreading centre has been derived from integrated structural data obtained in over a dozen ophiolite complexes (Fig. 1).

The main feature of this model is the diapiric uprise of asthenospheric mantle prior to its horizontal spreading in every direction. However, several other lines of information pertaining to ophiolites and oceanic-ridge structures were not incorporated in this qualitative model. Therefore, a more precise modelling using a numerical technique and taking into account these other data has been achieved (Rabinowicz *et al.* in press). The main features of this physical model will be recalled here together with critical assessments and discussions of its geological aspects.

The model

The numerical model (Fig. 2) prepared by Rabinowicz *et al.* (*in press*) is based on the assumption that the flow in the asthenospheric mantle beneath a spreading ridge results from the combination of two causes: (i) a regional cause responsible for a large-scale motion which is either the internally driven convection or, more probably, the drifting away of the growing lithosphere, and (ii) a local gravity instability developing in the rising diapir due to a difference in density with surrounding mantle as a consequence of partial melting. If the viscosity in the ridge vicinity where partial melting occurs is low enough (Fig. 3), the body force due to melting can create an active circulation which, superimposed on the general one leads to the complex flow pattern of Fig. 2. In this figure, the regional flow dominates away from the ridge in the lower right corner and the local flow, close to the ridge in the upper left corner.

The local flow is strongly channelled like in a rolling mill by the small cells centered at x and $z \sim 15$ km from the origin, on both sides of the ridge. The maximum velocity in the vertical channel attains 10.5 cm yr^{-1} at $z = 15$ km where the channel is narrowest (5–10 km radius). The mean velocity is however 2.5 cm yr^{-1} at $z = 30$ km

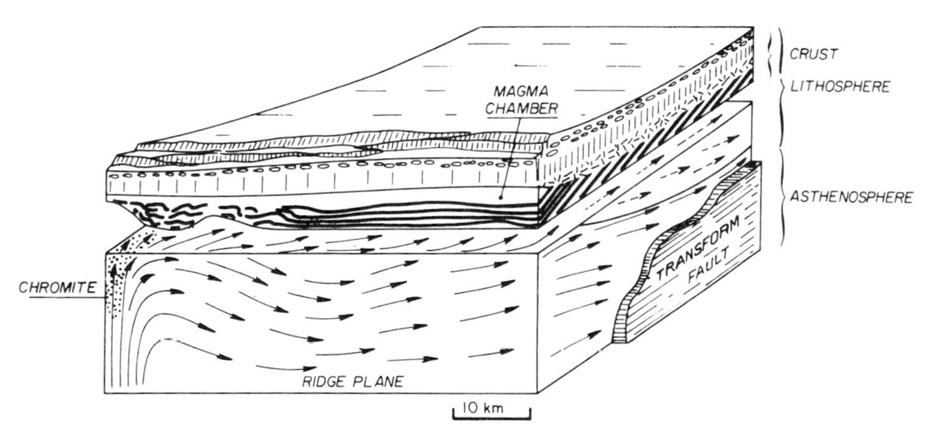

FIG. 1. Ridge model deduced from structural studies in ophiolites (after Nicolas & Violette 1982). Full arrows: flow lines in the asthenosphere; dashed arrows: fossil flow lines in the lithosphere.

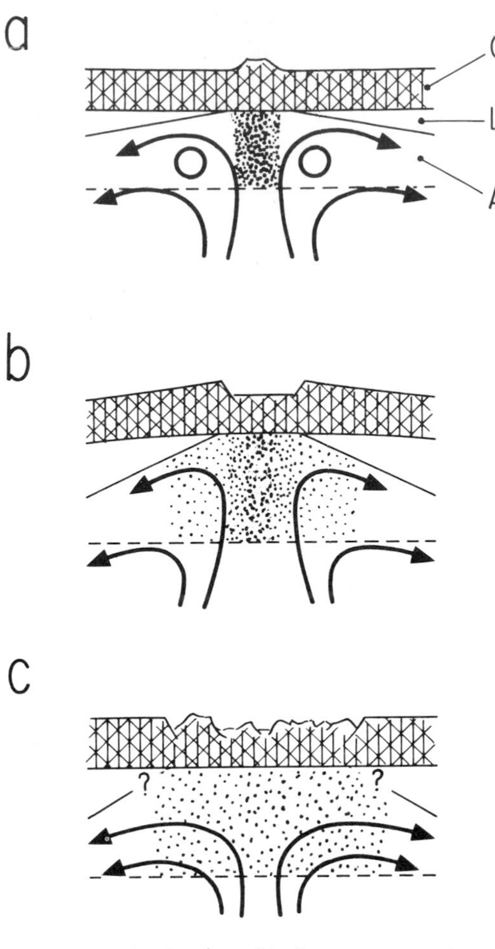

FIG. 4. Sketch of possible flow patterns and associated spreading activity as they could relate to spreading rates. Dots: zones of partial melting determining the area of active crustal generation. (a) fast; (b) moderate; (c) slow spreading rates.

threshold around 10^{20} P, choke this local circulation (Rabinowicz *et al.* in press). For high viscosities, only the large circulation can persist (Fig. 4c).

Assuming an inverse relation between viscosity and spreading rate (Lachenbruch 1976; Sleep & Rosendahl 1979), it can be predicted that below some critical spreading rate smaller than 5 cm yr^{-1} and corresponding to the critical viscosity, the rising flow is no longer channelled towards the ridge. As a consequence the magmatic delivery to the forming crust becomes more diffuse (Fig. 4c). Instead of being nearly totally created within a few kilometres of the ridge (Fig. 4a), the crust can now be created over distances up to several tens of kilometres, as shown in Fig. 2, by the lateral extent of the horizontal flow above the solidus horizon.

This speculation can be tested by reference to ophiolitic and oceanic structures. Ophiolites like the Xigaze massif in Tibet (Nicolas *et al.* 1981) illustrate a situation of diffuse crustal formation. The crust, which is thinner there than usual, is dominantly composed of diabase sills with only a few small gabbro patches. These diabase sills are also remarkably abundant and thick within the underlying peridotites. A colder thermal regime in the mantle is also suggested by a degree of partial melting smaller than usual and by the comparatively low-temperature plastic deformation of these peridotites.

Comparing oceanic ridges characterized by smaller spreading rates with ridges which are spreading at $\geqslant$5 cm yr^{-1}, it appears that the surface structures (mainly topographic and tectonic)* which can be related to the spreading activity are larger in the former. For the Mid-Atlantic Ridge (1–2 cm yr^{-1} rate) the tectonic activity extends over 10–15 km (Needham & Francheteau 1974) compared to 2–4 km in fast-spreading ridges (Ballard *et al.* 1982) (respectively 20 and 10 km in Macdonald (1982)); the rift floor attains 15 km (Tapponnier & Francheteau 1978) compared to 3–4 km in the Galapagos rift (Crane 1978) and the distance between the facing uplifted rift mountains, 30 km compared to 5–15 km for the central rise of fast-spreading ridges (Rea 1976; Van Andel & Ballard 1979). Off-axis volcanism is also more common and more scattered (5–20 km off-axis, compared to 4 km in fast-spreading ridges (Macdonald 1982)). Data are scarce on the extension of the spreading activity in marginal basins but altogether the spreading rates are usually small and the expansion considered as diffuse (Karig 1970). In the Mariana Trough, spreading rates vary from 1–2 cm yr^{-1} and the axial high is 60 km wide (to be compared with the 5–15 km horst of fast ridges), associated with a crustal shoaling over a similar distance (Amboss & Hussong 1982).

These data, showing that the spreading tends to become more diffuse when the spreading rate decreases, lend support to our hypothesis. When the viscosity increases, presumably due to spreading-rate decrease, the channelled asthenosphere flow becomes larger before vanishing (Fig. 4). It may be speculated that slow-spreading marginal basins with rugged topography and apparently diffuse spreading may illustrate the situation where the channelled flow is no longer active (Fig. 4c). Further modelling and more precise oceanographic data would be required to test this hypothesis.

ACKNOWLEDGMENTS: F. Boudier and J. L. Vigneresse are thanked for useful discussions. This work was supported by the Centre National de la Recherche Scientifique (ERA 547).

*In this analysis the volcanic activity and the magnetic anomalies are not considered because it is believed that, if there is a magma chamber at the ridge, their domain of extension will be essentially controlled by the mechanical aspects of fracturing the magma-chamber roof.

References

AMBOSS, E. L. & HUSSONG, D. M. 1982. Crustal structure of the Mariana Trough. *J. Geophys. Res.* **87**, 4003–18.

BALLARD, R. D., VAN ANDEL, T. H. & HOLCOMB, R. T. 1982. The Galapagos rift at 86°W. III. Variations in volcanism, structure and hydrothermal activity over a 30 km segment of the rift valley. *J. Geophys. Res.* **87**, 1149–61.

BOTTINGA, I. & ALLEGRE, C. J. 1978. Partial melting under spreading ridges. *Philos. Trans. R. Soc. London* **A 288**, 501–25.

BOUDIER, F. & NICOLAS, A. 1972. Fusion partielle gabbroique dans la lherzolite de Lanzo (Alpes Piémontaises). *Bull. Suisse Min. Petrol.* **52**, 39–56.

—— & —— 1977. Structural controls on partial melting in the Lanzo peridotites. *In*: DICK, H. J. B. (ed.) Magma genesis. *Oregon Dept. Geol. Miner. Ind. Bull.* **96**, 63–78.

CRANE, K. 1978. Structure and tectonics of the Galapagos inner rift, 86°10′W. *J. Geol.* **86**, 715–30.

FLOWER, M. F. J. 1981. Thermal and kinematic control on ocean ridge magma fractionation: contrasts between Atlantic and Pacific spreading axes. *J. Geol. Soc. London.* **138**, 635–712.

GIRARDEAU, J. & NICOLAS, A. 1981. The structure of two ophiolite massifs, Bay of Island, Newfoundland: a model for the oceanic crust and upper mantle. *Tectonophysics*, **77**, 1–34.

HALE, L. D., MORTON, C. J. & SLEEP, N. H. 1982. Reinterpretation of seismic reflection data over the East Pacific Rises. *J. Geophys. Res.* **87**, 7707–17.

HERRON, T. J., STOFFA, P. L., & BUHL, P. 1980. Magma chamber and mantle reflections—East Pacific Rise. *Geophys. Res. Lett.* **7**, 989–92.

KARIG, D. E. 1970. Ridges and basins of the Tonga-Kermadec island arc system. *J. Geophys. Res.* **75**, 239–55.

KUZNIR, N. J. & BOTT, M. H. P. 1976. A thermal study of the formation of oceanic crust. *Geophys. J.R. Astr. Soc.* **47**, 83–95.

LACHENBRUCH, A. H. 1976. Dynamics of a passive spreading centre. *J. Geophys. Res.* **81**, 1883–902.

MACDONALD, K. C. 1982. Mid-Ocean ridges: fine scale tectonic, volcanic and hydrothermal processes within the plate boundary zone. *Ann. Rev. Earth Planet. Sci.* **10**, 155–90.

NEEDHAM, H. D. & FRANCHETEAU, J. 1974. Some characteristics of the Rift valley in the Atlantic Ocean near 36°48′ north. *Earth Planet. Sci. Lett.* **22**, 29–43.

NICOLAS, A. & JACKSON, M. 1982. Dike structures in mantle peridotite: injection mechanism and relation to plastic flow. *J. Petrol.* **23**, 568–82.

—— & POIRIER, J. P. 1976. *Crystalline Plasticity and Solid State Flow in Metamorphic Rocks.* Wiley, New York, 444 pp.

—— & PRINZHOFER, A. 1983. Cumulative or residual origin for the transition zone in ophiolites: structural evidence. *J. Petrol.* **24**, 188–206.

—— & VIOLETTE, J. F. 1982. Mantle flow at oceanic spreading centres: models derived from ophiolites. *Tectonophysics* **81**, 319–39.

——, GIRARDEAU, J., MARCOUX, J., DUPRE, B., XIBIN, W., YOUGONG, C., HAIXIANG, Z. & XUCHANG, X. 1981. The Xigaze ophiolite (Tibet): a peculiar oceanic lithosphere. *Nature, Lond.* **294**, 414–7.

NISBETT, E. G. & FOWLER, C. M. R. 1978. The Mid-Atlantic Ridge at 37° and 45°N: some geophysical and petrological constraint. *Geophys. J.R. Abstr. Soc.* **54**, 631–60.

PRINZHOFER, A., NICOLAS, A., CASSARD, D., MOUTTE, J., LEBLANC, M., PARIS, J. P. & RABINOWICZ, M. 1980. Structures in the New Caledonia peridotites-gabbros: implications for oceanic mantle and crust. *Tectonophysics* **69**, 85–112.

RABINOWICZ, M., NICOLAS, A. & VIGNERESSE, J. L. A rolling mill effect in asthenosphere beneath oceanic spreading centres. *Earth Planet. Sci. Lett.* (In press).

REA, D. K. 1976. Analysis of a fast-spreading rise crest: the East Pacific Rise, 9° to 12° South. *Marine Geophys. Res.* **2**, 291–313.

SLEEP, N. H. 1975. Formation of ocean crust: some thermal constraints. *J. Geophys. Res.* **80**, 4037–42.

—— & ROSENDAHL, B. R. 1979. Topography and tectonics of mid-oceanic ridge axes. *J. Geophys. Res.* **84**, 6831–9.

TAPPONNIER, P. & FRANCHETEAU, J. 1978. Necking of the lithosphere and the mechanics of slowly accreting plate boundaries. *J. Geophys. Res.* **83**, 3955–70.

VAN ANDEL, T. H. & BALLARD, R. D. 1979. The Galapagos Rift at 86°W. II. Volcanism, structure and evolution of the Rift valley. *J. Geophys. Res.* **84**, 5390–406.

VIOLETTE, J. F. 1980. *Structure des ophiolites des Philippines (Zambales et Palawan) et de Chypre, écoulement asthénosphérique sous les zones d'expansion océaniques.* Thèse 3ème cycle, Nantes, France.

A. NICOLAS, Laboratoire de Tectonophysique, Université de Nantes, France.
M. RABINOWICZ, Groupe de Recherches de Géodésie Spatiale, CNES, Toulouse, France.

LAVAS AND SEDIMENTS

Petrology of the Upper Pillow Lava suite, Troodos ophiolite, Cyprus

J. Malpas & G. Langdon

SUMMARY: The recently redefined Upper Pillow Lava suite of the Troodos ophiolite in Cyprus is composed of a series of related rock types: ultrabasic rocks, komatiites, olivine basalts and aphyric basalts, derived from a highly depleted primary melt by dominant olivine fractionation. Differentiation within high-level magma chambers involving approximately 25% olivine fractionation, and subsequent tapping of the resultant magma types explains the field relationships observed.

Previous workers (e.g. Gass & Smewing 1973; Smewing 1975) divided the extrusive section of the Troodos ophiolite into: (i) Upper Pillow Lavas, described as generally undersaturated, often olivine-bearing basalts with more ultrabasic varieties occurring at the top of the sequence; and (ii) Lower Pillow Lavas defined as mainly oversaturated, often intensely silicified with common celadonite, aphyric basalts. Gass & Smewing (1973), Smewing (1975), and Smewing *et al.* (1975) concluded that the boundary between Upper and Lower Pillow Lavas was largely a metamorphic discontinuity and they reported geochemical overlap between the two suites.

Recent investigation on Troodos, accompanying a programme of deep crustal drilling, has confirmed the existence of two major magma suites forming the volcanic section of the ophiolite, which correspond to distinct stratigraphic intervals (Robinson *et al.*, 1983). The basal 400–500 m of the sequence consist of an andesite-dacite-rhyolite assemblage interpreted as an arc-tholeiite suite. The remainder of the section comprises a basalt-basaltic andesite assemblage having a characteristic high MgO, low TiO_2 composition suggestive of a boninite suite (Robinson *et al.*, 1983). The major compositional discontinuity now recognized in the extrusive section of the ophiolite lies at a much lower level than the division between Upper and Lower Pillow Lavas recognized previously. The following discussion is concerned with the petrology of the now-redefined Upper Pillow Lava suite.

Field characteristics

Extensive descriptions of the field characteristics of the Upper Pillow Lavas in their various areas of outcrop have been given by several authors (e.g. Wilson 1959; Bagnall 1960; Carr & Bear 1960; Gass 1960; Bear 1960; Pantazis 1967). Smewing (1975) described the general nature of the unit as seen in the field.

These lavas are considered the peripheral unit of the Troodos complex and are bounded above and below by Upper Cretaceous–Lower Tertiary sediments, and the Lower Pillow Lava sequence, respectively. They crop out around nearly 60% of the perimeter of the massif, being most continuously found on its northern and southeastern edges and absent in most of its western parts (Fig. 1).

The lavas range up to 750 m but average about 500 m in thickness (Robinson *et al.*, 1983). They occur as a variety of intrusive and extrusive types, including pillows, dykes, flows, sills, breccias and hyaloclastites. Pillow lavas are by far the most voluminous morphology, composing about 80% of the exposure, with the other types listed above in order of decreasing abundance. The sequence may be divided into four lava types based on texture and macroscopic phenocryst assemblages. These are: (i) picrite basalt and ultrabasic rocks; (ii) basaltic komatiites; (iii) olivine basalt; and (iv) aphyric basalts. Chemical compositions, presented later, support this fourfold classification.

The picrite basalts and ultrabasic rocks (here referred to together as ultrabasic rocks) have previously been described by Gass (1958, 1960) and Searle & Vokes (1969). They were generally considered as shallow intrusive or hypabyssal bodies but are clearly found also as pillow lavas and the basal portions of massive differentiated flows. The position of these rocks near the top of the pillow-lava sequence suggests that they were extruded at a late stage in the tapping of magma chamber(s).

Olivine basalts are found generally at or near the top of the Upper Pillow Lava succession and are identifiable in the field by their reddish-brown, pseudomorphed olivine phenocrysts, now completely replaced by iron oxide and carbonate. This field distinction is also borne out in thin section. These rocks are most commonly found as highly weathered pillows which are pervaded by carbonate- and zeolite-filled vesicles and veins.

In some of the pillows, a fresh black glassy margin is preserved which varies in width from 1–15 cm and has hyalopilitic texture. Smewing (1975) distinguished between two types of olivine

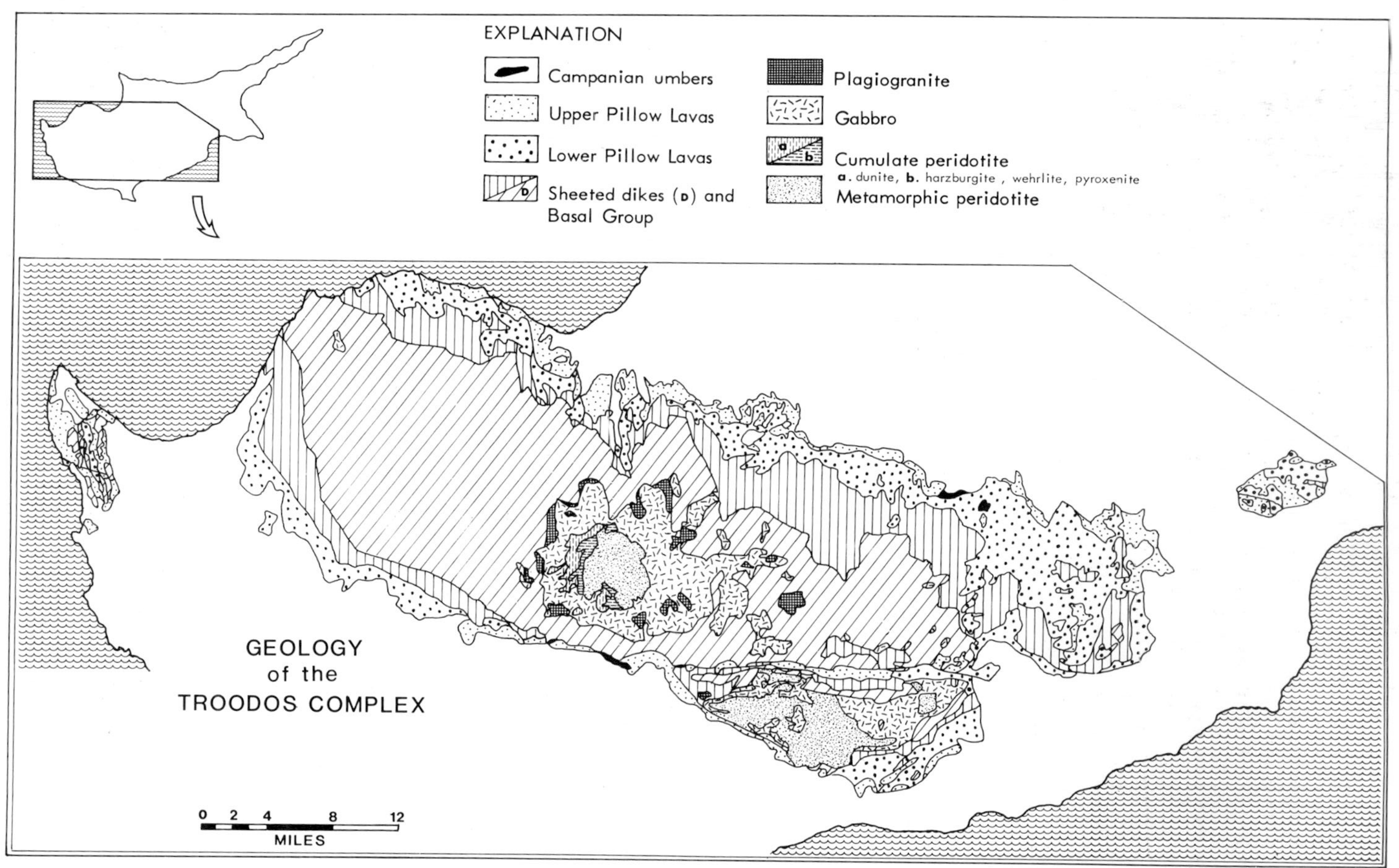

FIG. 1. Geology of the Troodos Complex, Cyprus, after Bear (1960).

basalts, based primarily on the presence or absence of this fresh black glassy margin which surrounds the altered cores of the pillows. These two types are clearly separated further on the basis of petrographical and geochemical data. Gass (1960) and Bear (1960) have referred to the olivine basalts containing the glassy margin as limburgites. These lavas are readily identified in the field; their black glassy edges stand out from the crumbly remains of the normal olivine basalts. Here, for reasons presented below, we refer to these basalts as basaltic komatiites, and distinguish them from the normal olivine basalts.

The aphyric basalts are by far the most abundant type of rock within the Upper Lavas, forming wide areas of outcrop where the sequence is thickest. Their main occurrence is as pillow lavas, with massive flows and sills occasionally being found. These lavas are perhaps best characterized in the field by their consistently altered state. Generally, whole pillows can be broken down by a few hammer blows, owing to a pervasive network of cracks and joints.

Petrography

Ultrabasic rocks

These rocks are found as both extrusive and intrusive types, this dichotomy leading to the development of two textures: vitrophyric and holocrystalline, respectively. The vitrophyric type has a dominance of olivine phenocrysts which are set in a glassy groundmass, and which compose up to 66% by volume of the rock. These unzoned olivine phenocrysts ($\sim Fo_{92}$) range up to 2.2×0.5 cm in size. They normally show euhedral but slightly corroded boundaries. The grains are often fresh but more commonly show alteration to antigorite along fracture planes. In some grains this alteration is accompanied by rounded patches of magnetite, which tend to occupy the corroded interiors.

Fresh clinopyroxene occurs also as a phenocryst phase but is much less abundant than olivine and forms much smaller grains (average 0.5 mm). These are commonly euhedral and occasionally show well-developed twinning along [100]. Euhedral crystals of chromite exist in small amounts and reach 0.6 mm in diameter. They are often present as inclusions within large olivine grains, where they appear to serve as loci for fracturing and serpentinization.

Partially devitrified glass of basaltic composition makes up the groundmass (determined by microprobe). Microlites of clinopyroxene are disposed parallel to phenocryst boundaries. Carbonates and zeolites occur both in amygdules and throughout the groundmass as deuteric patches.

Ultrabasic rocks with a holocrystalline texture show much higher degrees of alteration, a groundmass which is entirely crystalline and a reduction in the size of olivine phenocrysts (average 2 mm) as compared to the vitrophyric type. Small (0.8–1 mm), well-preserved, euhedral magnesian augite grains ($Wo_{41.3}En_{51.0}Fs_{7.8}$) occur as a phenocryst phase. Olivine euhedra are altered to fibrous antigorite, especially near grain boundaries and along fractures within the crystals. Chromite is present as intergranular euhedra and as inclusions within the olivine. Small magnetite inclusions within olivine appear to have developed as a byproduct of antigoritization. Small grains of orthopyroxene are sparse. Plagioclase is a fairly important constituent of the groundmass, being intergrown with augite in an intersertal texture. Much of the original holocrystalline aspect of the groundmass, however, has been lost during deuteric alteration. Primary pyroxenes and plagioclase have been extensively replaced by antigorite, magnetite, smectite and zeolites.

Basaltic komatiites

The basaltic komatiitic lavas are characterized by their very fresh condition and the well-preserved nature of both phenocryst and matrix phases. The rocks display hyalopilitic, variolitic, and rarely, intersertal and vitrophyric textures. Hyalopilitic types contain phenocrysts of olivine and orthopyroxene and microlites of subcalcic augite and pigeonite set in a glassy groundmass. Olivine forms euhedral phenocrysts which reach 2.4 mm and average 0.4 mm in their long dimension. They are occasionally completely fresh but more often occur as euhedral, skeletal forms in which alteration has attacked the interior of the grains along fracture planes. Antigorite is the common alteration product although some iron oxide is present along fractures.

The character of the pyroxenes is significantly different from that of olivine in that grains have developed forms and shapes indicative of rapid quenching. Orthopyroxene is the earliest pyroxene to nucleate as it forms stubby uniform grains averaging about 0.5 mm in length. Occasionally it attains an attenuate or acicular habit, more akin to that of the clinopyroxenes. Some grains have developed a mantle of clinopyroxene which presumably represents a lower-temperature reaction of Mg-Fe pyroxene with a liquid enriched in calcium.

Augite ($Ca_{26.6-42.6}$) is the most common monoclinic pyroxene in the rock. Subcalcic augite ($Ca_{20.8-21.3}$) and pigeonite ($Ca_{3.4-12.7}$) are present in subsidiary amounts, and are difficult to distinguish under the microscope. The absence of exsolved augite lamellae in the orthopyroxene, diagnostic of inverted pigeonite, suggests a rapid cooling history for these rocks.

The calcic pyroxenes are collectively characterized by their relatively fine-grained (0.2–0.4 mm) and quenched aspect—i.e. 'swallowtail', and occasionally, 'hourglass' structure.

Pyroxenes of the variolitic textured basalts differ markedly from those of the hyalopilitic basalts. They characteristically form highly skeletal or acicular grains which may congregate in radial growth. Average lengths are of the order of 1–1.2 mm but occasionally grains are up to 2 mm long. In most sections studied, dendritic, plumose or sheaflike varieties are present and in the better-developed specimens these grains form 'vertebrae' structures.

Olivine and orthopyroxene are considerably less abundant within rocks with this texture. Olivine tends to form skeletal grains altered to antigorite and iron oxide. Glass also forms a smaller proportion of the rock, mainly due to the interlocking mesh of pyroxene crystals. A higher degree of alteration is characteristic of this texture.

Opaques are very sparse in all the olivine-phyric rocks. Minute (0.1–0.2 mm) grains of chromite occur in clusters or as inclusions within larger pyroxene grains, and constitute a very small proportion of the groundmass.

Near the edge of pillows, hyalopilitic textures grade outward into vitrophyric textures with a decrease in crystalline pyroxene as a constituent of the groundmass. Near the centres of larger pillows, variolitic textures tend to grade inward towards intersertal textures, although are not found to be anywhere well developed.

The exclusive occurence of olivine as a phenocryst phase, and clinopyroxene as a groundmass phase has genetic implications. Olivine is always in an unquenched state and can be allocated to the pre-eruptive or intratelluric stage of the magma—i.e. olivine had begun to crystallize, presumably in a magma chamber, prior to extrusion and experienced little or no post-extrusive growth. There is clearly a continuum from obviously accumulative rocks to true liquids, and the degree to which the porphyritic rocks are accumulative is in some cases, difficult to ascertain, and has consequences for the interpretation of major-element geochemistry. Growth of clinopyroxene, however, is restricted to the extrusion stage, where rapid cooling produced an uninverted pigeonite/sub-calcic augite assemblage. Orthopyroxene may occupy an intermediate genetic position; nucleation and growth had been initiated in the intratelluric stage as some large grains display the typical chunky form of orthopyroxene. Others, however, show the quenched form noted for the monoclinic pyroxenes, and could only have formed after lava extrusion.

Olivine basalts

Almost without exception these rocks are more intensely altered than the quenched lavas. The great majority of specimens reveal a groundmass completely and penetratively altered to an assemblage of haematite, fibrous zeolites, and patches of carbonate. A few fresh specimens enable description of the primary texture of the rock, but even in the freshest specimens, olivine phenocrysts are totally pseudomorphed by haematite or calcite.

The most notable digression from komatiitic types, and the major reason for a separate grouping, is the absence of quench textures and the consequent appearance of plagioclase as a crystalline phase in the groundmass. Equigranular clinopyroxene and plagioclase average 0.2 mm in size. The occurrence of deuteric material and opaques in the interstices defines an intersertal texture. Clinopyroxene ranges from groundmass size up to microphenocrysts of 1.6 mm diameter. Such grains are always fresh, even in the more altered rocks, and tend to form glomeroporphyritic clusters.

Pseudomorphs after olivine form 0.4–2.6 mm grains. In some of the fresher specimens orthopyroxene is recognizable as small grains which occasionally are mantled by clinopyroxene.

In the outer parts of olivine basalt pillows, variolitic textures are suggested by the radial arrangement of pyroxenes. A gradual decrease in grain size is encountered from core to margin of pillows.

Aphyric basalts

These rocks are thoroughly crystalline and display intersertal or intergranular texture. Small patches of variolitic texture can occasionally be found near vesicles, where faster cooling has occurred. These rocks differ from the more mafic lavas in the absence both of olivine phenocrysts and a glassy mesostasis. Microphenocrysts of clinopyroxene are very common and range from groundmass size (0.1 mm) up to 1.5 mm. Plagio-

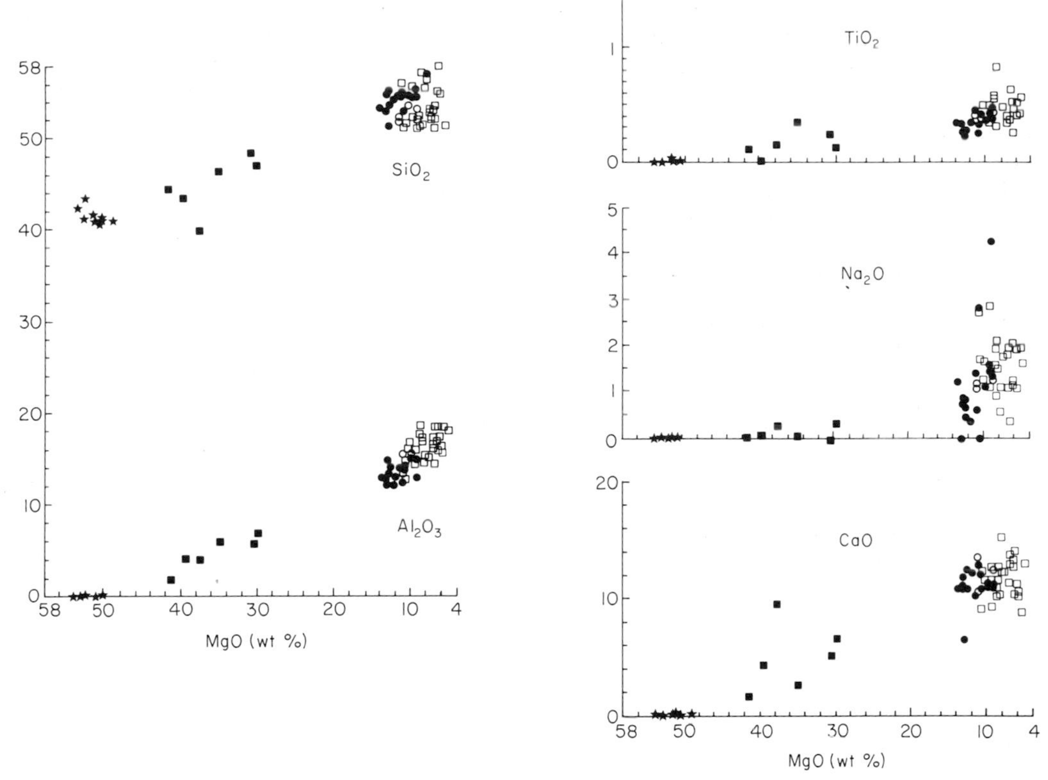

FIG. 2. Major elements plotted against MgO as a fractionation index. ★ = Olivine analyses, ■ = ultrabasic rocks, ● = komatiites, ○ = olivine basalts, □ = aphyric basalts.

clase normally forms laths in the groundmass (<0.4 mm) but occasionally occurs as microphenocrysts which reach 1.6 mm in length. Minute euhedra of magnetite generally constitute less than 5% of the rock.

A ubiquitous feature of this group of rocks is their intense and pervasive alteration. Celadonite in a common deuteric mineral, occurring as bright green patches replacing pyroxene or in amygdules. In many sections haematite has almost completely obscured the primary textures. Fine-grained pyroxene and magnetite are the phases most commonly affected by haematization. The succession also contains some coarse-grained bodies which presumably were intruded at shallow depths into warm pillow lavas. These rocks arc microgabbros consisting of 0.5 mm plagioclase and 0.4 mm pyroxene grains intergrown in an intergranular texture. Titanomagnetite is more common than within the fine-grained rocks. Pyroxene is similar in habit and abundance to its occurrence in fine-grained bodies, but the increase in abundance and grain size of plagioclase is notable.

Bulk-rock geochemistry

Major elements *v.* MgO

The trends of certain major elements *v.* MgO are depicted in Fig. 2. The plots also include olivine analyses from the ultrabasic rocks to show relationships between olivine compositions and the different lava types. The field and petrographic classifications made for the different types of Upper Pillow Lava are supported by the plots, primarily on the basis of MgO content alone. The ultrabasic rocks form a distinct group with MgO contents of 30–40%. Most of the major elements of the ultrabasic rocks show very small ranges against MgO; lime and silica show a wide scatter and probably indicate the effects of zeolite-facies metamorphism. Alumina shows a very broad range and may reflect the varying amounts of plagioclase in the different ultrabasic rocks. The basaltic komatiites also form a relatively distinct group with MgO ranging from 9.5 to 13.5%, and are characterized by very consistent values for the major oxides. Again SiO_2 and CaO show the

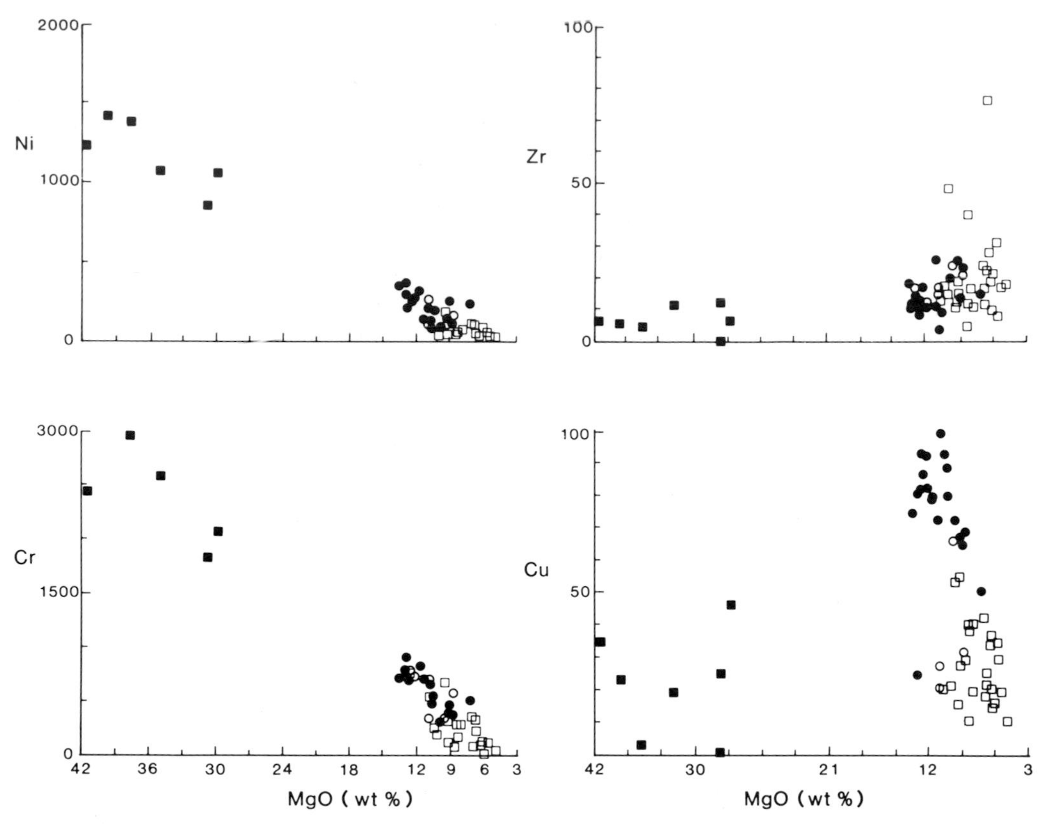

FIG. 3. Trace elements plotted against MgO as a fractionation index. Symbols as in Fig. 2.

widest absolute variation of 4 and 6%, respectively. Apparently the consistent results are produced by the very fresh condition of these glassy rocks. The field of basaltic komatiites is partially overlapped by that of olivine basalts and aphyric basalts in some of the plots; these show ranges of 8–11% MgO and 4.5–11% MgO respectively. These rocks also generally show broader ranges for the other oxides as well; again SiO_2 and CaO show especially large variations. The olivine basalts and aphyric lavas are chemically, as well as petrographically, similar. The olivine-phyric basalt plots define a field inside the aphyric basalt but intermediate between the komatiites and the bulk of the aphyric lavas. K_2O values in the ultrabasic rocks were too low to measure.

The most significant feature of the major oxides *v.* MgO plots is the colinear aspect of the fields defined by the different rock types. This is evident for all the plots with the possible exception of some of the fields defined by the highly weathered olivine basalts. Where the fields are broad, the points are dispersed with considerable symmetry around a line joining the fields. The dispersed nature of some plots, especially for SiO_2, Na_2O and CaO indicate possible alteration and element mobility. However, stable elements such as Al_2O_3 and TiO_2 show better cluster. The microprobe analyses of the olivine phenocrysts lie on an extension of the line connecting the fields of bulk-rock analyses. This colinear aspect, together with the petrographic data, is considered to demonstrate the importance of olivine as a control phase in the genesis of the different rock types.

Trace elements *v.* MgO

Generally the different rock types show the expected variation of trace elements with MgO. The incompatible elements all increase with decreasing MgO. Ni and Cr increase with MgO content, while Cu shows an increase from ultrabasic to komatiitic rock types and a decrease from komatiites to aphyric basalts (Fig. 3).

The very regular variation of Ni with MgO supports the major-element data, i.e. that these trends can be explained by olivine fractionation. Cr against MgO shows similar behaviour and

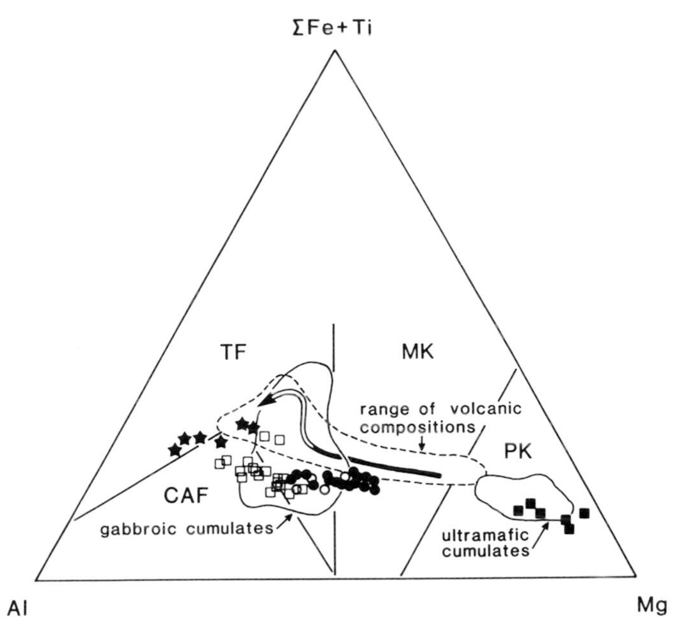

FIG. 4. Jensen Diagram (Fe + Ti—Al—Mg). Symbols as for Fig. 2 with the addition ★ = Lower Pillow Lava analyses. PK = peridotitic komatiite, MK = mafic komatiite, TF = tholeiitic field, CAF = calc-alkaline field. Solid curve represents olivine/pyroxene fractionation path, open curve represents feldspar/pyroxene fractionation path. Enclosed fields are derived from Chukotat Group (Francis & Hynes 1979).

suggests incorporation of this element into crystallizing clinopyroxene or chromian spinel; this possibility is confirmed by the petrography.

Zr contents are low (<75 ppm at 6% MgO) and together with low TiO_2 and, relatively high SiO_2 at given MgO contents, have led Robinson *et al.* (1983) to ascribe the Upper Pillow Lavas to a boninitic series, derived from a depleted mantle source.

Jensen diagram (Fig. 4 after Jensen 1976)

This plot uses Fe + Ti, Mg and Al and serves very well to characterize all the lavas concerned. The Upper Pillow Lava compositions as a whole show a continuous linear trend away from the Mg apex. The lack of a significant trend toward the Fe + Ti apex indicates the minor importance of clinopyroxene and plagioclase, and the predominance of olivine, as separating phases. The slight digression of some points away from both the Fe + Ti and Al apices may indicate small amounts of pyroxene and plagioclase fractionation, albeit this effect is deemed minimal when compared to the normal tholeiitic trend as delineated by the arrow in the Jensen diagram.

Most of the komatiite samples lie in the mafic komatiite field of the diagram. The derivative olivine-phyric and aphyric basalts trend away from the komatiite field toward the field of high-Al basalt. The rocks as a whole are enriched in aluminium with respect to the komatiitic-tholeiitic trend of Francis & Hynes (1979) for the lavas of the Chukotat Group, New Quebec. The more evolved members of the Upper Pillow Lavas are comparable in composition, however, to the gabbroic cumulates of Francis & Hynes (1979) which are in fact derived from residual liquids produced by the extraction of ultramafic cumulates. Although none of these basaltic rocks are cumulates, they are analogous to the Chukotat gabbroic cumulates in their status as a derived liquid.

TiO_2 v. SiO_2 (Fig. 5)

This diagram was used by Arndt *et al.* (1977) to chemically distinguish between tholeiites and komatiites from Munro Township, Quebec. It is considered less than ideal here, but serves to indicate both the extent of metasomatism which has occurred in the highly altered olivine and aphyric basalts, and the strikingly low TiO_2 contents of the Upper Pillow Lavas as a whole.

Except for the ultrabasic rocks, the TiO_2 values show no significant variation with SiO_2 content, in fact, the highest values are found where SiO_2 is lowest. This behaviour indicates the validity of

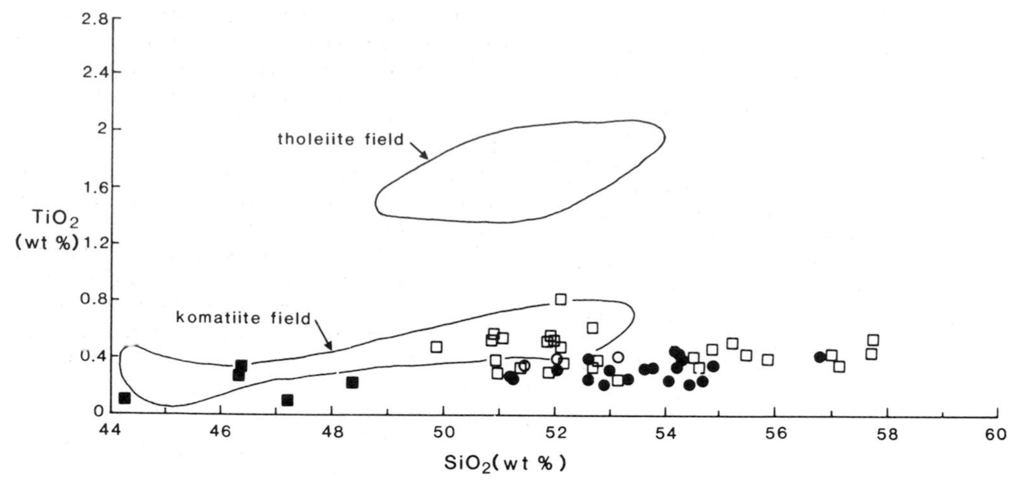

FIG. 5. TiO_2 *v.* SiO_2 (wt %). Symbols as in Fig. 2

using MgO rather than SiO_2 as a fractionation index. As TiO_2 is considered to be relatively immobile during low-temperature metamorphic processes, those lavas with the highest TiO_2 are likely to be the most evolved, and their lower SiO_2 contents probably represent the removal of silica by metasomatism. This consideration is further brought out by the large scatter of points for the olivine and aphyric basalts as compared to the komatiites.

A considerable proportion of the aphyric lavas plot inside the komatiite field as defined for the Munro Township rocks. The remainder plot below the field with TiO_2 values for some komatiites as low as 0.2%. As a group the aphyric lavas contain significantly less SiO_2 and more TiO_2 than the komatiites.

Mineral chemistry

Olivine

Olivine analyses were obtained by microprobe from ultrabasic rocks and komatiites. The complete alteration of olivine phenocrysts in the olivine basalts precluded their study. Results are presented in Table 1. Two ultrabasic rocks (KL-121, DL-45) have similar Fo contents (91.3 and 90.6) and both were collected from the bases of differentiated flows. A third, DL-36, collected from an ultrabasic pillow lava, is richer in the forsterite molecule (Fo_{93}). Olivines from the komatiitic lavas are markedly less magnesian varying from $Fo_{87.6}$ to $Fo_{89.6}$, but are within the range of komatiitic olivines from Munro Township reported by Arndt *et al.* (1977).

TABLE 1. *Mineral analyses from Upper Pillow Lavas*

		Olivine					Orthopyroxene		
		Ultrabasic rock			Komatiite		Ultrabasic rock	Komatiite	Olivine basalt
		KL-121(4)	DL-36(4)	DL-45(3)	DL-23(1)	*KL-33(3)	KL-121(3)	KL-34(1)	KL-104(1)
% wt	SiO_2	40.67	41.66	40.84	40.64	40.77	54.77	56.70	55.11
	Al_2O_3	0.07	0.04	0.02	0.19	0.02	3.74	1.81	1.76
	TiO_2	0.01	0.01	0.01	0.02	0.01	0.09	0.04	0.02
	Cr_2O_3	—	—	0.03	—	0.05	—	—	—
	FeO	8.48	6.90	9.20	12.12	10.03	11.26	7.64	6.74
	MnO	0.12	0.11	0.12	0.16	0.11	0.20	0.18	0.14
	MgO	50.40	52.63	50.57	47.88	48.51	28.60	31.04	31.99
	CaO	0.21	0.17	0.19	0.23	0.18	1.75	3.17	2.29
	Na_2O	0.01	0.01	0.04	—	—	0.02	—	0.01
	K_2O	0.01	—	—	—	—	0.01	—	—
	NiO	—	—	0.25	—	0.32	—	—	—
	Fo/En	91.3	93.0	90.6	87.6	89.6	81.5	87.6	89.2

*Nos in brackets = number of analyses.

Orthopyroxenes

The En content of orthopyroxenes increases from 81.6 in the ultrabasic rocks to 89.2 in the olivine basalts. The En/Fo ratio for the ultrabasic rocks is clearly not an equilibrium ratio which suggests that either: (i) olivine and enstatite are in marked disequilibrium due to rapid cooling of the lava; or (ii) the larger, and presumably earlier formed olivine grains precipitated out of a liquid of different composition than the pyroxene did. The texture of the rocks suggests the latter a more likely premise and that extrusion of an olivine-liquid mush occurred before orthopyroxene had begun to crystallize. The small proportion of orthopyroxene which then crystallized in the groundmass did so from a restricted batch of liquid already depleted in Mg^{2+} by the earlier crystallization of olivine.

The Mg^{2+} content of orthopyroxenes in the komatiitic lavas ($En_{87.6}$) is more comparable to that of the olivines, although it is again slightly low. The rapid quenching of this rock can account for the observed lack of equilibrium distribution of Mg^{2+} and Fe^{2+} between the two phases. The En content of orthopyroxene for the olivine basalts is still higher ($En_{89.2}$) and suggests conditions closer to equilibrium crystallization.

The molecular proportion of Ca^{2+} in orthopyroxene ranges from 0.065 in the ultrabasic lavas to 0.117 in the komatiitic lavas. The high value for komatiitic orthopyroxene may reflect the abundance of calcium in the magma, which may have entered the orthopyroxene lattice before clinopyroxenes began to crystallize. Atlas (1952) found that the amount of Ca^{2+} atoms accepted into enstatite in the synthetic system $MgSiO_3$–$CaMgSi_2O_6$ reached a maximum of 0.115 at a crystallization temperature of 1100°C, but decreased to 0.030 at 700°C. These data are in accord with the quenched nature of the komatiites and the preservation of mineral compositions achieved at high temperatures of equilibrium, such as the high Ca^{2+} content of orthopyroxenes.

Clinopyroxene

Clinopyroxenes from the ultrabasic rocks, olivine basalts and aphyric lavas (i.e. unquenched lavas) are all augites, although those from ultrabasic rocks are richer in both the Wo (39.6–41.2) and En (51.0–54.0) components (Fig. 6). Generally speaking, Ca^{2+} varies sympathetically with Mg^{2+} and antipathetically with Fe^{2+}, as there is an overall increase in the $MgSiO_3$ and decrease in the $FeSiO_3$ components going from ultrabasic to aphyric rocks. Cr_2O_3 was analysed from one ultrabasic rock and composes 1.8 wt% as compared with 0.10–1.47% from the aphyric lavas. TiO_2 shows no significant variation. Alumina in clinopyroxenes is consistently high in both the ultrabasic rocks and olivine basalts (2.01–3.16%), as compared with smaller values and a broader range for the aphyric basalts (1.62–3.02%).

Soda and potash contents are low in all clinopyroxenes analysed and reflect the extremely depleted nature of the magma.

A number of clinopyroxene grains from the most pristine hyalopilitic komatiites (i.e. quenched lavas) were studied to investigate the zoning already identified optically. There appeared no unique pattern of variation within the grains, the following cases being notable (see Fig. 6):

(i) Ca^{2+}-rich (augite) cores → rims lower in Ca^{2+} (augite or subcalcic augite) e.g. KL-33, Grain A, KL-34, Grains A, B and D.

(ii) Ca^{2+}-rich (augite) cores → rims much lower in Ca^{2+} (pigeonite) e.g. KL-33, Grain D.

(iii) Ca^{2+}-poor (pigeonite) cores → Ca^{2+}-rich rims (augite) e.g. KL-33, Grains B and C; KL-34, Grain C.

Such variations are undoubtedly the result of disequilibrium states produced by fast rates of cooling as suggested by Lofgren *et al.* (1974) and Donaldsen *et al.* (1975) on the basis of experimental investigations.

Plagioclase

Analyses of plagioclase were obtained only from the ultrabasic rocks and the aphyric lavas, as no free plagioclase was found in the komatiitic lavas, and the mineral within olivine basalts was invariably highly altered. Plagioclase grains are unzoned labradorites and bytownites (An_{60-77}). There is no significant compositional variation between plagioclase from ultrabasic and less basic rocks, however, as both the highest and lowest values of An are found in the aphyric basalts. The Or component is negligible in all analyses.

Petrogenesis

The heterogeneous nature of the upper mantle is intimately tied to the premise of complex melting/crystallization. Small degrees of partial melting have the effect of concentrating highly incompatible elements in the liquid phase with a concomitant depletion of these elements in the unmelted material. The immediate effect is to produce very local zones of depletion/enrichment in regions of high geothermal gradient.

Magnesian quartz tholeiites which have been derived by a second stage or advanced melting process (i.e. heterogeneous material) can be treated as primary liquids from which distinctive

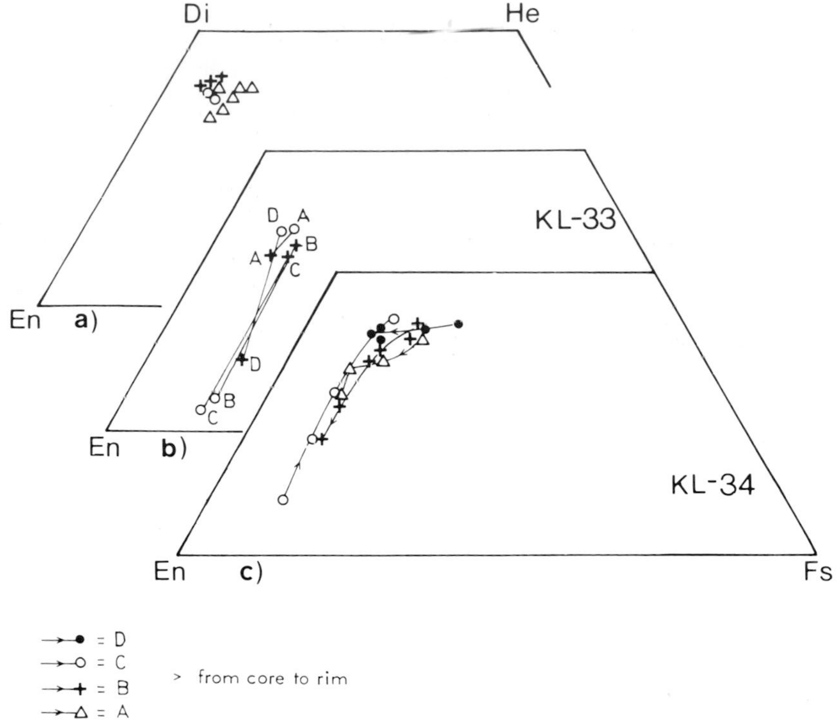

FIG. 6. Variation in clinopyroxene composition. (a) Unquenched lavas: + = ultrabasic rocks, ○ = olivine basalts, △ = aphyric basalts. (b) Quenched lava KL-33 (Komatiite): + = rim, o = core, (c) Quenched lava KL-34 (komatiite): traverses across four pyroxenes, arrows indicate direction from core to rim.

magma series may be produced by low-pressure fractionation. Duncan & Green (1980, figure 1) plotted the major-element compositions of olivines from the ultrabasic rocks, and the bulk-rock compositions of ultrabasic rock, olivine basalt and aphyric lavas of the Upper Pillow Lavas. They suggested that the exhibited linearity could be explained by the addition or removal of olivine, and using reported chemical data (Gass 1958; Searle & Vokes 1969; Kay & Senechal 1976; Simonian & Gass 1978) determined a parental-liquid composition, based on the principle that such a liquid must be in equilibrium with the most magnesian olivine observed. An olivine composition of $Fo_{91.7}$ implies a $Mg/(Mg+Fe^T)$ ratio of 0.77 (Roeder & Emslie 1970). This determined parental liquid is found to be more refractory than the least fractionated basalts from ocean-ridge and island-arc environments.

Here the concept of olivine extraction is tested. From the linearity expressed in the major- and trace-element variation diagrams, it can be empirically stated that addition and removal of olivine accounts for the composition of all the members of the Upper Pillow Lavas. Using a form of the equilibrium equation for the partitioning of iron and magnesium between olivine and liquid, with a partition coefficient (K_D) = 0.3 (Roeder & Emslie 1970) then, if for the average aphyric basaltic komatiite FeO = 4.76 mol% and MgO = 17.15 mol% (see Table 2).

$$\frac{Ol_{FeO}}{Ol_{MgO}} = \frac{K_D \times \text{liq FeO}}{\text{liq MgO}} = \frac{0.3 \times 4.76}{17.15} = 0.0833. \quad (1)$$

But Fo + Fa in olivine = 100. (2)

By substitution, $Ol_{MgO} = 92.31$, i.e. the olivine composition in equilibrium with the average komatiite is $Fo_{92.3}$. This can be directly compared with the olivine compositions from the ultrabasic rocks which range from 90.8 to 93.1.

Preliminary examination of the major-element data therefore suggests that the ultrabasic rocks, basaltic komatiites, olivine basalts and aphyric lavas of the Upper Pillow Lavas form a series which can be petrogenetically related by the fractionation of olivine from a parental-magma equivalent in composition to the basaltic komatiites. Trace elements may be used to provide an

TABLE 2. *Chemical analyses of parental and primary magmas of Upper Pillow Lavas, Cyprus; and average compositions of lava types.*

Lava type or magma type	Ol basalt (4)*	Aphyric basalt (20)	Komatiite (parent) (18)	'Boninite'† (primary) (1)
SiO_2 %wt	47.78	52.82	53.72	51.62
Al_2O_3	13.35	15.13	13.36	10.83
FeO } FeO	4.53	5.47	2.98 }	7.43
Fe_2O_3 }	3.17	3.16	5.57 }	
MnO	0.12	0.12	0.15	—
MgO	9.22	7.64	11.25	19.38
CaO	10.78	11.65	11.14	8.92
Na_2O	1.21	1.26	1.16	1.30
P_2O_5	—	—	0.10	—
H_2O+	8.81	2.98	recalculated to 100% anh.	nd‡
TOTAL	98.97	100.23		99.48
K (ppm)	7740	7305	1909	747
Rb	10	27	4	nd
Sr	100	91	104	nd
Ba	27	25	15	nd
Zr	19	21	16	nd
Cu	37	32	77	nd
Ni	159	62	217	nd
Ti	2336	2576	1977	1978
V	215	266	209	nd
Pb	1	3	1	nd
Cr	479	155	617	nd
Zn	60	60	57	nd
Nb	1	2	2	nd
La	20	16	20	nd
Y	12	17	12	nd
Cc	4	5	4	nd
Ga	11	12	11	nd

*No. of samples averaged;
†analysis of melt inclusion from A. Sobolev;
‡nd = not determined.

estimate of the amounts of olivine that may have been involved in this process. Elements with a high field strength (charge/radius ratio) such as Ti, Zr, Y and Nb are not likely to be transported in aqueous fluids (Pearce & Norry 1979). Since their values show a regular variation from komatiite to aphyric lava (except Nb) they provide a more reliable estimate of fractionation. Because of its very low concentrations, which are below the accepted detection limit for the method used, Nb is excluded from the calculations. The inclusion of olivine only in the model is designed to facilitate calculations although, in reality, the extraction of clinopyroxene or titanomagnetite will affect concentrations in residual liquids. However, olivine is by far the most important fractionating phase and for these calculations other effects are negligible.

Partition coefficients for Ti, Zr and Y are taken from Pearce & Norry (1979).

Titanium

$$D_{Ti}(Ol) = 0.02$$

$$F = (1-D)\sqrt{C_0/C_1},$$

where $C_0 = 1977$ ppm and $C_1 = 2576$ ppm (Table 2),

$$F = (1-0.02)\sqrt{\frac{1977}{2576}} = 0.7633,$$

or
~76% melt remaining, or ~24% of olivine fractionation.

Zirconium

$$D_{Zr}(Ol) = 0.01$$

where $C_0 = 15.8$ ppm and $C_1 = 21.1$ ppm (Table 2),

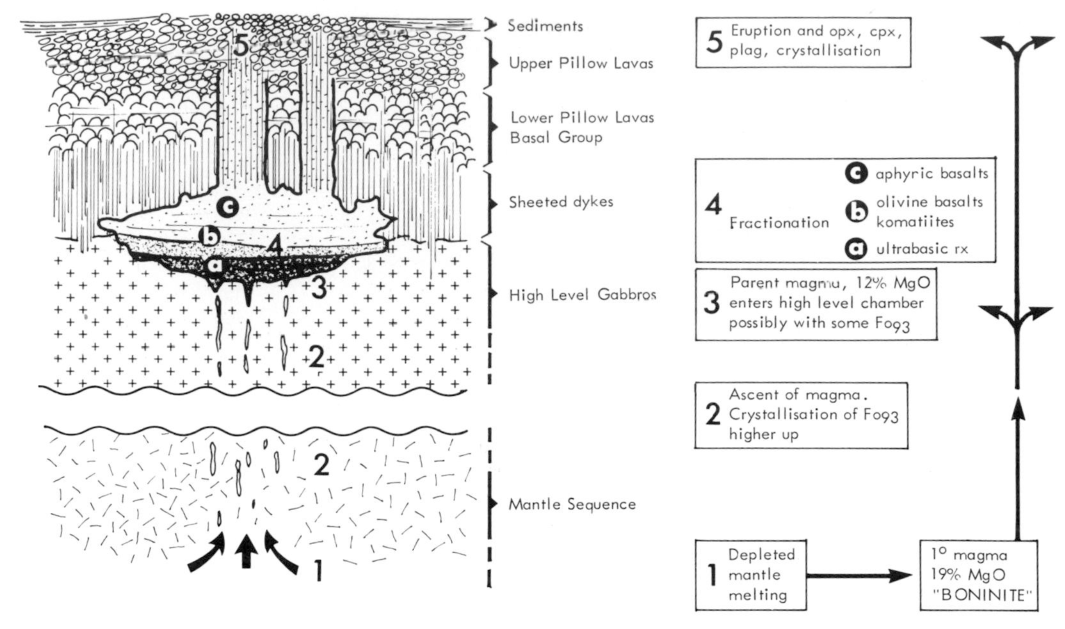

FIG. 7. Magma-chamber model and petrogenetic scheme for the derivation of the Upper Pillow Lavas.

$$F = (1-0.01)\sqrt{\frac{15.8}{21.1}} = 0.7466,$$

or

~75% melt remaining, or ~25% fractionation of olivine.

Yttrium

$D_Y(Ol) = 0.01$,
where $C_0 = 11.6$ ppm and $C_1 = 17.1$ ppm (Table 3),

$$F = (1-0.01)\sqrt{\frac{11.6}{17.1}} = 0.6757,$$

or

~68% melting remaining, or ~32% fractionation of olivine.

These values, generally in agreement, define a range in the amount of olivine crystallization required to derive the Upper Pillow Lava series from a parental magma of basaltic-komatiite composition. It is also worth emphasizing that the approximate proportion of rocks with accumulated olivine, compared to aphyric basalts, is 1 : 3 in the total exposure of Upper Pillow Lavas or within any one large differentiated flow, similar to results predicted above.

Although a basaltic-komatiitic composition thus appears a possible parent to the observed Upper Pillow Lava series, it is unlikely that it is a primary magma produced by melting of a depleted mantle source. More likely it is itself removed from the primary composition by olivine fractionation.

A closer approximation to the primary-magma composition is obtained by examination of melt inclusions within, and in equilibrium with, the most refractory olivine compositions observed, i.e. $Fo_{93.1}$. Examination of these olivines (A. V. Sobolev, pers. comm.) has shown that they crystallized from a silica-rich highly magnesian, but incompatible element depleted, liquid with boninitic affinities (Table 2). This liquid more likely approaches a primary-magma composition which on ascent to high-level magma chambers underwent olivine fractionation to produce the basaltic-komatiitic parental magma (Fig. 7).

Magma-chamber considerations

It is suggested that the parental magma (~basaltic komatiite) derived by olivine fractionation from a primary (boninitic) melt, accumulated in small magma chambers at a shallow crustal level. The continued high-temperature precipitation of olivine led to the accumulation of this mineral at the bottom of the chambers, with a portion of liquid being trapped between the crystals. The crystal/liquid mush formed in this manner is represented by the volumetrically small amount of ultrabasic rocks. The aphyric basalts thus represent magma in the upper parts of the chamber from which the olivine fraction has been removed. The gamut of compositions may thus represent the particular level in a magma chamber at which the magma batch attained its character. Olivine basalts, then, may represent a level intermediate between the olivine-rich and olivine-poor parts of the chamber. Extrusion of primitive magma soon after its accumulation in a chamber with minimal fractionation gave rise to the basaltic komatiites (Fig. 7).

The close association of the different rock types in the field, i.e. the common random interlayering of the different types, also lends support to their association in and extrusion from the same magma chamber or a number of small, consanguineous, closely spaced chambers. The relative abundance of the aphyric basalts again is in agreement with this model, as the derived aphyric liquid will be most preponderant in the magma chambers. Ultrabasic rocks occur at the top of the pillow lava sequence, suggesting their extrusion as a crystal/liquid mush upon advanced tapping of the magma chambers, subsequent to the extrusion of the aphyric lava.

The textures developed in the primitive liquid, i.e. the basaltic komatiite, are supportive of the above scheme. As pointed out above, olivine crystals were nucleated and experienced much of their growth in the intratelluric stage, while orthopyroxene may have undergone some post-extrusive growth. Clinopyroxene is restricted to the eruptive stage.

ACKNOWLEDGMENTS: Funding for this project was provided by an NSERC (Canada) operating grant to J. Malpas. We are grateful to P. Robinson and A. Sobolev for comments and the data they provided, the University of Bergen Geological Institute for manuscript typing and drafting, and Jim Scott for field assistance.

References

ARNDT, N. T., NALDRETT, A. J. & PYKE, D. R. 1977. Komatiitic and iron-rich tholeiitic lavas of Munro Township, northeast Ontario. *J. Petrol.* **18**, 319–69.

ATLAS, L. 1952. The polymorphism of $MgSiO_3$ and solid-state equilibria in the system $MgSiO_3$–$CaMgSi_2O_6$. *J. Geol.* **60**, 125.

BAGNALL, P. S. 1960. The geology and mineral resources of the Pano Lefkara—Larnaca Area. *Geol. Surv. Dept., Cyprus, Memoir No. 5*, 116 pp.

BEAR, L. M. 1960. The geology and mineral resources of the Akaki—Lythrodondha area. *Geol. Surv. Dept., Cyprus, Memoir No. 3*, 122 pp.

CARR, J. G. & BEAR, L. M. 1960. The geology and mineral resources of the Peristerona—Lagoudhera area. *Geol. Surv. Dept., Cyprus, Memoir No. 2*, 79 pp.

DONALDSEN, C. H., USSELMAN, T. M., WILLIAMS, R. J. & LOFGREN, G. E. 1975. Experimental modelling of the cooling history of Apollo 12 olivine basalts. *Proc. 6th Lunar Sci. Conf.* 843–69.

DUNCAN, R. A. & GREEN, D. H. 1980. Role of multistage melting in the formation of oceanic crust. *Geology*, **8**, 22–6.

FRANCIS, D. M. & HYNES, A. J. 1979. Komatiite-derived tholeiites in the Proterozoic of New Quebec. *Earth Planet. Sci. Lett.* **44**, 473–81.

GASS, I. G. 1958. Ultrabasic pillow lavas from Cyprus. *Geol. Mag.* **95**, 241–51.

—— 1960. The geology and mineral resources of the Dhali area. *Geol. Surv. Dept., Cyprus, Memoir No. 4*, 116 pp.

—— & SMEWING, J. D. 1973. Intrusion, extrusion and metamorphism at a constructive margin: evidence from the Troodos Massif, Cyprus. *Nature, Lond.* **242**, 26–9.

JENSEN, L. S. 1976. A new cation plot for classifying subalkalic volcanic rocks. *Ont. Dept. Mines Misc. Paper 66.*

KAY, R. W. & SENECHAL, R. G. 1976. The rare earth geochemistry of the Troodos ophiolite complex. *J. Geophys. Res.* **81**, 964–70.

LOFGREN, G., DONALDSEN, C. H., WILLIAMS, R. J., MULLINS, O. & USSELMAN, T. M. 1974. Experimentally-reproduced textures and mineral chemistry of Apollo 15 quartz normative basalts. Proc. 5th Lunar Conf., suppl. 5. *Geochim. Cosmochim. Acta* **1**, 549–67.

PANTAZIS, Th. M. 1967. The geology and mineral resources of the Pharmakas—Kalavasos Area. *Geol. Surv. Dept. Cyprus, Memoir No. 8*, 190 pp.

PEARCE, J. A. & NORRY, M. J. 1979. Petrogenetic implications of Ti, Zr, Y and Nb variations in volcanic rocks. *Contrib. Mineral Petrol.* **69**, 33–47.

ROBINSON, P. T., MELSON, W. G., O'HEARN, T. & SCHMINCKE, H.-U. (1983). Volcanic glass compositions of the Troodos ophiolite, Cyprus. *Geology* **11**, 400–4.

ROEDER, P. L. & EMSLIE, R. F. 1970. Olivine-liquid equilibrium. *Contrib. Mineral. Petrol.* **29**, 275–89.

SEARLE, D. L. & VOKES, F. M. 1969. Layered ultrabasic rocks from Cyprus. *Geol. Mag.* **106**, 513–30.

SIMONIAN, K. O. & GASS, I. G. 1978. Arakapas fault belt, Cyprus: A fossil transform fault. *Geol. Soc. Am. Bull.* **89**, 1220–30.

SMEWING, J. D. 1975. *Metamorphism of the Troodos Massif, Cyprus.* Unpub. Ph.D. Thesis, Open University, UK, 267 pp.

——, SIMONIAN, K. O. & GASS, I. G. 1975. Metabasalts from the Troodos Massif, Cyprus, Genetic implications deduced from petrography and trace element geochemistry. *Contrib. Mineral. Petrol.* **51**, 49–64.

WILSON, R. A. M. 1959. The geology of the Xeros—Troodos Area, with an account of the mineral resources (by F. T. Ingham). *Geol. Surv. Dept., Cyprus, Memoir No. 1*, 135 pp.

J. MALPAS, Department of Earth Sciences, Memorial University of Newfoundland, St John's, Newfoundland A1B 3X5, Canada.

G. LANGDON, Aramco, Exploration Division, Dhahran, Saudi Arabia.

Evolving metallogenesis at the Troodos spreading axis

J. F. Boyle & A. H. F. Robertson

SUMMARY: An area of the eastern Troodos Massif has been mapped in detail to identify stages of volcanism, faulting and metallogenesis which could be related to the evolution of the late Cretaceous Troodos spreading axis. The field relations and comparisons with modern spreading ridges suggest the following succession of events. Pillowed and massive flows making up most of the lava volume were erupted on a relatively flat sea floor at the axis of a well-defined median valley. Volcaniclastic sediments accumulated during a hiatus in volcanism. The Mathiati massive sulphide formed in the vicinity of ridge-parallel fractures towards the margins of the rift valley, whilst ferromanganiferous oxide-sediments were deposited in the surrounding area. Normal faulting, at first parallel to the rift axis, then at 30–40° to it, accompanied formation of the rift-valley walls, together with renewed eruption of lavas, equivalent to the Upper Pillow Lavas. Small grabens and half-grabens, which formed during this stage, were fringed with lava talus, then gradually filled, first with ferromanganiferous umbers precipitated from local hydrothermal vents, then, as the spreading axis migrated away, by more hydrogenous metal-oxide sediments and later deep-sea deposits. In Cyprus, metallogenesis persisted from the median valley to the flanks of particular regions of crust. Also the inferred site of formation of the Mathiati sulphide towards the margin of the rift valley differs from that of the known 'smokers' which are active closer to the spreading axis.

It is now firmly established that the various metalliferous sulphide- and oxide-deposits of the Troodos Massif were formed by hydrothermal processes at a late-Cretaceous oceanic spreading axis (Constantinou & Govett 1972, 1973; Robertson 1976; Spooner 1977; Heaton & Sheppard 1977). An important remaining question is how exactly the oxide- and sulphide-sediments can be related to the stages of construction of a spreading axis. The approach here has been to map the lava succession of a selected area in sufficient detail to identify the main phases of metalliferous accumulation, and to relate these to the structural and volcanic development of the ridge. This is timely in view of the recent upsurge of interest in the Troodos ophiolite, stimulated by the work of the International Deep Crustal Research Group (IDÇRG).

Over the last few years, new deep-ocean exploration techniques have immeasurably increased knowledge of hydrothermal processes at spreading ocean ridges (e.g. Ballard & van Andel 1977; Lonsdale 1977a, b). The extremely close parallels which can now be drawn with ophiolites like the Troodos facilitate interpretation of the latter; conversely, detailed mapping of ophiolites can add a depth dimension not readily available in the oceans.

Geological setting of the Mathiati–Margi area

The Troodos Massif has long been recognized as an intact ophiolite succession created at some form of Upper-Cretaceous spreading axis (Gass 1968; Moores & Vine 1971), although the exact tectonic setting remains controversial (e.g. Pearce 1980; Robertson & Woodcock 1980). The sheeted dykes, which constitute the key evidence of a spreading genesis, are orientated overall N–S in the Troodos (Kidd & Cann 1974), but complications exist, particularly in the east where the present study is located.

The extrusive succession has been subdivided into the Lower Pillow Lavas and the Upper Pillow Lavas (Wilson 1959), the former being largely basaltic andesites and the latter mostly olivine basalts. The boundary between these two units is uncomformable in some areas, but transitional in others (Wilson 1959; Bear 1960; Gass 1960), leading to uncertainty over the location and significance of this subdivision. More recent attempts to divide the lavas have been based upon other criteria, e.g. metamorphic grade (Gass & Smewing 1973; Smewing *et al.* 1975) and petrochemistry (current work by the International Deep Crustal Research Group). In the absence of a generally accepted definition of the Upper and Lower Pillow Lava units, we have simply mapped the lavas as local units and assigned letters to them.

An area on the eastern margin of the Troodos Massif in the vicinity of the well-known Mathiati mine (Fig. 1) was chosen for mapping because both sulphide- and oxide-metalliferous sediments are abundant and because regionally low dips and fault repetitions allow the individual units to be traced substantial distances laterally. In this area the dyke trend is consistently NW–SE deviating from the regional N–S direction. The sedimentary cover dips at 10–20° to the NE defining the mean dip of the lavas, but locally within the lavas dips range up to 30–40°, where rotation on normal faults has taken place. The distinctive ferromanganiferous oxide-sediments, the umbers, occur in small hollows along the contact between the

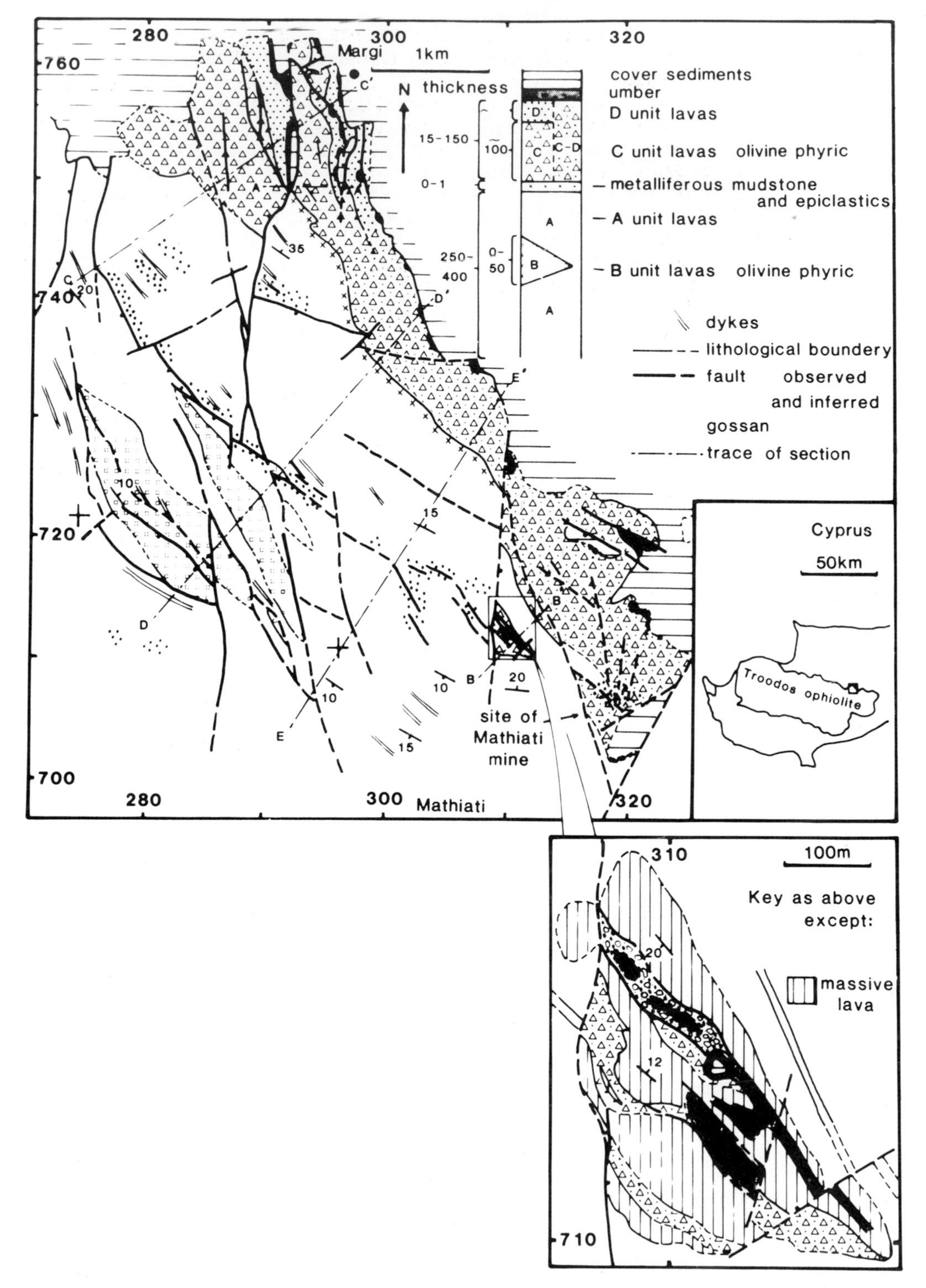

FIG. 1. Geological map of the Mathiati–Margi area, based on field mapping of selected areas using Cyprus series D.L.S. 17 1:5000 topographical maps, aerial photographs and the Cyprus Geological Survey Department 1:31,680 map (Gass, 1960). Inset: detail of an area 0.8 km NW of Mathiati mine.

stratigraphically highest lavas and the overlying sedimentary cover. Mostly ferruginous metal-oxide sediments are disseminated throughout the lava pile, largely in pillow interstices. The Mathiati massive sulphide orebody (Constantinou & Govett 1973) is located within the lava pile towards the southern end of the area mapped (Fig. 1).

The Mathiati–Margi area contains a higher proportion of both sulphide- and oxide-sediments (at all levels of the lava succession) than adjacent areas except the SE where there is the Sha mining area (1 km SE of Map, Fig. 1).

Lithostratigraphy

In the Mathiati–Margi area, Gass (1960) recognized a Lower Pillow Lava unit composed of pyroxene andesites, with subordinate basaltic, dacitic and keratophyric lavas, separated by a thin layer of metalliferous mudstones and volcaniclastic sediments from an Upper Pillow Lava unit composed of basalts, olivine basalts, limburgites and picrites. Both extrusives and high-level intrusives were identified. As our mapping was intended mainly to shed light on the metallogenesis in the area, we found it convenient to subdivide the succession along the laterally continuous interlava sediment horizon and then according to the presence or absence of olivine phenocrysts. A unit of olivine basalts was identified in the lower part of the succession in a region previously mapped as Lower Pillow Lavas (Gass 1960). Phenocrysts of pyroxene were found to be too variable in distribution to use in mapping. The stratigraphic distribution and relative abundance of each of the units mapped is shown in Fig. 1.

Lava morphology

The area mapped consists of *c.* 60–70% pillow lavas and 30–40% massive lavas. This is comparable with the proportions drilled in CY-I of the IDCRG (69.9% pillow lava, 28.4% massive lavas; Robinson & Gibson 1983). In both the area mapped and in CY-I, hyaloclastite is very subordinate, although in some exposures, e.g. the Akaki River section of North Troodos, hyaloclastite is relatively more abundant. In the Mathiati–Margi area dykes first appear in the lower part of the succession locally as swarms comprising >50% of the outcrop.

The original eruptive morphology of the extrusives in the area can be interpreted in the light of observations from modern spreading axes (Ballard *et al.* 1981, 1982; Ballard & Moore 1977). Steep-sided pillowed flows form topographic highs around eruptive centres which are located along fissures at the ridge axis. The early flows of each eruptive cycle tend to be sheet-like, travelling further than the pillowed flows (*c.* 1 km), pouring down slope and ponding in fault hollows. All except the most recent lavas are dissected by minor ridge-parallel fractures.

In the mapped area in Cyprus, differential erosion has tended to reverse the original sea-floor topography; the more resistant high-level intrusions and the massive flows form the high ground while the pillow lavas and the fragmented extrusives now form depressions (relief 20–50 m). The original relationships can however still be observed, e.g. near Mathiati (Fig. 1, inset; Fig. 3, Section B-B′), where massive flows of the C-D lava units dip in opposite directions on either sides of a valley. Older A-unit pillow lavas on the floor of the intervening valley are cut by a 4 m thick dyke which could have fed a pillow volcano originally located between the oppositely dipping massive flow units.

In outcrop most of the lava pillows are relatively uniform in size and shape (~1 m across), but locally throughout the succession some of the lava pillows are conspicuously elongated as 'bolsters', or larger still as 'mega-pillows'. The 'mega-pillows', which range up to 10 m across, commonly possess pillow-lobed lateral and upper contacts and are thought to represent sites of lava transport and extrusion (Schmincke & Rautenschlein 1982). The 'mega-pillows' often occur in groups capping small hills of more normal pillow lavas. Although hard to measure, the apparently radiating dips of these lavas suggest that they originated as small pillow-lava volcanoes.

Two types of fragmental extrusive rocks are recognized, hyaloclastites and pillow breccias. Small elongate lenses of hyaloclastite up to 5 m thick are present, associated with both massive and pillowed extrusives, throughout the lava pile and are best exposed in stream sections. The hyaloclastites are typically green to brown, weakly bedded to massive, consisting of poorly sorted glassy fragments and small intact pillows ~10 cm in diameter. In the oceans hyaloclastite is seen to form particularly on steep flow fronts (Ballard & Moore 1977), and a similar origin is favoured in Cyprus.

The pillow breccias are concentrated in the upper parts of the D lava unit. The breccias range from totally disorganized pillow segments, to some cases where the outlines of original pillows can still be recognized (Robertson 1975). These breccias contain little true hyaloclastite material and are interpreted as the result of collapse of pillow lavas along small submarine fault scarps (see below).

Interlava sediments

Excluding the fragmental extrusives mentioned above, three types of interlava sediments are present: (i) dispersed metal-oxide sediments; (ii) a horizon of ferromanganiferous mudstones and epiclastics; and (iii) massive sulphide and ochres.

Orange iron-rich oxide-sediments are dispersed throughout the volcanic succession but are concentrated in lava unit D and the upper part of lava unit A (Fig. 2). The oxide-sediment fills spaces between individual lava pillows and is often seen as small veins penetrating cooling fractures in the lavas. These field relations result from disruption of accumulating sediments by overriding laval flows. Compositionally, the oxide-sediments are largely ferruginous with variable amounts of ferromanganiferous oxides, clays, carbonates and spalled volcanic glass. Faecal pellets, coccolith guards and poorly preserved radiolaria are also present. It is unclear as to what extent the sediments were originally ferromanganiferous but were leached to leave a ferruginous residue during the later stages of hydrothermal activity. The dispersed metal-oxide sediment has previously been interpreted regionally as a hydrothermal precipitate ponded in the pillow lavas during the later stages of lava eruption (Robertson 1976; Robertson & Fleet 1976). The present work now shows that in the Margi area the dispersed sediment is more widespread within the lava pile, defining broad horizons extending several kilometres around Margi and to the SE (Fig. 2). A possible source for this oxide-material would be oxidation of sulphide precipitates drifting away from more axial vents, and this would adequately explain their widespread dispersed character. This process has been observed at Pacific spreading axes (Edmond *et al.* 1982). The downward decrease in abundance of dispersed oxide-sediment within the lava pile could result either from reduced sedimentation rates, or more probably from more rapid eruption of the earlier pillow lavas.

At the contact between the A and C lava units there is a laterally continuous horizon of ferromanganiferous mudstones underlain by epiclastic sediments. The epiclastics are fine to medium grained, poorly bedded, generally massive and contain altered lava fragments in a muddy matrix which is commonly iron- and manganese-rich. Thickness varies from 0 to 1 m. The ferromanganiferous horizon above is 0–1 m thick, very fine grained, either massive or finely laminated and generally resembles the umbers above the highest lavas. This sediment is interpreted as a mixture of background pelagic and hydrothermal oxide-sediment which accumulated above the epiclastics, during a pause in lava eruption. The laterally continuous, unponded character of these sediments indicates that the sea floor was relatively flat and had not undergone significant tilting at this stage.

The formerly economic Mathiati sulphide orebody and stockwork mineralization in the S of the area mapped (Fig. 1) now exists as a tilted horst overlain by unmineralized lavas (Constantinou & Govett, 1972). Until recent quarrying the massive ore was seen to be overlain by ferruginous ochres. These formed by a combination of submarine oxidation of massive sulphide (submarine gossans) and as hydrothermally precipitated ferruginous metal-oxide sediment; ochre of erosive origin is minimal (Robertson 1976). Constantinou (1980) reports that the sulphide occurs at the contact between the Lower and Upper Pillow Lavas, which places it at the same stratigraphic level as the ferromanganiferous mudstone horizon described above. Later faulting has obscured the sea-floor topography in the vicinity of the mineralization, but a stratigraphic association with the sediment horizon would indicate that it formed prior to the major faulting. The inferred depositional geometry of the orebody suggests it was probably formed in a fault-controlled bathymetric depression (Constantinou 1980).

Supra-lava sediments

In the area mapped, the stratigraphically highest lavas are overlain by ferromanganiferous oxide-sediments (umbers), metal-enriched clays, radiolarites and mudstones of the Perapedhi Formation, dated as Campanian. In turn these are overlain by Maastrichtian pelagic chalks of the Lefkara Formation (Robertson & Hudson 1973, 1974). The umbers commonly occur in faulted depressions on the original lava surface where underlying lavas have locally undergone extensive low-temperature hydrous alteration indicating a local source for the hydrothermal fluids. Robertson (1976) interpreted the umbers as hydrothermal precipitates ponded in hollows during and after the waning stages of volcanism. The pelagic chalk cover accumulated later in gradually shallowing seas up to a Mid-Tertiary emergence.

Fault stages

A major objective of this work has been to determine the relationships of metalliferous sediment accumulation to faulting of the extrusive sequence and to use this as a basis for comparisons with modern spreading axes.

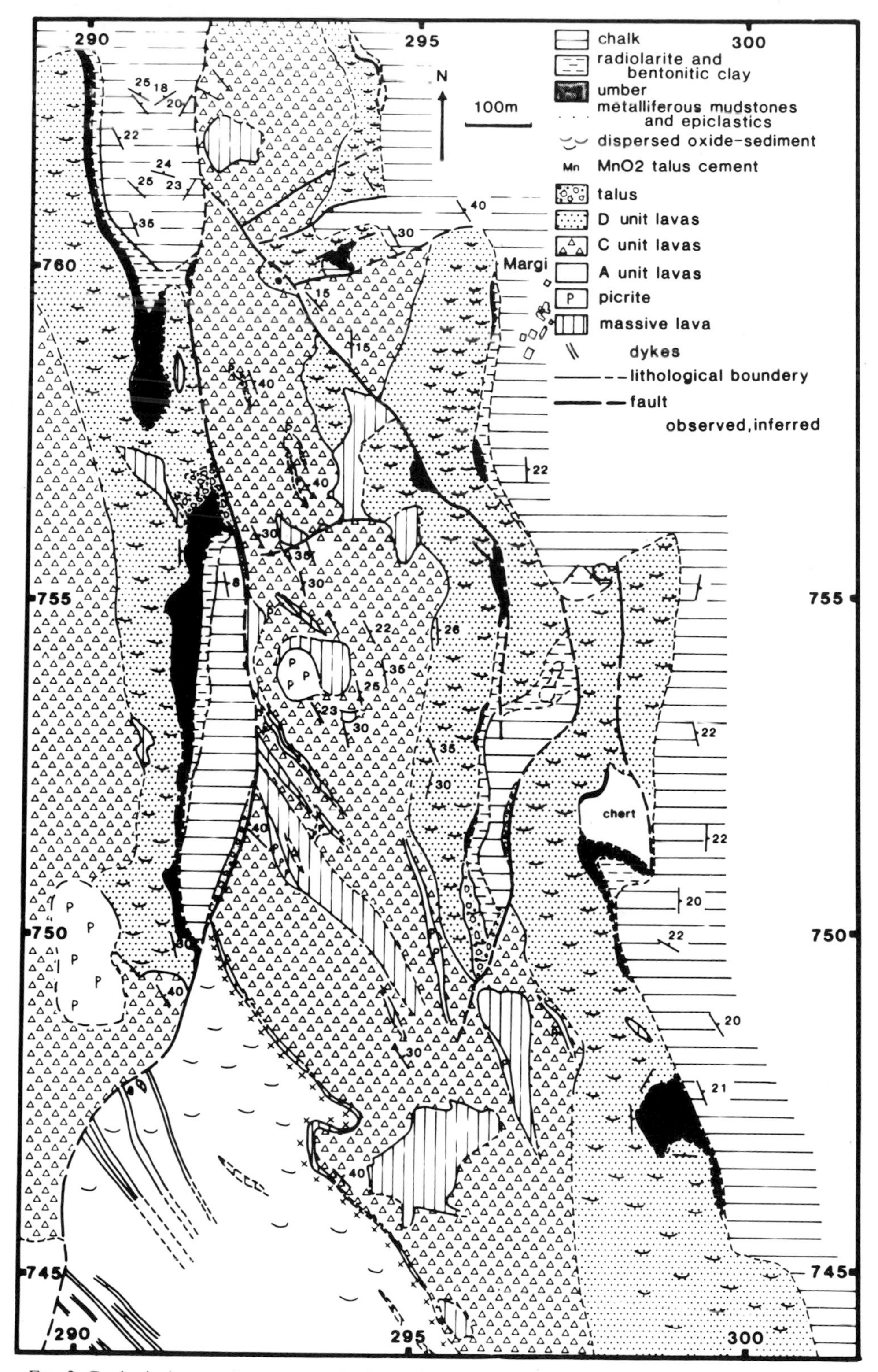

FIG. 2. Geological map of an area south of Margi village to show the relationships of the metalliferous sediments to the lava units and faults. See Fig. 1 for location (grid) and details of the volcanic succession. Base map: 1:5000 topographical series D.L.S. 17 sheets 39/V, 30/XXXIX.

In the mapped area, a major transverse fault and two generations of normal faults predate pelagic sediments above the umbers. The transverse fault is at the SE end of the area and trends NE–SW, perpendicular to the dykes (Fig. 1). An earlier generation of normal faults trends parallel to the dykes (NW–SE) and a later one trends both N–S and NW–SE (Fig. 1). The earliest sediments, the umbers, are found ponded in grabens and half-grabens created by this faulting. The umbers, after allowing for differential compaction, have the same dip as the overlying chalks, showing that faulting and tilting was mostly before umber accumulation.

The transverse fault

The most prominent fault in the area runs close to the southern edge of the Mathiati opencast at a high angle to both generations of normal faults and is also perpendicular to the orientation of dykes in the area mapped. None of the normal faults can be traced across this fault. To the S outside the area mapped, the dykes swing round to parallel this NE–SW-trending fault suggesting that it may have been active early. Since this fault is at right angles to the local dyke trends, we suggest it may have operated as a minor transform fault.

The normal faults

The normal faults are of two generations here termed *type 1* and *type 2*. The type 1 faults displace only the A and B lava units. The type-2 faults displace the type-1 faults and were coeval with eruption of the C and D lava units.

Type 1: earlier normal faults

The earlier normal faults, which cut the A and B lava units, are parallel to the NW–SE dyke trend (Fig. 1). The maximum observed vertical-fault offset is *c.* 160 m, determined by displacement of the B lava unit (Fig. 3, Section D-D′), but offsets could possibly have been as great as 400 m. These earlier faults are parallel to those which bound the Mathiati sulphide body and have often been extensively gossanized. Shearing, which has taken place along these fault planes, is compatible with post-mineralization movement. The field relations discussed above show that the laterally continuous horizon of ferromanganiferous and epiclastic sediment above the A lava unit accumulated on an almost flat sea floor, thus the type-1 faulting cannot have involved significant differential vertical motion prior to this stage. Although the type-1 faults may have existed as fractures from a very early stage, major movement on them is constrained to an interval following the deposition of the sediment blanket, and prior to eruption of the C and D unit lavas.

Type 2: later normal faulting

The C lavas and the type-1 faults are displaced by later faults, mostly trending N–S, but some, particularly in the S of the area, follow the pre-existing NW–SE trend. Several lines of evidence suggest that the N–S faulting took place over a short time interval during eruption of the C and D lava units. Firstly, there is a marked upward decrease in dip and change in direction of strike through the C and D lava units (Fig. 2). Secondly, thin sheet and pillowed flows of the lower part of the *c.* 150 m thick succession give way upward to disrupted pillow lavas intercalated with lava breccia and hyaloclastite, indicating a steeper palaeoslope. Thirdly, upfaulted C unit lavas are locally overlain by the D unit with an unfaulted extrusive contact (Fig. 3). The D lavas did not entirely fill the half-grabens, leaving fault hollows which subsequently became sites of umber accumulation.

There are also the type-2 faults which cut the C–D lava units but are parallel to the NW–SE trend of the earlier type-1 faults. These are best exposed in an outlier of C–D unit lavas, umbers and cover sediments located in the S of the area mapped, 1 km N of Mathiati (Fig. 1, inset; Fig. 3). These faults form small grabens in the C–D unit lavas which also later became sites of umber accumulation. No cross-cutting relations with the N–S trending type-2 faults have been observed.

It is important to note that, although there is a clear relative age difference between the type-1 and the type-2 normal faults, this need not imply a large time separation. The volcaniclastic and metal-oxide blanket beneath the C unit lavas could well record a significant time interval, followed by the normal faulting and extrusion of the C and D lava units.

Stages of spreading-axis construction

The field relations of the Mathiati–Margi area can now be considered in the context of new crust forming and moving away from a spreading axis. The following stages are recognized:

(i) Most of the lava volume was rapidly extruded as pillowed and massive flows very close to the spreading axis. The rapidity of this earlier eruption restricted interlava sediment to minor dispersed oxide-sediment. During a subsequent hiatus, volcaniclastic material was eroded, presumably by sea-floor currents, and accumulated on a more or less flat sea floor.

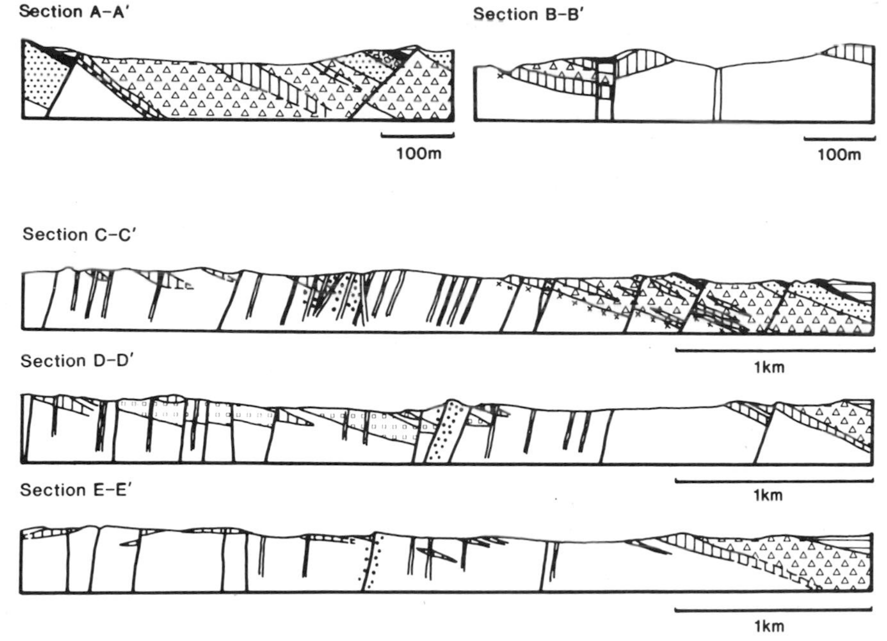

FIG. 3. Geological cross-section to show the relationships of faults to the lava succession. See Fig. 1 for lines of section and the key to symbols. Vertical scale = horizontal scale.

(ii) The sulphide sediments and a halo of metal-oxide sediments were then deposited in an area close to the minor transform fault. Numerous fractures parallel to the spreading axis were mineralized, and the sulphide probably formed in a related shallow graben.

(iii) Blocks of lava 0.5–3 km wide were rapidly rotated parallel to the spreading axis, with vertical throws in the order of 160 m. The sulphide orebody was also dissected and rotated at this stage.

(iv) Basaltic eruptions resumed partially smoothing the topography. At the same time N–S faults became active and blocks up to 0.5 km across were rotated up to 30°. Metal-oxide sediments accumulated during eruption and were dispersed throughout the upper part of the lava succession. Restricted to the south of the area mapped, the later faulting continued after the end of volcanism forming a series of grabens.

(v) The newly formed grabens, and the half-grabens of the north of the area which were not completely filled by the later lavas, were flanked by lava talus and then progressively filled by umber. The solutions from which the umbers were precipitated caused extensive hydrous alteration of the subjacent lavas. During the waning stages of hydrothermal discharge the umbers and the associated lavas were veined by palygorskite and poorly crystalline MnO_2.

(vi) Later, after the spreading axis had migrated away, metalliferous clays accumulated with a greater hydrogenous component (unpublished chemical data), then the remaining hollows were progressively filled with radiolarites, mudstones and pelagic carbonates, although the basement remained exposed in places for up to 20 Ma.

The question which then arises is how typical these stages are of the Troodos Massif as a whole. This excludes the Arakapas fault zone, which is interpreted as a fossil transform fault (Simonian & Gass 1978), with its own distinctive hydrothermal metalliferous oxide-sediments (Robertson 1978). The gross lava succession, the position of the sulphide orebodies and the style of faulting in the Mathiati–Margi area are very similar to many other parts of the Troodos Massif. There is often an abrupt upward change to more mafic compositions in the lavas (Gass 1980). The major sulphide orebodies are mostly, if not all, located beneath these more mafic lavas (Constantinou & Govett 1973). Also Adamides (1980) has demonstrated that the major sulphide mineralization of Kalavasos was controlled by early faulting, although this area is complicated by the close proximity to the Arakapas fault zone. The Mathiati–Margi area is peculiar in its richness of mineralization and metalliferous sediments throughout the lava succession. Away from mining districts hydrothermal sediments are generally scarce and are completely absent from some areas, e.g. the Akaki river section on the northern margin of the Troodos Massif. The implication of the tendency of interlava metalliferous sediments, umbers and sulphides to form at different levels in the lava stratigraphy, but in the same general area, is that zones of elevated hydrothermal activity must have persisted on particular segments of the crust throughout and for some time after lava eruption.

One remaining puzzle is why some of the later (type-2) faults were orientated *c.* 40° clockwise from the trend of the earlier (type-1) faults which had paralleled the inferred spreading axis. This change took place during the later stages of the principal tectonic event, which corresponds to the construction of the rift-valley walls, and was coincident with eruption of the later more mafic lavas.

Metallogenesis at modern spreading axes

We now review the main morphological, structural and metallogenic features of the Atlantic and Pacific spreading axes, which are relevant to interpretation of the Troodos ophiolite.

Atlantic

The Mid-Atlantic Ridge in the TAG area (26°N) is topographically rugged with a median valley *c.* 20–30 km across and 1–2 km deep. Normal faults along the walls of the median valley are inferred to possess throws up to 1 km, locally exposing gabbros (Rona *et al.* 1976). The TAG hydrothermal field comprising manganese deposits, is situated on the inner wall of the rift, 10–12 km from the axis.

In the FAMOUS area (37°N) the axial graben is still 20–30 km across, but shallower (1 km relief). The topography of the inner rift-valley floor is subdued and is controlled by eruption with relatively little faulting. The walls of the median valley are dissected by normal faults spaced 0.5 to 8 km apart with inferred displacements up to 500 m (Ballard & van Andel 1977). Only in transform fault 'A' have hydrothermal deposits been found, these comprising nontronite and manganese oxides (Hoffert *et al.* 1978).

Pacific

At the Galapagos rift, which has an intermediate spreading rate (60 mm yr^{-1} total), there is a narrow median valley 3–8 km across with a total relief of 100–250 m (Lonsdale 1977a; Ballard *et al.* 1982). Normal faults spaced 1–6 km apart, with inferred throws of *c.* 100 m, are formed in a tectonically active zone at the margin of the median valley (*c.* 4 km from the axis). Inactive sulphide deposits are located as 85° 49–50′W along a fault associated with the southern boundary of the north rift valley (Skirrow & Coleman 1982) and at 86° 0–15′W located along both the northern and southern scarps of the rift valley. Active hydrothermal fields, none of high-temperature type, are located immediately on the axis. The hydrothermal mounds, composed of MnO_2 and nontronite, are situated on the sedimentary cover above basement fractures, 20–30 km south of the axis.

The Juan de Fuca ridge, which is also an intermediate-rate spreading axis (60 mm yr^{-1} total), is morphologically similar to the Galapagos rift, with a well-defined median valley *c.* 3 km across and *c.* 100 m deep (Normark *et al.* 1983). The inner walls are composed of low terraces bounded by steep normal faults (throws *c.* 30 m) and define an axial volcanic floor about 1 km across. Along the centre of the valley, which is largely flat and unfaulted, there is a nearly continuous depression 50–200 m across and about 25 m deep. Four of the known sulphides occur in this median depression, and one is located against the scarp at the foot of the eastern inner wall. Low-temperature active vents occur along the median depression.

The much faster-spreading East Pacific Rise at 3°25′S (150 mm yr^{-1} total) differs from the Galapagos by the absence of a well-defined axial graben. Most of the faulting takes place in a zone of extension 1–15 km from the ridge axis (Lonsdale 1977b). In contrast to the Atlantic and the Galapagos spreading axes, the fault blocks are symmetrical horsts and grabens rather than asymmetrical half-grabens (Lonsdale 1977b). Numerous sulphide deposits of high-temperature origin (*c.* 370°C) have been discovered on the EPR at 21°N. The active vents are located in a linear zone 200–300 m wide right on the spreading axis. The sulphides sampled by Francheteau *et al.* (1979) were found in small faulted depressions about 700 m from the axis, with notional ages of 10–20,000 years.

Discussion

Profiles of basement topography on the rift flanks are compared in Fig. 4. The slow-spreading TAG area, with huge fault throws and exposed plutonic rocks, and the fast-spreading EPR, with symmetrical block-faulting are not compatible with the area mapped. The FAMOUS area, and the intermediate spreading axes of the Galapagos

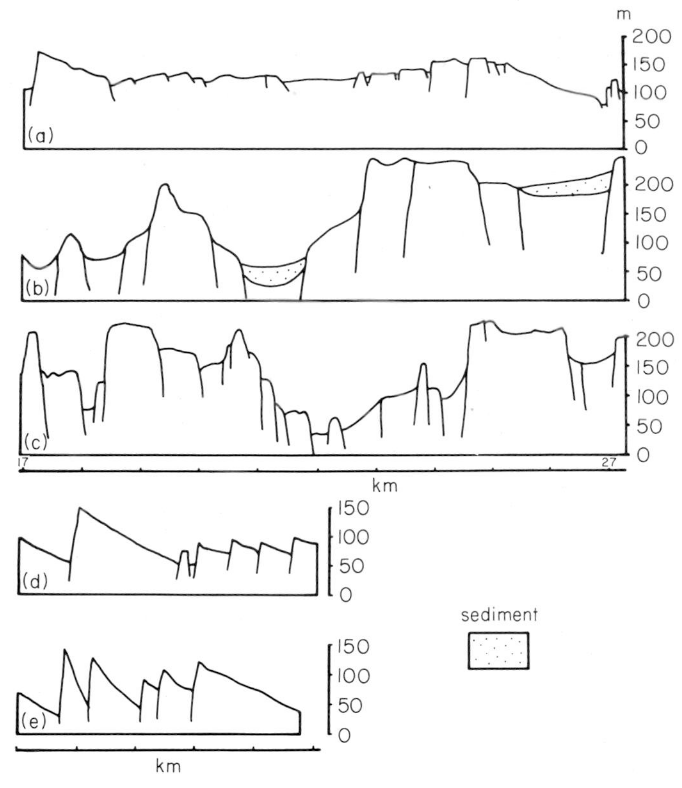

FIG. 4. Comparisons of the topography of modern spreading axes with those inferred from the Mathiati–Margi area of Cyprus. (a) Galapagos (Lonsdale 1977a); (b) FAMOUS area (Macdonald & Luyendyk 1977); (c) East Pacific Rise (Lonsdale 1977b). A–C are shown for 17–27 km from the spreading axis on the ridge flanks. D–E style of basement topography inferred for Cyprus at an equivalent distance from the axis. In each case the ridge is to the left.

and Juan de Fuca ridges which have asymmetrical block-faulting and lower relief compare more closely. The maximum estimated fault-controlled relief of *c.* 160 m for the mapped area compares with 500 m for the FAMOUS area, *c.* 100 m for the Galapagos and 30 m for the Juan de Fuca spreading axes. In modern spreading axes the median-valley walls are uplifted on ridge-parallel faults, apparently in contrast with the two directions of faulting in the Mathiati–Margi area. It is however clear that the area mapped is not large enough to allow a more confident comparison with modern-ridge morphology to be made.

On modern spreading axes the bulk of the lava pile is produced in a narrow relatively unfaulted volcanically active zone around 2 km across where topography is largely controlled by volcanism (Lonsdale 1977a, b; Ballard & van Andel 1977; Ballard *et al.* 1981, 1982). Consistent with earlier interpretations (Smewing *et al.* 1975; Robertson & Fleet, 1976) we envisage the A and B lava units of the Mathiati–Margi area (Lower Pillow Lavas) as having been erupted in a similar narrow, volcanically controlled, axial zone. The massive and pillowed flows were erupted from fissures with little vertical relief. Eruption rates were high enough to preclude much accumulation of pelagic sediments, in contrast, for example, with the higher levels of the extrusive succession of the Semail ophiolite, Oman, which contain pelagic chalks in addition to metalliferous oxide-sediments (Fleet & Robertson 1980). The dispersed sediments in Cyprus consist of mixed-background pelagic sediment and metalliferous oxide-sediment which was hydrothermally precipitated from vents, possibly originating as fine sulphide particles. Volcanism ceased and, as the area moved off the axis still within the axial valley, currents were sufficiently active to erode lava to form a laterally continuous thin blanket of volcaniclastic sediment over a wide area of relatively flat sea floor.

Massive sulphides in the ocean have to date

only been discovered on the Pacific spreading axes. All of the high-temperature vents (*c.* 370°C) are located on the youngest lavas on the spreading axis. Some inactive sulphides are found near major scarps at the edges of the median valley as in the Galapagos rift (Skirrow & Coleman 1982) and the Juan de Fuca ridge (Normark *et al.* 1983). These could either be fossil sulphides transported away from the axis, or else could have formed near the walls of the median valley where greater fracturing and permeability might have favoured hydrothermal discharge (Normark *et al.* 1983). Sulphides formed at the spreading axis are likely to be rapidly buried by later lavas and thus would be expected to be located low in a hypothetical vertical succession of axial lavas. This contrasts with the Mathiati–Margi area, where the massive sulphides are located on top of the lavas interpreted as having formed in the axial valley prior to major faulting. In this regard the sulphides seem to differ somewhat in location from the known 'black smokers' which are more axial. Constraints on the actual distance from the axis that the Cyprus massive sulphides were deposited can be inferred from modern examples, assuming that they were deposited beyond the axial volcanic zone but inside the axial valley. Maximum development of the rift-valley walls is achieved at 10 km from the axis at the FAMOUS area, *c.* 4 km at the Galapagos rift, and *c.* 1.5 km at the Juan de Fuca rift. No sulphides have been found on the slow-spreading axes, so a realistic maximum distance for sulphide formation is *c.* 4 km.

The later C and D unit lavas in the Mathiati–Margi area were erupted during a period of intense fault rotation and differential uplift which can be intrepreted as the construction of the rift-valley walls. The site of eruption would thus be at the margins of the axial graben rather than for example some distance out on the flanks of the ridge. To date, off-axis extrusion has been reported only in the FAMOUS area, but these extrusions are more evolved than their axial counterparts (Bryan & Moore 1977), in contrast with the Troodos. In many parts of the Troodos Massif the traditional Upper Pillow Lava–Lower Pillow Lava break is less marked or apparently absent which could imply either that the later lavas were more nearly axial or that the median valley was less well defined.

Off-axis hydrothermal activity is known from the TAG and the Galapagos mounds area. The TAG hydrothermal vents, now inactive, produced MnO_2 deposits which are located on the inner walls of the rift, 10–12 km from the axis, on crust *c.* 1 m.y. old (Rona *et al.* 1976). In the Galapagos area, the well-known hydrothermal mounds are sited along fractures 20–30 km away from the spreading axis on crust 500,000–700,000 years old, outside the main tectonic zone (Lonsdale 1977a). Unlike the Cyprus umbers, the hydrothermal solutions which formed the mounds percolated through 20–30 m of pelagic carbonate formed beneath a high productivity zone. The umbers are chemically rather less akin to the Galapagos mounds than they are to the EPR crestal sediments (Boström & Peterson 1969), but this may be due in part to the modifying effects of the underlying pelagic sediment at Galapagos (Corliss *et al.* 1978). However, the location is analogous, suggesting that active hydrothermal exhalation precipitating the umbers could have operated in crust up to 0.5–1 m.y. old and up to 30 km from the axis. A minimum distance, given that formation post-dated rift-valley construction, is 4 km. The overlying metal-enriched pelagic clays then formed further from the spreading axis with a hydrogenous component related to the spreading axis.

The question remains as to why the Mathiati–Margi area is one of a number of areas of the Troodos which are metal-enriched at all levels of the extrusives. Although these sediments accumulated close to a major fault, there are plenty of other highly faulted areas of the Troodos which are devoid of mineralization. Modelling of convective circulation by Spooner (1977) has demonstrated the need for local thermal anomalies to impose spatial stability, and he pointed out that large magma chambers would act in this way. Theoretical considerations of the thermal budget of the hydrothermal 'smokers' indicate that the heat source needed to sustain hydrothermal discharge at 370°C must involve magma crystallization (Strens & Cann 1982), which thus implies the existence of magma chambers under the 'smokers'. Even after freezing, magma chambers would remain thermal anomalies which could localize cooler hydrothermal circulation after they had moved off-axis. Unfortunately, no relationship between the plutonic rocks and extrusive sequence can be established in the area mapped. Regardless of the mechanism, however, the localization of hydrothermal sediments at particular sites in the Troodos ophiolite indicates a local persistence of hydrothermal metallogenesis not yet demonstrated at modern spreading axes.

A further point is that the close integration now possible of all the Troodos lavas and sediments into the evolution of spreading axis offers no obvious support for earlier suggestions that the Upper Pillow Lavas formed in some separate tectonic setting from the Lower Pillow Lavas, e.g.

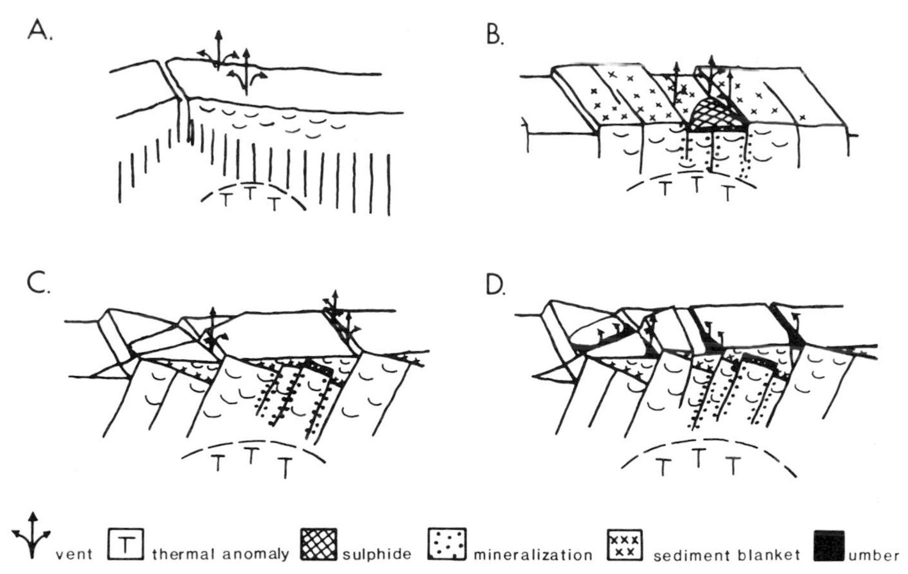

FIG. 5. Summary of evolving Troodos metallogenesis. (a) Axial volcanic zone. Rapid lava extrusion dilutes hydrothermal iron-oxide and background pelagic sediments. (b) Details of part of rift-floor towards the rift walls. Sulphides form together with volcaniclastic sediment during a pause in volcanism. (c) Rift valley wall. Here normal faulting and rotation is accompanied by renewed volcanism. (d) The ridge flanks. After cessation of faulting, the umbers accumulate in fault hollows, which later fill with metal-rich clays and more normal pelagic sediments.

as an off-axis seamount (Gass & Smewing 1973) or as an incipient island arc (Pearce 1975).

Conclusions

The combined field observations and the comparisons with the modern oceans show that the volcanic and structural development of the Mathiati–Margi area of the eastern Troodos Massif (Upper Cretaceous) is compatible with formation at an intermediate spreading rate axis with a well-defined axial graben, as illustrated in Fig. 5. The bulk of the lava pile was erupted on an area of relatively flat sea floor in the axial part of the median valley (traditional Lower Pillow Lavas). During a pause in volcanism submarine erosion produced a thin layer of volcaniclastic sediments. The massive sulphides accumulated in small fault-controlled depressions towards the walls of the axial graben, with a halo of Fe-Mn oxide-sediments. Ferruginous oxides dispersed throughout the lava pile are possibly drifted products of high-temperature sulphide vents.

The walls of the median valley were constructed by two phases of normal faulting, of which the second accompanied the eruption of the later more mafic lavas (traditional Upper Pillow Lavas). The Fe-Mn oxide-sediments, the umbers, were then precipitated in small fault-controlled hollows near the crest of the ridge flanks. Later, as the spreading axis migrated away, pelagic clays accumulated which were enriched in metals by hydrogenous processes, followed by pelagic chalk deposition.

Analogy with modern intermediate-rate spreading axes suggests an axial volcanic zone *c.* 2 km across within an 3–8 km wide rift valley. The inferred origin of most, if not all, the Cyprus massive sulphides above the highest axial lavas differs somewhat from the 'smokers' of modern Pacific spreading axes which mostly appear to have formed immediately on the axis. By analogy with the Galapagos mounds the umbers accumulated on crust of less than 1 Ma, up to several tens of kilometres from the axis. The persistence of both sulphide- and oxide-sediment formation throughout the lava pile in some areas of Cyprus implies a record of continuing hydrothermal activity from the axis to ridge flanks on particular areas of crust. This feature has not yet been reported in the oceans.

ACKNOWLEDGMENTS: The authors wish to thank G. Constantinou and the members of the International Deep Crustal Research Group for helpful discussion in the field. JFB was supported by an NERC studentship; AHFR by Edinburgh University.

ISOTOPE STUDIES
& METAMORPHISM

Modelling of oxygen-isotope data from the Sarmiento ophiolite complex, Chile

D. Elthon, J. R. Lawrence, R. E. Hanson & C. Stern

SUMMARY: The results of an oxygen-isotope investigation of the Sarmiento ophiolite complex show that the extrusive rocks are isotopically enriched in ^{18}O compared to fresh basalts as a consequence of low-temperature interaction with seawater. The basaltic sheeted dykes, dykes cross-cutting plutonic rocks, and the plutonic rocks themselves are depleted in ^{18}O compared to fresh basalts as a consequence of high-temperature interaction with seawater. K_2O contents of pillow lavas indicate minimum water/rock ratios of 15 to 90. Within the sheeted dyke complex and the plutonic section, the inferred water/rock ratios become <0.6. Modelling of basalt–seawater interaction indicates that hydrothermal fluids in equilibrium with actinolite-facies assemblages (525–700°C) have $\delta^{18}O = +5$ to $+7‰$, whereas hydrothermal fluids in equilibrium with greenschist-facies assemblages (230–525°C) have $\delta^{18}O = 0$ to $+5‰$. It appears that an inverse correlation exists between the temperature of metamorphism and the water/rock ratio. This would be anticipated from the cooling effect that circulating seawater has on the oceanic crust.

The recent discovery of hydrothermal vents along the Galapagos Rift (Corliss *et al.* 1979) and the East Pacific Rise (Macdonald *et al.* 1980) supports the hypothesis that active hydrothermal circulation of heated seawater through the oceanic crust is an integral aspect of crustal formation along spreading ridges. Prior to the discovery of these vents, the existence of extensive hydrothermal circulation of seawater in the crust had been inferred from features such as near-ridge metalliferous deposits (e.g. Bonatti & Joensuu 1966), heat-flow anomalies along oceanic ridges (e.g. Williams *et al.* 1974), and the abundant recovery of metamorphic rocks from the oceanic basins (e.g. Melson & van Andel 1966; Bonatti *et al.* 1975; Humphris & Thompson 1978).

It is now generally accepted that most ophiolite complexes represent uplifted sections of oceanic or marginal basin crust (Coleman 1977). Due to the technological difficulties involved in sampling of the crust in the present-day oceanic basins, petrological and geochemical investigations of ophiolite complexes have often been undertaken in an effort to develop or refine models for various aspects of crustal formation in the oceanic basins. In this paper we present results from an oxygen-isotope investigation of metamorphic rocks from the Sarmiento ophiolite complex in southern Chile and we develop a water/rock interaction model for oxygen-isotope changes based on observed mineral assemblages in the rock. The oxygen-isotope results are discussed in the framework of this model.

Geology of the Sarmiento complex

A discontinuous belt of mafic rocks, interpreted as fragments of a Mesozoic marginal basin, extends along the spine of the Andean Cordillera in southern Chile (Katz 1972; Dalziel *et al.* 1974). The Sarmiento ophiolite complex, at latitude 51°S, is the northernmost-known exposure of these ophiolite bodies. The general field relationships and petrology of the Sarmiento complex have been described elsewhere (deWit & Stern 1981; Dalziel 1981, and references therein). In the present study, the rocks selected for analysis were collected from the region adjacent to and between Fjordo Lolos and Fjordo Encuentros (see map in deWit & Stern 1981), where the field relationships and petrology are best documented and where the sheeted dykes and plutonic rocks crop out.

In this region, the internal stratigraphy of the Sarmiento complex is characterized by a thick (>2 km) extrusive section, which contains pillowed and massive basalts and locally abundant hyaloclastites. This extrusive section grades downward in an irregular manner into a sheeted dykes section (~300 m thick) that, in turn, grades progressively downward into the plutonic section (>1 km thick). The plutonic section consists of gabbros, gabbronorites, norites and rare trondhjemites that are cross-cut by basaltic and trondhjemitic dykes (Stern & Elthon 1979). The deeper levels of the plutonic section, such as the transition zone and cumulate ultramafic assemblages that are commonly observed in most ophiolite complexes, are not observed in the Sarmiento complex. The absence of these rocks is probably due to insufficient uplift of the ophiolite during closure of the marginal basin (Saunders *et al.* 1979).

Petrography and metamorphic assemblages

Previous studies on metamorphosed rocks from the present-day oceanic basins and ophiolites (Melson & van Andel 1966; Gass & Smewing 1973; Bonatti *et al.* 1975; Humphris & Thompson 1978; Stern & Elthon 1979; and many others) have noted that there are many general similarities between the mineralogical assemblages produced by metamorphism in continental terrains and those assemblages produced during metamorphism within the oceanic crust. Elthon & Stern (1978) and Elthon (1981) have proposed the following facies classification for ocean-floor metamorphism:

(i) Lower greenschist facies: (230–320°C) epidote + chlorite + albite + sphene ± calcite ± quartz.
(ii) Upper greenschist facies: (320–525°C) actinolite + epidote + chlorite + albite + sphene ± quartz.
(iii) Lower actinolite facies: (525–600°C) actinolite + calcic plagioclase + sphene.
(iv) Upper actinolite facies: (600–700°C) actinolite + calcic plagioclase + magnetite + ilmenite.

The amphibole in the Sarmiento metabasalts and metagabbros and in many other examples of high-temperature ocean-floor metamorphism is generally a fibrous, green actinolite rather than the brown hornblende that is present in the classically defined amphibolite facies from continental terrains (e.g. Miyashiro 1973). Consequently, this calcic plagioclase and actinolite assemblage has been termed the 'actinolite facies' in order to avoid confusion with the 'amphibolite facies' (hornblende + calcic plagioclase). Because this actinolite-facies assemblage has been reported from other, low-pressure contact metamorphism terrains (e.g. Shido 1958; Seki 1961; Loomis 1966), it is clear that it is not restricted to the oceanic basins, even though it is more common there than in continental terrains. It should also be noted that there are examples of true amphibolites or amphibolite-facies rocks from the oceanic basins and ophiolites.

The complex petrographic features that characterize these rocks from the Sarmiento complex have been discussed by Elthon & Stern (1978) and Stern & Elthon (1979). For the sheeted dykes, dykes that cross-cut plutonic rocks, and the extrusive rocks, all of the original igneous minerals that were present in these rocks have been replaced by metamorphic minerals. There are, however, relict cores of unaltered ortho- and clino-pyroxenes in some of the plutonic samples.

As a rule, most of these Sarmiento rocks are characterized by complex retrograde metamorphic textures that we have interpreted to result from the incomplete transformation of actinolite-facies assemblages to greenschist-facies assemblages in a decreasing geothermal gradient during transport away from the spreading axis. The preservation of these retrograde textures is attributed to the limited access of water at increasing distances from the spreading axis, rather than dramatic differences in temperature at the centimetre scale (Elthon & Stern 1978; Stern & Elthon 1979).

Analytical techniques

The oxygen from these samples was extracted for isotopic analysis by the technique of Clayton & Mayeda (1963) without conversion of the oxygen gas to carbon dioxide. The $^{18}O/^{16}O$ ratio of the extracted gas was analysed with a Nuclide 6″ mass spectrometer. Reproducibility of $\delta^{18}O$ in numerous analyses of our laboratory isotope standard, a powdered sample of the Palisades Sill diabase (PS-1), was better than ±0.2‰. All data are reported in Table 1 and Fig. 1 as the $\delta^{18}O$‰ difference from standard mean ocean water (SMOW).

Results

All of the samples from the *extrusive* section of the ophiolite complex that were analysed in this study (see Table 1 and Fig. 1) are enriched in $\delta^{18}O$ relative to fresh, unaltered oceanic basalts, which have $\delta^{18}O = 5.8 \pm 0.3$‰ (Taylor 1968; Muehlenbachs & Clayton 1972). These results are similar to those reported from other ophiolite complexes—E. Liguria (Spooner *et al.* 1974), Troodos (Spooner *et al.* 1974, 1977; Heaton & Sheppard 1977), Semail (Gregory & Taylor 1981)—and samples from the present-day oceanic basins (Muehlenbachs & Clayton 1976; Muehlenbachs 1980; Lawrence & Drever 1981; Stakes & O'Neil 1982). In all of these cases, the observed ^{18}O enrichment has been interpreted as the result of low-temperature (<300°C) interaction of basalts with seawater.

Those samples from the sheeted dykes, the dykes that cross-cut plutonic rocks, and the plutonic rocks themselves are, with few exceptions, depleted in $\delta^{18}O$ relative to unaltered rocks (see Table 1 and Fig. 1). The exceptions to this rule are, for the most part, trondhjemites and other silica-rich rocks. The analytical results on gabbroic rocks from Troodos (Spooner *et al.* 1977; Heaton & Sheppard 1977), Semail (Gregory & Taylor 1981), and Sarmiento are all similar, with $\delta^{18}O$ between +4.0 and +6.0‰ for most samples. (In the Sarmiento complex, all of the samples with $\delta^{18}O$ <3.8 have abundant (≳25%) titanomagnetite grains). In ophiolites, this transition from $\delta^{18}O$-enriched rocks in the extrusive section to the $\delta^{18}O$-depleted rocks generally occurs in the vicinity of the sheeted dykes unit. In the Sarmiento and Troodos ophiolites the sheeted dykes, for the most part, are depleted in $\delta^{18}O$ whereas the sheeted dykes in the Semail ophiolite are enriched in $\delta^{18}O$.

TABLE 1. *Oxygen-isotopic analyses*

Sample No.	Description	Facies*	Composition†	$\delta^{18}O$
Extrusives				
PA 56D	pillow	G	basic	+ 8.5
PA 56E	pillow	G	intermediate	+10.3
PA 55	pillow	G	acid	+ 7.5
PA 56	dyke	G	basic	+ 6.4
FL 70A	flow	G	basic	+ 7.4
FL 75E	flow	Z	acid	+10.4
PA 57	tuff	G	basic	+ 9.6
Sheeted dykes				
PA 28V	dyke	G	basic	+ 5.1
PA 28B	dyke	G	basic	+ 4.7
PA 28T	dyke	G	basic	+ 4.8
PA 23I	dyke	G	basic	+ 5.0
PA 23H	dyke	A	basic	+ 5.2
PA 23R	dyke	G	basic	+ 5.4
PA 28P	dyke	A	basic	+ 4.0
PA 23U	dyke	A	basic	+ 3.3
PA 23Y	dyke	G	basic	+ 5.4
PA 28K	dyke	G	basic	+ 4.7
PA 28AA	dyke	G	acid	+ 6.2
FL 43E	dyke	G	acid	+ 7.9
PA 28Z	dyke	G	acid	+ 8.1
Dykes cross-cutting plutonic rocks				
PA 32C	dyke	A	basic	+ 4.7
PA 32A	dyke	A	basic	+ 4.3
PA 32I	dyke	A	basic	+ 4.2
FL 2	dyke	A	basic	+ 4.3
PA 32H	dyke	A	basic	+ 4.1
PA 43H	dyke	A	basic	+ 3.8
PA 32G	dyke	G	basic	+ 5.6
PA 32J	dyke	G	basic	+ 4.6
PA 43F	dyke	A	basic	+ 3.0
PA 51A	dyke	A	basic	+ 4.7
PA 32F	dyke	A	basic	+ 4.4
PA 32K	dyke	G	basic	+ 5.3
PA 43A	dyke	G	basic	+ 3.9
PA 51C	dyke	G	intermediate	+ 5.3
PA 32E	dyke	A	acid	+ 7.9
Plutonic rocks				
PA 31K	metagabbro	A	basic	+ 6.1
PA 31P	metagabbro	A	basic	+ 4.8
PA 31N	metagabbro	G	basic	+ 5.0
PA 31A	metagabbro	A	basic	+ 3.1
PA 31B	metanorite	A	basic	+ 2.6
PA 25E	metagabbro	A	basic	+ 5.0
PA 24A	metagabbro	A	basic	+ 4.0
PA 31E	metanorite	A	basic	+ 5.6
PA 25B	metagabbro	A	basic	+ 4.3
PA 30E	metagabbro	A	basic	+ 4.1
PA 30H	metagabbro	A	basic	+ 3.8
PA 25A	metagabbro	G	basic	+ 4.0
PA 49G	trondhjemite	G	acid	+ 7.5
PA 42B	trondhjemite	G	acid	+ 7.6
PA 47C	trondhjemite	G	acid	+ 7.3
PA 34A	trondhjemite	G	acid	+ 7.1
PA 25D	crustal xenolith		acid	+ 5.6
PA 42A	crustal xenolith		intermediate	+ 6.4
PA 24A	crustal xenolith		intermediate	+ 4.0

*Facies: A – actinolite; G – greenschist; and Z – zeolite.
†Composition: basic $<53\%$ SiO_2; intermediate 53–66% SiO_2; and acid $>66\%$ SiO_2.

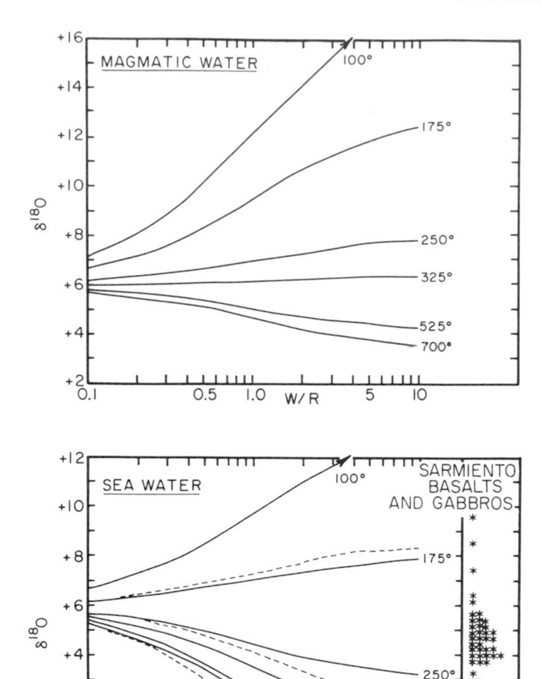

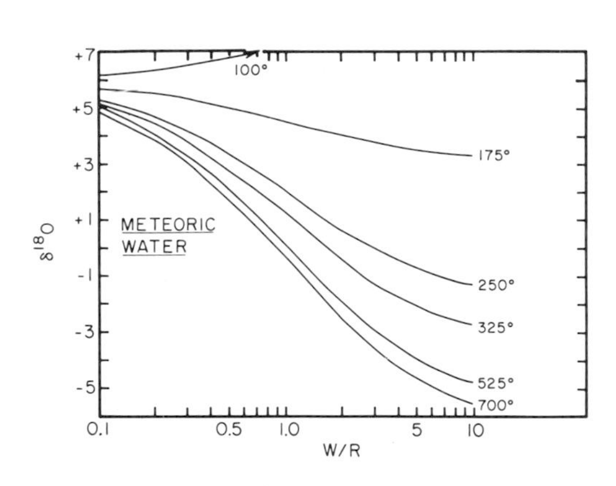

FIG. 2. Modelling of the $\delta^{18}O_{rock-final}$ *v.* water/rock ratio for basalts interacting with hydrothermal fluids (magmatic water, $\delta^{18}O_{water-initial} = +5‰$; seawater, $\delta^{18}O_{water-initial} = 0‰$; meteoric water, $\delta^{18}O_{water-initial} = -5‰$) at various temperatures. The solid curves shown are for closed-system interaction. Curves for open-system interaction can be calculated from $W/R_{open} = \log_e[(W/R)_{closed} + 1]$. The dashed curves are examples at 175, 250 and 700°C for seawater interaction.

The comparison of the observed $\delta^{18}O$ values in the metabasalts and metagabbros with those predicted by the model indicates that either seawater or magmatic water are the likely hydrothermal fluids (see Fig. 2). The interaction of large volumes of meteoric water with the Sarmiento complex during hydrothermal metamorphism would tend to produce basalts that are extremely depleted in $\delta^{18}O$. Rocks of this type have not been found within the Sarmiento complex and it is unlikely that meteoric water was a major component in the hydrothermal fluid.

Seawater is believed to be the dominant hydrothermal fluid accompanying metamorphism in the Sarmiento complex. The wide extent of the hydrothermal metamorphism in the Sarmiento complex indicates that magmatic water played only a minor role in this metamorphism due to the small quantities of water in oceanic basalts ($\lesssim$0.30 wt%: Moore 1972; Delaney *et al.* 1978). The amount of water in the greenschist-facies metabasalts from Sarmiento is generally 2–4%, with the actinolite-facies rocks containing 1–2% water (our unpublished data). Consequently, the amount of magmatic water that could have been derived from the solidification of basaltic magma is probably <10% of the total water presently bound within the metamorphic rocks themselves.

The very high K_2O (average of 2.14%) and Rb (average of 77 ppm) contents of the extrusive section (Stern & Elthon 1979) require that the hydrothermal fluid is a source of these two elements. These high K_2O and Rb contents in the extrusive rocks are manifested in the development of potassium feldspars and clay minerals in the metabasalts. It is likely that this cation-bearing fluid was seawater because very similar effects have been noted in experimental studies of basalt–seawater interaction and studies of metamorphosed extrusive rocks (Hart *et al.* 1974; Donnelly *et al.* 1979).

In terms of the volumetric abundances of water that are required, the K_2O and Rb abundances, and the predicted shifts in $\delta^{18}O$ of the metabasalts, seawater best fits the requirements for the hydrothermal fluid that accompanied metamorphism in the Sarmiento complex. Nonetheless, it must be considered that the areal extent of the marginal basin in which the Sarmiento complex formed is poorly known (Dalziel 1981), although it is believed to be small. It may be that the isotopic composition of the water in this marginal basin was slightly lower than seawater in the open oceanic basins due to the influx of fresh water from the drainage basins in the adjacent island arc and continental margin. This effect should not lower the $\delta^{18}O$ of the water in the marginal

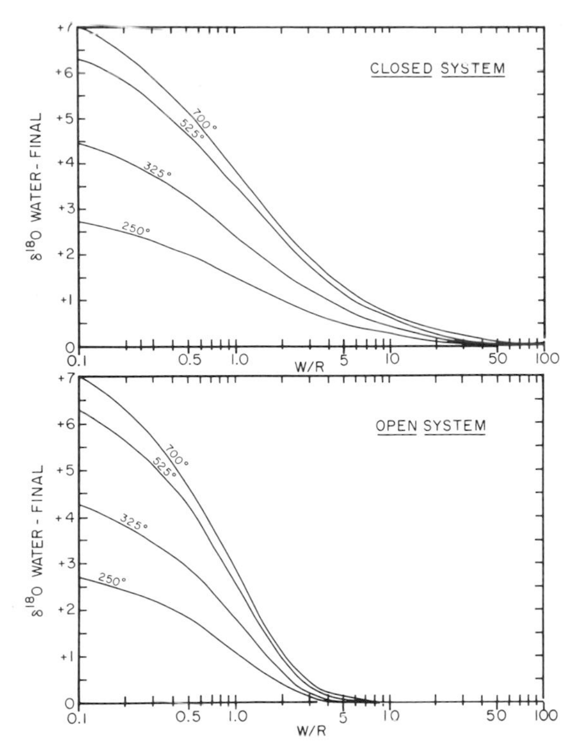

FIG. 3. The modelling of $\delta^{18}O_{water-final}$ *v.* water/rock in seawater interacting with basalts at various temperatures (°C). Note that these curves for open- and closed-system interaction are calculated by the model described in the text ($\delta^{18}O_{rock-initial} = +6‰$, $\delta^{18}O_{water-initial} = 0‰$). (See text for details.)

basin by more than 1.5‰, if one assumes hydrological conditions similar to those in the Long Island Sound (Riley 1956). For the following evaluation of the effects of rock–water interactions, this possible effect is ignored.

Estimates of the water/rock ratio

One might anticipate that the $(W/R)_A$ is quite variable in both time and space during metamorphism of the oceanic crust. In the proposed models for ocean-floor metamorphism, where seawater penetrates into the crust, causing chemical and mineralogical alteration, and is vented onto the floor of the oceans, the $(W/R)_A$ (integrated over the life of the system) would, in a general sense, be related to the stratigraphic level within the crust, i.e. $(W/R)_A$ is expected to be high within the upper portions of the crust, but decreases with depth as the effects of metamorphism diminish. Below, we present evidence indicating that the $(W/R)_A$ varies from 15 to 90 in the extrusive section where low-temperature metamorphism occurs to <0.6 in the sheeted dykes and plutonic section where high-temperature metamorphism occurs. In detail, however, the situation is probably very complicated, with the effects of metamorphism locally controlled by the permeability and thermal history of the crust. Especially important in this regard are fault zones and other areas of (assumed) high permeability where the penetration of water and its movement within the crust are particularly effective in cooling and altering the crust.

Large values of $(W/R)_A$ during metamorphism are inferred for the extrusive section of the Sarmiento ophiolite on the basis of potassium contents and the presence of calcite and quartz veins. As noted above, the K abundances in oceanic and ophiolitic basalts may be severely modified by interaction with seawater (Hart 1970; Hart *et al.* 1974; Coish 1977). Recent investigations of the hydrothermal alteration of basalts show that at high temperatures (>250°C) K is removed from the basalt (Mottl & Holland 1978), whereas K is enriched in the basalt at lower temperatures (Donnelly *et al.* 1979). Stern & Elthon (1979) suggest that the K_2O contents of basalts from the Sarmiento complex averaged 0.30% prior to metamorphism. Assuming that seawater has 380 ppm (Brewer 1975) and that all of the K is retained by the metabasalt, the $(W/R)_A$ for these extrusive ranges from 15 to 90. This

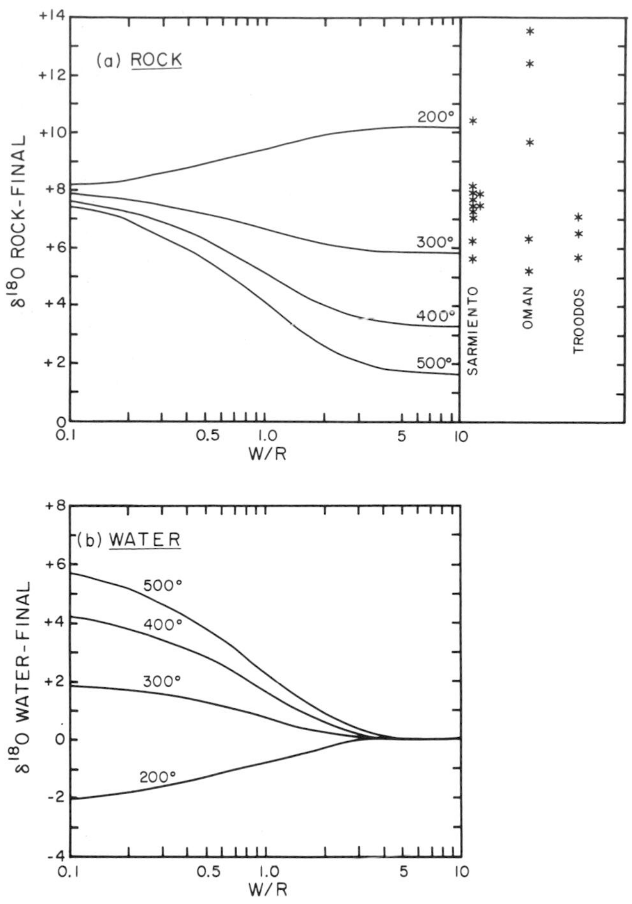

FIG. 5. Modelling of trondhjemite–seawater (open-system) interaction at various temperatures. These curves are calculated by the method described in the text assuming that the $\delta^{18}O_{trondhjemite\text{-}initial}$ was +8‰. Figure 5a compares the analytical results on trondhjemites from the Sarmiento, Oman, and Troodos ophiolites. Figure 5b shows the isotopic composition of seawater after equilibration under these conditions. These figures suggest that most of the trondhjemites were hydrothermally metamorphosed at >250°C and that the isotopic composition of the fluid was >0‰ (and probably less than +5‰). (See text for details.)

Contrary to the situation in the Peninsular Ranges Batholith, it is likely that the ^{18}O of the magma in Sarmiento *increased* during differentiation. In the Sarmiento complex, many of the plutonic rocks contain >25% magnetite-ilmenite crystals indicating that large quantities of magnetite-ilmenite formed during the final stages of crystallization of ferrobasaltic magmas. Because magnetite has much lower $\delta^{18}O$ than co-precipitating pyroxenes, amphiboles, feldspars, or quartz (Friedman & O'Neil 1977), crystallization of significant amounts of magnetite will increase the $\delta^{18}O$ of the fractionating magma. We suggest, therefore, that in the Sarmiento complex the abundant magnetite ± ilmenite in the plutonic rocks implies significant enrichment of ^{18}O during the final, extreme stages of magmatic differentiation. The extent of this ^{18}O enrichment during differentiation is dependent upon the parameters one chooses to model these processes, but we suggest a range of ~+7 to +8.5‰ for most petrologically reasonable alternatives. In fact, just such an enrichment from +5.7 to +7.1‰ by magmatic differentiation has been documented by Muehlenbachs & Byerly (1982) in unmetamorphosed volcanic rocks from the Galapagos Spreading Centre.

In accordance with the above discussion, we present an alternative scenario for trondhjemite–seawater interaction in Figs 5a and b. For this model $\delta^{18}O_{rock\text{-}initial}$ is assumed to be +8.0‰, with the mineral proportions and fractionation factors that are used listed in Table 2. Figures 4a and 5a are similar to each other, as are 4b and 5b. In

Fig. 5a, the temperatures of interaction for most of the Sarmiento (as well as Semail and Troodos) trondhjemites are >250°C, i.e. their $\delta^{18}O_{rock-final}$ is <+8.0. These trondhjemites would be in equilibrium with hydrothermal fluids having $\delta^{18}O$ ranging from 0 to ~+5, similar to the hydrothermal fluids in equilibrium with the abundant basaltic rocks (~1 to +7‰).

Summary and conclusions

1. The imprint of ocean-ridge metamorphism on the Sarmiento ophiolite complex is characterized by large variations in metamorphic grade over short distances. This variation is reflected in metamorphic mineral assemblages, oxygen-isotopic ratios, and bulk-rock chemistry. The oxygen-isotopic variations reflect the interaction of seawater with basaltic and gabbroic rocks over a wide range of temperature and water to rock ratios.
2. A model that accounts for the individual preference of minerals to concentrate ^{18}O during metamorphism of basalts is presented. This model describes the oxygen-isotope shifts that accompany hydrothermal ocean ridge metamorphism more adequately than a model using a single mineral such as plagioclase to represent basalts.
3. This modelling of basalt–seawater interaction at elevated temperatures indicates that the modified seawater in equilibrium with greenschist-facies metabasalts would be similar to those hydrothermal fluids collected from vents on the East Pacific Rise.
4. Modelling of basalt–seawater interaction at elevated temperatures also indicates that the modified seawater in equilibrium with actinolite-facies metabasalts or metagabbros would have a similar $\delta^{18}O$ (+5 to +7‰) as 'magmatic water'. The notion that hydrothermal fluids identified as 'magmatic water' must be directly derived from the mantle is thereby questioned, particularly when only oxygen-isotope data are available.
5. We consider it likely that the $\delta^{18}O$ of the magmas increased during the extreme differentiation processes that produced the trondhjemites. This increase in $\delta^{18}O$ is probably similar to that noted by Muehlenbachs & Byerly (1982).
6. Although the pattern of oxygen-isotopic variation within the Sarmiento ophiolite has certain common characteristics with those patterns measured in other ophiolites, there are some differences. These differences may contain valuable information about the intrusive history, water circulation patterns, tectonics and ocean-floor spreading rates in the particular ophiolite being studied.

ACKNOWLEDGMENTS: Discussions with Ian Dalziel, Ian Ridley and Maarten deWit on the petrology and the geology of the Chilean ophiolites are appreciated. We would like to thank J. Stroup, M. A. Skewes and J. Logsden for assistance in the field and/or laboratory. This work has been supported by National Science Foundation Grants OCE-75-02968, OCE-76-81952, OCE-77-24819, OCE-80-24044, EAR-76-82456, and OPP-74-21415, and EAR-80-26445. Travel to London to present these results was supported by the University of Houston and OCE-81-21232. L-DGO Contribution No. 3469.

References

BONATTI, E. & JOENSUU, O. 1966. Deep sea iron deposits from the South Pacific. *Science* **154**, 643–5.

——, HONNOREZ, J., KIRST, P. & RADICATI, F. 1975. Metagabbros from the Mid-Atlantic Ridge at 06°N: contact-hydrothermal-dynamic metamorphism beneath the axial valley. *J. Geol.* **83**, 61–78.

BOTTINGA, Y. & JAVOY, M. 1975. Oxygen isotope partitioning among the minerals in igneous and metamorphic rocks. *Rev. Geophys. Space Phys.* **13**, 401–8.

BREWER, P. G. 1975. Minor elements in sea water. *Chemical Oceanography* **1**, 415–96.

CLAYTON, R. N. & MAYEDA, T. K. 1963. The use of bromine pentafluoride in the extraction of oxygen from oxides and silicates for isotopic analysis. *Geochim. Cosmochim. Acta* **27**, 43–52.

——, O'NEIL, J. R. & MAYEDA, T. K. 1972. Oxygen isotope exchange between quartz and water. *J. Geophys. Res.* **77**, 3057–66.

COISH, R. A. 1977. Ocean floor metamorphism in the Betts Cove ophiolite, Newfoundland. *Contrib. Mineral. Petrol.* **60**, 255–70.

COLEMAN, R. G. 1977. *Ophiolites: Ancient Oceanic Lithosphere?* Springer-Verlag, New York, 229 pp.

CORLISS, J. B., DYMOND, J., GORDON, L. I., EDWARD, J. M., VON HERZEN, R. P., BALLARD, R. D., GREEN, K., WILLIAMS, D., BAINBRIDGE, A., CRANE, K. & VAN ANDEL, T. H. 1979. Submarine thermal springs on the Galapagos Rift. *Science* **203**, 1073–83.

CRAIG, H. 1980. Geochemical studies of the 21°N. EPR hydrothermal fluids. *Trans. Am. Geophys. Un.* **61**, 992.

DALZIEL, I. W. D. 1981. Back-arc extension in the southern Andes: a review and critical reappraisal. *Phil. Trans. R. Soc. Lond.* **A300**, 319–35.

——, DEWIT, M. J. & PALMER, K. F. 1974. Fossil marginal basin in the southern Andes. *Nature, Lond.* **250**, 291–4.

DELANEY, J. R., MUENOW, D. W. & GRAHAM, D. G. 1978. Abundance of water, carbon, and sulfur in

Oxygen-isotope and geochemical characterization of hydrothermal alteration in ophiolite complexes and modern oceanic crust

D. S. Stakes, H. P. Taylor, Jr & R. L. Fisher

SUMMARY: Stable isotopic, geochemical, and mineralogic variations in plutonic and hypabyssal rocks from oceanic crust (mainly from the Indian Ocean) and ophiolitic terranes (principally the Semail complex, Oman) are very similar. Several stages of seawater–oceanic crust interaction are recognized in these gabbros and diabases. Isotopic and chemical compositions of secondary mineral assemblages reflect changing temperatures and water-rock ratios, and record the effects of (i) pervasive seawater-hydrothermal circulation associated with the main stage of crustal formation at an oceanic spreading centre, (ii) subsequent alteration associated with off-axis volcanism (upper pillow-lava sequences), and (iii) progressively lower-temperature alteration associated with hydrothermal 'ageing' of the oceanic crust. Along the contact between the high-level, isotropic gabbro of the ophiolite, and the overlying sheeted dyke complex, repeated stoping of hydrothermally altered roof rocks into the magma chamber appears to be a ubiquitous process. Stoping is a maximum where the roof is intruded by large (off-axis) gabbro-diorite-plagiogranite bodies which may be 60% xenoliths. Abundant quartz-epidote-sulphide veins originate near these silicic intrusions and alter the overlying sheeted dyke complex. Diabase from the sheeted dyke complex in Oman, typically much more altered than dredged oceanic rocks of similar texture, exhibits the integrated effects of both the axis and off-axis hydrothermal systems.

The aim of this paper is to directly compare the textural, mineralogical and isotopic characteristics of rocks from ophiolite complexes with analogous rocks collected from the modern sea floor. The data presented here are the initial results of a detailed, integrative study of various phenomena associated with the hydrothermal alteration of both oceanic and ophiolitic plutonic rocks. These data provide insight into: (i) the depth to which seawater penetrates into various kinds of oceanic crust. (ii) the resulting profile of $\delta^{18}O$ and secondary minerals in such oceanic crust. (iii) the effect of tectonic deformation and/or proximity to possible higher permeability zones (e.g. transform faults) on the pattern of seawater alteration. (iv) The bulk chemical and isotopic changes that take place as a result of alteration. (v) The possible importance of stoping of hydrothermally altered roof-rocks in modifying late-stage, magmatic differentiates in the oceanic crust.

Direct comparisons between the ophiolite complexes and 'normal' sections of oceanic crust are somewhat ambiguous, because many of the ophiolitic terranes may have formed in a back-arc-spreading environment or in narrow, intracratonal rift zones such as the one found today in the Red Sea. By concentrating our studies on the Semail ophiolite in Oman, we feel we are working on probably the best-exposed intact piece of obducted oceanic lithosphere. The thickness, stratification, lithology, and geologic setting of the Semail ophiolite all correspond extremely well with 'normal' oceanic crust (as documented by the various papers in the Oman Ophiolite Volume of the *J. Geophys. Res.* **86**, 2495–782). However, we note that Pearce *et al.* (1981) and Alabaster *et al.* (1982) believe that the Semail complex formed in a back-arc oceanic environment.

Even direct sampling of modern oceanic crust does not necessarily give a complete picture of what a 'normal' section of MORB-type crust looks like. Most available oceanic samples of plutonic rocks have been recovered by dredging the steep scarps of cross-cutting fracture systems. Such tectonically active regions may involve hydrothermal alteration patterns that are different from 'normal', intact oceanic crust. In some occurrences the crustal section exposed in fracture zones may also be much thinner than intact crust (Fox & Gallo 1982).

Previous work

In this study emphasis is placed on characterizing the mineralogical and isotopic features of crystalline rocks—mostly gabbros, diabases and associated silicic differentiates—from the modern ocean basins. Such data are sparse because of limited sample availability. Most samples we have worked on were dredged from large fracture zones, taken as marking transform faults that cross-cut the Indian Ocean Ridge System (Fig. 1). These samples and the associated localities were previously described in detail in a petrographic-

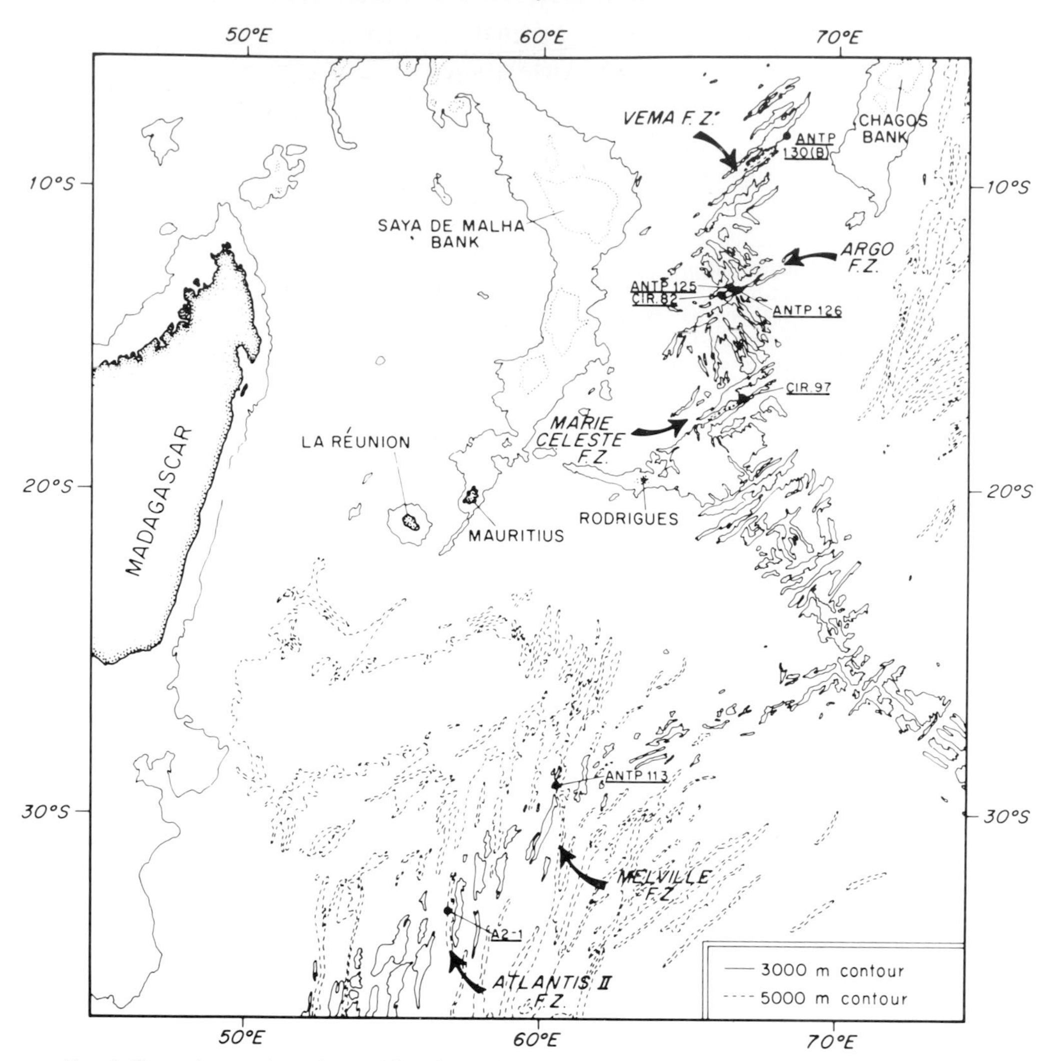

FIG. 1. General tectonic setting and locations of samples dredged from the western Indian Ocean. CIR represents CIRCE Expedition 1968: ANTP represents ANTIPODE Expedition 1970–71.

chemical-mineralogic paper by Engel & Fisher (1975): the Vema Fracture Zone at 9°S, the Argo Fracture Zone near 13°30′S, and the Marie Celeste Fracture Zone near 17°30′S, all trending ENE on the moderately slowly spreading Central Indian Ridge; and the Melville and Atlantis II Fracture Zones that trend N–S on the slowly spreading SW Indian Ridge (Fig. 1). Locally detailed and well-located dredging in this structurally complex area has yielded abundant plutonic rocks representing all the various rock types characteristic of tholeiitic magmas including lherzolite and minor harzburgite, orthopyroxenite, olivine-, clinopyroxene- and two-pyroxene gabbro, Ti-rich-ferrogabbro, norite, and anorthosite. Associated diabase is granophyric and cut by late-stage dykelets of granodiorite and, more commonly, plagiogranite. This bountiful collection of varied plutonic rocks was supplemented by two samples from the Marianas Trough, described by Bloomer & Hawkins (1983), and several from the Mid-Atlantic Ridge (MAR). The isotopic and chemical systematics observed in this sample set are further supplemented by references to previous data on samples from the East Pacific Rise (EPR) and the MAR described by Stakes & O'Neil (1982).

By transporting heat and mobilizing chemical constituents, seawater cools newly formed oceanic crust, chemically interacts with this crust, and finally discharges on the sea floor to form metal-rich deposits. Deposits of hydrothermal origin include Fe- and Mn-rich sediment mounds (e.g. near the Galapagos, Corliss *et al.* 1979), pyrrhotite and talc-rich sediments—e.g. in the Guaymas Basin, Lonsdale *et al.* (1980)—and zinc- and copper-rich sulphides that build high-temperature vents along the axial faults of active spreading centres (Hekinian *et al.* 1980; Hayman & Kastner 1981) and on off-axis seamounts (Lonsdale *et al.* 1982). Such interactions have also been shown to be responsible for the formation of similar metal-rich deposits observed in many ophiolite complexes (Coleman 1977; Heaton & Sheppard 1977; Alabaster *et al.* 1980; Fleet & Robertson 1980).

Geochemical and field observations from the Cretaceous Semail ophiolite (Oman) provide one of the best working models of a sub-sea-floor hydrothermal system (Alabaster *et al.* 1980; Gregory & Taylor 1981). Previous oxygen-isotopic data from the southern half of the ophiolite document a pervasive subsolidus exchange between the rocks and seawater, at a range of temperatures and water–rock ratios (Gregory & Taylor 1981). In this southern region, the base of the gabbro section is intact, and the maximum depth of seawater penetration could be mapped on the basis of the oxygen-isotopic compositions of the rocks. These results show that locally seawater penetrated below the Moho into the tectonized peridotite, but more commonly, only very small quantities of heated seawater reached the base of the layered gabbros. In the northern half of the Semail ophiolite, the basal part of the section is commonly disrupted by large faults, further disrupted by a fossil transform zone in the Wadi Ragmi section (Smewing 1980) and also intruded by large gabbro-diorite-plagiogranite bodies that are thought by Alabaster *et al.* (1980) to be associated with off-axis volcanism (sea-mount or island-arc complex?). The upper part of the ophiolite section, which includes the upper layered gabbros, the isotropic gabbro that crystallized next to the roof of the magma chamber, the sheeted dyke sequence, and the pillow lavas, is very well preserved in the northern part of the Semail ophiolite. Thus, samples from three transects in the northern part of the ophiolite (Fig. 2) complement the previous studies (*J. Geophys. Res.* **86**, 2495–782) in the southern sector. These three sections include a probable fossil transform fault (W. Ragmi), a section cut by a silicic centre (W. Shafan) and an intact region between silicic centres (W. Hilti).

Oxygen- and hydrogen-isotope systematics

The $\delta^{18}O$ and δD data are reported in the standard delta notation, relative to SMOW (standard mean ocean water). The δD values are measured with a precision of $\pm 1‰$ and the $\delta^{18}O$ values with a precision of $\pm 0.15‰$. The oxygen- and hydrogen-isotopic compositions of hydrothermally exchanged minerals are controlled by (i) the formation temperature; (ii) the isotopic composition of the fluid phase, which in the cases described below is either seawater ($\delta^{18}O \approx 0$) or an ^{18}O-shifted, modified seawater-hydrothermal solution; (iii) the effective water-rock ratio; and (iv) the degree to which isotopic equilibrium is attained. Over a wide temperature range, at equilibrium the common hydrous minerals all have D/H ratios about 30–60‰ lower than their coexisting water, and thus most minerals equilibrated with ocean water will have $\delta D = -30$ to -60. The fractionation of oxygen isotopes between minerals and water (or between two minerals) is strongly dependent on the individual characteristics of the mineral phase(s), as well as on the temperature, and varies for different mineral species at a single temperature. Thus the $\delta^{18}O$ values of minerals deposited from ocean water could vary from values as low as zero at high T to values as high as +35 at low T. These various principles of stable-isotope geochemistry are described in several standard references (Clayton & Epstein 1958; Epstein & Taylor 1967; Taylor 1968, 1977; Suzuoki & Epstein 1976).

Presently available data on the variation of δD and $\delta^{18}O$ values in hydrous minerals from ophiolites and oceanic rocks are illustrated in Fig. 3, together with our new data on amphibole separates from Indian Ocean gabbroic rocks. The fields shown on Fig. 3 were delineated by Stakes & O'Neil (1982), and represent data for chlorite separated from oceanic greenstones, smectite separated from saponite-rich altered pillow basalt, and amphibole separates from amphibolized rocks from the Mid-Atlantic Ridge and from three ophiolites (Troodos, Semail and San Luis Obispo; see Stakes & O'Neil, 1982). It should be noted that these rocks probably all started out with primary, whole-rock $\delta^{18}O$ values of about +5.8, which is the uniform, accepted value for MORB; also, except for the high-level amphibole-bearing gabbros and plagiogranites, these rocks probably originally contained no primary hydrous minerals (see Gregory & Taylor 1981).

The amphiboles from the MAR, the Indian Ocean, and the ophiolite sections all have very similar $\delta^{18}O$ and δD values (Fig. 3). Variations in the $\delta^{18}O$ values are a function of temperature, water/rock ratio and extent of metamorphic recrystallization. All of the oceanic and ophiolite amphiboles have $\delta D = -35$ to $-60‰$, compatible with a seawater origin. Small differences between the δD values of amphiboles from the modern

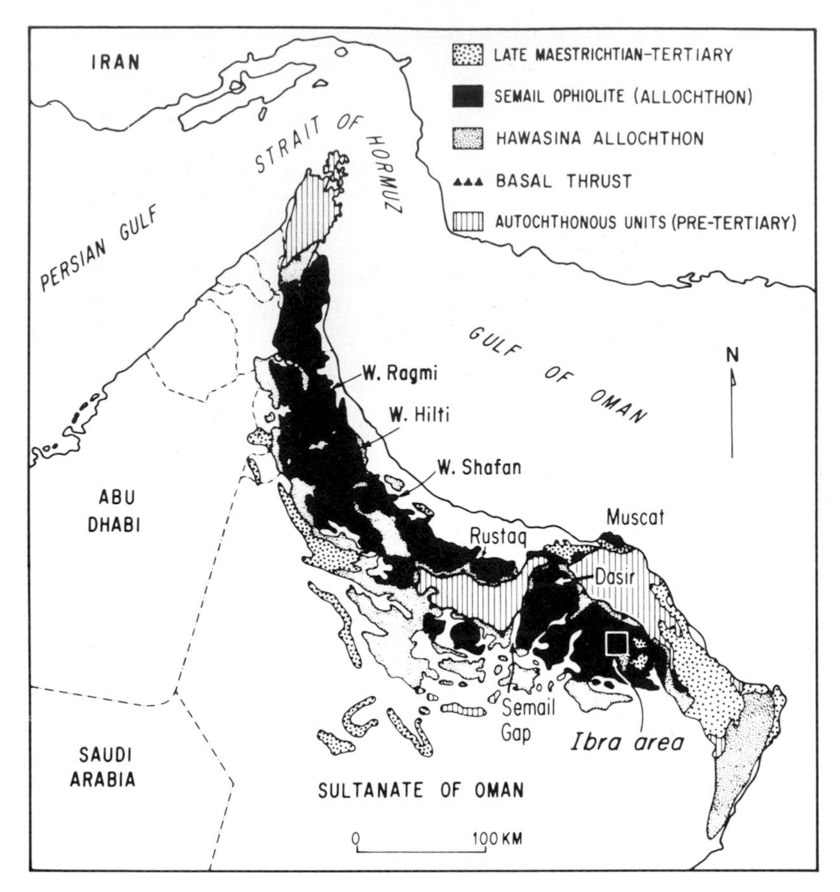

FIG. 2. Location map of the Semail ophiolite. Samples discussed in text are from (i) a fossil transform zone at W. Ragmi, (ii) near a large diorite-plagiogranite body at the gabbro-sheeted dykes contact in W. Shafan, (iii) from an intact section through W. Hilti, and (iv) earlier data of Gregory & Taylor (1981) from the Ibra area.

ocean and the ophiolitic rocks can be attributed to a difference in the δD value of Cretaceous seawater, which was about 10‰ lower in δD than present-day seawater as a result of the absence of the polar ice sheets (see Gregory & Taylor 1981).

The oceanic saponites on Fig. 3 define a field of low-temperature alteration associated with the formation of abundant Mg-rich secondary clay minerals (smectites). This type of alteration is characteristic of glassy basalts that were altered at low to moderate temperatures (30–200°C). At higher temperatures (200–350°C), basaltic rocks are altered to greenschist assemblages of sodic plagioclase-chlorite-epidote-quartz. Relative to the saponites, the chlorites characteristic of these sea-floor greenstones have lower values of δ^{18}O and higher values of δD, compatible with a higher-temperature origin. Increasing amounts of amphibole and decreasing amounts of chlorite are associated with metamorphic conditions transitional between upper greenschist facies and the lower amphibolite facies (Liou *et al.* 1974). The amphiboles separated from the gabbroic rocks reflect formation over this broad range of temperatures, but at generally lower water–rock ratios than the shallower extrusive rocks.

Table 1 provides a brief summary of the isotopic characteristics of a small set of samples that are representative of the major types of alteration observed in the crystalline rocks. These types include: (i) high-temperature interactions at low water–rock ratios, associated with minimal mineralogical changes (the mineralogically fresh gabbros); (ii) interactions at high to moderate temperatures and higher water–rock ratios, associated with significant secondary mineral formation under static conditions (amphibolized gabbros); (iii) extensive alteration and recrystallization associated with tectonic deformation (hornblende-rich gabbros); and (iv) metamorphic

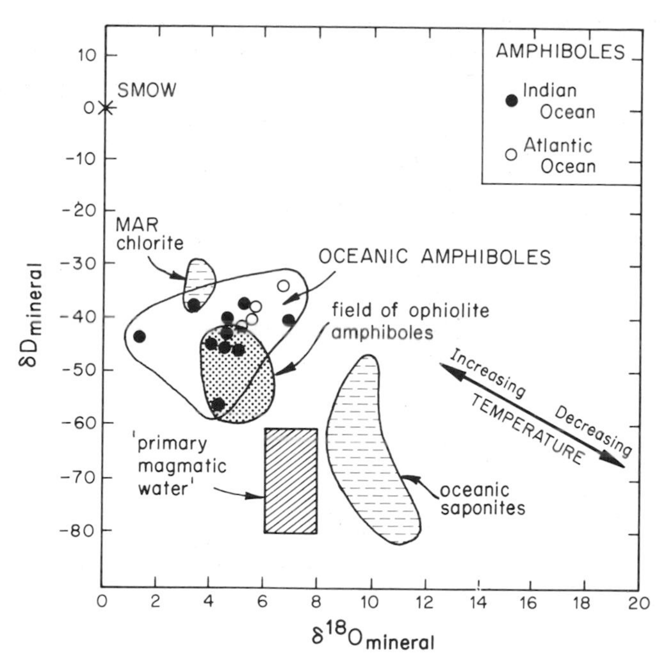

FIG. 3. Plot of δD *v.* $\delta^{18}O$ for amphiboles from the Indian Ocean compared to those from ophiolite complexes; also shown are values for chlorite and smectite from oceanic metabasalts. The arrow indicates the general direction of change of δD and $\delta^{18}O$ in hydrous minerals equilibrated with a reservoir of H_2O of constant isotopic composition at various temperatures (after Stakes & O'Neil 1982).

reworking associated with both axis and off-axis hydrothermal events and cumulative alteration effects as the crust ages (the silicic rocks, diabase dykes, xenoliths, and quartz-epidote veins).

High-temperature seawater interactions associated with minimal mineralogic changes

The freshest-appearing gabbros from the Indian Ocean are olivine gabbros largely composed of clear, twinned plagioclase, (An_{65-75}), fresh olivine ($\sim Fo_{78}$), clinopyroxene, and minor amounts of intercumulus orthopyroxene. Microfractures in the plagioclase grains are the locus of turbid zones (plane light) and patchy extinction (crossed nicols); however, numerous microprobe analyses failed to detect any chemical changes associated with those textural features. The olivine (and orthopyroxene) is partially replaced by talc and magnetite. The clinopyroxenes locally display zones of patchy extinction and incipient replacement by a magnesio-hornblende along cleavage planes, giving the grains a cross-hatched appearance.

The apparently pristine petrological condition of these rocks is misleading because the plagioclase grains in all but a couple of samples have undergone thorough isotopic exchange that has lowered the $\delta^{18}O$ of the feldspar from its primary value of +6.1 to +6.4 to values of +5.0 to +6.0‰. Some apparent lowering of the values of $\delta^{18}O$ of clinipyroxene actually reflects incipient replacement of the clinopyroxene by hornblende. The minor, but ubiquitous, partial replacement of olivine by talc (not serpentine) and the replacement of clinopyroxene by aluminous hornblende, suggest that temperatures of alteration extended to well above 500°C. The general paucity of hydrous minerals, together with the total absence of any low-temperature hydrothermal minerals, indicates that the alteration of these rocks occurred either under very high initial temperatures or relatively low water–rock ratios (or both) and that it largely ceased at these high temperatures, preventing retrograde alteration.

The isotopic perturbations observed in these freshest gabbros are illustrated in Fig. 4, which shows a plot of bulk rock $\delta^{18}O$ *v.* $^{87}Sr/^{86}Sr$. Of the four analysed gabbros originally selected by petrographic examination as the freshest, least-altered samples from the Indian Ocean dredge collection by Engel & Fisher (1975), only one retains the isotopic composition characteristic of fresh, MOR-type gabbro. The remaining three samples exhibit clear-cut depletion in ^{18}O and enrichment in $^{87}Sr/^{86}Sr$ (seawater is more radio-

TABLE 1. *Oxygen- and hydrogen-isotopic compositions of some typical plutonic and hypabyssal rocks from oceanic fracture zones and the Semail ophiolite, Oman. The ophiolitic rocks are indicated by an asterisk*

Rock type	$\delta^{18}O$ values (‰) plag	cpx	amph	qtz	WR	δD_{amph}	Location
A. *Silicic rocks*							
1. qtz-monzonite	7.3		4.7	8.7	7.5		Argo
2. granophyric diabase	6.4		4.4		5.4	−57	Argo
3. plagiogranite dykelet	5.9			7.0			MAR
4. axis plagiogranite*	9.7			7.7	8.5		Oman
5. high-level gabbro*	6.8			7.3	6.0		Oman
B. *Amphibolized gabbro*							
6. cpx-gabbro	4.5		4.6		3.8	−43	Melville
7. ol-cpx gabbro	4.0	5.1	4.1		4.9		Vema
8. cpx gabbro*	5.0		0.6		2.4		Oman
C. *Hornblende-rich gabbroic rocks*							
9. >60% hb	4.5		4.1		4.1	−45	Marie Celeste
10. >50% hb*	3.4		5.0		5.0		Oman
D. *Mineralogically fresh gabbro*							
11. ol-cpx gabbro	6.3	5.7			6.2		Vema
12. ol-cpx gabbro	5.4	5.5			4.9		Argo
13. ol-cpx gabbro*	5.8	5.7	5.7		5.5		Oman
E. *Diabase dykes*							
14. axis dyke*					9.6		Oman
15. axis dyke*					7.8		Oman
16. axis dyke*					6.3		Oman
17. island-arc dyke*					6.4		Oman
18. island-arc dyke*					7.0		Oman
19. cpx-phyric upper dyke*					5.8		Oman
20. sea-mount dyke*					2.9		Oman
21. trench diabase			2.9		3.4	−44	Marianas
F. *Quartz-epidote veins*							
22. adjacent to axis plagiogranite*				7.4			Oman
23. adjacent to silicic centre*				7.3			Oman
24. cuts MAR greenstone				7.1			MAR
G. *Xenoliths of diabase*							
25. in axis plagiogranite*					5.7		Oman
26. in axis plagiogranite*					2.8		Oman

1. ANTP 125-4c. Quartz monzonite dykelet in granophyric diabase (No. 2) from Argo Fracture Zone.
2. ANTP 125-4b. Horblende-rich granophyric diabase from Argo Fracture Zone.
3. AII60-12-32W. White aplite (plagiogranite) dyke cutting ferrodiorite from 22°S MAR. Qtz + plag (An_2) + minor amphibole. Sample from G. Thompson.
4. D81-25. Axis plagiogranite from near Musafiyah, N Semail ophiolite.
5. D81-29. High-level gabbro with granophyric texture and 'sweats' of axis plagiogranite; adjacent to sample D81-25.
6. ANTP 113-1(1). Originally an augite-gabbro from the Melville Fracture Zone. Turbid but twinned plagioclase is An_{48-53}. Clinopyroxene is completely replaced by actinolitic hornblende and plagioclase and clinopyroxene boundaries are replaced by edenitic hornblende.
7. ANTP 130B-1(1). An ol-cpx gabbro from the Vema Fracture Zone. Olivine (Fo_{68}) is replaced by talc-magnetite pseudomorphs. Plagioclase varies from An_{58-52} and is turbid and twinned. Clinopyroxene is replaced by magnesio-hornblende. Plagioclase and clinopyroxene boundaries are replaced by Fe-rich chlorite and edenitic hornblende.
8. D81-38. Gabbro pegmatite from W Ragmi south. Shear zone near bottom of section. Plagioclase is fresh and

genic than fresh MOR basalt, 0.7094 *v.* 0.7028). The combination of ^{18}O depletions and ^{87}Sr enrichments occurring together in the same gabbros is proof that these rocks were subjected to *high-temperature* ($T \geqslant 400°C$) hydrothermal alteration. The fresh mineralogy of the gabbros and the lack of OH-bearing minerals implies even higher temperatures of alteration (>500°C). The observed isotopic and mineralogical characteristics in these oceanic-dredge samples are identical to those described for the layered cumulate gabbros in the southern Semail ophiolite (Gregory & Taylor 1981; McCulloch *et al.* 1981). In particular, note that although it differs in detail, the central V-shaped pattern indicated in Fig. 4 for the oceanic samples is the same established by McCulloch *et al.* (1981) in the Semail ophiolite: (i) The lowest $^{87}Sr/^{86}Sr$ values (0.7028) are associated with the freshest gabbros, which have MORB-type $\delta^{18}O$ values (+5.8). (ii) The lowest $\delta^{18}O$ values, indicative of high-temperature seawater-hydrothermal alterations, are associated with somewhat more radiogenic $^{87}Sr/^{86}Sr$ attributable to the same process. (iii) The most radiogenic $^{87}Sr/^{86}Sr$ values are associated with higher $\delta^{18}O$ values indicative of lower-temperature alteration or 'ageing' of the oceanic crust.

The most sensitive indicator of the above-described, high-temperature hydrothermal exchange event is the preferential ^{18}O-depletion of plagioclase compared to clinopyroxene. This exchange results in a reversal of the normal, equilibrium sequence of $\delta^{18}O$ variation between these two mineral phases, and has been observed in all meteoric-hydrothermal systems (Taylor 1977), as well as the Semail ophiolite (Gregory & Taylor 1981). Fig. 5 is a compilation of oxygen-isotope data for plagioclase-clinopyroxene pairs from a number of fossil hydrothermal centres (Taylor 1983), including the samples from the Indian Ocean. Most of the plagioclase-clinopyroxene pairs fall steeply below the equilibrium $\Delta\ ^{18}O$ plag-cpx line as a result of this subsolidus

---

twinned. Clinopyroxene is replaced by hornblende with hornblende replacement of plagioclase-clinopyroxene contacts.

9. CIR 97F(2). Foliated amphibolite from Maric Celeste Fracture Zone. Plagioclase is somewhat recrystallized, An_{45-50} cores and An_{8-23} rims. Amphibole is red-brown or green-brown pargasitic or edenitic hornblende.
10. D81-18. Coarse hornblende gabbro from small, later-stage, zoned intrusion, W Kanut, N Semail ophiolite.
11. ANTP 130-1(2A). Isotopically and petrographically fresh gabbro from Vema Fracture Zone. Plagioclase is An_{65-72}, olivine is Fo_{78}.
12. ANTP 126-1(4). Plagioclase is An_{64-75}; traces of talc replacing olivine (Fo_{79}), and magnesio-hornblende replacing rims of clinopyroxene. Argo Fracture Zone.
13. D81-41. Layered gabbro from W Ragmi. Clinopyroxene and olivine are fresh. Plagioclase is slightly turbid.
14.+ OM 5085. Axis dyke from W Kanut, northern Semail ophiolite. Pervasively altered to actinolite, prehnite, pumpellyite(?), chlorite, sphene, and cut by many thin quartz veins. Sample (and classification) from A. W. Shelton.
15.+ OM 5080B. Axis dyke from W Sarami, northern Oman ophiolite. Sample (and classification) from A. W. Shelton.
16.+ OM 5066. Axis dyke from sheeted dyke complex, W Ragmi. Sample (and classification) provided by A. W. Shelton.
17.+ OM 5060. Off-axis (island-arc) dyke from N of Ajib Fort. Sample (and classification) supplied by A. W. Shelton.
18.+ OM 5089. Off-axis (island-arc) dyke from sheeted dyke complex at Hawquayn NW. Sample (and classification) from A. W. Shelton.
19.+ OM 7124. Off-axis dyke (cpx-phyric upper) associated with Upper Pillow Lavas. Sample (and classification) supplied by T. Alabaster).
20. D81-15. Classified by J. Smewing as an off-axis dyke intruding composite sheeted-dyke complex, Wadi Kanut, Northern Semail ophiolite. Associated with small wehrlite-gabbro pluton.
21. MARA 05 D25-10. Diabase dredged from Marianas trough south of Guam. Clinopyroxene replaced by fibrous amphibole (supplied by S. Bloomer).
22. D81-74. Quartz separated from a quartz-epidote vein near axis plagiogranite body, W Ragmi.
23. D81-147. Quartz vein in shear zone near silicic centre at W Shafan, Northern Semail ophiolite.
24. GS2405. Euhedral quartz crystals in greenstone, 36°N, Mid-Atlantic Ridge. Data from Stakes & O'Neil (1982).
25. D81-26B. Dark diabase xenolith from axis plagiogranite body, Musafiyah, Northern Semail ophiolite.
26. D81-30. Dark, partially resorbed xenolith from axis plagiogranite body, Musafiyah, Northern Semail ophiolite.

Abbreviations: ol = olivine, cpx = clinopyroxene, plag = plagioclase, amph = amphibole, qtz = quartz, WR = whole rock, hb = hornblende.

+Dyke samples with OM-prefix were obtained from the Open University collection. The classification as axis or off-axis is theirs, and based on field relations and trace-element chemistry (Alabaster *et al.* 1980).

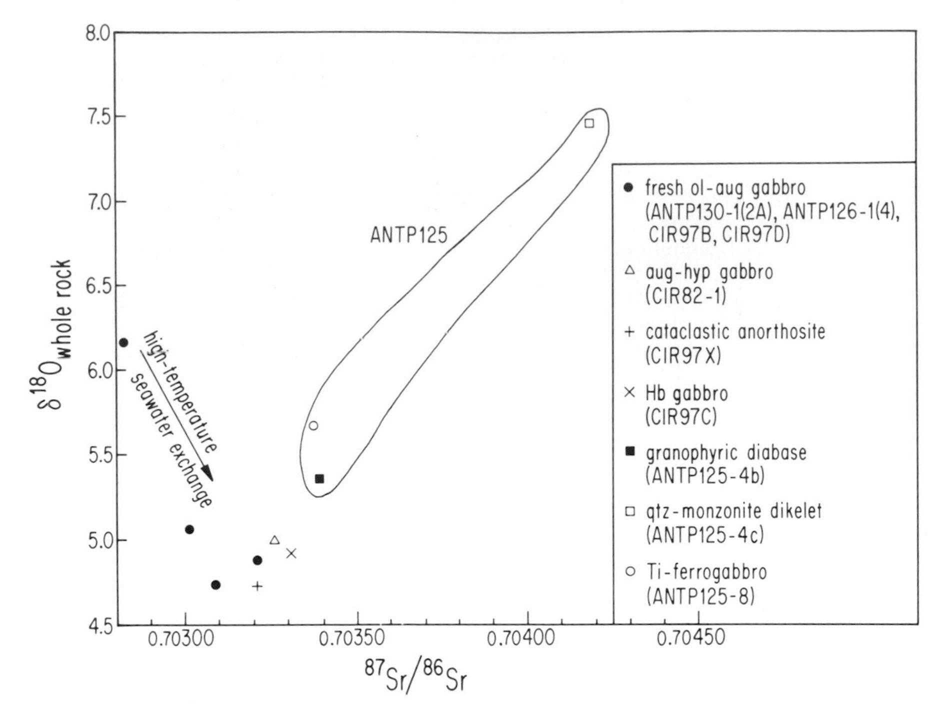

FIG. 4. A plot of $\delta^{18}O$ *v.* $^{87}Sr/^{86}Sr$ for whole-rock samples from the Indian Ocean. $^{87}Sr/^{86}Sr$ values are from Hedge *et al.* (1979). High-temperature subsolidus seawater interactions progressively decrease the $\delta^{18}O$ (through exchange with igneous plagioclase) and increase the $^{87}Sr/^{86}Sr$. ANTP 125 is a suite of unusual rocks that includes small amounts of evolved quartz monzonite and plagiogranite dyke rocks, characterized by strong Fe-enrichment in the gabbros and diorites.

hydrothermal exchange, just as do the previously analysed samples from Oman, the Skaergaard, and Jabal at Tirf. A detailed discussion of these types of non-equilibrium $\delta^{18}O$ trajectories is given by Gregory & Taylor (1981). Note that the non-equilibrium pyroxene-plagioclase pairs from the Indian Ocean and the southern Semail ophiolite in Oman plot in similar positions on Fig. 5, but the least-squares lines through the data-points are quite different, as follows:

Oman (Gregory & Taylor, 1981): $\delta^{18}O$ plag = 2.3, $\delta^{18}O$ cpx - 6.7.

Indian Ocean (this work): $\delta^{18}O$ plag = 4.8, $\delta^{18}O$ cpx -20.3.

The intercept and slope of the Indian Ocean least-squares line are both similar to the values obtained by Taylor & Forester (1979) for the Skaergaard intrusion, formed 55 Ma during the initial stages of opening of the North Atlantic Ocean, and also to the values obtained by Taylor & Coleman (1977) for the Jabal at Tirf complex, a subaerial spreading centre formed about 22 Ma during the opening of the Red Sea. The Jabal at Tirf complex is in many respects geologically similar to an ophiolite complex, and it was subjected to a similar type of hydrothermal activity. However, because it was subaerial, the hydrothermal fluids were derived from meteoric ground waters (Tertiary lake waters?) that are estimated by Taylor & Coleman (1977) to have had an initial (unshifted) $\delta^{18}O$ value of about −5 or −6 (i.e. 5‰ lower than ocean water). We can at present only speculate about the reasons for the slight differences in slope and intercept for the various clinopyroxene-plagioclase isotopic pairs shown on Fig. 5. The steep slopes for the Skaergaard, Jabal at Tirf, and Indian Ocean pairs may be due to the fact that the analysed pyroxene mineral separates at these localities were very pure and selected to be as free of amphibole as possible. In less pure separates, the pyroxene grains that have suffered incipient alteration to amphibole are physically 'opened up' and partially disrupted by the accompanying volume changes, thus allowing the hydrothermal fluids to readily penetrate the pyroxene grains along cleavage cracks and therefore to exchange more easily with the oxygen in the clinopyroxene lattice. A process analogous to this (inversion of β ferrowollastonite to a very fine-grained aggregate of ferrohedenbergite grains) in the Upper Zone of the Skaergaard intrusion is certainly the explanation

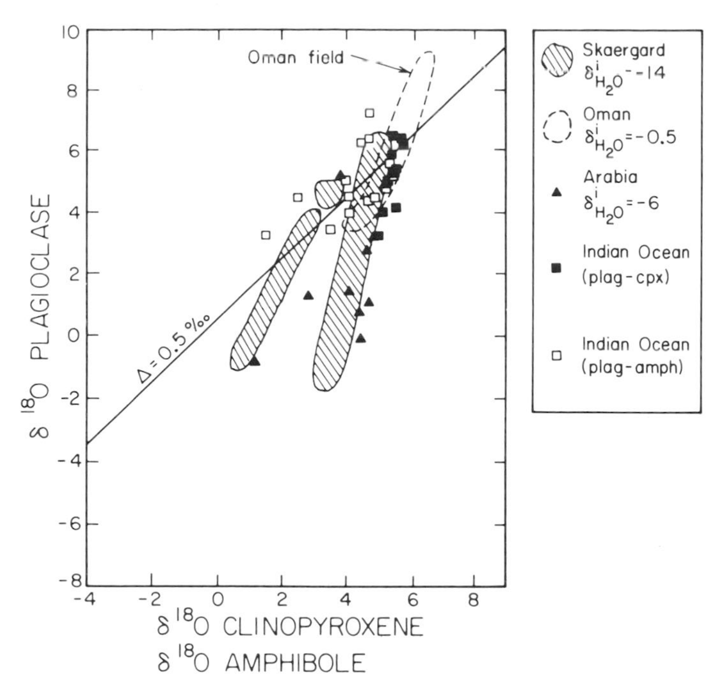

FIG. 5. Comparison of $\delta^{18}O$ data on plagioclase-clinopyroxene pairs from layered gabbros in the Semail ophiolite complex, Oman, the Skaergaard intrusion, East Greenland, and the Jabal at Tirf complex, Saudi Arabia (after Taylor 1983) with the gabbroic rocks from the western Indian Ocean. The steep slopes of the plagioclase-clinopyroxene $\delta^{18}O$ arrays in most of these localities indicate marked isotopic disequilibrium, typically with negative $\Delta^{18}O$ plagioclase-pyroxene values. The effects are a result of high-temperature, subsolidus isotopic exchange with either ocean waters or meteoric ground waters.

of that part of the Skaergaard pattern indicated as a separate data-point envelope on the left-hand side of Fig. 5 (Taylor & Forester 1979).

Seawater alteration associated with substantial mineralogical changes

Petrographic features

Increasing degrees of alteration in the gabbroic rocks at temperatures below 550°C and higher water/rock ratios result in extensive mineralogical and metamorphic changes, in addition to the isotopic effects. These changes probably took place either upward in the crustal section near the spreading axis, or perhaps laterally on approaching a transform fault zone. In such rocks the olivine is totally replaced; the only evidence of its existence is the presence of talc-magnetite pseudomorphs. Clinopyroxene grains are replaced along their edges by actinolite or magnesio-hornblende; zones along clinopyroxene-plagioclase grain contacts are commonly replaced by edenitic hornblende, and chlorite replaces plagioclase along grain boundaries and corroded margins. Talc fills microfractures and grain boundaries near relict olivine. Where these tiny veins cross plagioclase, the phyllosilicate phase that fills them is Fe-rich chlorite.

Amphibole compositions

Amphibole compositions vary from pargasitic hornblende, edenite and edenitic hornblende through magnesio-hornblende to actinolite (after Leake 1978). In the amphibolized gabbros, pargasitic and edenitic hornblendes characteristically replace the assemblage plagioclase-clinopyroxene, and magnesio-hornblendes replace individual clinopyroxene grains; such samples typically contain the lowest-^{18}O amphibole and the most complete replacement of clinopyroxene of any of the specimens analysed in this study. This replacement probably occurs at temperatures in the mid to lower amphibolite facies (450–550°C). Chlorite is typically present in all rocks in which actinolite or actinolitic hornblende substantially replaces clinopyroxene. The association of chlorite and Al-poor actinolite reflects temperatures falling into the greenschist range (~350–450°C).

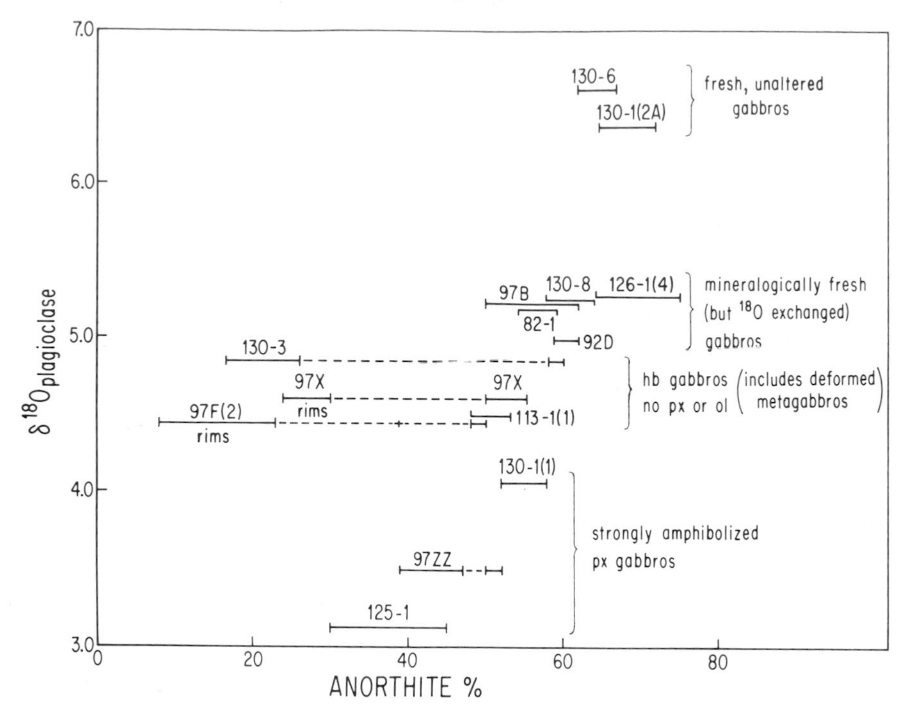

FIG. 8. Values of $\delta^{18}O$ for plagioclase separates from samples from the Indian Ocean, plotted against the range of anorthite compositions as determined by microprobe analyses. Although the total range of An content in any one sample is indistinguishable from values typical of normal, unexchanged igneous plagioclases, the An contents *do* decrease with decreasing values of $\delta^{18}O$. Cataclastic rocks contain clear, sodic plagioclase rims on twinned calcic cores. The range of compositions between the cores and the (smaller volume) rims are shown connected by a dashed line.

produced by the interaction of crystalline, hydrous, amphibolized gabbro or diabase blocks with the underlying basaltic magma. Silicic stringers of small volumes of plagiogranite liquid appear to have been 'sweated' from these blocks and injected upward into the base of the dyke complex (e.g. at Musafiyah, samples 4 and 5, Table 1). Repeated stoping events and dehydration of xenoliths would result in highly variable H_2O concentrations in the contaminated magma; these extremely variable P_{H_2O} values can produce the textural and mineralogical heterogeneity that typifies the isotropic gabbro (Gregory & Taylor 1981).

Silicic dykes broadly similar to those observed in ophiolite complexes have been dredged from the Argo Fracture Zone (Engel & Fisher 1975, see samples 1 and 2 in Table 1) and at 22°S on the MAR (sample 3, Table 1). However, one of the Argo samples is a very unusual dyke rock classified as a quartz monzonite by Engel & Fisher (1975). This hypabyssal rock, which is actually closer to a granodiorite in composition, has a *much* higher K/Na ratio than the typical plagiogranite of an ophiolite complex. Oxygen- and strontium-isotope data for this set of Argo Fracture Zone samples are shown in Fig. 4. The combination of a fairly radiogenic $^{87}Sr/^{86}Sr$ value in the 'quartz monzonite', together with a relatively high $\delta^{18}O$ value of +7.5, is compatible with an origin involving partial melting of incipiently altered oceanic crust (see McCulloch *et al.* 1981), but such crust obviously cannot have undergone any significant Na metasomatism or spilitization prior to the melting event.

The most dramatic evidence of stoping in the northern part of the Semail ophiolite is associated with large gabbro-diorite-plagiogranite bodies. These intrusions have in some places completely engulfed the high-level gabbro and the lower portions of the sheeted-dyke complex (e.g. the Suhaylah and Lasail localities). The largest silicic bodies may contain up to 60% xenoliths that appear as a jumble of recrystallized mafic blocks with pegmatitic borders surrounded by diorite. Such silicic centres are thought to be associated with off-axis intrusive activity during sea-mount formation (Alabaster *et al.* 1980), or possibly as a major episode of stoping and lateral intrusion of late-stage silicic liquid into the 'wings' of an 'aircraft-carrier-shaped' axis magma chamber (Gregory & Taylor 1981; Hopson *et al.* 1981).

Regardless of how these larger silicic bodies formed along the gabbro–sheeted-dyke contact zone, (e.g. Wadi Shafan) they represent a plausible source zone for the Ca- and Si-rich hydrothermal fluids that alter the overlying dyke complex, precipitating quartz and epidote and sulphides in shear zones and along dyke margins. Smaller (axis) plagiogranite bodies are associated with smaller metasomatic zones of quartz and epidote enrichment (e.g. Wadi Hilti).

We infer from this that the entire gabbro–sheeted-dyke contact zone may involve active mobilization of metals at and near the mid-ocean-ridge axis. Thermodynamic constraints on fluid buoyancy, coupled with the experimentally determined requirements for metal solubility, predict that the zone of maximum metal solubilization should occur at temperatures of about 400–450°C and hydrostatic pressures of about 500 bars (Bischoff 1980). In a mid-ocean-ridge environment, such conditions are likely to occur only in the vicinity of the roof of the axis magma chamber. Note that numerical modelling of the meteoric-hydrothermal system associated with the Skaergaard intrusion also suggests that hydrothermal solutions having temperatures greater than 400°C originated within or below the roof zone of that intrusion (Norton & Taylor 1979). Also note that in Oman the zone between the sheeted dykes and the diffuse boundary between the high-level gabbro and the layered gabbro is also the zone of maximum ^{18}O depletion by seawater (see Gregory & Taylor 1981).

Quartz veins from all of the above-described environments, namely quartz separates from axis plagiogranites (Musafiyah), quartz veins from a metasomatic zone adjacent to a small plagiogranite body (W. Ragmi), and quartz from a quartz-epidote-pyrite vein adjacent to the large silicic centre at W. Shafan, all have uniform $\delta^{18}O$ values of +7.1 to +7.7 (Table 1). This is the same compositional range observed in quartz-epidote veins in an epidote-rich greenstone from the MAR that apparently formed at temperatures of about 350–400°C (Stakes & O'Neil 1982).

The diabasic rocks of the sheeted-dyke complex are cooled by hydrothermal circulation near the axis of the spreading centre resulting in hydrothermally altered diabasic rocks with low $\delta^{18}O$ values and upper greenschist mineralogies (Gregory & Taylor 1981). These initial high-temperature alteration effects were later partially masked by subsequent off-axis hydrothermal events, resulting in diabasic rocks with $\delta^{18}O = +6.8$ to $+10.9$ in the southern part of the Semail ophiolite (Gregory & Taylor 1981). Note that Gregory & Taylor (1979) found xenoliths of diabase with $\delta^{18}O = +3.7$ to $+6.3$ that had escaped this off-axis hydrothermal reworking by being incorporated into plagiogranite bodies at the ridge axis. Diabase samples, obtained from several localities in the northern part of the Semail ophiolite and previously classified by A. W. Shelton (pers. comm.) and Alabaster *et al.* (1980) as (i) axis dykes; (ii) off-axis dykes; and (iii) off-axis clinopyroxene-phyric dykes, show a wide range of $\delta^{18}O$, from values as low as +2.9 to values as high as +9.6. There is also a very good correlation between the $\delta^{18}O$ values and the geologic setting of the dykes.

The isotopic data presented in Table 1 are representative of a much larger data-set, and demonstrate that the dykes that Alabaster *et al.* (1980) consider to have been emplaced near the spreading axis (14, 15, 16) are richer in ^{18}O than the dykes emplaced during later, off-axis events (17, 18, 19, 20). The off-axis diabasic dykes are also much less altered than the axis dykes, and they differ markedly in trace-element contents (Alabaster *et al.* 1980). The axis dykes tend to be highly oxidized, and are completely replaced by alteration assemblages that include actinolite, prehnite, abundant sphene, and some chlorite and zoisite; they are cut by myriads of tiny quartz stringers. The off-axis dykes are also altered to similar mineral assemblages, but they typically display much better preservation of their original textures and primary mineralogy.

These data are explicable as follows: the sheeted dykes formed in or near the axis spreading centre, and since then they have been subjected to the cumulative effect of all of the hydrothermal events that later affected the oceanic crust. Thus, they tend to be more enriched in ^{18}O than later off-axis dykes. Such 'hydrothermal ageing' and ^{18}O-enrichment may be associated with the 'lower convective system' of Gregory & Taylor (1981), which involves the highly ^{18}O-shifted seawater that produced the ^{18}O depletion observed in the layered gabbros. It could also be associated with continued circulation (and alteration) in young, warm oceanic crust that has been sealed by an impermeable sediment cover, or with the Ca- and Si-rich hydrothermal solutions generated by the off-axis intrusive events. Each of the off-axis intrusive events can be expected to generate its own hydrothermal circulation, as demonstrated convincingly by a small 'upper' intrusive body of wehrlite and hornblende gabbro that appears within the sheeted dyke complex in W. Kanut. Here the hornblende gabbro (No. 10, Table 1) bears the imprint of high-temperature hydrothermal alteration ($\delta^{18}O_{plag} = 3.4$) as do the

—— & Coleman, R. G. 1977. Oxygen isotopic evidence for meteoric-hydrothermal alteration of the Jabal at Tirf complex, Saudi Arabia. *EOS, Trans. Amer. Geophys. Union*, **58**, 516.

—— & Forester, R. W. 1979. An oxygen and hydrogen isotope study of the Skaergaard intrusion and its country rocks: A description of a 55 m.y. old fossil hydrothermal system. *J. Petrology* **20**, 355–419.

D. S. Stakes & H. P. Taylor Jr, Division of Geological and Planetary Sciences, California Institute of Technology, Pasadena, CA 91125, USA.
R. L. Fisher, Scripps Institution of Oceanography, La Jolla, CA 92093, USA.

Sr–Nd isotope and chemical evidence that the Ballantrae 'ophiolite', SW Scotland, is polygenetic

M. F. Thirlwall & B. J. Bluck

SUMMARY: Sr-Nd isotope and major- and trace-element analyses, including REE, are presented for a suite of basaltic lavas from the Ballantrae complex, SW Scotland. The chemistry of the lavas is discussed with reference to 'spider' diagrams, using normalization of incompatible element abundances to those estimated in the primordial earth, rather than conventional element–element plots. These diagrams facilitate assessment of element mobility during alteration, and some samples are shown to have close to original igneous concentrations of the mobile elements Rb, K, Sr etc. Clinopyroxene-phyric basalts from Games Loup and Mains Hill have chemical affinities with modern intra-oceanic arc volcanics, and their acid-leached augite phenocrysts have more radiogenic Sr than the mantle array, defined using clinopyroxene separates from Ordovician oceanic basalts from the Southern Uplands. Basalts from Pinbain, Bennane Head and Downan Point are in contrast typical ocean-island tholeiites in both chemistry and Nd-Sr isotopes. At least four independent volcanic sequences can be recognized from the chemical data: these do not form a continuous 5 km thick sequence but rather represent tectonically juxtaposed units generated in very different environments. The 'ophiolite' can not therefore be regarded as a single entity, produced in a single tectonic environment. Sm-Nd mineral ages have been determined for samples from Games Loup, Mains Hill and Downan Point in an attempt to provide a chronology for the development of the different units, but ages obtained are indistinguishable from the zircon age reported by Bluck *et al.* (1980) for a Ballantrae trondhjemite. Evaluation of the significance of the complex to Caledonian plate models will only be possible with more precise age determinations on the different members of the complex.

The Ballantrae complex, on the SW margin of the Midland Valley of Scotland (Fig. 1), has been an important item of evidence in reconstructions of the Caledonian development of Britain. It consists of two major bodies of serpentinized ultramafic rocks with minor gabbro and trondhjemite intrusions, associated with volcanic units comprising mainly pillow lavas and volcaniclastic sediments. The volcanic rocks are associated with radiolarian cherts, black graptolitic shales and olistostromes containing blocks of blueschist, ariegite and all the other lithologies. Church & Gayer (1973) compared the complex with the Ordovician ophiolites of Newfoundland, and recognized that most of the components of the standard ophiolite sequence were present at Ballantrae, although no stratigraphy remains as the complex is greatly dissected by faults. They noted that the ultramafic rocks were associated with a high-temperature metamorphic aureole, implying that ophiolite emplacement occurred while the ultramafic rocks were still hot, a feature ascribed to their generation close to the site of obduction, in a marginal basin or a small ocean basin. The presence of clasts of these metamorphic rocks in olistostromes associated with the Bennane Head volcanic sequence (Fig. 1) was used as evidence that at least some of the Ballantrae volcanism post-dated ophiolite emplacement, and Church & Gayer (1973) suggested that these parts of the Bennane sequence were erupted in an arc environment. The present authors have however so far been unable to find metamorphic clasts in the Bennane Head olistostromes.

Recent mapping has shown that volcaniclastic material forms a far greater proportion of the volcanic units than would be expected for rocks generated at a spreading ridge in a major ocean. Bluck *et al.* (1980) therefore suggested that most of the volcanic units were generated in an arc environment, separated from the continent further north by a marginal basin in which the plutonic rocks of the complex were formed. This was supported by the concordance of a zircon U-Pb age of 483±4 Ma for a trondhjemite body, dating ophiolite genesis, with a K-Ar age of 478±4 Ma on amphibole from the metamorphic aureole, dating ophiolite emplacement. Bluck (1981) demonstrated that most of the volcaniclastic sediments at Pinbain (Fig. 1) were deposited in shallow water, which he concluded provided further evidence for an arc environment. These interpretations of the complex were disputed by Barrett *et al.* (1982), who noted the absence of arc material along strike from Ballantrae and suggested that the complex represented an obducted seamount terrain on the northern margin of the Iapetus Ocean.

Several workers have published chemical data on the Ballantrae lavas, but either the elements analysed have been limited to small groups supposedly diagnostic of tectonic setting, or the sampling techniques have been suspect. Wilkinson & Cann (1974) published Ti, Zr, Y and

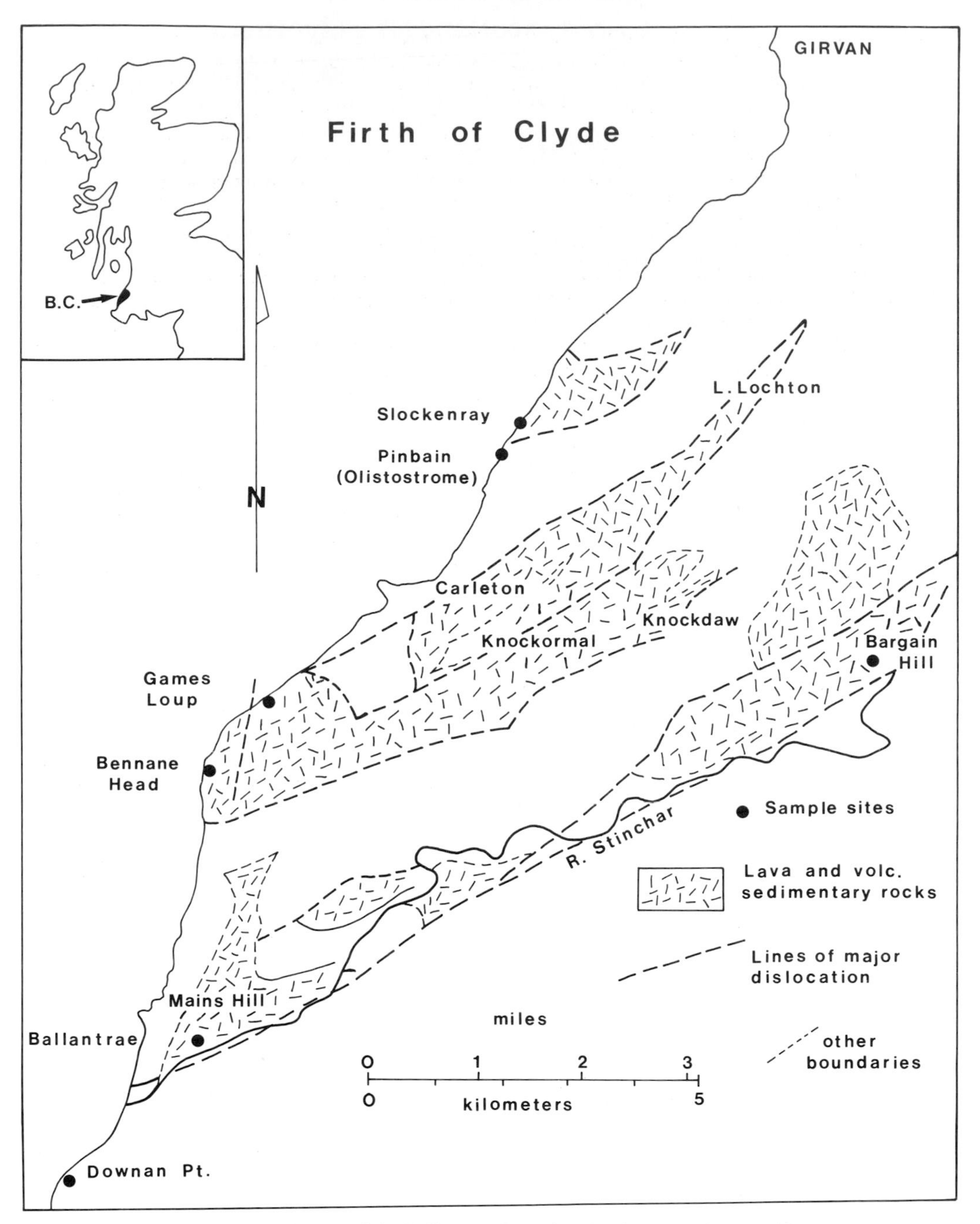

FIG. 1. Sketch map of the Ballantrae Complex showing sampling localities.

Nb concentrations for lavas from Games Loup, Bennane Head, Downan Point and from blocks in the Pinbain Beach olistostrome (Fig. 1), for blueschists and amphibolites near Knockormal and for dykes cutting the ultramafic rocks. The discriminant diagrams of Pearce & Cann (1973) were used to identify tectonic settings, which ranged from island arc at Games Loup and possible ocean floor at Knockormal to within-plate elsewhere. The highly varied chemistry was ascribed either to major time differences between the different volcanic blocks or to the tectonic juxtaposition of geographically widely separated volcanic units. Later workers have sought to relegate all the Ballantrae volcanics to a single tectonic environment and have either ignored the complexity suggested by Wilkinson & Cann (1974) or have suggested that Zr, Y and Nb concentrations have been modified by alteration and 'crustal contamination' (Jelinek *et al.* 1980).

Lewis & Bloxam (1977) presented rare earth element (REE) data for Arenig lavas from Bennane Head, Games Loup, Knockdaw and Bargain Hill (Fig. 1) and also from the Downan Point, Currarie and Portandea lavas south of the Stinchar Valley Fault, which they regarded as Caradocian. However, the relationship between the Downan Point lavas and those associated with the Caradocian fauna (Currarie) is not clear (J. K. Leggett, unpublished mapping), and the Downan Point lavas are not demonstrably Caradocian. With the exception of the Bennane Head sample, all Arenig lavas analysed by Lewis & Bloxam (1977) are light-REE depleted, from which they inferred volcanism at either a spreading ridge or an island arc. They suggested that the lavas formed a continuous 5 km thick sequence, noted that this was excessive for ocean-floor basalts and therefore proposed that all the Arenig lavas formed part of an island-arc pile. Unfortunately, they failed to analyse samples from Pinbain, which they relegated to the status of 'other Arenig volcanics'. The Bennane Head sample was labelled as an arc basalt because of its occurrence within the Arenig sequence, thus ignoring the data of Wilkinson & Cann (1974) suggesting a within-plate origin for the Bennane lavas.

Jones (1977) made a careful and detailed mapping and chemical study of the Ballantrae complex. He showed that the Arenig lavas did not form a continuous 5 km thick sequence and emphasized that there were at least seven volcanic blocks, distinct in chemistry, petrography and structural history. Samples from the Pinbain, Bennane Head, Knockormal and Knockdolian blocks were identified as within-plate basalts while samples from the southeastern blocks were thought to be characteristic of ocean-floor basalts. Samples from Mains Hill (Fig. 1) were anomalous and Jones (1977) tentatively attributed this to accumulation of clinopyroxene phenocrysts. Jones (1977) therefore regarded the 'ophiolite' as generated at a spreading ridge, with an ocean island sequence developed en route to obduction. This is very similar to the ideas recently suggested by Barrett *et al.* (1982), but Jones (1977) unfortunately did not analyse samples from Games Loup.

These conflicting interpretations of the chemistry of Ballantrae lavas have led to considerable confusion in palaeoenvironmental models for the complex. This was compounded by the work of Jelinek *et al.* (1980) who concluded that the Ballantrae lavas had been affected by crustal contamination and that the diagnostic elements used in identifying tectonic setting by previous authors (particularly Wilkinson & Cann 1974) were mobile during alteration and 'metasomatism'. Such conclusions are the results of suspect sampling procedure and misunderstanding of the effects of alteration processes. The idea of crustal contamination was based solely on higher mean Rb/Sr ratios for the Downan Point, Pinbain and Bennane Head lavas than observed in metadolerite dykes ('beerbachites') traversing the complex. Apart from the fact that the Rb/Sr ratios reported for lavas and dykes show massive variation and overlap, the Rb/Sr ratio in rocks as altered as these provides little information on igneous processes, and no information whatsoever on 'crustal contamination'. Their conclusion that Ti, Y, Zr, etc., had been mobile during alteration was based on comparison between cores and margins of four pillows and the scatter of the lavas on a Ti-Zr diagram. The chemistry of the core/margin pairs reported in their Table 5 bears little relation to the plotted position of the analyses on their fig. 9b, while Pearce & Cann (1973) state clearly that the Ti-Zr diagram is to be used for rocks *not* falling within the within-plate basalt field on their Ti-Zr-Y diagram. Further, pillow lavas from Downan Point are described as very rich in amygdales, which increase markedly towards the pillow margin: the very high $H_2O + CO_2$ content (up to 15%) indicates that a substantial part of the 'rock' analysed was in fact amygdale.

The purpose of this study is to present comprehensive chemical and Nd isotope data for some of the freshest available Ballantrae lavas, and to discuss these primarily with regard to their petrogenesis. This is a prerequisite for satisfactory comparison with modern volcanic rocks, for simple processes such as alteration, phenocryst separation or accumulation can produce erroneous results in many of the diagrams used to identify tectonic environments.

Sampling and analytical techniques

Concentrations of major and trace elements, including REE, in eighteen Ballantrae lavas are given in Table 1, with REE concentrations for additional rocks and mineral separates in Table 2. Sample localities and brief petrographies are also given in these Tables. Samples were collected from massive parts of outcrops or pillow cores

TABLE 1. *Major- and trace-element compositions of Ballantrae lavas*

	PB1	PB2	PB3	PB4	PB5	PB6	PB7	PO3	PO5	PO7	PO9	BH1	BH2	GL1	GL2	MH1	MH2	MH3
SiO_2	49.29	48.70	49.87	49.90	51.26	49.58	51.66	48.12	49.94	49.39	49.46	53.93	56.55	52.24	54.10	52.99	53.32	52.37
Al_2O_3	17.69	16.54	17.60	17.88	15.96	17.14	13.04	15.64	18.49	18.38	15.94	15.94	13.35	15.66	16.50	12.71	12.80	16.63
Fe_2O_3	11.05	13.23	10.58	9.87	11.65	11.70	12.59	12.98	11.48	11.50	15.86	6.03	9.84	10.29	8.80	8.95	8.65	10.93
MgO	3.40	4.00	3.52	4.49	5.21	5.44	4.19	3.94	4.25	4.53	4.06	3.69	5.60	10.44	8.44	9.04	8.59	6.22
CaO	11.81	10.68	11.24	7.81	7.57	7.80	10.99	9.07	8.44	8.66	4.11	14.52	7.45	5.07	5.59	13.76	13.66	9.82
Na_2O	3.50	3.42	3.68	5.05	5.65	4.89	3.43	5.90	4.79	5.16	6.25	3.12	3.43	5.22	5.60	2.38	2.44	2.52
K_2O	0.362	0.392	0.309	1.159	0.292	0.672	0.351	0.430	0.303	0.366	0.062	1.353	1.309	0.513	0.584	0.386	0.368	1.288
TiO_2	2.79	2.90	2.69	2.16	2.22	2.34	3.37	3.11	1.658	1.599	3.20	1.912	2.052	0.528	0.456	0.479	0.493	0.618
MnO	0.241	0.334	0.280	0.164	0.170	0.200	0.331	0.160	0.298	0.437	0.179	0.143	0.095	0.158	0.126	0.192	0.170	0.125
P_2O_5	0.391	0.405	0.377	0.288	0.291	0.311	0.476	0.634	0.161	0.163	0.707	0.254	0.320	0.074	0.020	0.078	0.101	0.138
LOI	1.06	1.14	0.75	2.98	2.88	2.78	1.25	6.58	4.28	6.22	5.05	4.97	5.23	3.40	2.70	2.44	2.56	1.79
Ni	34	33	33	44	49	49	31	144	67	59	228	58	36	98	65	69	67	32
Cr	33	35	31	47	41	55	39	164	195	184	207	214	34	211	181	504	499	102
V	365	371	344	271	310	326	432	222	423	381	179	313	305	264	322	327	322	426
Sc	37.7	38.0	34.9	29.1	35.2	35.0	42.0	21.8	76	77	20.8	35.8	31.3	50.3	47.2	58	59	48.5
Cu	104	108	103	72	57	83	130	18	65	60	12	78	85	66	83	104	101	29
Zn	106	108	96	81	96	85	133	132	98	93	206	82	87	95	74	119	132	61
Pb	1.9	2.3	2.0	2.8	1.5	0.8	2.4	1.3	0.3	0.5	0.6	0.9	0.7	4.9	3.5	1.8	1.9	3.3
Sr	432	397	428	662	141	604	332	317	209	213	85	455	257	272	279	267	270	348
Rb	3.6	2.6	2.2	11.9	4.0	5.6	3.7	7.3	5.6	6.5	1.4	14.9	11.9	5.2	5.5	4.9	4.9	25.7
Zr	252	260	239	183	186	193	306	326	96	93	354	160	163	24	23	22	23	33
Nb	12.4	12.5	12.2	11.8	12.2	13.2	15.1	54	5.5	3.9	57	11.9	11.0	1.9	1.2	1.8	0.9	2.2
Ba	74	82	76	297	53	241	84	134	70	84	16	378	124	58	41	71	72	154
Th	n.d.	n.d.	n.d.	n.d.	n.d.	n.d.	2	6	n.d.	n.d.	4	n.d.	n.d.	n.d.	n.d.	n.d.	n.d.	3
Nd*	27	26	23	22	18	19	31	41	6	9	40	21	19	n.d.	n.d.	n.d.	n.d.	5
Y	35.0	36.9	35.9	25.8	27.6	29.9	43.0	34.7	31.3	29.9	33.0	23.0	24.4	17.4	15.4	13.0	13.1	17.8
La	22*	18.7	15*	13*	14*	13*	18*	41.9	13.05	16*	41*	13.23	10*	0.64	0.934	n.d.*	3.00	6.72
Ce	39*	44.8	36*	33*	32*	33*	53*	94.0	18.90	17*	86*	31.4	31*	1.95	2.48	n.d.*	7.10	14.73
Nd	—	28.65	—	—	—	—	—	45.57	12.019	—	—	18.75	—	2.030	2.349	—	4.839	8.910
Sm	—	7.060	—	—	—	—	—	9.305	3.619	—	—	4.530	—	0.927	1.005	—	1.402	2.284
Eu	—	2.284	—	—	—	—	—	3.34	1.206	—	—	1.580	—	0.369	0.422	—	0.487	0.760
Gd	—	7.63	—	—	—	—	—	8.93	4.65	—	—	4.75	—	1.58	1.74	—	1.65	2.56
Dy	—	7.17	—	—	—	—	—	6.88	5.39	—	—	4.24	—	2.44	2.38	—	1.82	2.62
Er	—	4.01	—	—	—	—	—	3.23	3.37	—	—	2.268	—	1.784	1.620	—	1.134	1.616
Yb	—	3.58	—	—	—	—	—	2.55	3.08	—	—	1.940	—	1.892	1.501	—	1.904	1.531
Pheno-crysts '%'	pl-ol 35	pl-ol 35	pl-ol 35	pl-ol 40	pl-ol 40	pl-ol 40	pl-ol 35	ol 1	ol 1	none —	ol 1	pl-ol 3	pl-ol 3	cpx-pl 10	cpx-pl-ol 15	cpx-pl-ol 60	cpx-ol-pl 60	cpx-pl-ol 80
Grid ref.	NX141 920	NX141 920	NX141 920	NX141 920	NX141 920	NX141 920	NX139 918	NX136 913	NX136 913	NX136 913	NX136 913	NX092 869	NX092 869	NX105 881	NX105 881	NX088 825	NX088 825	NX088 825

Oxides in wt %, trace elements in ppm n.d. = not detected. Fe_2O_3 = total iron as Fe_2O_3, LOI = loss on ignition: concentrations are reported on a volatile-free basis. * = La, Ce, Nd by XRF. Localities are: PB = Pinbain, PO = Pinbain olistostrome, BH = Bennane Head, GL = Games Loup and MH = Mains Hill. Phenocrysts are listed in approximate order of abundance. Abbreviations: ol = olivine, pl = plagioclase, cpx = calcic clinopyroxene. Phases underlined are present as pseudomorphs only. '%' refers to the approximate volume % phenocrysts.

TABLE 2. *REE concentrations (ppm) for mineral separates and for whole rocks for which XRF chemical data are not available.*

	GL3	BG2	DP1	GL2-cpx	GL3-cpx	GL3-pl	MH2-glass	MH2-cpx	MH3-cpx	DP1-cpx
La	0.964	3.39	10.92	0.051	0.0607	0.054	4.79	0.179	0.53	2.24
Ce	2.66	9.24	27.2	0.221	0.298	0.1287	10.93	0.760	1.97	9.28
Nd	2.365	7.280	17.14	0.4781	0.5992	0.0949	7.379	1.110	2.370	11.79
Sm	0.9362	2.573	4.268	0.3446	0.3869	0.0353	2.025	0.4953	0.992	4.261
Eu	0.365	1.055	1.496	0.1473	0.1658	0.0283	0.703	0.1769	0.306	1.395
Gd	1.489	3.74	4.48	—	0.73	0.0571	2.31	0.708	1.421	5.49
Dy	2.09	4.86	4.10	—	1.241	—	2.43	0.838	—	5.11
Er	1.432	3.33	2.157	—	0.856	0.0586	1.540	0.497	1.048	2.53
Yb	1.468	3.35	1.812	0.877	0.811	0.0619	1.492	0.435	0.979	1.941

GL3—cpx-pl-ol-phyric basalt, Games Loup (NX104880). *c.* 15% phenocrysts, cpx and pl fresh.
BG2—aphyric basalt, Bargain Hill (NX198888).
DP1—aphyric, coarsely ophitic basalt, Downan Point (NX070806); cpx fresh.
cpx = calcic clinopyroxene, pl = plagioclase, ol = olivine.

N.B. Concentrations in Table 2 are *not* reported on a volatile-free basis.

with the fewest possible amygdales or veins. As these are dominantly of calcite, their presence is reflected in high loss on ignition (LOI) values, e.g. for the samples from the Pinbain olistostrome (PO). The poor exposure and very altered nature of the volcanic rocks inland mean that the collection is biased towards rocks exposed on the coast and should not be considered as representative of the complex as a whole. Analytical techniques, precision and accuracy were the same as those of Thirlwall (in press). Pb concentrations are much lower, but the XRF calibration included several samples within the range reported here and Pb concentrations are considered accurate to ± 0.3 ppm (2σ).

Results

Introduction

Major- and compatible trace-element (Si, Al, Fe, Mg, Ca, Na, Mn, Cr, Ni, Sc and V) concentrations in basaltic lavas principally reflect their low-pressure fractional crystallization history. Low-pressure processes can show significant differences between different tectonic regimes, and many chemical diagrams supposedly discriminant of tectonic setting have been based on these differences. We consider that these can not be satisfactorily applied to ancient rocks unless full information is available on the phase relationships and P, T and a_{H_2O} conditions during their crystallization history. Such information is not available for the Ballantrae lavas.

In contrast, the isotope systematics and incompatible-element concentrations and ratios of basaltic lavas can be used to provide information on mantle melting processes and on the compositions of the mantle-source regions. As basalt generation processes differ markedly between tectonic regimes, incompatible-element ratios can be satisfactorily used for discrimination processes, as demonstrated by Pearce & Cann (1973) and other workers. High-level processes can modify such relationships, however; for example on the Zr-Ti diagram of Pearce & Cann (1973), extensive low-pressure fractional crystallization can increase Zr and Ti concentrations at constant Zr/Ti, while titanomagnetite extraction can lead to higher Zr/Ti ratios. More complex open-system fractional crystallization can produce large variation in incompatible-element ratios (e.g. O'Hara & Mathews 1981), but this seems either unable to modify incompatible-element ratio tectonic setting relationships or it is in part responsible for them. Hydrothermal alteration will modify concentrations and ratios of the elements most readily transported in hydrous fluids (Ca, Na, K, Rb, Sr, Ba, U and perhaps Pb) while a much wider range of elements are mobile in CO_2 rich fluids at high temperatures and pressures. Such effects are often associated with mineralogical alteration and can be detected by wide scatter in mobile-element concentrations uncorrelated with more stable elements.

These processes may be effectively discussed with reference to 'spider' diagrams (e.g. Fig. 2). These are extensions of the chrondrite-normalized REE patterns to other incompatible elements. Normalization is to chondrites for refractory elements and to estimated bulk earth for volatile elements (after Sun 1980). Elements are arranged in an estimated left-to-right order of decreasing incompatibility in normal lherzolite-phase assemblages. This order is not known precisely, but is unimportant for comparative purposes. To facilitate comparison, all diagrams presented here have the same vertical scale. Examples of typical spider diagrams of modern volcanic rocks are given in Fig. 2. The concave downwards curve for N-type mid-ocean ridge basalts (N-MORB) reflects their derivation by large degrees of melting of depleted mantle, a mantle source from which a melt rich in incompatible elements has previously been extracted. The characteristic curves for a Hawaiian tholeiite and an alkali basalt show enrichment in more incompatible elements, though in both cases the slope flattens out at the left-hand side. These probably represent small partial melts of a slightly depleted mantle source although other more complicated petrogeneses can be suggested. Crystal fractionation of mafic minerals causes general translation of the curve to higher levels, and slight clockwise rotation relative to the heavy REE, while fractionation of plagioclase causes translation with development of Sr- and Eu-negative anomalies.

Figure 2b shows typical spider diagrams of arc volcanic rocks. Although these again show a range from depletion to enrichment in LREE relative to HREE, they have marked positive 'anomalies' in Pb, Rb, Ba, U, K and Sr (hydrophile elements), and a strong negative Nb 'anomaly'. The hydrophile elements are believed to be preferentially added to the mantle source of arc magmas through their much greater solubility in hydrous fluids released from subducted oceanic crust (e.g. Hawkesworth & Powell 1980), or through their relative enrichment by weathering in subducted oceanic sediments (e.g. Thirlwall, in press). The origin of the negative Nb anomaly is unclear but it is an almost universal feature of arc magmas.

The depletion in incompatible elements of the MORB source is known to be ancient because MORB have higher $^{143}Nd/^{144}Nd$ than chondrites (e.g. O'Nions *et al.* 1977). This reflects the existence of higher than chondritic Sm/Nd in the MORB source for a long period of time. Similarly, the low $^{87}Sr/^{86}Sr$ of MORB reflects long-term depletion of Rb relative to Sr in the MORB source. O'Nions *et al.* (1977) and other workers have shown that $^{87}Sr/^{86}Sr$ and $^{143}Nd/^{144}Nd$ are strongly anti-correlated in modern oceanic basalts (the mantle array), with MORB having isotopic compositions corresponding to the most extreme time-integrated source depletion in incompatible elements. The extreme depletion of the MORB source has been related to the progressive extraction of the continental crust through time from mantle initially chondritic in Sm/Nd (e.g. O'Nions *et al.* 1979). Because of growth in chondrite $^{143}Nd/^{144}Nd$ over geological time comparison with ancient rocks is best expressed in terms of $\varepsilon_{Nd}(t)$, the deviation in parts per 10,000 of the initial isotope ratio from chondrites at time t (DePaolo & Wasserburg 1977). ε_{Sr} is similarly defined for Sr isotopes, but is relative to a hypothetical bulk-earth model (UR). In principal, the high ε_{Nd} of MORB could be used to identify MORB components in ophiolite complexes, as modern MORB have ε_{Nd} from +8 to +14 (White & Hofmann 1982; Machado *et al.* 1982), mostly greater than +10. However, if the MORB source has developed its depletion progressively, whether by crustal extraction or by some other process, ancient MORB will be expected to have lower ε_{Nd} than modern MORB. Thus Jacobsen & Wasserburg (1979) reported $\varepsilon_{Nd}(t)$ from +6.5 to +8.1 (average +7.6) for members of the early Ordovician Bay of Islands Complex, Newfoundland which they interpreted as a characteristic value for early Ordovician MORB.

Modern ocean-island lavas typically show lower ε_{Nd} and higher ε_{Sr} than MORB, and define the mantle array extending from MORB towards the model bulk-earth isotopic composition. Many modern arc lavas are offset to the right of the mantle array, and usually have somewhat lower ε_{Nd} than MORB. The higher ε_{Sr} has been attributed to contamination of the mantle-source regions by Sr derived from subducted oceanic lithosphere altered by seawater (e.g. Hawkesworth *et al.* 1977), or to mixing with Sr derived from subducted sediment (e.g. Thirlwall, in press). These are similar processes to those thought to give rise to the hydrophile element enrichment of arc lavas.

Ballantrae isotope data

Nd-Sr istope data for eleven Ballantrae lavas are reported in Table 3. ε parameters are calculated at 490 Ma, based on the zircon age reported by Bluck *et al.* (1980), but Sm/Nd and Rb/Sr are sufficiently close to CHUR and UR values (DePaolo & Wasserburg 1977) respectively that an error of ±20 Ma in the age assumed makes little difference. The data is discussed in detail relative to the chemistry of the individual volcanic blocks, but some general comments are relevant here. $\varepsilon_{Nd}(490)$ ranges from +4.9 to +7.8, lower than modern MORB and mostly lower than data from the Bay of Islands Complex (Jacobsen & Wasserburg, 1979) and from Arenig N-MORB in the Southern Uplands accretionary prism (Thirlwall & Leggett, unpublished data). Despite evidence for addition of LREE to basalts during prolonged sea-floor weathering (Ludden &

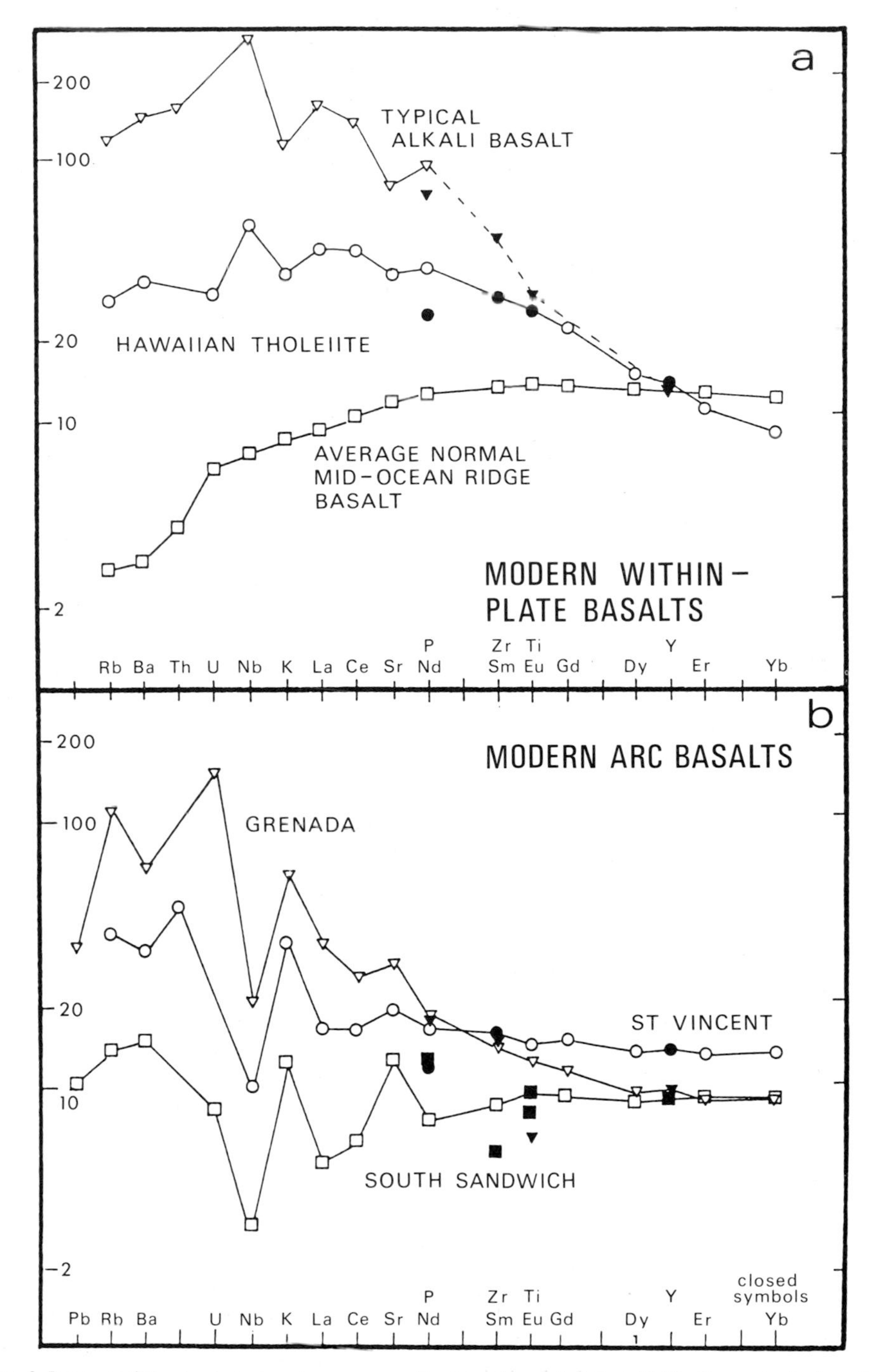

FIG. 2. Incompatible-element abundances normalized to 'chondrite' (Sun 1980) of some typical modern volcanic rocks. Closed symbols are for P, Zr, Ti and Y. Data sources are: alkali basalt from Cantal, France (H. Downes, unpublished data); Hawaiian tholeiite (BHVO-1) and average N-MORB from Sun *et al.* (1979); Grenada basic andesite 310A from Graham (1980) and Thirlwall (unpublished data); St Vincent basic andesite 7060 from Graham & Thirlwall (1981) and Thirlwall (unpublished data); South Sandwich Island basalt from Luff (1982).

TABLE. 3. *Isotope data and Rb, Sr concentrations (XRF) for Ballantrae lavas and mineral separates.*

	Rb (ppm)	Sr (ppm)	$^{87}Sr/^{86}Sr$	Sm/Nd	$^{143}Nd/^{144}Nd$	$^{148}Nd/^{144}Nd$	ε_{Nd}	ε_{Sr}
PB2	2.6	397	0.70311±5	0.2464	0.512834±7	0.241610±8	+6.6	−13.3
PB3	2.2	428	0.70304±3	—	—	—	—	−13.9
PO3	7.3	317	0.70642±5	0.2042	0.512715±9	0.241566±5	+5.9	+29.0
PO5	5.6	209	0.70477±4	0.3011	0.512950±7	0.241584±7	+6.8	+4.5
				0.3030				
BH1	14.9	455	0.70476±2	0.2417	0.512788±8	0.241606±4	+5.9	+2.6
GL1	5.4	272	0.70740±4	0.4569	0.513264±10	—	+7.0	+43.8
GL2	5.5	279	0.70720±20	0.4278	0.513218±15	—	+7.2	+41
GL2-cpx	—	2.8*	0.70348±3	0.7208	0.51374±4	0.24157±2	—	−6.1
GL3	—	—	—	0.3959	0.513184±9	0.241573±11	+7.8	—
GL3-cpx	0.043*	4.5*	0.70406±8	0.6457	0.513655±10	0.241573±14		−0.6
GL3-pl	—	—	—	0.3718	0.51311±6	—		
MH2	4.9	270	—	0.2898	0.512838±8	0.241560±12	+5.0	
				0.2896	0.512834±11	0.241544±14	+4.9	
MH2-glass	—	—	—	0.2745	0.512805±8	0.241563±8		
				0.2743	0.512801±13	—		
MH2-cpx	0.02*	21.5*	0.70332±4	0.4463	0.513145±8	0.241570±11		−8.7
	0.02*	21.6*	0.70332±5	0.4469	0.513149±11	—		−8.7
MH3	25.7	348	0.70540±7	0.2563	0.512787±11	0.241587±7	+5.3	−0.1
				0.2565				
MH3-cpx	0.02*	19.2*	0.70338±3	0.4185	0.51308±5	—		−7.9
BG2	—	—	—	0.3534	0.513000±8	0.241567±7	+5.8	—
DP1	—	—	—	0.2490	0.512789±8	0.241574±9	+5.6	—
					0.512803±7	0.241590±9		
DP1-cpx	—	30.8*	0.70283±10	0.3612	0.513005±7	0.241577±7		−15.4

ε parameters are calculated at 490 Ma, using $(^{143}Nd/^{144}Nd)_{CHUR,\,O} = 0.51264$, $(^{87}Sr/^{86}Sr)_{UR,\,O} = 0.7045$, $(^{147}Sm/^{144}Nd)_{CHUR} = 0.1936$, $(^{87}Rb/^{86}Sr)_{UR} = 0.0839$, $\lambda^{147}Sm = 6.54\times10^{-12}\ yr^{-1}$, $\lambda^{87}Rb = 1.42\times10^{-11}\ yr^{-1}$.
Errors quoted on the isotope analyses are 2 s.e. on the mean, and apply to the last digit(s) quoted. * = Rb, Sr analysed by isotope dilution.
Nd isotope ratios are normalized to $^{146}Nd/^{144}Nd = 0.7219$, and $^{87}Sr/^{86}Sr$ to $^{86}Sr/^{88}Sr = 0.1194$. Duplicate analyses are of completely separate dissolutions.

Sm–Nd mineral ages: Regression of mineral Sm-Nd data using the method of York (1969) gives the following results:
GL3: age = 476±14 Ma (2 points); DP1: Age = 468±22 Ma (2 points).
MH2: age =501±12 Ma (6 points) MSWD = 0.23.
Data for GL2 and MH3 are not sufficiently precise.

Thompson 1979), all studies so far have failed to show significant mobilization of Nd isotopes during alteration (e.g. McCulloch *et al.* 1980; Cohen & O'Nions 1982). Except for one very altered basalt from the Pinbain olistostrome (PO5), none of the samples analysed here shows the Ce anomaly found by Ludden & Thompson (1979) in basalts showing LREE alteration. The coherent behaviour of LREE with the more stable elements (Zr, Nb, etc.) is strong evidence that LREE have not been mobile in most analysed Ballantrae lavas, as is the closely similar LREE content of three spilite samples from Pinbain, despite wide variation in hydrophile element content. It is therefore thought that the measured $^{143}Nd/^{144}Nd$ represents the erupted composition modified since only by ^{147}Sm decay, which can be easily corrected for.

Except for fresh basalts from Pinbain, all whole-rock samples have much higher $\varepsilon_{Sr}(490)$ for their ε_{Nd} than modern oceanic basalts. This is a function of hydrothermal exchange of Sr and Rb during sea-floor alteration and later weathering (cf. Jacobsen & Wasserburg 1979). Fresh clinopyroxenes are known to retain magmatic Sr isotopes (e.g. Hawkesworth & Morrison 1978), and they have therefore been separated from several Ballantrae samples by standard magnetic and heavy liquid techniques. Prior to Sr isotope analysis, the pyroxenes were leached several times in hot 6 M HCl with a few drops of HF added, and were subsequently washed carefully. The Sr-Nd isotope data is plotted in Fig. 3: pyroxenes separated from Southern Uplands basalts (Thirlwall & Leggett, unpublished data) and from a Downan Point lava plot close to the centre of the modern mantle array, as does the fresh Pinbain basalt. In contrast, pyroxenes from

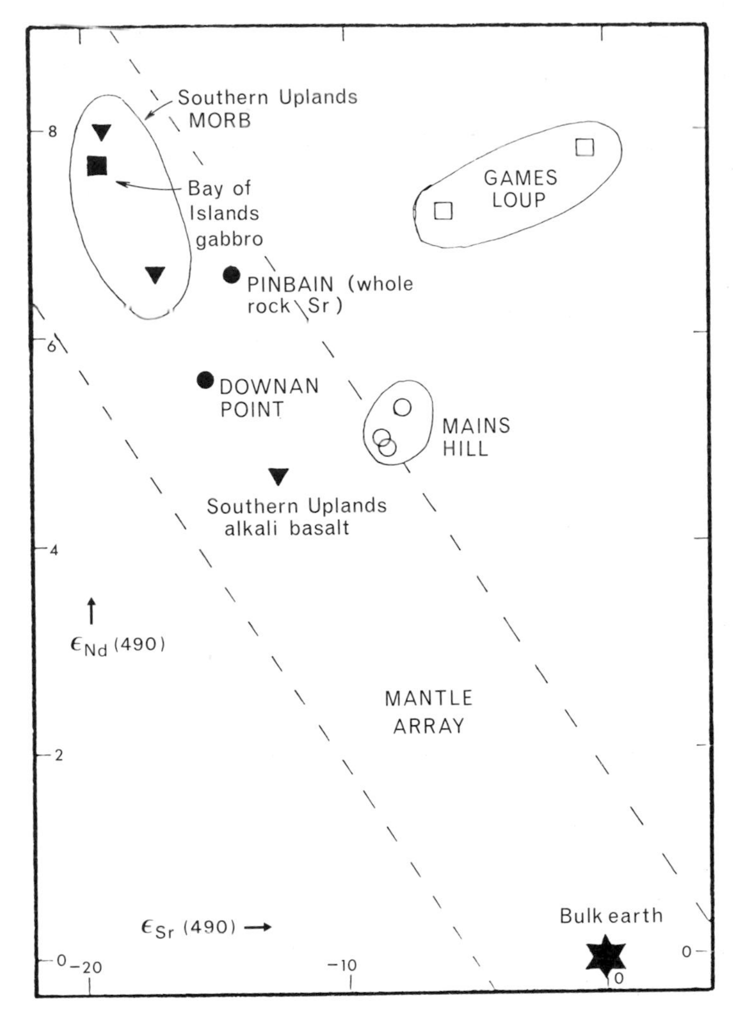

FIG. 3. ε_{Nd}–ε_{Sr} isotope data for Ballantrae samples. Sr data is from leached separated clinopyroxene except for the Pinbain sample. Southern Uplands data (also pyroxenes) from Thirlwall & Leggett (unpublished data). Bay of Islands data from Jacobsen & Wasserburg (1979). The latter extend to higher ε_{Sr} at constant ε_{Nd} through alteration (not shown).

Mains Hill and Games Loup have more radiogenic Sr than the mantle array. That this is not the result of ineffective leaching is suggested by the duplicate data for MH2, which were leached separately, and the fact that the Mains Hill and Games Loup pyroxene phenocrysts are appreciably less altered than the groundmass pyroxene separated from the Southern Uplands basalts.

Pinbain block

Volcanic rocks of the Pinbain block (Fig. 1) are of shallow-water origin and include spilite lava flows and lava clasts in tuffs which have remained unspilitized (Bluck 1981). All samples analysed are rich in plagioclase phenocrysts and have subordinate olivine pseudomorphs. Plagioclase is albite-oligoclase in the spilites but is fresh labradorite in the unspilitized basalt clasts. To assess alteration effects, three separate basalt clasts (PB1–3) and three samples of spilitic lava flow (PB4–6) have been analysed. It should be noted that the clasts are unlikely to have been derived from this lava flow, as there is a small fault between tuff and lava, and the samples are petrographically distinct.

The clasts are typical tholeiitic basalts with high Fe/Mg, V and Sc and low Ni and Cr. The high V and Sc suggest that fractional crystallization of olivine and plagioclase has been the most important high-level process. Spider diagrams for the three clasts are given in Fig. 4a and

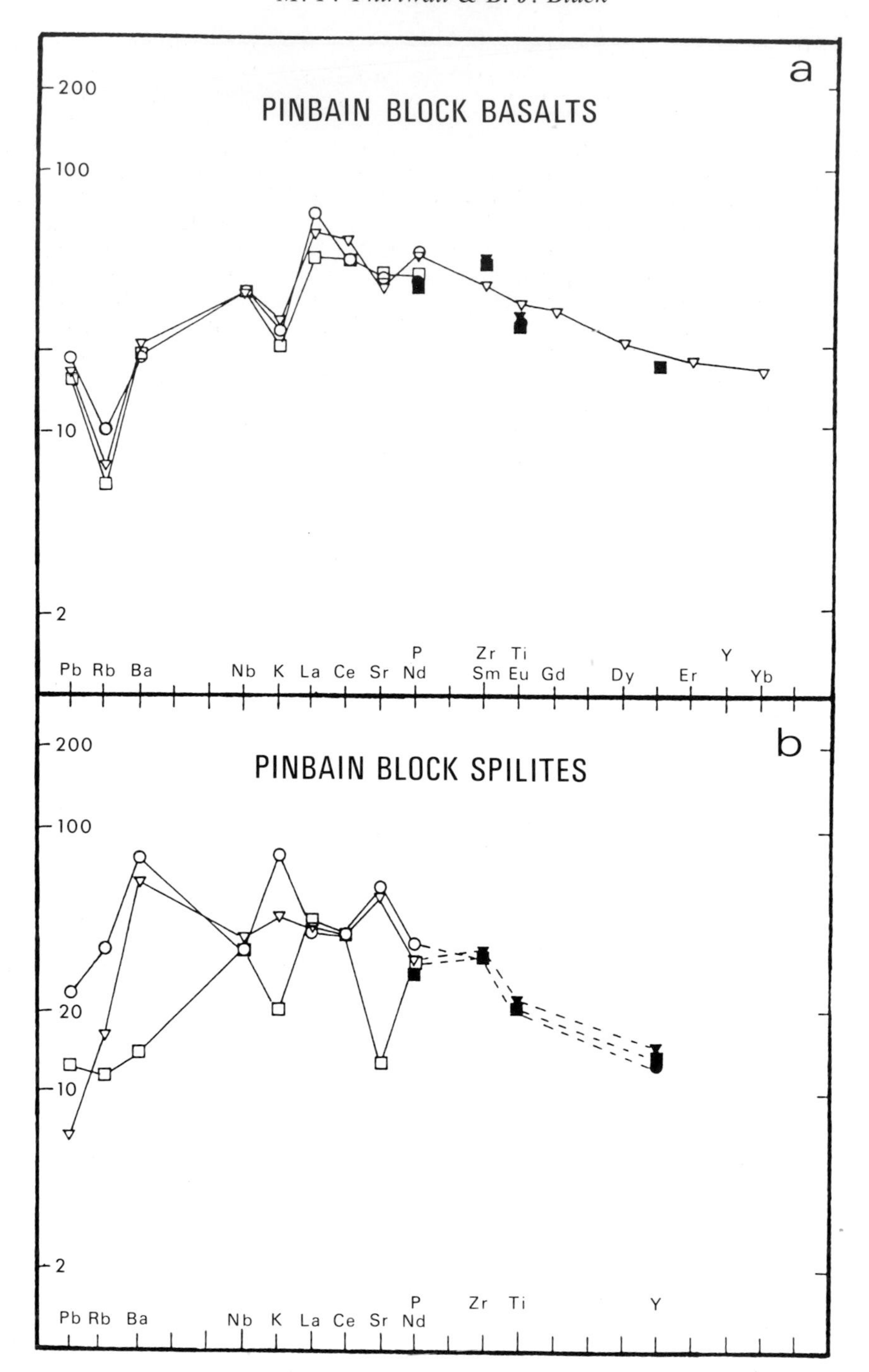

FIG. 4. Spider diagrams for Pinbain block samples (a) basalt clasts PB1–3 (b) samples PB4–6 of a single spilitized lava flow. Note the effect of alteration on incompatible-element concentrations. Normalization and closed symbols as Fig. 2.

are almost coincident. The patterns are smooth except for K and Rb depletion. The close similarity in composition and the smoothness of the patterns strongly implies that no elements have been significantly mobile during alteration of these clasts. Further evidence for the lack of alteration is provided by the low loss on ignition (<1.5%), and the low Sr isotope ratios, plotting within the mantle array. The chemical variation shown by the three clasts can be attributed to variable plagioclase accumulation or separation: thus the lower Al and Ca of PB2 are accompanied by slight negative anomalies in Sr and Eu, and higher incompatible-element concentrations. A fourth clast from lower in the sequence, PB7, also appears to have retained its original chemistry but is somewhat more fractionated than the other three.

The spilite samples show characteristically high Na_2O and low CaO. Spider diagrams (Fig. 4b) are highly scattered for the elements expected to be mobile in hydrous fluids (Pb, Rb, Ba, K and Sr), while other more stable elements have very similar concentrations in the three samples. The contrast between Figs 4a and b strongly emphasizes the effect of alteration on hydrophile-element concentrations. The spilites have substantially lower concentrations of immobile incompatible elements (Zr, Ti, Nb etc), and higher Ni and Cr than the basalt clasts, implying that they are somewhat less fractionated.

No Pinbain sample shows the characteristic Nb depletion of arc magmas, nor do fresh samples show relative hydrophile-element enrichment. The spider diagrams are very similar to that of the Hawaiian tholeiite (Fig. 2a), and the Sr-Nd isotopes are typical of ocean-island basalts, in that they plot a little lower in the mantle array than contemporary MORB. It is most probable that the Pinbain block lavas are ocean-island tholeiites, although it is possible that they were generated at a highly anomalous spreading ridge. This confirms the conclusion reached by Jones (1977) about the Pinbain block.

Blocks in the Pinbain Beach olistostrome

Four highly altered clasts from the Pinbain olistostrome have been analysed (PO3, 5, 7 and 9). All four have high Na_2O suggesting spilitization, and high LOI reflecting substantial secondary calcite. The samples form two pairs: PO3 and PO9 have high concentrations of the stable incompatible elements with marked enrichment in Nb and light REE, while PO5 and PO7 have essentially flat normalized patterns (Fig. 5). Negative anomalies for many hydrophile elements in PO3 are most probably the result of alteration: this is supported by the high ε_{Sr} reflecting Sr exchange with hydrothermal fluids. The stable element pattern of PO3 resembles those of modern alkali basalts, particularly in the Nb enrichment. Such a composition is most likely to be generated in an ocean-island environment, which is also suggested by the relatively low ε_{Nd}. The flat pattern of PO5 is interrupted by an anomaly at La; this is not the result of analytical error (cf. Sm and Gd 'anomalies' of Lewis & Bloxam 1977) as it was precisely duplicated in a second dissolution of this sample. This suggests it is an alteration effect, and that the otherwise flat pattern is in part a chance product of alteration. The higher ε_{Nd} of this sample indicates greater time-integrated LREE depletion in its source, although such an interpretation is tentative because of the evidence for La alteration. A composition such as PO5 could reasonably be generated at a spreading ridge or ocean island, and these data therefore confirm the interpretation of Wilkinson & Cann (1974).

Bargain Hill

One somewhat sheared aphyric basalt has been collected from Bargain Hill, within the region of LREE-depleted basalts interpretated as island-arc material by Lewis & Bloxam (1977), and as MORB by Jones (1977). REE data for this sample are plotted in Fig. 5, and compare closely with mean N-MORB (Fig. 2a), but chemical data are at present insufficient to demonstrate that this represents MORB rather than primitive intra-oceanic island-arc material (cf. the South Sandwich Islands basalt, Fig. 2b). The low ε_{Nd} of the Bargain Hill sample is most interesting (Table 3). The mantle-source region of this lava has clearly not had a long history of LREE depletion relative to the source of the ocean-island-type basalts discussed earlier: it seems unlikely that the Bargain Hill sample could have been generated at a long-lived spreading centre. Further data are required for basalts from this area.

Bennane Head

The two samples analysed from Bennane Head are mildly plagioclase-phyric basalts with substantial secondary carbonate, reflected in the high LOI of both and the high CaO of BH1. Spider diagrams are very similar in stable elements to those of the Pinbain spilites (Fig. 4), and are nearly parallel to the Pinbain basalt clasts, but hydrophile elements show considerable scatter. Again no negative Nb anomaly is present. The strong similarity with the Pinbain rocks implies a closely similar petrogenesis, and therefore the Bennane Head rocks were probably also

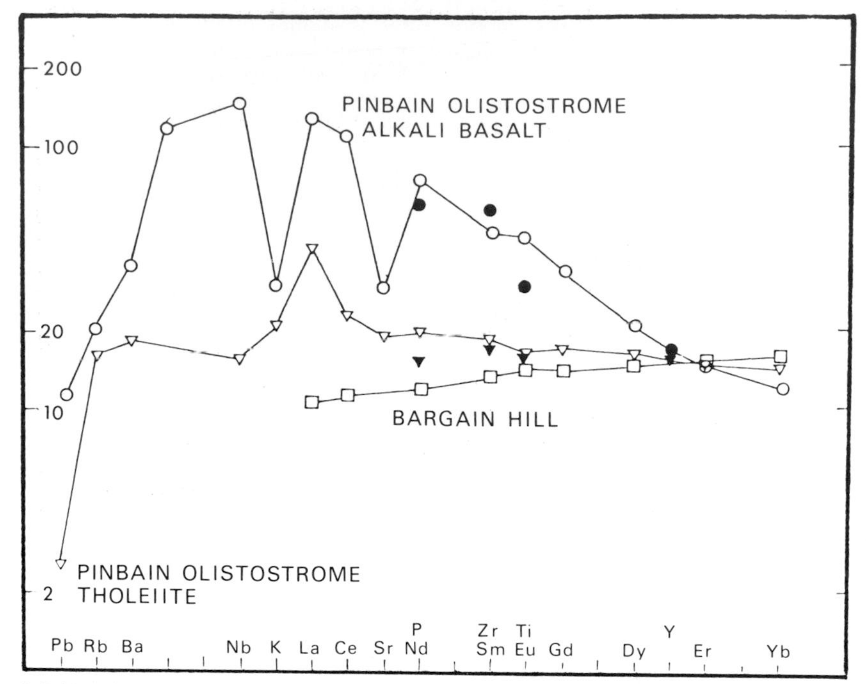

FIG. 5. Spider diagrams for an alkali basalt (PO3) and a tholeiite (PO5) from blocks in the Pinbain Beach olistostrome, and REE data for BG2 (Bargain Hill). Normalization and closed symbols as Fig. 2.

generated in an ocean-island setting. This confirms the data of Wilkinson & Cann (1974). The lower ε_{Nd} than the Pinbain samples could suggest a slightly different source region, however.

Mains Hill

Three basalt clasts have been collected from volcaniclastic units at Mains Hill. Two of these have abundant large euhedral fresh augite phenocrysts with subordinate sericitized plagioclase and pseudomorphed olivine, while the third (MH3) is rich in much smaller phenocrysts of the same minerals. Jones (1977) suggested that the unusual chemistry of Mains Hill lavas might be explained by augite accumulation. MH1 and MH2 are chemically almost identical, and the very high Cr, Sc and CaO most probably result from augite accumulation. The much lower concentrations of these elements in MH3 suggest that this is close to a liquid composition, and that the small phenocrysts in MH3 are merely microphenocrysts grown rapidly during quenching. Spider diagrams for MH2 and MH3 are given in Fig. 6a and are very distinctive subparallel patterns with marked enrichment in hydrophile elements and depletion in Nb. The rocks are mildly LREE-enriched but show marked depletion in Zr, Ti and Nb relative to the REE. These features are characteristic of modern arc lavas, and the Mains Hill rocks closely resemble the St Vincent sample shown in Fig. 2b.

The REE concentrations in augite phenocrysts separated from MH2, MH3 and two Games Loup lavas (with $<10\%$ phenocryst augite), and in groundmass material ('glass') separated from MH2, are given in Table 2. The ratios (REE in augite/REE in whole rock) show little variation between samples (from 0.06 to 0.08 for La, and from 0.41 to 0.65 for Yb) and are within the range of augite-melt distribution coefficients (Irving 1978). The ratio (REE in augite/REE in groundmass) for MH2 gives inordinately low 'distribution coefficients' implying that the augite can be only partly accumulative. The concentrations of LREE, Sr and presumably Nb in the augite are so low that the incompatible-element ratios of the sample are dominated by the groundmass, and hence augite accumulation can not explain the distinctive chemistry of these samples. It can not be proved that the positive hydrophile-element anomalies are primary magmatic features rather than the products of alteration, but the almost identical chemistry of MH1 and MH2 and the parallel spider diagrams of MH2 and MH3 strongly suggest that they have a primary magmatic origin, perhaps enhanced a little by alteration. The rocks are not spilitized and have relatively low LOI.

The $\varepsilon_{Nd}(490)$ of the Mains Hill rocks is the

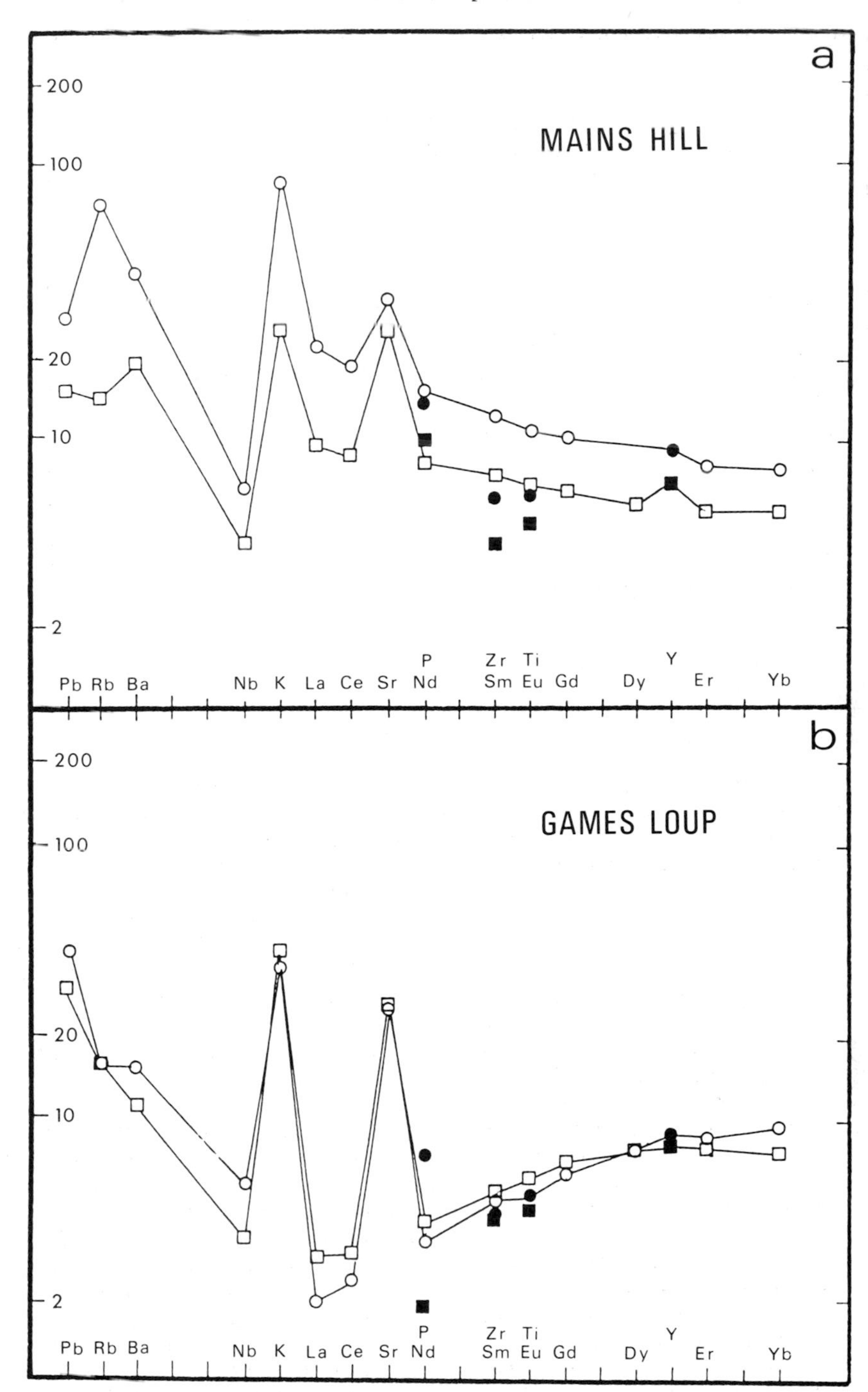

FIG. 6. Spider diagrams for samples from (a) Mains Hill ○ = MH3, □ = MH2 and (b) Games Loup ○ = GL1, □ = GL2. Normalization and closed symbols as Fig. 2.

lowest so far analysed at Ballantrae; combined with Sr isotope data on separated augite phenocrysts the samples are offset slightly to the right of the mantle array defined by modern and Ordovician oceanic basalts (Fig. 3). As discussed earlier, this is most unlikely to be the result of alteration of the pyroxene. A comparable offset is often seen in modern arc rocks, and arc lavas with similar chemistry to the Mains Hill rocks tend to plot in a similar position on the Sr-Nd isotope diagram relative to contemporary MORB—e.g. St Kitts (Hawkeswort & Powell 1980); St Vincent (Thirlwall, unpublished data). The lower ε_{Nd} of the Mains Hill rocks requires a different source region to other samples in the Ballantrae complex, and it seems highly likely that the rocks originally formed part of a fairly mature island arc.

Games Loup

This small region is separated from Bennane Head by a fault that passes through Balcreuchan Port. The volcanics are easily distinguished from those of Bennane Head: volcaniclastic material is sparse and the lavas lack vesicles, suggesting a relatively deep-water origin, while all samples so far examined in thin section have some 5–15% fresh augite and plagioclase phenocrysts. Analyses of two samples are presented in Table 1 and REE concentrations in a third are given in Table 2. The samples are primitive basalts with high MgO, Ni and Cr, but their high Na_2O and low CaO reflect spilitic alteration. The rocks have very distinctive incompatible-element patterns (Fig. 6b) with extremely low concentrations of LREE, Zr, Nb, Ti and P and very low LREE/HREE ratios. This implies that the magmas represent greater degrees of partial melting of the MORB source than MORB or are melts of a mantle more depleted than the MORB source. Such extreme incompatible-element depletion is only common in modern primitive intra-oceanic island arcs, e.g. in the South Sandwich Islands (Fig. 2b). This forms the basis of discrimination between MORB and island-arc tholeiites using the Zr-Ti diagram of Pearce & Cann (1973), but conditions can be envisaged in which such depletion can be produced outside the island arc environment. The rocks do not show Nb depletion relative to LREE as observed in other arc rocks, but the Nb depletion is also not very convincing in South Sandwich basalts (Fig. 2b). This could be due to Nb concentrations being close to the XRF detection limit, or to the introduction of small amounts of Nb during grinding in a tungsten-carbide mill. The Games Loup lavas show marked hydrophile-element enrichment (Fig. 6b), which is fairly consistent in the two samples analysed, but much more extreme than observed in South Sandwich lavas (Fig. 2b). The high whole-rock $\varepsilon_{Sr}(490)$ (Table 3) suggests hydrophile-element alteration, however, and the hydrophile-element enrichment can not confidently be identified as primary.

The $\varepsilon_{Nd}(490)$ for Games Loup lavas ranges from +7.0 to +7.8, considerably higher than any other Ballantrae sample. This must reflect greater time-integrated LREE depletion in the source, but, if the lava Sm/Nd ratios are similar to those of their sources, the Games Loup lava source only needs to have remained distinct from other Ballantrae magma sources for *c.* 100 Ma. Combined with Sr isotopes determined on separated augite phenocrysts, the samples are offset far to the right of the mantle array on the ε_{Nd}–ε_{Sr} diagram (Fig. 3). Unfortunately, it is just possible that this reflects inadequate leaching of altered Sr-rich material coating the pyroxenes, especially as their Sr contents are very small (Table 3). However, the Sr-Nd isotopes of Games Loup pyroxenes are very similar to those of South Sandwich lavas (Hawkesworth *et al.* 1977) in that they are strongly offset to the right of the mantle array but have comparable ε_{Nd} to contemporary MORB. We believe that the similarities are sufficiently strong for the Games Loup lavas to be regarded as primitive island-arc tholeiites.

Downan Point

One ophitic basalt from Downan Point has been analysed for REE and Nd isotopes. As noted earlier, it is not clear that the Downan Point lavas are Caradocian, as suggested by Lewis & Bloxam (1977). The sample is fairly strongly LREE-enriched and compares closely with the analyses of Lewis & Bloxam (1977). $\varepsilon_{Nd}(490)$ is relatively low, and combined with Sr isotopes determined on leached separated groundmass augite, falls within the region of Fig. 3 typical of modern ocean-island lavas. The high REE content of the augite (Table 3) is a function of its quench origin: REE are not expected to show equilibrium distribution between augite and host magma. Both chemistry and isotopes support the views of Lewis & Bloxam (1977) and Wilkinson & Cann (1974) that the Downan Point basalts were erupted in an ocean-island setting.

Discussion and conclusions

The comprehensive chemical and Nd-isotope data presented here confirm the conclusions of Wilkinson & Cann (1974) that basalts generated in a wide variety of tectonic environments are present within the Ballantrae 'ophiolite', and that individual blocks of volcanic rocks represent

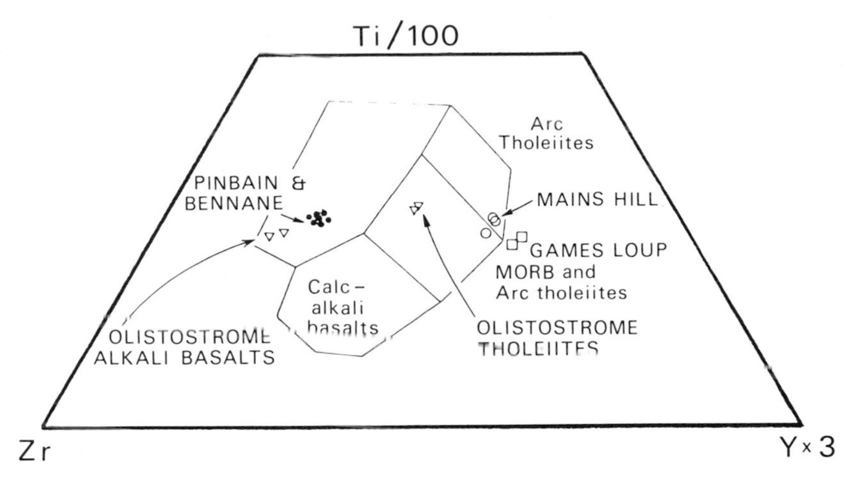

FIG. 7. Ti-Y-Zr diagram for Ballantrae samples. Fields from Pearce & Cann (1973). Note close coherence of samples from a single locality.

tectonically juxtaposed material which may be unrelated in time or space. The simpler models advanced by Lewis & Bloxam (1977), Jones (1977) and Jelinek *et al.* (1980) were principally the result of not taking into account data from all volcanic units. Ti-Y-Zr data presented here are plotted in Fig. 7 and closely resemble the data published by Wilkinson & Cann (1974).

As observed by Wilkinson & Cann (1974) the close proximity of such a wide variety of basalt types requires major tectonic shortening or age differences between the volcanic units, or both. An attempt has been made to date volcanism in the Mains Hill, Games Loup and Downan Point units by means of Sm-Nd internal isochrons on lavas and their separated clinopyroxene phenocrysts. Unfortunately, this does not provide much spread in Sm/Nd, and the ages obtained (Table 3) are just within error of each other, although the Mains Hill age is significantly greater than the zircon age for the Byne Hill trondhjemite reported by Bluck *et al.* (1980). A study of microfossils in interbedded sediments may prove more effective in establishing age differences. The variations in $\varepsilon_{Nd}(490)$ probably imply considerable initial geographic separation between the volcanic units, although their sources need not have remained distinct for more than some 100 m.y. prior to volcanism. The Games Loup rocks are unlikely to have formed in the same arc environment as those of Mains Hill.

Although it is not the purpose of this contribution to assess plate tectonic models in the light of data from such a small area, the confirmation of the existence of intra-oceanic arc lavas within the Ballantrae complex does tend to favour the idea that the complex was related to closure of a small marginal basin, as proposed by Bluck *et al.* (1980).

ACKNOWLEDGMENTS: We are grateful to Drs J. K. Leggett and A. H. F. Robertson for helpful discussion, to Messrs W. Wilkinson and T. Oddy for performing the mineral separations, and to Dr J. G. Fitton, Mrs D. E. James and Mr G. R. Angell for help with the XRF analyses in Edinburgh. Isotope and REE research at Leeds are supported by NERC and the Royal Society.

References

BARRETT, T. J., JENKYNS, H. C., LEGGETT, J. K. & ROBERTSON, A. H. F. 1982. Comment on Bluck, Halliday, Aftalion & MacIntyre 1980, Age and origin of Ballantrae ophiolite and its significance to the Caledonian orogeny and Ordovician timescale. *Geology* **10**, 331.

BLUCK, B. J. 1981. Hyalotuff deltaic deposits in the Ballantrae ophiolite of SW Scotland: evidence for crustal position of the lava sequence. *Trans. R. Soc. Edinb. Earth Sci.* **72**, 217–28.

——, HALLIDAY, A. N., AFTALION, M. & MACINTYRE, R. M. 1980. Age and origin of Ballantrae ophiolite and its significance to the Caledonian orogeny and Ordovician time scale. *Geology* **8**, 492–95.

CHURCH, W. R. & GAYER, R. A. 1973. The Ballantrae ophiolite. *Geol. Mag.* **110**, 497–510.

COHEN, R. S. & O'NIONS, R. K. 1982. The lead, neodymium and strontium isotopic structure of ocean ridge basalts. *J. Petrology* **23**, 299–324.

DEPAOLO, D. J. & WASSERBURG, G. J. 1977. The sources of island arcs as indicated by Nd and Sr isotopic studies. *Geophys. Res. Lett.* **4**, 465–8.

GRAHAM, A. M. 1980. *Genesis of the igneous rock suites of Grenada, Lesser Antilles.* Thesis, PhD, Univ. Edinburgh (unpubl.).

—— & THIRLWALL, M. F. 1981. Petrology of the 1979 eruption of Soufriere Volcano, St Vincent, Lesser Antilles. *Contrib. Mineral. Petrol.* **76**, 336–42.

HAWKESWORTH, C. J. & MORRISON, M. A. 1978. A reduction in $^{87}Sr/^{86}Sr$ during basalt alteration. *Nature, Lond.* **276**, 381–3.
—— & POWELL, B. M. 1980. Magma genesis in the Lesser Antilles island arc. *Earth Planet. Sci. Lett.* **51**, 297–308.
——, O'NIONS, R. K., PANKHURST, R. J., HAMILTON, P. J. & EVENSEN, N. M. 1977. A geochemical study of island-arc and back-arc tholeiites from the Scotia Sea. *Earth Planet. Sci. Lett.* **36**, 253–62.
IRVING, A. J. 1978. A review of experimental studies of crystal/liquid trace element partitioning. *Geochim. Cosmochim. Acta* **42**, 743–70.
JACOBSEN, S. B. & WASSERBURG, G. J. 1979. Nd and Sr isotopic study of the Bay of Islands ophiolite complex and the evolution of the source of mid-ocean ridge basalts. *J. Geophys. Res.* **84**, 7429–45.
JELINEK, E., SOUCEK, J., BLUCK, B. J., BOWES, D. R. & TRELOAR, P. J. 1980. Nature and significance of beerbachites in the Ballantrae ophiolite, SW Scotland. *Trans. R. Soc. Edinb. Earth Sci.* **71**, 159–79.
JONES, C. M. 1977. *The Ballantrae Complex as compared to the ophiolites of Newfoundland.* Thesis, PhD, Univ. Wales, Cardiff (unpubl.).
LEWIS, A. D. & BLOXAM, T. W. 1977. Petrotectonic environments of the Girvan-Ballantrae lavas from rare-earth element distributions. *Scott. J. Geol.* **13**, 211–22.
LUDDEN, J. N. & THOMPSON, G. 1979. An evaluation of the behaviour of the rare earth elements during the weathering of sea-floor basalt. *Earth Planet. Sci. Lett.* **43**, 85–92.
LUFF, I. W. 1982. *Petrogenesis of the island arc tholeiite series of the South Sandwich Islands.* Thesis, PhD, Univ. Leeds (unpubl.).
MACHADO, N., LUDDEN, J. N., BROOKS, C. & THOMPSON G. 1982. Fine-scale isotopic heterogeneity in the sub-Atlantic mantle. *Nature, Lond.* **295**, 226–8.
MCCULLOCH, M. T., GREGORY, R. T., WASSERBURG, G. J. & TAYLOR, H. P. JR 1980. A neodymium, strontium and oxygen isotopic study of the Cretaceous Samail ophiolite and implications for the petrogenesis and seawater-hydrothermal alteration of oceanic crust. *Earth Planet. Sci. Lett.* **46**, 201–211.
O'HARA, M. J. & MATHEWS, R. E. 1981. Geochemical evolution in an advancing, periodically replenished, periodically tapped, continuously fractionated magma chamber. *J. geol. Soc. London* **138**, 237–77.
O'NIONS, R. K., EVENSEN, N. M. & HAMILTON, P. J. 1979. Geochemical modelling of mantle differentiation and crustal growth. *J. Geophys. Res.* **84**, 6091–101.
—— , HAMILTON, P. J. & EVENSEN, N. M. 1977. Variations in $^{143}Nd/^{144}Nd$ and $^{87}Sr/^{86}Sr$ ratios in oceanic basalts. *Earth Planet. Sci. Lett.* **34**, 13–22.
PEARCE, J. A. & CANN, J. R. 1973. Tectonic setting of basic volcanic rocks determined using trace element analysis. *Earth Planet. Sci. Lett.* **19**, 290–300.
SUN, S.-S. 1980. Lead isotopic study of young volcanic rocks from mid-ocean ridges, ocean islands and island arcs. *Phil. Trans. R. Soc. London* **A297**, 409–46.
——, NESBITT, R. W. & SHARASKIN, A. Y. 1979. Geochemical characteristics of mid-ocean ridge basalts. *Earth Planet. Sci. Lett.* **44**, 119–38.
THIRLWALL, M. F. Isotope geochemistry and origin of calc-alkaline lavas from a Caledonian continental margin volcanic arc. *J. Volcanol. Geotherm. Res.* (in press).
WHITE, W. M. & HOFMANN, A. W. 1982. Sr and Nd isotope geochemistry of oceanic basalts and mantle evolution. *Nature, Lond.* **296**, 821–5.
WILKINSON, J. M. & CANN, J. R. 1974. Trace elements and tectonic relationships of basaltic rocks in the Ballantrae igneous complex, Ayrshire. *Geol. Mag.* **111**, 35–41.
YORK, D. 1969. Least squares fitting of a straight line with correlated errors. *Earth Planet. Sci. Lett.* **5**, 320–4.

M. J. THIRLWALL, Department of Earth Sciences, University of Leeds, Leeds 2, UK.
B. J. BLUCK, Department of Geology, University of Glasgow, Glasgow, Scotland.

Chemical and isotopic heterogeneities in orogenic and ophiolitic peridotites

M. A. Menzies

SUMMARY: Rare-earth element (REE) and isotopic analysis of clinopyroxenes from orogenic and ophiolitic peridotites reveal a complex multi-stage origin resulting in chemical and isotopic heterogeneity in the upper mantle. The bulk of the pyroxenes plot within the 'mantle array' in a region characterized by mantle depleted in light REE for much of Earth's history. Interestingly the majority of the pyroxenes separated from orogenic and ophiolitic peridotites have ε_{Nd} and ε_{Sr} = MORB. As a consequence of such isotopic (and REE) characteristics the mantle sections of ophiolites are not believed to be related to the overlying plutonic and volcanic rocks by a simple parent–daughter melting relationship.

Ultramafic rocks found in orogenic belts are fragments of the upper mantle emplaced by obduction processes or diapirism. Trace-element and isotopic analyses of these orogenic and ophiolitic peridotites provide an invaluable insight into mantle heterogeneity and mantle evolutionary processes. In this context neodymium (Nd) isotopes are particularly useful in conjunction with rare-earth elements (REE) since these elements appear to be unaffected by serpentinization. Furthermore, the unique isotopic signature of magmas erupted at mid-ocean ridges may assist in determining the relationship between ophiolites and underlying metamorphic peridotites and the origin of orogenic and ophiolitic peridotites.

This preliminary study of diopsides in orogenic and ophiolitic peridotites was undertaken (i) to ascertain whether variations in REE and Sr-Nd isotopic composition exist in diopside separates; (ii) to shed some light on the origin of these variations, and (iii) to stimulate further research into the origin of orogenic and ophiolitic lherzolites and harzburgites. Clinopyroxenes were separated from six ophiolitic and orogenic peridotites and analysed for Sm, Nd, $^{143}Nd/^{144}Nd$ and $^{87}Sr/^{86}Sr$ isotopes by methods to be described elsewhere. These data are listed in Table 1. In the case of the Trinity peridotites from N. Carolina more complete REE data are also available (Menzies *et al.* 1977c).

Rare-earth geochemistry

Previous investigations of orogenic and ophiolitic peridotites established their light rare-earth element (LREE)-depleted character (Haskin *et al.* 1966; Frey 1969). Moreover these authors proposed that loss of an alkali earth and LREE-enriched fluid from a mantle with an initial heavy REE content of 1.5–3.0 × chondrite produced this depletion. Since plagioclase and spinel lherzolites have a relatively undepleted major- and minor-element content (Boudier 1972, 1978), it was further suggested that these peridotites may be a potential source for MORB-type liquids (Menzies & Murthy 1978). Some of the heterogeneity in REE is shown in Fig. 1. Clinopyroxenes from orogenic and ophiolitic peridotites exhibit a considerable range in $(Ce/Yb)_N$ ratio and the lowest $(Ce/Yb)_N$ ratio was found in pyroxenes from the Lizard peridotite (Frey 1969).

New REE analyses of clinopyroxenes from orogenic and ophiolitic peridotites display the same range in Sm/Nd as was previously reported (Fig. 1). These variations point to a mantle extremely heterogeneous for the LREE where $(La/Yb) < 1$. This characteristic depletion in LREE differs markedly from that observed in fragments of mantle disrupted by alkaline and kimberlitic magmas.

Diopsides occurring in lherzolites from Beni Bouchera, Morocco and Trinity, California are considerably richer in Σ REE than equivalent phases in the harzburgites underlying the Troodos and Othris ophiolites. The lherzolites are known to be significantly richer in clinopyroxenes (5–15%) relative to the harzburgites (<1%). Although complete REE analyses are not available, it can be seen that harzburgites below the Othris and Troodos ophiolitic complexes have an extreme depletion in LREE somewhat akin to the Lizard Complex (Frey 1969) (Fig. 1). One can infer from the modal analyses and clinopyroxene REE analyses that the whole-rock harzburgites would contain $(Sm)_N \sim 0.009$ and $(Nd)_N \sim 0.0039$ whereas the lherzolites would have a rare-earth content of 1–2 × chondrite. The extremely low relative abundance of LREE in the Troodos and Othris harzburgites is compatible with the chemically refractory nature of these rocks (Menzies & Allen 1974) and their petrographic and mineralogic continuity over several tens of kilometres. Moreover, major- and trace-element analyses have shown the plagioclase and spinel lherzolites to be CaO- and Al_2O_3-rich and relatively undepleted in major- and minor-

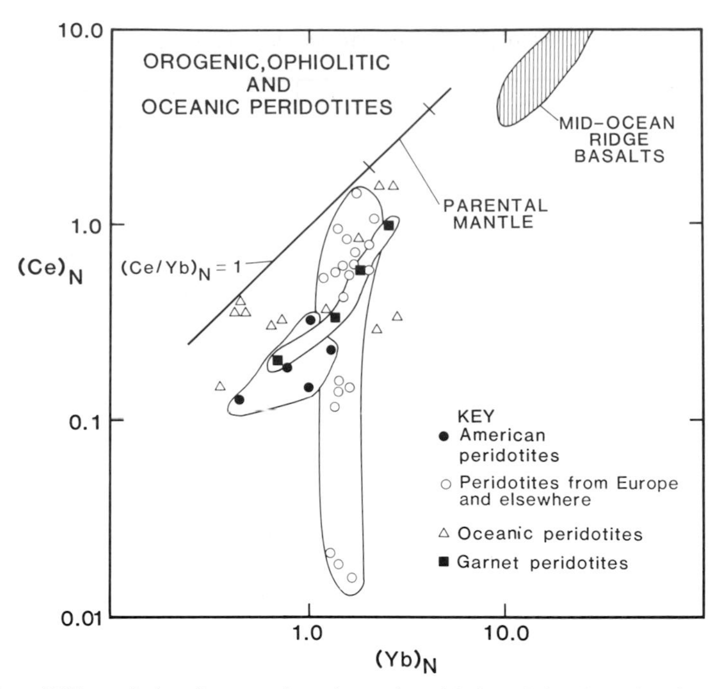

FIG. 2. $(Ce)_N$ *v.* $(Yb)_N$ variations in orogenic and oceanic peridotites. Other data taken from Shih (1972), Loubet *et al* (1975), Helmke (1983, unpubl. data) and Frey (1983).

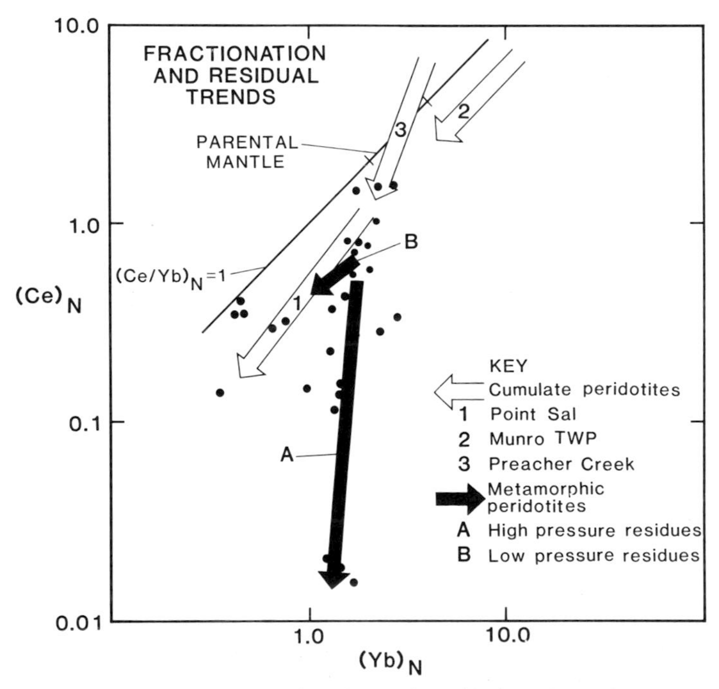

FIG. 3. $(Ce)_N$ *v.* $(Yb)_N$ variations in orogenic and oceanic peridotites (data points and references as in Fig. 2). Trend A and B represent the interpretations of Loubet *et al* (1975). Fractional crystallization trends are based on natural cumulus peridotites and pyroxenites (Potts & Condie 1971; Menzies *et al.* 1977a; Whitford & Arndt 1977). Note that natural garnet peridotites do not coincide with trend A, as would be expected, and that the low-pressure residual trend B parallels low-pressure fractionation trends.

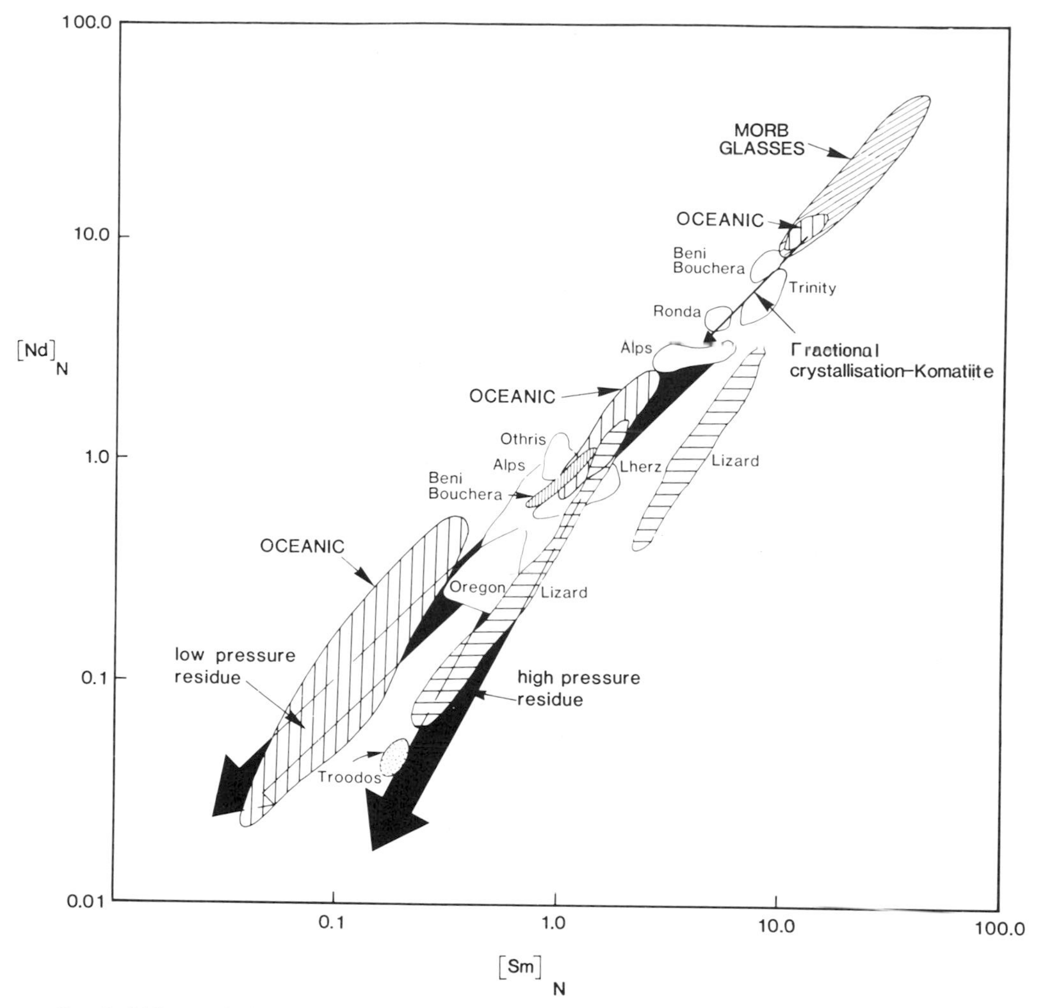

FIG. 4. $(Nd)_N$ *v.* $(Sm)_N$ variations in orogenic, ophiolitic and oceanic peridotites. Data sources are MORB glass (Cohen *et al.* 1980) and peridotite data (Loubet *et al.* 1975 and references therein; Ottonello *et al.* 1979). Note that the Lizard and Troodos tectonite peridotites appear to lie along a steeper trend perhaps indicative of formation in the garnet stability field.

Sm/Nd data are not compatible with melting in either the plagioclase or spinel stability field (Fig. 4). The extreme depletion in LREE relative to heavy REE (HREE) implies that the HREE were retained during melting in the garnet stability field. Frey (1969) proposed a similar origin for the Lizard metamorphic peridotites on the basis of a detailed REE study. These data have important implications since the Troodos, Othris and perhaps the Lizard metamorphic peridotites form a relatively inert basement for the accumulation of plutonic and volcanic members of an ophiolite suite. In the past it has been generally accepted that this metamorphic basement of refractory harzburgite is a residue left after removal of tholeiitic melts similar to those found in the volcanic and plutonic suite. If indeed the harzburgites formed during melting events in the garnet stability field, it is highly unlikely that the melt extracted was tholeiitic in composition. Consequently, a simple genetic relationship between mantle harzburgites and the overlying plutonic-volcanic suite seems unwarranted. This is compatible with the limited isotopic data to be presented next. Several other authors have suggested that the metamorphic peridotites underlying ophiolites and the volcanic-plutonic rocks are unrelated (Suen *et al.* 1979; Pallister & Knight 1981). For the mantle peridotite to represent a residue produced by removal of ophiolitic lavas, they must be characterized by a LREE-depleted profile. This is not the case in the

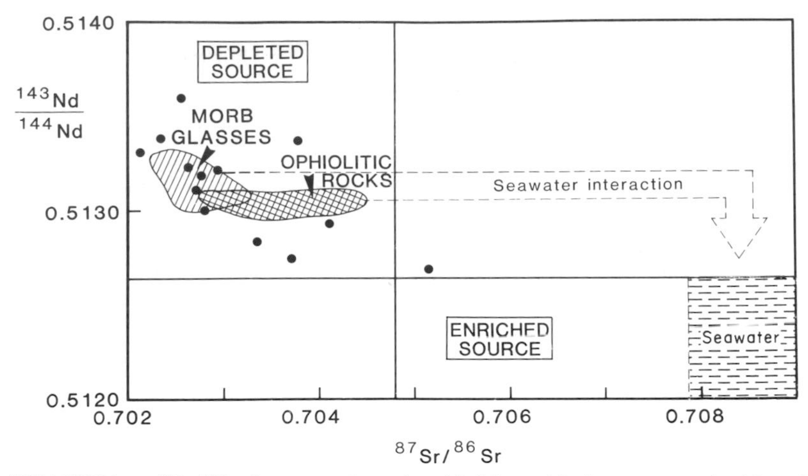

FIG. 6. $^{143}Nd/^{144}Nd$ *v.* $^{87}Sr/^{86}Sr$ for orogenic and ophiolitic peridotites compared with ophiolitic volcanic and plutonic rocks (Richard & Allegre 1980) and MORB glasses (Cohen *et al.* 1980). Note the overall isotopic heterogeneity in orogenic lherzolite massifs.

brought up by alkaline basalts (Menzies *et al.* 1982). However, there appears to be a fundamental difference between orogenic-ophiolitic peridotites and xenolithic peridotites. While both groups display near identical heterogeneity in Sr and Nd isotopes the xenolithic peridotites are both LREE-enriched and -depleted. To date a marked enrichment in the LREE $(Ce/Yb)_N \gg 1$ has not been reported in any orogenic or ophiolitic metamorphic peridotites. Consequently, there is either a fundamental difference between mantle sampled by diapirism-obduction and volcanism, or, more likely, lherzolite inclusions are modified by the very processes transporting them to the surface.

(ii) Mantle heterogeneity is apparent in all the orogenic lherzolite massifs, in particular Beni Bouchera, Morocco (Figs 5b and 6). The considerable range in Nd- and Sr-isotopic composition can result from a complex multi-stage history occurring over several billion years. Model ages calculated for all the orogenic lherzolites vary from 0.5 to 2.4 billion years supporting a long complex evolutionary history. Quick (1981) established a record of complex mantle processes in the Trinity body, California. These included melt segregation, migration and wall rock reaction. This may in part explain the isotopic heterogeneity observed in the two Trinity lherzolites that occur within 100 m of each other (Table 1). These preliminary data indicate that isotopic heterogeneities occur on a local and regional scale. However it should be stressed that some of this isotopic variation may be the result of ^{147}Sm decay after emplacement.

(iii) Chemical and isotopic heterogeneity at the level observed in orogenic and ophiolitic peridotites is also apparent in terrestrial basalts and, by inference, their source regions. Flood basalts may have retained some of the isotopic variability inherent in the mantle, prior to limited crustal contamination. Furthermore, a similar degree of Nd- and Sr- isotopic heterogeneity to that of the peridotites is also apparent in the source regions of successive episodes of volcanism erupted on Oahu, Hawaii (Stille *et al.* 1982).

(iv) A diopside separated from the tectonite harzburgite that underlies the Troodos ophiolite has a Sr-isotopic composition within the range of MORB-type liquids. However, this differs from the isotopic composition of the overlying volcanic and plutonic rocks where studies of fresh feldspar separates from the plutonic suite produces a minimum $^{87}Sr/^{86}Sr \sim 0.70344$ (Spooner *et al.* 1977). If this is an accurate representation of the minimum isotopic composition of the parental magma that formed the plutonic suite and overlying volcanic suite, then it is rather unlikely that it equilibrated with the underlying harzburgite ($^{87}Sr/^{86}Sr \sim 0.70283$) (Table 1). Consideration of both the isotopic and REE data indicates that the mantle sequence attached to ophiolite complexes is not simply related to the overlying tholeiitic basalts.

Conclusions

The upper mantle, as represented by plagioclase and spinel-facies lherzolites, is heterogeneous both for rare earths and isotopes, on a regional and local scale. Orogenic and ophiolitic peridotites exhibit a considerable variation in Sr- and Nd-isotopic composition comparable with that observed in Type-1A xenoliths (LREE-depleted lherzolites and harzburgites) and volcanic rocks erupted in oceanic and continental environments. Furthermore, the similar Sr- and Nd-isotopic composition of most orogenic peridotites and N-type MORB liquids appears to indicate that the peridotites are genetically related to the production of MORB-type magmas. The melt segregations within the Lanzo massif are derivatives of N-type MORB that have interacted with MORB-type residual wall rock.

Mantle sampled by obduction-diapirism (ophiolites-orogenic lherzolites) and that entrained in alkaline magmas (ultramafic xenoliths) differ in their degree of heterogeneity for both the REE and isotopes. Since a fundamental difference in mantle-type is highly unlikely, the inhomogeneities evident in ultramafic xenoliths *must* in part record the very magmatic process that eventually transport them to the surface. Ultramafic xenoliths are disrupted mantle conduits whose wall rocks record a varied history of fluid and magma transport, whereas orogenic lherzolites are random mantle slabs that are less chemically and isotopically heterogeneous, and thus record a less complex history.

In conclusion, one can speculate that perhaps much of the earth's upper mantle has a Sr- and Nd-isotopic signature similar to MORB, a time-integrated response to an overall depletion in LREE. The isotopic and trace-element heterogeneity observed in mantle xenoliths is possibly a time-integrated manifestation of interaction between ancient MORB-type (depleted component) and upwelling alkaline-kimberlitic magmas (enriched component).

ACKNOWLEDGMENTS: The author appreciates the invaluable assistance of Peter van Calsteren and Andrew Gledhill in the isotope laboratory at the Open University. John Taylor, Neil Mather and Helen Boxall are thanked for the artwork and Donna Evans is thanked for typing the manuscript.

References

BOUDIER, F. 1972. *Relations lherzolite-gabbro-dunite dans le massif de Lanzo (Alpes piemontaisses): Example de fusion partielle.* These presentee a l'institut des Sciences de la Nature de l'Universite de Nantes, janvier 29.

—— 1978. Structure and petrology of the Lanzo peridotite (Piedmont Alps). *Bull. Geol. Soc. Am.* **89**, 1574–91.

COHEN, R. S., EVENSEN, N., HAMILTON, P. J. & O'NIONS, R. K. 1980. U-Pb, Sm-Nd and Rb-Sr systematics of mid-ocean ridge basalt glasses. *Nature, Lond.* **283**, 149–53.

DICKEY, J. S. 1975. A hypothesis of orogin for podiform chromite deposits. *Geochim. Cosmochim. Acta*, **39**, 1061–1074.

FREY, F. A. 1969. Rare earth abundances in a high temperature peridotite intrusion. *Geochim. Cosmochim. Acta* **33**, 1429–77.

—— 1983. Rare earth abundances in upper mantle rocks. *In*: HENDERSON, P. (ed.). *Rare earth element geochemistry.* Elsevier, (in press.).

GARMANN, L. B., BRUNFELT, A. O., FINSTAD, K. G. & HEIER, K. S. 1975. Rare earth element distribution in basic and ultrabasic rocks from west Norway. *Chemical Geology* **15**, 103–16.

HASKIN, L. A., FREY, F. A., SCHMITT, R. A. & SMITH, R. H. 1966. Meteoritic, solar and terrestrial rare-earth distributions. *Phys. Chem. Earth* **7**, 169–321.

HOFMANN, A. W. & HART, S. R. 1978. An assessment of local and regional isotopic equilibrium in the mantle. *Earth Planet. Sci. Lett.* **38**, 44–62.

JAGOUTZ, E., CARLSON, R. W. & LUGMAIR, G. 1980. Equilibrated Nd-unequilibrated Sr isotopes in mantle xenoliths. *Nature, Lond.* **286**, 708–710.

KEMPTON, P., DUNGAN, M. & MENZIES, M. A. 1982. Petrology and geochemistry of ultramafic xenoliths from the Geronimo volcanic field. *Terra Cognita* **2**, 222.

LOUBET, M., SHIMIZU, N. & ALLEGRE, C. J. 1975. Rare earth elements in alpine peridotites. *Contr. Mineral. Petrol.* **53**, 1–12.

MENZIES, M. A. 1976. Rare earth geochemistry of fused ophiolitic and alpine lherzolites. I. Othris, Lanzo and Troodos. *Geochim. Cosmochim. Acta* **40**, 645–56.

—— & ALLEN, C. A. 1974. Plagioclase lherzolite residual mantle relationships within two eastern Mediterranean ophiolites. *Contr. Mineral. Petrol.* **45**, 197–213.

—— & MURTHY, V. R. 1978. Strontium isotope geochemistry of alpine tectonite lherzolites: data compatible with a mantle origin. *Earth Planet. Sci. Lett.* **38**, 346–54.

——, BLANCHARD, D., BRANNON, J. & KOROTEV, R. 1977(a). Rare earth and trace element geochemistry of a fragment of Jurassic seafloor Point Sal, California. *Geochim. Cosmochim. Acta* **41**, 1419–30.

——, ——, —— & —— 1977(b). Rare earth Geochemistry of fused ophiolitic and alpine lherzolites. *Contr. Mineral. Petrol.* **64**, 53–74.

——, —— & JACOBS, J. 1977(c). Rare earth geochemistry of alpine tectonite lherzolites from northern California. Geol. Soc. Abs. (abs) **9**, 465.

——, KEMPTON, P. & DUNGAN, M. 1982. Nature of the

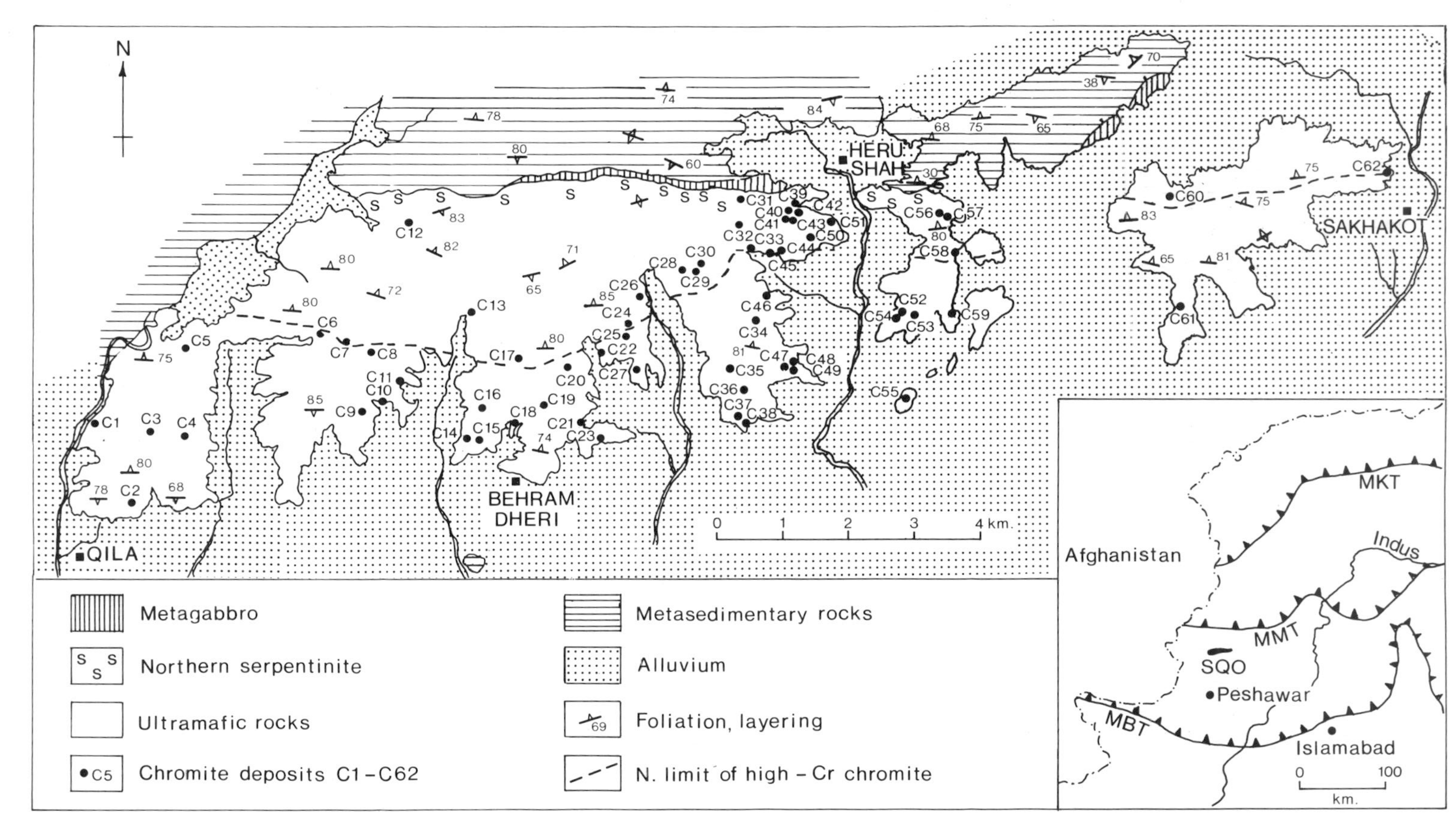

FIG. 1. Geological map of the Sakhakot-Qila ophiolite, Pakistan, simplified from the more detailed maps by Ahmed (1982a). Inset: the position of the SQO in relation to the major thrusts of the India–Eurasia collision zone of northern Pakistan, after Tahirkheli (1980). MKT—Main Karakoram Thrust, MMT—Main Mantle Thrust, MBT—Main Boundary Thrust.

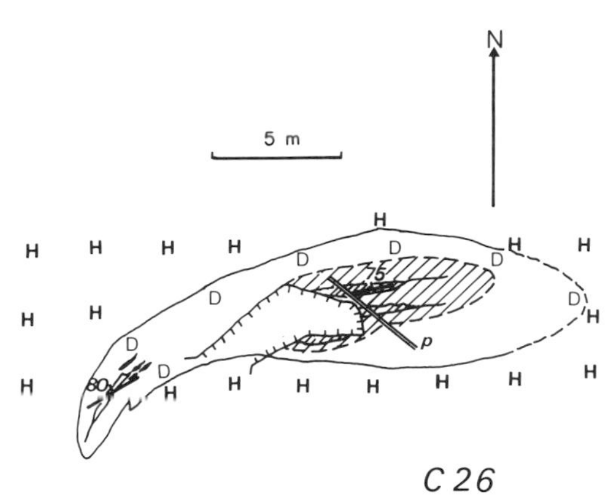

FIG. 2. A podiform chromite ore-body (C 26), from the northern part of the complex. Chromitite is shaded. The country rocks are dunite (D) and harzburgite (H), and the ore-body is cut by a pyroxenite dyke (P).

extent are more usual in a more southerly zone extending subparallel to the elongation of the complex, e.g. deposit C27 (Fig. 3). The layer-like bodies often contain banded ores.

Petrology

Harzburgite is the commonest rock type in the complex and is easily recognizable in the field by its 'hob-nail' outcrops. Generally it shows only a small amount of serpentinization. The chemical compositions of harzburgite show a limited variation with 40–46% MgO and very low contents of Na_2O, K_2O and Al_2O_3 (Table 1). Some of the harzburgites contain a minor amount (0–10%) of diopside. The SQO harzburgites show very low contents of incompatible elements but high levels of Cr, Ni and Co, in common with other ophiolitic harzburgites. The Cr content in harzburgite from near the chromite deposits usually exceeds 2500 ppm, but elsewhere it varies from 970 to 1900 ppm.

Dunites and *wehrlites* occur throughout the ultramafic part of the complex but do not outcrop in well-defined, separately mappable units. The dunite forms layer-like, lensoid and irregular-shaped minor bodies surrounded by either harzburgite or wehrlite. They are concordant as well as discordant to the harzburgite foliation, and exhibit sharp contacts. Dunite is the immediate host for most of the chromite concentrations, and may cut across them as dykes a few centimetres to a metre across. The dunites may contain minor amounts of chrysotile, antigorite, magnesite, magnetite, diopside and chromite. The wehrlites have conspicuous white specks of coarse clinopyroxene on weathered surfaces and are easily recognized in the field, but the amount of clinopyroxene is small (<10%) and the wehrlites do not differ greatly in chemical composition from the dunites.

Pyroxenite dykes, like their ultramafic host rocks, are highly magnesian. Clinopyroxenites are the most abundant kind, websterites are less abundant and orthopyroxenites are quite rare. In the pyroxenites, minor amounts of olivine, serpentine and chlorite are often present with chromite, Ni sulphides and awaruite as accessories. Most of the pyroxenites are very coarse grained, with crystals up to several centimetres in diameter. At least one 'Fe-websterite' dyke is observed on the ridge W of Bada Sar. It has a finer grain size than most other pyroxenites, and is richer in Fe, Al, Na and K, with trace-element contents different from the rest of the clinopyroxenites (Table 1, No. 9).

Among the *gabbros* the western metagabbro is extensively metasomatized, containing clinopyroxenes of two generations, tremolite, chlorite, clinozoisite and rare albite. The chemical analyses (e.g. Table 2, No. 1) show it to be desilicated and enriched in Ca and Al, i.e. changes similar to rodingitization. The eastern metagabbro is also metamorphosed but not so severely metasomatized. It contains abundant tremolite accompanied by albite, clinozoisite, epidote, chlorite, quartz and rarely sphene. Hornblende occurs in the western part of the eastern metagabbro.

Metadolerites in the SQO do not form a sheeted complex but occur either as swarms of dykes and lenses, or as satellite dykes below metagabbro near the northern contact of the ultramafics. Ophitic texture is sometimes observed. Metamorphism has changed the original doleritic mineral assemblage to the present one consisting of high-Ti brown hornblende, low-Ti green amphibole, clinopyroxene, clinozoisite, albite, grossular, sphene, ilmenite, chlorite and chlorapatite. Most of the dykes have been metasomatized, resulting in depletion in SiO_2 (41.1–43.8%) and enrichment in Na_2O (3.6–4.6%). The least altered rock is an ophitic metadolerite (analysis No. 12 in Table 1), which contains abundant hornblende accompanied by albite, clinopyroxene and epidote.

Secondary alteration, metasomatism and veining are widespread in the SQO. The dunites, wehrlites and harzburgites are in some places completely serpentinized, and some of the serpentinites are tremolite- and actinolite-bearing. Chrysotile and tremolite asbestos veins are commercially exploited. Magnesite veins are abundant, and occur in three ways: (i) as sheared magnesite-serpentine veins, (ii) as metre-scale

TABLE 2. *Chemical analyses of metasomatized basic rocks and rodingites*

	1	2	3	4	5	6	7A	7B	8A	8B
SiO_2	40.27	47.52	34.24	41.88	41.53	43.77	36.83	31.37	40.03	34.94
TiO_2	0.04	0.04	2.01	2.20	2.16	2.18	0.02	0.01	0.12	0.04
Al_2O_3	26.40	24.35	16.31	15.63	15.06	14.52	18.49	18.53	13.66	11.67
Fe_2O_3	1.17	1.08	2.23	2.13	1.54	1.32	1.77	2.22	0.96	3.09
FeO	0.80	1.35	12.31	14.11	13.77	13.89	0.65	3.53	2.32	2.92
MnO	0.05	0.03	0.26	0.26	0.25	0.24	0.07	0.07	0.10	0.06
MgO	5.91	5.37	11.73	6.87	6.52	6.29	3.85	31.88	13.72	36.17
CaO	20.68	14.49	15.86	11.54	12.32	11.52	36.55	0.16	26.90	2.78
Na_2O	0.11	2.80	1.28	3.82	3.59	4.55	0.02	0.02	0.08	0.02
K_2O	0.00	0.00	0.25	0.18	0.20	0.08	0.02	0.02	0.02	0.04
P_2O_5	0.06	0.05	0.35	0.27	0.16	0.16	0.04	0.05	0.08	0.09
H_2O^+	4.10	2.09	3.60	1.92	2.38	0.87	0.90	10.62	2.82	8.83
H_2O^-	0.22	0.08	0.09	n.d.	0.52	0.07	0.40	0.55	0.00	n.d.
Total	99.81	99.25	100.52	100.81	100.00	99.46	99.61	99.03	100.81	100.65
Trace elements (ppm)										
Ba	10	4	18	27	19	7	1	1	14	10
Rb	5	3	6	n.d.	7	7	5	5	5	4
Sr	219	127	29	74	74	60	17	6	12	7
Zr	43	39	34	35	82	67	5	n.d.	7	5
Co	20	16	54	64	59	60	11	70	34	56
Cr	460	147	178	100	78	49	280	5920	2725	6504
Cu	22	39	783	190	82	155	4	n.d.	54	152
Li	30	25	8	10	9	10	4	0	5	3
Nb	n.d.	n.d.	8	n.d.	n.d.	n.d.	n.d.	n.d.	11	13
Ni	470	350	273	363	160	190	210	1190	504	900
Sc	n.d.	n.d.	42	n.d.	n.d.	n.d.	n.d.	n.d.	29	10
V	27	48	506	504	577	570	2	80	132	76
Y	2	2	33	37	29	30	2	3	3	2
Zn	72	88	95	n.d.	140	146	65	80	31	66

1. Metagabbro, western outcrop (Z233),
2. Metagabbro, eastern outcrop (Z394),
3. Metadolerite, cut by grossular veins (Z371),
4. Metadolerite (Z370),
5. Metadolerite (Z369),
6. Metadolerite (Z372).

7A. Rodingite dyke (Z399A).
7B. Chloritic wall-rock of rodingite (Z399B).
8A. Rodingite dyke (Z361A).
8B. Chloritic wall-rock of rodingite (Z361B).

Mineral chemistry

Olivine

Microprobe analyses of olivines (Table 3) show that they are highly magnesian. Those from chromitites and their dunitic matrix are Fo_{92-97}, with a maximum frequency at Fo_{95}. Away from chromitites, the dunites have olivines with lesser Fo. In harzburgites and orthopyroxenites, olivines range from Fo_{88-92}, in clinopyroxenites Fo_{83-89}, and in Fe-websterite from Fo_{74-75}. TiO_2, Cr_2O_3 and CaO contents in olivine are negligible, but Ni is usually high and increases with Fo content. In chromitites, the olivine is most rich in Fo where the chromite is coarse-grained or there is a high chromite : silicate ratio. This is comparable to stratiform complexes, where olivines are known to display higher Fo in high-chromite, low-olivine cumulates than in adjacent low-chromite, high-olivine cumulates (Cameron & Desborough 1969; Jackson 1969; Hamlyn & Keays 1979).

Pyroxene

The orthopyroxenes are low in Al_2O_3, rarely exceeding 3%. The most magnesian orthopyroxenes (Fs_6) occur as inclusions in chromite crystals. Those of the harzburgites show a narrower range and slightly more magnesian composition (Fs_{8-10}) than those of the pyroxenites (Fs_{8-15}). The most Fe-rich orthopyroxene occurs in the Fe-websterite dyke (Fs_{21-22}). The primary clinopyroxenes are of similarly high Mg content, but show increasing amounts of Fe in the rock sequence; chromitite-harzburgite-wehrlite-clinopyroxenite-websterite-metagabbro-Fe websterite. Some of the clinopyroxenes are not, however, magmatic. In the metasomatized metagabbro there are two types of clinopyroxene: early magmatic and later replacive (e.g. Table 3, analyses 23 and 24). Some rodingitized chromitites consist of chromite grains matrixed by metasomatic clinopyroxene. The secondary clinopyroxenes in the metasomatized rocks are relatively Fe-rich, and some show Wo contents exceeding 50%. Cr_2O_3 is present in the clinopyroxenes up to a maximum of 1.77%.

Amphibole

Ultramafic rocks sometimes contain amphibole of secondary origin, e.g. tremolite asbestos veins. In some chromitites, amphibole inclusions are present in chromite. Anthophyllite is rarely seen. Amphibole is more abundant in the pyroxenites than in the other ultramafic rocks. The metagabbros contain both tremolite and green hornblende. In metadolerites, Ti-rich brown hornblende and Ti-poor green amphibole occur together.

Others

Chlorite occurs as a secondary mineral in all the rock types of the SQO. Microprobe analyses of chlorites show large, unsystematic variations in FeO, MgO and Al_2O_3 contents. The chlorites in chromitites are always of low-Fe, high-Mg type, but in mafic rocks they are more Fe-rich. Clinozoisite and albite occur in metagabbros and metadolerites, and ilmenite, sphene and chlorapatite are usually present as accessories in the metadolerites. Grossular and hydrogrossular are the most abundant minerals of the rodingites. Where they cross chromitites, a green coloration develops adjacent to chromite grains due to Cr-enrichment, and rarely uvarovite is formed. Metasomatized metadolerites contain abundant grossular similar to the rodingite dykes crossing them. Perovskite was observed in a rodingitized chromitite. Euhedral vesuvianite prisms form on hydrogrossular in the rodingites. Clintonite and corundum have not previously been reported from rodingites, and their presence in the SQO rodingites reflects strong Al-enrichment. Native Cu present in grossular (rodingitic) veins cutting across metadolerites is sometimes accompanied by nantokite, indicating the presence of both Cu and Cl in the rodingitizing solutions.

Ore mineralogy

The chromite-rich rocks exhibit a great variety of primary textures: disseminated, massive, banded, graded-layered, nodular, porphyritic nodular, orbicular, occluded silicate, chromite net, 'speckled', 'augen-disseminated' and 'pseudo clastic' (Ahmed 1982b). The chromite is characteristically of coarse grain size (average 2–3 mm), with anhedral, rounded, shattered crystals. Rarely, severe cataclastic effects, resulting in pulverized chromite with strong lineation, are observed. Massive and disseminated textures may occur in single specimens, with gradational or sharp contacts. Ores with alternate chromite-rich and silicate-rich rhythmic bands occur dominantly in chromite deposits that are elongated roughly parallel to the elongation of the complex. The banded ores sometimes show graded layering. Typical occluded silicate and chromite net textures may occur in single specimens, with gradational or sharp contacts with each other; in the latter case, the coarser chromites may carry abundant olivine inclusions in a row along the sharp contact. 'Speckled' chromitites contain irregular-shaped patches or 'speckles' of coarse-grained massive chromite enclosed in, or in contact with, finer-grained disseminated chromite. 'Pseudoclastic' ores are made of coarse, pebble-like chromite crystals, sometimes size-sorted, and usually in high concentrations like massive ores. They occur only in deposits with high-Al, low-Cr chromites. Unaltered chromite contains crystal inclusions of olivine, usually with a thin jacket of serpentine. Rarely, sieve-texture chromite grains occur. Sometimes, pyroxenes and/or amphibole inclusions are also observed. Rarely, clinopyroxene-free dunite contains chromite crystals with clinopyroxene inclusions.

A detailed electron-microprobe study was made of over 500 polished sections, and a very great number of chromites was analysed (Ahmed 1982a). There are three major parageneses of chromite: accessory chromite in the ultramafic rocks; segregated chromite, forming ore-bodies; and altered chromite (ferritchromite).

The primary chromite compositions show a typical alpine-type trend in the spinel prism plot of Irvine (1965). Many analyses plot in the 'metallurgical grade' chromite field of Stevens (1944), but others are either 'aluminian chromite' or 'chromian spinel'. The chromites exhibit a large range of reciprocal Cr-Al variation with Cr/(Cr + Al) mostly in the range 0.5–0.8. Total Fe (as FeO) content varies from 13.7 to 22.1%, which is independent of Cr_2O_3 content, and is much lower than that of stratiform chromites (Thayer 1970). Fe_2O_3 is below 8%, also lower than that found in stratiform deposits and does not vary systematically. TiO_2 rarely exceeds 0.3% and does not correlate with major elements. The Cr/Fe ratio of chromites varies from 1.4 to 4.5.

TABLE 3. *Representative electron-microprobe analyses of olivines (1–14), clinopyroxenes (15–27), orthopyroxenes (28–34) and amphiboles (35–41)*

	Olivines													
	1	2	3	4	5	6	7	8	9	10	11	12	13	14
SiO_2	41.62	40.91	41.00	40.63	40.65	39.79	40.66	40.78	40.88	40.61	40.47	40.31	38.65	38.17
FeO	4.01	5.88	5.83	8.08	8.97	9.35	9.19	8.48	9.62	11.09	9.47	10.53	14.20	22.90
MnO	0.09	0.15	0.05	0.12	0.11	0.12	0.10	0.05	0.14	0.11	0.16	0.09	0.26	0.26
MgO	53.86	51.86	52.04	51.08	49.11	49.24	49.23	50.25	50.07	48.32	48.67	48.24	45.28	38.76
NiO	0.57	0.33	0.41	0.32	0.36	0.35	0.47	0.40	0.30	0.17	0.38	0.27	0.35	0.06
Total	100.15	99.13	99.33	100.23	99.20	98.85	99.65	99.96	101.01	100.30	99.15	99.44	98.74	100.15
Fo (%)	95.98	94.03	94.04	91.73	90.65	90.27	90.48	91.32	90.13	88.55	90.01	88.99	85.05	74.88

	Clinopyroxenes												
	15	16	17	18	19	20	21	22	23	24	25	26	27
SiO_2	53.20	53.96	52.77	53.20	53.82	54.52	53.44	52.20	51.24	54.36	53.33	51.17	51.94
Al_2O_3	2.28	1.05	1.83	2.59	2.04	0.40	1.45	2.24	3.81	1.37	0.38	1.26	2.56
TiO_2	0.36	0.37	0.10	0.07	0.23	0.00	0.06	0.24	0.31	0.00	0.00	0.21	0.16
Cr_2O_3	1.77	0.42	0.91	1.05	0.98	1.61	0.63	0.39	1.22	0.13	0.00	0.00	1.09
FeO	1.75	0.91	1.10	2.24	2.50	2.34	2.65	5.49	3.69	2.12	9.14	12.83	2.59
MnO	0.00	0.00	0.00	0.08	0.00	0.07	0.09	0.13	0.10	0.14	0.30	0.37	0.08
MgO	16.17	16.79	18.84	16.98	17.10	16.99	17.01	15.47	14.95	15.61	14.27	11.63	15.97
NiO	0.06	0.00	0.00	0.00	0.00	0.00	0.00	0.00	0.00	0.06	0.00	0.00	0.00
CaO	23.57	25.27	23.93	23.83	23.91	22.11	24.24	22.93	23.53	25.55	22.44	21.63	23.63
Na_2O	0.00	n.d.	0.00	0.17	n.d.	1.15	n.d.	0.57	0.39	0.57	0.00	0.21	0.33
Total	99.16	98.77	99.48	100.21	100.58	99.19	99.57	99.66	99.24	99.91	99.86	99.31	98.35

	Orthopyroxenes							Amphiboles						
	28	29	30	31	32	33	34	35	36	37	38	39	40	41
SiO_2	57.39	56.01	56.96	56.84	57.35	56.04	54.07	46.72	52.41	54.28	45.09	55.16	49.71	43.13
Al_2O_3	1.55	2.11	1.24	0.99	0.00	1.40	1.79	10.77	3.69	2.26	11.05	1.59	4.89	9.54
TiO_2	0.08	0.04	0.00	0.07	0.04	0.07	0.16	0.88	0.00	0.00	1.08	0.03	0.19	2.20
Cr_2O_3	1.03	0.63	0.52	0.46	0.17	0.34	0.16	2.50	0.09	0.67	1.01	0.21	0.10	0.06
FeO	4.20	6.02	5.97	5.84	5.78	8.44	13.90	2.37	3.02	4.52	7.93	7.53	16.28	17.54
MnO	0.14	0.14	0.12	0.13	0.17	0.23	0.30	0.00	0.11	0.23	0.00	0.15	0.31	0.24
MgO	35.30	33.39	33.87	33.44	34.23	32.55	28.15	19.56	23.29	23.07	16.42	17.80	11.27	10.00
NiO	0.08	0.88	0.09	0.09	0.11	0.12	0.06	0.16	0.30	0.06	0.00	0.08	0.00	0.00
CaO	0.42	0.89	1.25	1.25	0.80	0.66	0.71	12.37	11.39	11.42	12.40	13.11	10.15	10.88
Na_2O	0.19	0.11	n.d.	n.d.	0.28	n.d.	0.09	2.43	n.d.	0.22	2.26	0.23	3.55	3.79
Total	100.38	99.42	100.02	99.11	98.93	99.85	99.39	97.76	94.30	96.73	97.24	95.89	96.45	97.38

Olivines from:
1. Chromitite, Cr-rich lower part of graded layer (Z40A)
2. Chromitite, Cr-poor upper part of graded layer (Z40C)
3. Disseminated ore with about 70% chromite (ZA228A)
4. Dunite dyke in chromitite (ZA228B)
5. Harzburgite with minor clinopyroxene (Z222)
6. Harzburgite (Z216)
7. Harzburgite (Z105)
8. Dunite (Z142)
9. Wehrlite (Z147)
10. Wehrlite (Z183)
11. Pyroxenite dyke (Z104)
12. Pyroxenite dyke (Z188B)
13. Pyroxenite dyke (Z70B)
14. Fe-websterite dyke (Z36)

Orthopyroxenes from:
28. Chromitite (pyroxene inclusions in chromite) (Z277)
29. Harzburgite with minor clinopyroxene (Z222)
30. Harzburgite (Z105)
31. Pyroxenite dyke (Z104)
32. Pyroxenite (ZA275)
33. Websterite (Z30)
34. Fe-websterite (Z36)

Clinopyroxenes from:
15. Chromitite (pyroxene inclusions in chromite) (Z277)
16. Rodingitized chromitite (Z41A)
17. Dunite (pyroxene inclusions in chromite) (Z327)
18. Harzburgite (Z222)
19. Wehrlite (Z183)
20. Pyroxenite (ZA275)
21. Websterite (Z30)
22. Fe-websterite (Z36)
23. Metasomatized gabbro, primary pyroxene (Z233)
24. Metasomatized gabbro, secondary pyroxene (Z233)
25. Subophitic metadolerite (ZB182)
26. Metasomatized metadolerite (Z372)
27. Rodingite dyke (Z361A)

Amphiboles from:
35. Chromitite (amphibole inclusions in chromite) (Z277)
36. Harzburgite with minor clinopyroxene (Z326)
37. Pyroxenite dyke (Z54)
38. Fe-websterite (Z36)
39. Hornblende gabbro (Z368A)
40. Metasomatized metadolerite, green amphibole (Z372)
41. Metasomatized metadolerite, brown amphibole (Z372)

The accessory chromites have a generally finer grain size than the segregated chromites, and differ in composition. In general, the accessory chromites are richer in Fe and Al, and the segregated chromites are richer in Mg and Cr. The atomic ratio $Mg/(Mg+Fe^{2+})$ varies from 0.38 to 0.64 for the accessory chromites and from 0.57 to 0.75 for segregated chromites.

There is a stratigraphic variation in the chromite compositions. Analyses from the coarsest-grained available chromite samples, often of massive texture and containing very little silicate matrix, were averaged separately for each chromite deposit. Such samples would be little affected by post-crystallization re-equilibration with adjacent silicates. Two groups of chromite bodies emerge—those with $Cr/(Cr + Al)$ ratio >0.61 and those <0.61. The former samples, with high-Cr chromites, occur in the more southerly (stratigraphically lower) deposits, and the latter in the more northerly (stratigraphically high) deposits. The latter usually form dykes, lenses or podiform bodies; the former tend to form layer-like and tabular bodies, although more typically podiform bodies are also present. Within segregated chromite samples and ore bodies, there is often a bimodal grain size distribution, and the finer-sized chromite populations possess higher Fe^{2+} and lower Mg than the coarser-sized chromite populations.

Chromite is usually altered to the extent of about 5% although individual samples may vary from almost unaltered to completely altered. Paired analyses of fresh and altered chromite in individual grains are given by Ahmed & Hall (1981). Ferritchromite shows features indicating a volume decrease, and formed by residual enrichment of Cr, Fe and Mn, and strong depletion of Al and Mg. In Stevens' (1944) triangular plot, it trends towards the Cr-richer 'ferrian-chromite' field. When ferritchromite develops along chromite grain margins, chlorite haloes develop around them, and chlorite inclusions grow inwards towards fresh chromite cores along opened-up cleavages (Ahmed & Hall 1981). The chlorite associated with ferritchromite is bluish-grey, and is believed to result from reaction between the enclosing serpentine and Al liberated from the original chromite during its alteration.

Accessory amounts of nickeliferous opaque minerals consisting, in order of decreasing abundance, of awaruite-heazlewoodite-pentlandite frequently occur in the partly or completely serpentinized ultramafic rocks. The usual grain size of these minerals is 10–50 μm, although sometimes finer particles of 1–3 μm diameter are abundant, and awaruite grains 4–5 mm across may sometimes occur. The grains and specks are preferentially concentrated in chromitites and adjacent to chromite grains. They are also found on chromite grain boundaries or fractures, and as inclusions in ferritchromite. Any two of the three minerals may occur intergrown. The Fe-websterite contains about 2% of troilite and pentlandite. A hydrated oxide of Ni was observed in a serpentinite from near Qila, accompanied by heazlewoodite grown in contact with carbonate, serpentine and chlorite grains. The awaruites are richer in Ni than most known occurrences, and contain Cu up to 5.5% by weight, whereas their Co content is very low. In pentlandites, Ni varies from 22.8 to 34.8, Fe from 19.0 to 28.4, S from 45.1 to 47.7, and Co from zero to 1.1 atomic %. In pentlandites, S is usually below the stoichiometric content. In heazlewoodite, Ni varies from 52.8 to 60.8 atomic %. Rare specimens of two new mineral phases, 'iridian awaruite' and a Ru-Os-Ir-Ni-Fe alloy have been recorded (Ahmed & Bevan, 1981).

The nickeliferous opaque minerals occur widely distributed without recognizable relation to the chloritic or serpentinous nature of the chrome ores. The evidence supports the formation of the common awaruite-heazlewoodwite-pentlandite assemblage during serpentinization of highly magnesian rocks under low fO_2-fS_2 (Ahmed & Hall 1982). Serpentinizing solutions liberated Ni and Fe from the olivine structure, which combined with sulphur particularly in the reducing environment adjacent to the ferritchromite alteration sites. The anomalous troilite-pentlandite assemblage in the Fe-websterite results from the scarcity of Ni in its olivine, the presumed source of the Ni and Fe in the secondary sulphide minerals. The presence of troilite rather than pyrrhotite in this rock indicates a low fS_2 in common with the nickeliferous assemblage of the other rocks.

Rodingite formation

Rodingites are generally considered to have formed by serpentinization-related Ca-rich fluids acting metasomatically on a variety of rocks such as gabbro, diabase, dacite, granite, greywacke and shale. They may form during all the four phases of ophiolite metamorphism, i.e. oceanic hydrothermal, subduction, obduction and regional metamorphism, of protolith peridotites (Coleman 1977). Rodingitization at SQO seems to have acted independently of serpentinization, The latter is usually a more pervasive alteration, with mesh serpentine-veinlets developed on the scale of grains, whereas the former is restricted to relatively large fractures. The serpentinite on the northern upper margin of the ultramafic unit contains more rodingites, but on the other hand, bigger rodingite dykes are developed in very weakly serpentinized ultramafic rocks.

Enrichment in Al is not a conspicuous feature of most rodingites, and the presence of clintonite and corundum in the SQO rodingites is probably the first report of such Al-rich minerals in these rocks. The Al may not have come from the chromite or pyroxenes in the ultramafic rocks, as the Al released by chromite during serpentinization does not move great distances but is absorbed in the chlorite haloes around individual chromite grains or segregations (Ahmed & Hall 1981). Thick chlorite walls around rodingite lenses form by Al-addition from strongly Al-rich rodingitizing solutions to the already-serpentinized ultramafic rocks. The dispersed ferrit-chromite suggests rodingite formation from an ultramafic predecessor or in its fractures, and does not necessarily require the prior occurrence of diabase or gabbro to be replaced. Some of the rodingites in the SQO occur in chromitite, harzburgite and serpentinite.

At SQO, the rodingites seem to suggest formation by the action of seawater channelled through fractures and cracks in rocks during the spreading-centre stage. A sea-floor origin for rodingites has previously been suggested by Hall (1979) for those in the Lizard complex, and rodingites are known to occur on the present-day sea floor (Honnorez & Kirst 1975). The more frequent rodingite lenses in the northern serpentinite of the SQO may be caused by its relatively high position in the ophiolite sequence, where the seawater would penetrate most readily; the Al content may have been acquired from the immediately overlying gabbros. Crystallization of nantokite in the grossular veins in the metadolerites suggests Cl-enriched rodingitizing solutions. This is also supported by the chlorapatite present in the metasomatized metadolerites. Cu is also rather high in metadolerites with abundant hydrogrossular veins, e.g. Table 2, No. 3, with 783 ppm Cu. This favours the metasomatic introduction of Cu during rodingitization, in contrast to the post-igneous hydrothermal alteration reported from other ophiolites during which Cu is removed instead.

ACKNOWLEDGMENTS: The authors thank Drs R. M. F. Preston, J. V. P. Long, P. Henderson, R. F. Symes, I. Young, and J. C. Bevan for the electron-microprobe facilities at University College London, Cambridge University and the British Museum (Natural History). Rock analyses were carried out using the inductively coupled plasma spectrometer at King's College, London, which is supported by NERC under grants to Professor R. A. Howie and Dr J. N. Walsh. The Government of Pakistan provided financial support to the first author. The manuscript has benefited from comments by Professor R. A. Howie, Professor F. A. Shams and Dr T. P. Thayer.

References

AHMED, Z. 1982a. *Chromite deposits of the Sakhakot-Qila ultramafic complex, Pakistan.* Unpublished Ph.D. thesis, University of London.

—— 1982b. Porphyritic-nodular, nodular, and orbicular chrome ores from the Sakhakot-Qila complex, Pakistan, and their chemical variations. *Mineralog. Mag.* **45**, 167–78.

—— & BEVAN, J. C. 1981. Awaruite, iridian awaruite and a new Ru-Os-Ir-Ni-Fe alloy from the Sakhakot-Qila complex, Malakand Agency, Pakistan. *Mineralog. Mag.* **44**, 225–30.

—— & HALL, A. 1981. Alteration of chromite from the Sakhakot-Qila ultramafic complex, Pakistan. *Chem. Erde* **40**, 209–39.

—— & —— 1982. Nickeliferous opaque minerals associated with chromite alteration in the Sakhakot-Qila complex, Pakistan, and their chemical variation. *Lithos* **15**, 39–47.

CAMERON, E. N. & DESBOROUGH, G. A. 1969. Occurrence and characteristics of chromite deposits—Eastern Bushveld complex. *Econ. Geol. Monograph* **4**, 23–40.

COLEMAN, R. G. 1977. *Ophiolites—Ancient Oceanic Lithosphere?* Springer-Verlag, Berlin, 229 pp.

HALL, A. 1979. Rodingites in the Lizard complex. *Proc. Ussher Soc.* **4**, 269–73.

HAMLYN, P. R. & KEAYS, R. R. 1979. Origin of chromite compositional variation in the Panton Sill, Western Australia. *Contrib. Mineral. Petrol.* **69**, 75–89.

HONNOREZ, J. & KIRST, P. 1975. Petrology of rodingites from the Equatorial and Mid-Atlantic fracture zones and their geotectonic significance. *Contrib. Mineral. Petrol* **49**, 233–57.

IRVINE, T. N. 1965. Chromian spinel as a petrogenetic indicator. I. Theory. *Can. J. Earth Sci.* **2**, 648–72.

JACKSON, E. D. 1969. Chemical variation in coexisting chromite and olivine in chromite zones of the Stillwater complex. *Econ. Geol. Monograph* **4**, 41–71.

STEVENS, R. E. 1944. Composition of some chromites of the Western Hemisphere. *Am. Mineral.* **29**, 1–34.

TAHIRKHELI, R. A. K. 1980. Major tectonic scars of Peshawar Vale and adjoining areas, and associated magmatism. *In*: TAHIRKHELI, R. A. K., JAN, M. Q. & MAJID, M. (eds) *Proceedings of the International Committee on Geodynamics, Group 6 meeting at Peshawar 23–29 November, 1979.* Special Issue of the Geological Bulletin, University of Peshawar, vol. 13, pp. 39–46.

——, MATTAUER, M., PROUST, F. & TAPPONNIER, P. 1979. The India–Eurasia suture zone in northern Pakistan: synthesis and interpretation of recent data at plate scale. *In*: FARAH, A. & DE JONG, K. (eds) *Geodynamics of Pakistan*. Geological Survey of Pakistan, Quetta, pp. 125–30.

THAYER, T. P. 1960. Some critical differences between alpine-type and stratiform peridotite-gabbro complexes. *21st Int. Geol. Congr. Copenhagen*, **13**, 247–59.

—— 1970. Chromite segregations as petrogenetic indicators. *Geol. Soc. S. Africa, Spec. Publ.* **1**, 380–90.

ZULFIQAR AHMED, Department of Geology, University of the Punjab, New Campus, Lahore-20, Pakistan.

A. HALL, Department of Geology, University of London King's College, Strand, London WC2R 2LS.

II. EMPLACEMENT (OBDUCTION) OF OPHIOLITES

OPHIOLITE EMPLACEMENT AND OBDUCTION

Possible causes and consequences of upper mantle decoupling and ophiolite displacement

J. G. Spray

SUMMARY: The possible causes and consequences of tectonic decoupling within the oceanic upper mantle and the subsequent displacement of ophiolites are considered with specific reference to the petrology, chemistry and structure of the basal mantle tectonites of ophiolites and their metamorphic soles. Investigations of the basal peridotites provide information on high temperature–high pressure events that have occurred in the upper mantle and on its mechanical properties at depths <15 km, while metamorphic soles reveal the pressure–temperature regimes operating during and as a direct result of overthrusting. Currently available age data from several well-documented ophiolites reveal that ophiolite igneous crystallization and basal metamorphism are broadly contemporaneous, in keeping with the localized occurrence of granulite-facies sole assemblages. Theories on the formation of sub-ophiolite metamorphic rocks are reviewed and the process of ophiolite inception is reappraised. It is concluded that many ophiolites are tectonically decoupled along the lithosphere–asthenosphere boundary and initially displaced as complete slices of hot, thin, immature lithosphere in close proximity to spreading centres.

The evolution of ophiolites as fragments of oceanic crust and upper mantle can be considered as a series of distinct tectonic events which may finally result in their emplacement onto continental margins. While the allochthonous nature of many ophiolites is manifest from studies of their field relations, the initial stages of their history remain poorly understood. Insight into the efficacy of the displacement process can be gained from studies of the approximate physical dimensions of ophiolites: reconstruction of numerous cross-sections has revealed that they tend not to be thicker than 15 km and most are substantially less than this (probably in the order of 5–10 km). While estimates of total original thicknesses may be inaccurate due to the effects of tectonic thickening or thinning, dismembering and erosion, it is clear that ophiolites are not of any thickness, but rather there is an upper limit to their vertical extent (Fig. 1). With regard to their other dimensions, ophiolites characteristically form linear belts extending for up to hundreds of kilometres along recent or ancient plate sutures. In contrast, their width rarely exceeds 100 km. These approximate dimensions suggest that there has been a relatively consistent ophiolite-producing process operating throughout the Phanerozoic. Furthermore, a number of important questions arise from this apparent consistency:

(i) What facilitates decoupling within the oceanic upper mantle at <15 km?
(ii) How is only an apparently limited width of oceanic crust and upper mantle displaced?
(iii) How are ophiolites produced as linear belts?

In order to provide constraints on these questions, investigations must necessarily focus on the base of ophiolites, particularly with regard to the lithologies and styles of deformation found in the basal mantle tectonites and underlying metamorphic soles. This paper is primarily concerned with appraising these basal relationships with the aim of revealing how ophiolites may be initially decoupled and displaced from the uppermost mantle.

Ultramafic and mafic mantle tectonites

Ultramafic tectonites

Beneath the cumulate ultramafic-mafic and crustal intrusive-extrusive mafic components of ophiolites lie peridotites which show the development of pronounced tectonic fabrics acquired during subsolidus recrystallization under upper mantle conditions (Fig. 1). The formation of mineral and compositional foliations, mineral lineations and folds testifies to a complex deformational and metamorphic history distinct from the predominantly magmatic character of the overlying crustal units. Although mantle tectonites have received significantly less attention than supra-Moho sequences, their investigation provides important insight into tectonic, metamorphic and igneous processes operating within the upper mantle. Their structural analysis, led by researchers such as Nicolas and co-workers, reveals at least two dominant types of metamorphic texture (Fig. 2). Most of the sequence shows evidence of high temperature–low stress (200–400 bars deviatoric) deformation resulting in the formation of a coarse-grained equigranular mineral foliation. This recrystallization fabric is especially prevalent near the top of the tectonites, but becomes progressively less pronounced towards the base where instead a medium-grained porphyroclastic texture predominates

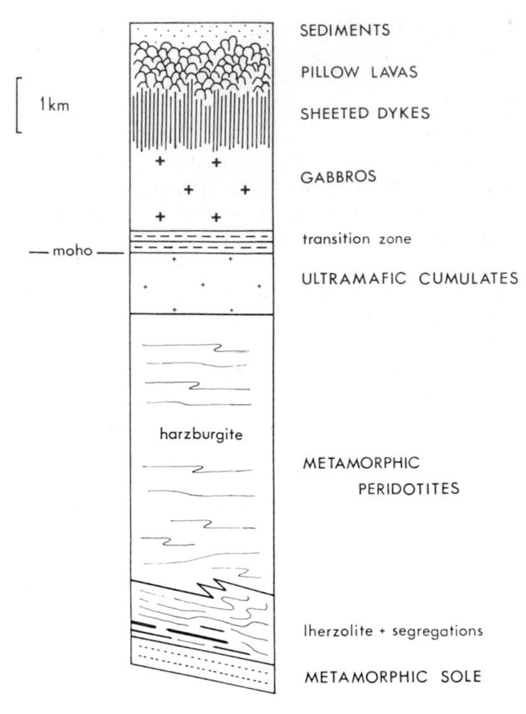

FIG. 1. Simplified and idealized cross-section of an ophiolite showing internal components, including positions of lherzolites and mafic segregations in the mantle sequence (metamorphic peridotites) and metamorphic sole.

(Nicolas *et al.* 1980; Nicolas & Le Pichon 1980). These fabrics are considered to result from deformation occurring at or near the peridotite solidus during the plastic flow of upwelling mantle beneath a spreading centre (e.g. Juteau *et al.* 1977; Boudier & Coleman 1981). The second type of metamorphic texture is restricted to the basal 1–2 km of the tectonites where peridotites exhibit a superimposed mylonitic fabric indicative of large strain–high stress (1–2 kb deviatoric) deformation occurring at temperatures of *c.* 700°C. Significantly, the peridotite mylonites commonly possess a similar orientation to fabrics developed in underlying metamorphic soles. Taken together, the basal tectonite–metamorphic sole structures are considered to result from deformation occurring during initial detachment and thrusting of the ophiolite (e.g. Boudier *et al.* 1982).

Compositionally, mantle tectonites usually comprise harzburgite with lesser amounts of dunite, lherzolite and pyroxenite. However, in many cases lherzolite rather than harzburgite is prevalent in the basal mylonitized zone of the tectonites (e.g. Jones 1977; Malpas 1979; Jamieson 1981; Spray 1982) and in other examples lherzolite may be the dominant peridotite and constitute the major part of the teconites, as in the central ophiolite zone of Yugoslavia (e.g. Pamic & Majer 1977; Karamata *et al.* 1980).

Mafic segregations

Mafic segregations occur as a minor but significant component of many ophiolite mantle sequences (Spray 1982). Most occur as thin layers (<10 cm thick) aligned parallel to peridotite mylonite fabrics and less commonly as pods and discordant veins and dykelets. They may be wholly monomineralic, comprising pyroxene (usually aluminous varieties of salite and augite) or amphibole (kaersutite and pargasite) or mica (phlogopite), or polymineralic combinations of pyroxene, amphibole, garnet, spinel, phlogopite and olivine. Similar segregations are found in lherzolite massifs (such as the type Lherz complex of southern France and the Ronda peridotite of southern Spain) where they are classified as varieties of ariegite (e.g. Conquere 1971; Obata *et al.* 1980). The segregations from ophiolites studied to date possess a basaltic chemistry and high temperature-high pressure parageneses: e.g. 900–1400°C and 17 kb were obtained from mafic segregations in the Zlatibor ophiolite (Pamic 1977; Popevic 1977), 900–950°C and 7–10 kb at the contact between the peridotite and metamorphic sole of the White Hills Peridotite (Jamieson 1980), 700–850°C and 15–20 kb from mafic segregations in the Othris and Ballantrae ophiolites (Spray 1980). Treloar *et al.* (1980) obtained 900 ± 70°C and 10–11 kb (possibly 14–15 kb) for similar lithologies from the Ballantrae ophiolite. These pressures indicate depths of equilibration ranging from 23 to 66 km which are in excess of the known thicknesses of their respective host ophiolites (i.e. <15 km), thus reflecting events which occurred at considerable depth *prior* to their juxtapositioning at higher levels. Significantly, those mafic segregations with the highest pressure parageneses show the greatest deformation, possess high MgO contents (*c.* 15–20%) and appear more 'primitive', in keeping with the proposed depths of generation of hypersthene- or nepheline-normative picritic melts. Those with the lower-pressure assemblages are usually less deformed and possess a more evolved MORB-like chemistry (Spray 1982). There are several possible explanations for the origin of mafic segregations: they may have originally been formed by partial melting and/or mantle metasomatism or they may be exotic inclusions. Early textural relationships within host peridotites are invariably obscured by later mylonitization so that it is presently unclear whether they were originally discordant intrusions (dykes and veins), concordant bands within the tectonite fabric or exotic lithologies incorporated within the basal peridotites during mylonitization. However, the chemistry of the segregations is in keeping with them being partial melts derived by the fusion of undepleted mantle. It is also likely that the supposed high-pressure segregations were rapidly injected from depth and subsequently sheared (transposed) during mylonitization, while their lower-pressure equivalents probably originated synkinematically within the tectonites during high-level adiabatic decompression and asthenospheric flow. Further work on these mafic segregations is in progress.

In addition to the mafic segregations, a variety of post-tectonite and post-mylonite dykes and sills commonly intrude the mantle peridotites. These typically comprise gabbro, dolerite or pyroxenites and are considered to represent either (i) highly differentiated melts which have been intruded *downwards* into the tectonites from an overlying magma chamber (e.g. Reuber *et al.* 1982), or (ii) fractionated partial melts (pyroxenite and gabbro) and their residua (dunite) derived from the fusion of underlying fertile peridotite (e.g. Quick 1981; Nicolas & Jackson 1982). These late-stage intrusions contrast with mafic segregations in that they retain their igneous textures, generally occur undeformed and have crystallized within the uppermost mantle rather than under high-pressure conditions (Fig. 2).

Metamorphic soles

Occurrences and diagnostics

It is a decade since metamorphic soles beneath ophiolites (also known as sub-ophiolite metamorphic rocks and dynamothermal aureoles) were first specifically described by Williams & Smyth (1973) using examples from Newfoundland. Since then numerous occurrences of sub-ophiolite metamorphic rocks have been discovered and documented from all but one of the continents with ages spanning the Phanerozoic eon (Table 1). As a result, there is now a substantial literature concerned with their description testifying to their common rather than exceptional association with transported ophiolite complexes. In well-preserved examples, soles normally form a coherent basal sequence or rind (usually <500 m thick) to the overlying mantle peridotites of the ophiolite (Figs. 1 & 2), but in many cases the relationship is obscured by the effects of subsequent tectonism and metamorphism. This has restricted investigations of type metamorphic soles to a handful of relatively intact ophiolites (e.g. the Bay of Islands, Newfoundland and the Semail, Oman) with which less complete examples have been compared. In other cases, intact metamorphic soles may be overlain by lherzolite bodies bereft of a crustal mafic unit (e.g. the Ronda peridotite of southern Spain) which do not fit the rather specific Penrose definition of an ophiolite currently in use. Recently, more practical classifications have been proposed (Moores 1982). Nevertheless, despite variations in the nature of the overlying peridotite the metamorphic soles themselves possess a number of features in common (Fig. 2):

(i) They are of limited thickness; usually ranging from a few metres up to 500 m, but thicknesses of over 1 km have been reported (e.g. MacKenzie 1960). However, tectonic thinning and thickening may have occurred both during and after sole formation.

(ii) Despite their relatively limited vertical thickness they may be of great lateral extent. The Trinity ultramafic massif in the Klamath Mountains of northern California possesses a metamorphic sole approximately 35 km in length (Davies *et al.* 1965) and the Semail ophiolite of Oman has a metamorphic sole discontinuously exposed at constant stratigraphic position beneath mantle peridotites for 450 km (Searle 1980).

(iii) Most intact soles show a sharp decrease in metamorphic grade from top to base (i.e. in the opposite sense to that operating during regional metamorphism) with isograds lying parallel or sub-parallel to the contact with peridotite. Granulite facies and even anatexis have been reported from the contact between peridotite and sole (MacKenzie 1960; Green 1964a; Challis 1965; Pamic 1977; Jamieson 1979; Searle & Malpas 1982), while only incipient metamorphism may be registered at the base.

(iv) Metamorphic soles are normally highly deformed and show folding and the development of a pronounced tectonic foliation. Typically, this decreases in intensity from schistose and gneissose fabrics at the contact with peridotites to more phyllitic, slaty and relatively undeformed textures nearer the base.

(v) The original lithological nature of metamorphic soles is not easy to establish. Many comprise predominantly mafic rocks now represented by amphibolites. Geochemical studies of these rocks indicate a tholeiitic or alkalic character. Metasediments may be present in subordinate

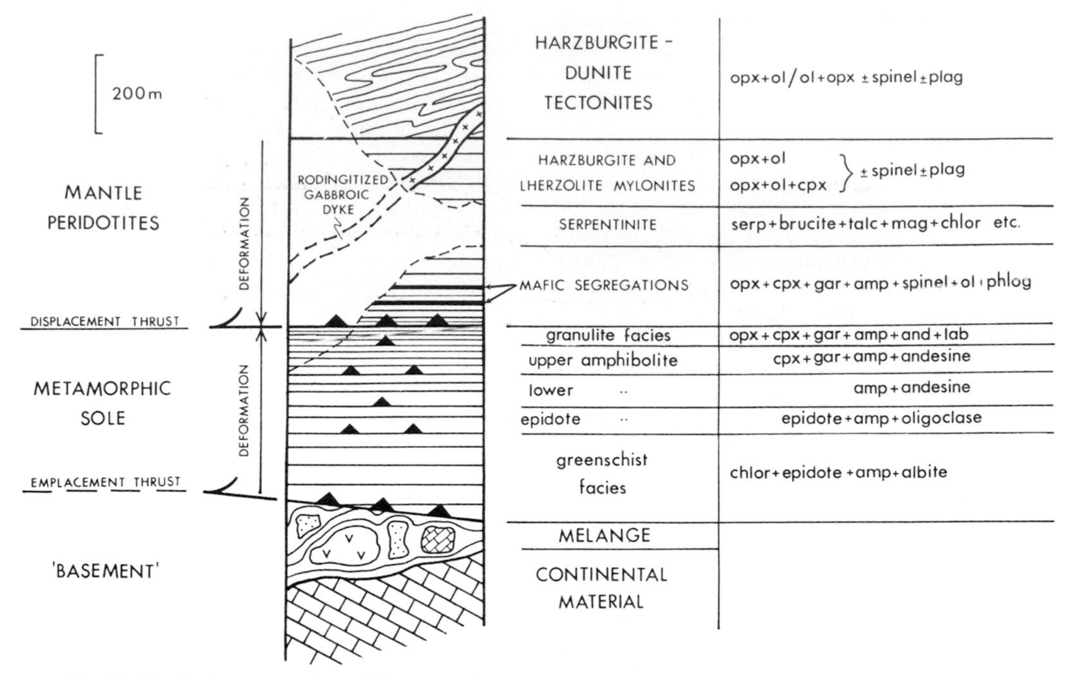

FIG. 2. Idealized cross-section of the base of an ophiolite showing relationship between metamorphic sole and overlying mantle peridotites. Variation in metamorphic grade and diagnostic parageneses for a mafic protolith for the sole are also shown. If present, metasedimentary lithologies would normally be restricted to the lower parts of the sole and yield, for example, quartz-albite-epidote-mica schists (metapelites), quartzites (meta-arenites) or calc-silicate assemablages (impure carbonates). Overlying mantle peridotites show a 'mirror' style of deformation at their base which overprints earlier tectonite fabrics. Lherzolite containing mafic segregations is also prevalent in the basal region. Varieties of pyroxenite, dunite and gabbroic dykes and sills may intrude the main lithologies. (The granulite-facies unit of the sole is shown enlarged for clarity, in reality it is likely to be ⩽10 m thick.)

amounts and usually towards the base of the sole. Often they are varieties of metapelites, metashales and metacherts. Occasionally, soles may predominantly comprise metasediments (e.g. beneath the Brezovica ophiolite, Yugoslavia; Karamata 1968), although certain micaceous zones have been interpreted as the products of metasomatism (Jamieson & Strong 1981).

The above features indicate that metamorphic soles are not aureoles in the conventional sense. 'Aureole' is normally used to denote relatively undeformed thermally metamorphosed rocks formed at constant pressure around hot igneous bodies. In contrast, metamorphic soles are schistose, polydeformed and occur in association with tectonically transported igneous bodies. Furthermore, it is likely that many metamorphic soles are tectonically assembled through both time and space and, as a result, their present profiles represent a composite assemblage of metamorphosed slivers formed under different P–T conditions which have become juxtaposed during overthrusting (Fig. 3). In this respect soles may form by successive underplating and welding onto the hanging wall (base of ophiolite) as it moves from the upper mantle to the surface. This indicates that the metamorphic-facies boundaries within soles may represent discrete tectonic contacts (Figs 2 & 3). The tectonic assembling of soles also permits the inclusion of unusual lithologies into the sequence: e.g. jacupirangites and related alkaline pyroxenites occur at the base of the White Hills Peridotite (Jamieson & Talkington 1980) and Semail ophiolite (Searle & Malpas 1982).

Some of the early descriptions interpreted soles as slices of regionally metamorphosed basement which had become detached and trapped along the base of ophiolites during overthrusting (e.g. Brunn 1956). This may be the case for certain exotic blocks found in sub-ophiolite mélanges. However, the steep inversion of metamorphic isograds shown by intact soles, their relatively constant thickness, the development of internal tectonic fabrics parallel or sub-parallel to basal

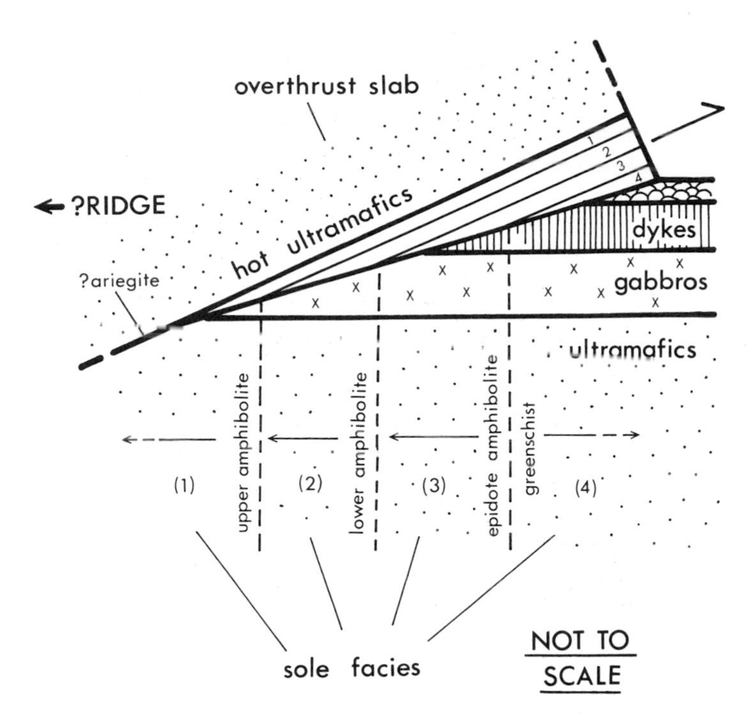

FIG. 3. Simplified model for the formation of a metamorphic sole along a thrust fault penetrating oceanic crust and upper mantle. The sole is shown divided into four arbitrary '*PT* planes' ranging from greenschist to upper amphibolite-granulite facies. The sole is only complete when all four *PT* planes are tectonically juxtaposed. This obviates the necessity of having extremely high thermal gradients existing through the whole thickness of the complete sole. As thrusting and metamorphism continue, lower-grade rocks are progressively 'welded' onto the hanging wall such that the locus of shearing moves downwards and away from the peridotite.

thrusts and overlying peridotite mylonites, and their predominantly mafic chemistry together indicate a close genetic relationship with ophiolites.

Overall, the evidence indicates that metamorphic soles are the products of dynamothermal metamorphism associated with the tectonic displacement of ultramafic bodies. One of the major problems has been to establish the heat source for this metamorphism; the two main contenders being residual and shear heating.

Formation by residual heat

Carslaw & Jaeger (1959) and Jaegar (1961) have shown that for the linear flow of heat from a hot upper slab to a lower cooler one (via a plane of contact on which there is no movement) the maximum contact temperature attainable is about 0.5 T (where T is the temperature difference between the base of the hot slab and the rocks below). Maximum temperatures are achieved immediately after emplacement at the contact and are reached further away from it later. Thus immediately beneath a peridotite slab at 1000°C the maximum temperature of the country rock will be about 500°C if the initial temperature of the country rock is 0°C (T_{ic}0°), 600°C at T_{ic}200°C and 700°C at T_{ic}400°. So that, in the case of ophiolites, the *static* juxtaposition of a hot peridotite slab on cooler basic rock imposes severe constraints on the contact temperatures that can be realized at the sole.

In order that an ophiolite may be tectonically displaced it must behave as a rigid body and consequently be at a temperature below its solidus. Figure 4 shows the melting relations for mantle peridotite and reveals the likely maximum temperatures available to produce a metamorphic sole. Hydrous and anhydrous peridotites commence melting in the range 950–1200°C at low (<5 kb) pressures (Mysen & Boettcher 1975a, b), so the base of an ophiolite could impart a maximum contact temperature of 550–650°C if juxtaposed with basalt at T_{ic}0–100°C. Therefore, contact temperatures in excess of 650°C are unlikely to be achieved by the effects of residual

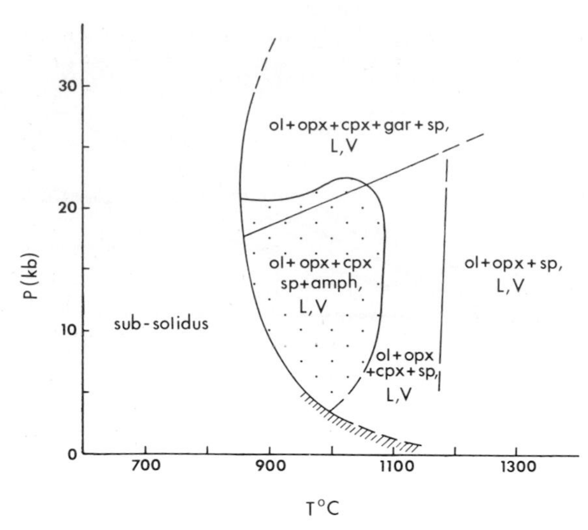

FIG. 4. Phase relations of hydrous and anhydrous spinel lherzolite. Diagram modified after Cox *et al.* (1979) compiled from the data of Mysen & Boettcher (1975a, b). Amphibole-bearing assemblages above the solidus are shown stippled.

heat alone unless the sole protoliths themselves initially possessed temperatures above 100°C. This limitation does not present problems for the greater part of metamorphic soles because they comprise low-pressure mineral assemblages formed below 650°C in the greenschist and amphibolite facies. However, exceptions include the presence of granulite-facies assemblages and evidence of partial melting.

Granulite-facies mafic assemblages have been reported from the top of several soles where they occur vertically contiguous with amphibolite-facies lithologies below (e.g. Green 1964a; Challis 1965; Jamieson 1979; Malpas 1979). They are recognized by the association of orthopyroxene (hypersthene) and clinopyroxene (augite-salite) and the predominance of brown pargasitic amphibole rather than green hornblende. Jamieson (1979), using a two-pyroxene geothermometer, calculates that temperatures of not less than 850°C were reached within 25 m of the peridotite contact beneath the White Hills massif and that localized partial melting of the amphibolites, resulting in the production of quartz-feldspathic segregations, occurred between 700–800°C. McCaig (1983) obtains temperatures of 750–850°C for the peridotite–sole contact of the Bay of Islands complex at pressures of 7–11 kb. Similar temperatures have been obtained elsewhere, mainly by using garnet-clinopyroxene pairs from granulite and upper amphibolite-facies garnet-pyroxene amphibolites: *c.* 850°C and approximately 7 kb from beneath the Ballantrae and Othris ophiolites (Spray & Williams 1980; Spray & Roddick 1980), 756–875°C at 5 kb (Ghent & Stout 1981) and 670–750°C at 4 kb (Searle & Malpas 1982) from beneath the Semail ophiolite, and *c.* 780°C at 5 kb beneath the Thetford Mines ophiolite (Feininger 1981). These temperatures were obtained using the geothermometers of Raheim & Green (1974), Perchuk (1969), Ellis & Green (1979), Ganguly (1979), Saxena (1979) or Dahl (1980), which in some cases resulted in ranges of up to 250°C for the same pyroxene–garnet pair (e.g. Feininger 1981). However, results obtained using the works of Ellis & Green, Dahl, Ganguly and Saxena (op. cit.) are preferred as they have the advantage of compensating for certain compositional variations, such as Ca in garnet. Notwithstanding these variations, it is clear the upper amphibolite- and granulite-facies temperature estimates from the highest-grade components of metamorphic soles are beyond the capacity of residual heating caused solely by the static juxtaposition of a hot peridotite. While static heating could provide the major portion of heat for sub-ophiolite metamorphism, other sources must be sought to account for the highest grades attained.

A variant of the static residual model involves a moving heat source (Jaegar 1942). In this case a hot body is continuously displaced over the contact rocks such that high temperatures are maintained at any given point at the contact for longer periods than could be achieved statically. This allows the contact to approach the temperature of the overriding material provided that large displacements of the hot body occur relatively rapidly. Significantly, the tectonic displacement and overthrusting of a hot ophiolite could satisfy the conditions of the dynamic residual heat model. However, evaluation of the heat transfer from a moving hot body is complicated by the possible effects of shear heating occurring along the plane of contact.

Formation by shear heating

Shear heating is the process by which heat is generated by the mechanical work of tectonic movements. One of the most concentrated sources is from the motion of faults which can produce heat from friction at shallow depths or from plastic deformation at deeper levels (Sibson 1977). Scholz *et al.* (1979) invoke shear heating in the Alpine continental transform fault of New Zealand to account for an increase in metamorphic grade from prehnite-pumpellyite facies at distances of 9–12 km from the fault, to low amphibolite facies within 4 km of the fault. It may also boost temperatures during faulting within oceanic ridge–ridge transform systems (Spray & Roddick 1981).

In an ideal situation, where all the mechanical work is converted to heat, the amount of heat produced is given by:

$$q = \tau \nu$$

where τ is the shear stress on the fault and ν the velocity of fault motion. The temperature increase at the contact is given by:

$$\Delta T_{x=0} = \frac{q}{\rho Cp}\left(\frac{t}{\pi k}\right)^{1/2}$$

where x is the distance from the contact, ρ is the rock density, Cp is the specific heat of the medium, t the duration of motion and k the thermal diffusivity. Unlike residual heating, the contact temperature increases with time as shear heating continues. In addition, shear heating may cause softening of the rock leading to strain concentration and maximum temperatures being attained in the centre of the shear zone which will subsequently widen by thermal conduction (Yuen *et al.* 1978; Fleitout & Froidevaux 1980).

Reitan (1968a, b, 1969) was one of the first workers who seriously attempted to model the possible effects of shear heating on rocks. He concluded that frictional heating may account for the *local* steepening of temperature gradients of up to 100°C if deformation is concentrated in time and space, i.e. for geologically short time periods (e.g. <1 m.y.) in zones of limited thickness (e.g. <1 km). However, Graham & England (1976) estimate that temperatures of *c.* 300°C greater than those produced by thermal relaxation of an overthrust sheet were created by shear heating in the Pelona Schist–Vincent thrust system of southern California. They showed that the extra heat could be produced by shear stresses of between 7.5 and 0.75 kb at the base of the upper slab for movements of between 1 and 10 cm yr^{-1} respectively in order to yield a suitable q value of 6–8 hfu. From these estimates it is clear that if relative displacements across such faults occurred at plate tectonic rates or slower, then shear stresses of several kb must have existed on the faults to produce the observed temperature rise. Alternatively, the faults could have moved at several tens of centimetres per year, but this would be significantly higher than supposed past or present plate tectonic rates (Scholz 1980).

To obtain temperatures of *c.* 600°C for the production of metamorphic soles by shear heating alone, it is apparent that very high rates of movement and/or high shear stresses are required. McKenzie & Brune (1972) conclude that moderate to large earthquakes can melt rock within up to about 1 cm of fault planes by frictional heat. This requires fault velocities in the order of 10^{-3} cm s^{-1} (about 1 cm 17 min^{-1}!) for estimated stress drops of 30–200 bars. For earthquakes such velocities are short lived and involve distances of centimetres rather than kilometres. In contrast, the overthrusting of ocean crust and upper mantle is more likely to occur at plate tectonic rates (<10 cm yr^{-1}) whereby shear stresses would have to be unrealistically high to generate temperatures in excess of 300°C. For these reasons, it appears unlikely that shear heating would be the major heat source for sub-ophiolite metamorphism. However, modelling the effects of residual and/or shear heating is besieged by several additional factors. Both the conduction and shear-produced heat budget may be increased or decreased if exothermic or endothermic reactions take place during metamorphism. The breakdown of pyroxene to amphibole and the process of serpentinization are both exothermic reactions and may be important processes operating during sub-ophiolite metamorphism. Graham & England (1976) showed the importance of dehydration reactions in buffering the maximum temperatures during shear heating and in particular noted the breakdown of chlorite, epidote and dehydration of serpentinite, all occurring at between 500–600°C. These dehydration reactions may further weaken the rock and help concentrate movement into a narrow zone of shear (Fleitout & Froidevaux 1980; Jamieson & Strong 1981)).

Résumé

The above discussions show that it is unlikely for either residual heating or shear heating alone to provide a sufficient temperature increase for cases of sub-ophiolite metamorphism which attain granulite facies. Contact temperatures

TABLE 1. *Some global occurrences, characteristics and age relations of metamorphic soles*

Country	Ophiolite massif	Approx. sole thickness	Main sole lithologies	Metamorphic facies	Ophiolite age (period or Ma)	sole age (Ma)	References*
Canada	Thetford Mines	<800 m	Amphibolite	Granulite to greenschist	Ordovician	488–494	Clague *et al.* (1981) Feininger (1981)
Greece– Yugoslavia	Pindos Vourinos Othris, Euboea —— Brezovica Krivaja-Konjuh	Generally <100 m but 1000 m recorded at Krivaja	Mainly metabasic schistose amphibolites with minor metasediments, latter predominante at Brezovica	Granulite to incipient	160–180	160–180	Spray *et al.* (in press)
Newfoundland	Bay of Islands —— White Hills	<1000 m	Basic amphibolites + minor metaseds.	Granulite to incipient	475–515	454–485†	Archibald & Farrar (1976) Dallmeyer (1977) Dallmeyer & Williams (1975) Jacobsen & Wasserburg (1979) Jamieson (1979, 1980, 1981) McCaig (1983) Malpas (1979) Williams & Smyth (1973)
New Zealand	Red Hills	<800 m	Metaspilites + metatuffs	Pyroxene hornfels to incipient	Permian	?	Challis (1965)

Oman-UAE	Semail	Units of 150 m tectonically thickened to several km	Amphibolite + metaseds.	Granulite to greenschist	94–95	76–93	Ghent & Stout (1981) Searle & Malpas (1980, 1982)
Papua-New Guinea	Papua	100–500 m	Amphibolites	Granulite to greenschist	Cretaceous	?	Davies (1971)
South America	Tinaquillo	1000–1500 m	Gneissose and schistose amphibolites	Granulite to greenschist	Cretaceous	?	MacKenzie (1960)
Syria	Baër-Bassit	<100 m	Amphibolite + metaseds.	Upper amphibolite to greenschist	Cretaceous	86–100	Thuizat *et al.* (1981) Whitechurch & Parrot (1974)
Turkey	Pozanti-Karsanti	<100 m	Amphibolite + metaseds.	Amphibolite to greenschist	Cretaceous	92–101	Cakir *et al.* (1978) Thuizat *et al.* (1981)
United Kingdom	Ballantrae	<50 m	Amphibolite + metaseds.	All granulite to greenschist	479–487	474–482	Bluck *et al.* (1980) Spray & Williams (1980)
	Lizard	<100 m	Basic granulite + amphibolite	All granulite to greenschist	Silurian–Devonian	337–377	Green (1964a & b)
	Shetland	<50 m	Amphibolite + metaseds.	All granulite to greenschist	Ordovician	465–479	Prichard & Spray (in press)
USA	Trinity	<500 m	Amphibolite	Granulite to greenschist	450–480	?	Davies *et al.* (1965)

*Only major or review references are cited.
†Note that the spread of sole ages for the Newfoundland ophiolites has been interpreted as indicating diachronous obduction (Dallmeyer 1977).

>650°C are beyond the capacity of static residual heating by a hot solid peridotite, and it is unlikely that temperatures >300°C can be achieved by shear heating. If metamorphic soles show no evidence of granulite-facies metamorphism then residual heat alone may account for grades up to amphibolite facies and shear heating for grades up to greenschist facies. However, for soles possessing uppermost amphibolite-granulite facies mineral assemblages a combination of dynamic residual and shear heating remains the most plausible explanation. Even with the additional heat supplied by shearing, high grades of metamorphism and anatexis necessitate the juxtaposition of a hot body. Significantly, this limits a site of initial tectonic decoupling of the ophiolite to the proximity of a spreading centre (hot spot).

Age relations

A close association between the ages of ophiolite crystallization and sub-ophiolite metamorphism is further supported by geochronological evidence (Table 1). The Appalachian–Caledonian ophiolites and soles (Quebec, Newfoundland, Ballantrae and Shetland) possess ages which either overlap or occur within 10 Ma of each other. The same appears to be the case for the Greek–Yugoslavian examples (Spray *et al.* in press). The best case of virtually synchronous ophiolite igneous generation and sole formation is afforded by the Semail ophiolite, a factor which led Coleman (1981) and Boudier & Coleman (1981) to conclude that overthrusting was initiated in close proximity to the spreading centre.

At present, the available data constraining the igneous ages of ophiolites is scant and even when both sole and igneous ages are known it is often necessary to compare results using different dating techniques. This may have disadvantages: for example, the blocking temperatures of the U-Pb and K-Ar systems differ by several hundreds of degrees. Nevertheless, Table 1 indicates that a broad synchronicity exists between ophiolite crystallization and sole formation and it is predicted that other, as yet undated ophiolites, possessing high temperature sole components will have similar age relations. By the same reasoning, ophiolites possessing lower-grade soles in the greenschist to low amphibolite facies may show significant igneous–metamorphic age differences, indicating that initial decoupling occurred well away from a spreading centre.

Upper mantle decoupling

If ophiolites are tectonically displaced in close proximity to spreading centres then a knowledge of the deep structure beneath ridges is required in order to understand the decoupling process more fully. As yet only approximate reconstructions can be attempted:

The base of the oceanic lithosphere is defined by the top of the low-velocity zone (LVZ) where shear-wave velocities reach a minimum. The LVZ becomes deeper with age so that in young lithosphere (<5 Ma old) it occurs at <25 km, whereas in mature lithosphere (>40 Ma old) it is reached at about 90–100 km (e.g. Green & Liebermann 1976; Forsyth 1977), although this is clearly dependent on the spreading rate. The LVZ is probably caused by a small degree of partial melting in a zone of low viscosity which extends to a depth of about 175 km allowing rigid lithospere to passively decouple and shear over the underlying more mobile asthenosphere. At spreading centres, the asthenosphere virtually breaches the surface and facilitates shearing at very high levels. It is proposed that it is this shallowness of the lithosphere–asthenosphere boundary at spreading centres which is critical for understanding how ophiolites may be generated (Fig. 5). Clearly, if mature lithosphere was decoupled and overthrust it would produce ophiolites of 100 km thickness; an unrecorded and unlikely event, but decoupling in the proximity of a spreading centre allows thin, immature lithosphere to be overthrust with the characteristics of an ophiolite (many of which deviate in detailed structure from that of mature ocean crust and upper mantle). Many ophiolites may therefore represent intact but young and consequently thin lithosphere rather than the upper part of mature lithosphere. The most likely cause of overthrusting near spreading centres is compression of the ridge. How this takes place depends on the location of the spreading centre; whether it be mid-ocean, back-arc or intracontinental. Furthermore, on closure of an oceanic basin, collision between ridge and trench is a likely eventuality (Christensen & Salisbury 1975). As young lithosphere is relatively buoyant and not easily subducted, this presents a hot, low-density topographic high for the trench to consume which is more likely to be overthrust than subducted. A consequence of this process is a ridge flake of intact lithosphere becoming detached from the shallow asthenosphere to potentially form an ophiolite (Fig. 5). In addition, the lower limit of the lithosphere is considered to approximate the 1200°C isotherm (peridotite solidus): suitable for the subsequent production of high-temperature metamorphic soles when tectonically displaced as an ophiolite and accounting for the presence of more fertile lherzolitic peridotites and related partial melts at the very

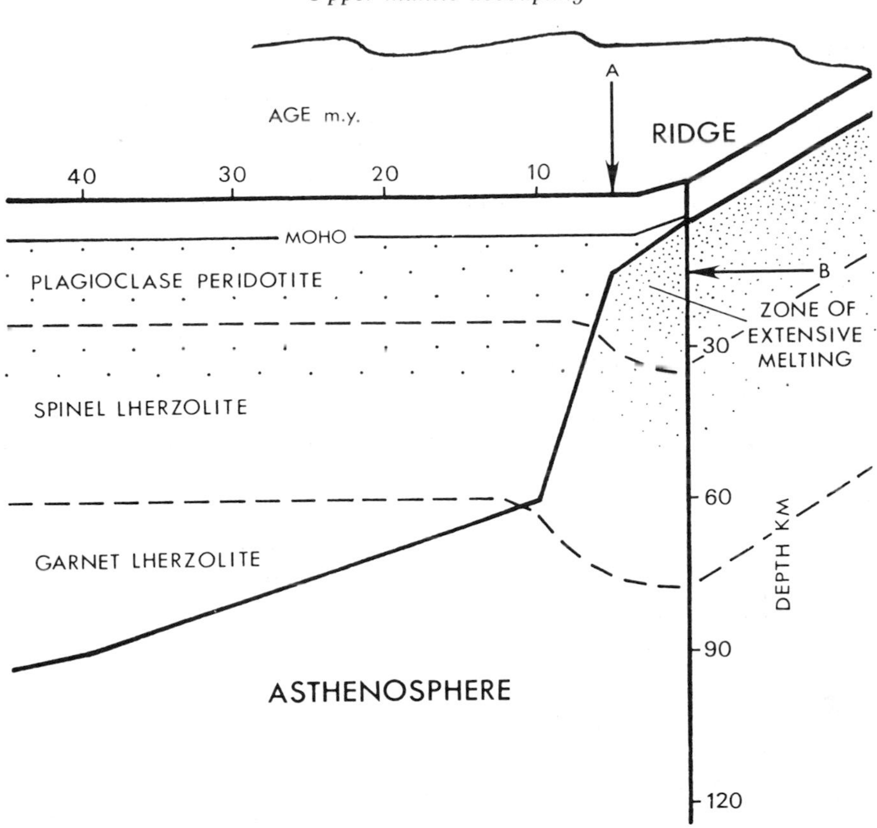

FIG. 5. Profile of oceanic lithosphere overlying asthenosphere. Coarse stipple represents depleted mantle (mainly harzburgite-dunite) which overlies relatively undepleted mantle (lherzolite). Fine stipple corresponds to area of extensive melting beneath ridge. Arrows A and B indicate likely maximum depths and age of lithosphere in order to facilitate ophiolite inception. Horizontal scale much reduced relative to the vertical. Modified after Forsyth (1977) and Green & Liebermann (1976).

base of ophiolites. This model has been applied to the Greek–Yugoslavian ophiolites whereby major transcurrent movements, as well as ridge decoupling, are considered important processes for the tectonic release of the ophiolites (Smith & Spray, in press).

Conclusions

The early stages of ophiolite evolution have been considered with the aim of more fully understanding how ophiolites are initiated. It is concluded that many ophiolites are tectonically decoupled and displaced as complete slices of hot, thin, immature lithosphere in close proximity to spreading centres undergoing compression. This conclusion has been derived using the following arguments:

(i) Ophiolites are generally <15 km in thickness. This implies that there is a fundamental discontinuity at this depth which, when under compression, may facilitate ophiolite displacement.

(ii) Metamorphic soles possessing granulite-facies components require a hot body in addition to shear heating to account for this formation, thus necessitating peridotites as hot as 1000°C. This temperature approximates the isothermal boundary between oceanic lithosphere and asthenosphere.

(iii) In several well-documented cases, geochronological studies confirm that the age differences between ophiolite igneous crystallization and metamorphic sole formation are <10 m.y.

(iv) The occurrence of lherzolite containing recrystallized primitive mafic segregations at the very base of ophiolites testifies to decoupling and overthrusting occurring between relatively depleted peridotite and relatively undepleted peridotite containing partial melts above high-level asthenosphere.

(v) The lithosphere–asthenosphere boundary represents the transition from elastic to ductile behaviour and at shallow depths (i.e. beneath a hot spot) provides a mechanically weak horizon suitable for overthrusting and ophiolite release.

ACKNOWLEDGMENTS: This work was written during tenure of an Earth Sciences Research Fellowship from King's College Cambridge. Cambridge Earth Science Contribution 286.

References

ARCHIBALD, D. A. & FARRAR, E. 1976. K-Ar ages of amphiboles from the Bay of Islands ophiolite and the Little Port Complex, western Newfoundland, and their geological implications. *Can. J. Earth Sci.* **12**, 520–9.

BLUCK, B. J., HALLIDAY, A. N., AFTALION, M. & MACINTYRE, R. M. 1980. Age and origins of the Ballantrae ophiolite and its significance to the Caledonian orogeny and Ordovician time scale. *Geology*, **8**, 492–5.

BOUDIER, F. & COLEMAN, R. G. 1981. Cross section through the peridotite in the Samail ophiolite, southeastern Oman mountains. *J. geophys. Res.* **86**, 2573–92.

——, NICOLAS, A. & BOUCHEZ, J. L. 1982. Kinematics of oceanic thrusting and subduction from basal sections of ophiolites. *Nature, Lond.* **296**, 825–8.

BRUNN, J. H. 1956. Contibutions à l'étude géologique du Pinde Septentrional et d'une partie de la Macédoine occidentale. *Ann. géol. Pays Helleniques* **7**, 358 pp.

CAKIR, U., JUTEAU, T. & WHITECHURCH, H. 1978. Nouvelles preuves de l'écaillage intro-océanique précoce des ophiolites téthysiennes: les roches métamorphiques infra-péridotitiques du massif de Pozanti-Karsanti (Turquie). *Bull. Soc. geol. France* **20**, 61–70.

CARSLAW, H. S. & JAEGER, J. C. 1959. *Conduction of Heat in Solids*. Oxford University Press, London. 510 pp.

CHALLIS, G. A. 1965. High-temperature contact metamorphism at the Red Hills ultramafic intrusion—Wairau Valley—New Zealand. *J. Petrol.* **6**, 395–419.

CHRISTENSEN, N. I. & SALISBURY, M. H. 1975. Structure and constitution of the lower oceanic crust. *Rev. geophys. Space Phys.* **13**, 57–86.

CLAGUE, D., RUBIN, J. & BRACKETT, R. 1981. The age of the garnet amphibolite underlying the Thetford Mines ophiolite, Quebec. *Can. J. Earth Sci.* **18**, 469–86.

COLEMAN, R. G. 1981. Tectonic setting for ophiolite obduction in Oman. *J. geophys. Res.* **86**, 2497–508.

CONQUERE, F. 1971. Les pyroxénolites à amphibole et les amphibolotites associées aux lherzolites du gisement de Lherz (Ariège, France): un example du rôle de l'eau au cours de la crystallization fractionée des liquids issus de la fusion partielle de lherzolites. *Contrib. Mineral. Petrol.* **32**, 32–61.

COX, K. G., BELL, J. D. & PANKHURST, R. J. 1979. *The Interpretation of Igneous Rocks*. George Allen & Unwin, London. 450 pp.

DAHL, P. S. 1980. The thermal-compositional dependence of Fe^{2+}-Mg distributions between coexisting garnet and pyroxene: applications to geothermometry. *Am. Min.* **65**, 852–66.

DALLMEYER, R. D. 1977. Diachronous ophiolite obduction in western Newfoundland: evidence from $^{40}Ar/^{39}Ar$ ages of the Hare Bay metamorphic aureole. *Am. J. Sci.* **277**, 61–72.

—— & WILLIAMS, H. 1975. $^{40}Ar/^{39}Ar$ ages from the Bay of Islands metamorphic aureole: their bearing on the timing of Ordovician ophiolite obduction. *Can. J. Earth Sci.* **12**, 1865–90.

DAVIES, G. A., HOLDAWAY, M. J., LIPMAN, P. W. & ROMEY, W. D. 1965. Structure, metamorphism and plutonism in the south-central Klamath Mountains, California. *Bull. geol. Soc. Am.* **76**, 933–66.

DAVIES, H. L. 1971. Peridotite-gabbro-basalt complex in eastern Papua: an overthrust plate of oceanic mantle and crust. *Australian Bur. Min. Res. Geol. & Geophys.* **128**, 48 pp.

ELLIS, D. J. & GREEN, D. M. 1979. An experimental study of the effect of Ca upon garnet-clinopyroxene Fe-Mg equilibria. *Contrib. Mineral. Petrol.* **71**, 13–22.

FEININGER, T. 1981. Amphibolite associated with the Thetford Mines ophiolite complex at Belmina Ridge, Quebec. *Can. J. Earth Sci.* **18**, 1878–92.

FLEITOUT, L. & FROIDEVAUX, C. 1980. Thermal and mechanical evolution of shear zones. *J. Struct. Geol.* **2**, 159–64.

FORSYTH, D. W. 1977. The evolution of the upper mantle beneath mid-ocean ridges. *Tectonophysics* **38**, 89–118.

GANGULY, J. 1979. Garnet and clinopyroxene solid solutions and geothermometry based on Fe-Mg distribution coefficients. *Geochim. cosmochim. Acta* **43**, 1021–9.

GHENT, E. D. & STOUT, M. Z. 1981. Metamorphism at the base of the Samail ophiolite. *J. geophys. Res.* **86**, 2557–71.

GRAHAM, C. M. & ENGLAND, P. C. 1976. Thermal regimes and regional metamorphism in the vicinity of overthrust faults: an example of shear heating and inverted metamorphic zonation from southern California. *Earth planet. Sci. Lett.* **31**, 142–52.

GREEN, D. H. 1964a. The metamorphic aureole of the peridotite at the Lizard, Cornwall. *J. Geol.* **72**, 543–63.

—— 1964b. The petrogenesis of the high-temperature peridotite intrusion in the Lizard area, Cornwall. *J. Petrol.* **5**, 134–88.

—— & LIEBERMANN, R. C. 1976. Phase equilibria and elastic properties of pyrolite model for the oceanic mantle. *Tectonophysics* **32**, 61–92.

JACOBSEN, S. B. & WASSERBURG, G. J. 1979. Nd and Sr isotopic study of the Bay of Islands ophiolite complex and the evolution of the source of midocean ridge basalts. *J. geophys. Res.* **84**, 7429–45.

JAEGAR, J. C. 1942. Moving sources of heat and the temperatures at sliding contacts. *J. proc. R. Soc. NSW* **76**, 203–24.

—— 1961. The cooling of irregularly shaped igneous bodies. *Am. J. Sci.* **259**, 721–34.

JAMIESON, R. A. 1979. The St Anthony Complex, northwestern Newfoundland: petrological study of the relationship between a peridotite sheet and its dynamothermal aureole. Thesis, PhD, Memorial Univ. Newfoundland (unpubl.).

——1980. Formation of metamorphic aureoles beneath ophiolites—evidence from the St Anthony Complex, Newfoundland. *Geology* **8**, 150–4.

—— 1981. Metamorphism during ophiolite emplacement—the petrology of the St Anthony Complex. *J. Petrol.* **22**, 397–449.

—— & STRONG, D. F. 1981. A metasomatic mylonite zone within the ophiolite aureole, St Anthony Complex, Newfoundland. *Am. J. Sci.* **281**, 264–81.

—— & TALKINGTON, R. W. 1980. A jacupirangite-syenite assemblage beneath the White Hills peridotite, NW Newfoundland. *Am. J. Sci.* **280**, 459–77.

JONES, C. M. 1977. The Ballantrae complex as compared to the ophiolites of Newfoundland. Thesis, PhD, Univ. Wales (unpubl.).

JUTEAU, T., NICOLAS, A., DUBESSY, J., FRUCHARD, J. C. & BOUCHEZ, J. L. 1977. Structural relationships in the Antalya ophiolite complex, Turkey: possible model for an oceanic ridge. *Bull. geol. Soc. Am.* **88**, 1740–8.

KARAMATA, S. 1968. Zonality in contact metamorphic rocks around the ultramafic mass of Brezovica (Serbia, Yugoslavia). *23rd Int. geol. Congress I, Beograd*, 197–207.

——, MAJER, V. & PAMIC, J. 1980. Ophiolites of Yugoslavia. *Ofioliti (Spec. Iss. Tethyan Ophiolites Vol. 1)*, 105–25.

MCCAIG, A. M. (1983). P–T conditions during emplacement of the Bay of Islands ophiolite complex. *Earth planet. Sci. Lett.* **65**, 459–73.

MACKENZIE, D. B. 1960. High temperature Alpine-type peridotite from Venezuela. *Bull. geol. Soc. Am.* **71**, 303–18.

MCKENZIE, D. P. & BRUNE, J. N. 1972. Melting on fault planes during large earthquakes. *Geophys. J.R. astr. Soc.* **29**, 65–78.

MALPAS, J. 1979. The dynamo-thermal aureole of the Bay of Islands ophiolite suite. *Can. J. Earth Sci.* **16**, 2086–101.

MOORES, E. M. 1982. Origin and emplacement of ophiolites. *Rev. geophys. Space Phys.* **20**, 735–60.

MYSEN, B. & BOETTCHER, A. L. 1975a. Melting of hydrous mantle. I. Phase relations of natural peridotite at high pressures and temperatures with controlled activities of water, carbon dioxide and hydrogen. *J. Petrol.* **16**, 520–48.

—— & —— 1975b. Melting of hydrous mantle. II. Geochemistry of crystals and liquids formed by anatexis of mantle peridotite at high pressures and high temperatures as a function of controlled activities of water, hydrogen and carbon dioxide. *J. Petrol.* **16**, 549–93.

NICOLAS, A. & JACKSON, M. 1982. High temperature dikes in peridotites: origin by hydraulic fracturing. *J. Petrol.* **23**, 568–82.

—— & LE PICHON, X. 1980. Thrusting of young lithosphere in subduction zones with special reference to structures in ophiolite peridotites. *Earth planet. Sci. Lett.* **46**, 397–406.

——, BOUDIER, F. & BOUCHEZ, J. L. 1980. Interpretation of peridotite structures from ophiolitic and oceanic environments. *Am. J. Sci.* **280A**, 192–210.

OBATA, M., SUEN, J. & DICKEY, J. S. 1980. The origin of mafic layers in the Ronda high-temperature peridotite intrusion, south Spain: an evidence of partial fusion and fractional crystallization in the upper mantle. *Coll. Internat. CNRS.* **272**, 257–68.

PAMIC, J. 1977. Variation in geothermometry and geobarometry of peridotite intrusions in the Dinaride central ophiolite zone, Yugoslavia. *Am. Mineral.* **62**, 874–86.

—— & MAJER, V. 1977. Ultramafic rocks of the Dinaride central ophiolite zone in Yugoslavia. *J. Geol.* **85**, 553–69.

PERCHUK, L. I. 1969. The effect of temperature and pressure on the equilibrium of natural iron-magnesium minerals. *Int. Geol. Review*, **11**, 875–901.

POPEVIC, A. 1977. Zicni eklogit Bistrice. Juzni deo Zlatiborskog ultramafitskog masiva. *8th Yugoslavian geol. congress, Ljubljana 1976*, pp. 245–60.

PRICHARD, H. M. & SPRAY, J. G. (in press). The metamorphic sole of the Shetland ophiolite, NE Scotland. *J. geol. Soc. London.*

QUICK, J. 1981. Petrology and petrogenesis of the Trinity peridotite, an upper mantle diapir in the eastern Klamath mountains, northern California, *J. geophys. Res.* **86**, 11827–63.

RAHEIM, A. & GREEN, D. H. 1974. Experimental determination of the temperature and pressure dependence of the Fe-Mg partition coefficient for coexisting garnet and clinopyroxene. *Contrib. Mineral. Petrol.* **48**, 179–203.

REITAN, P. H. 1968a. Frictional heat during metamorphism: quantitative evaluation of concentration of heat generation in time. *Lithos* **1**, 151–63.

—— 1968b. Frictional heat during metamorphism: quantitative evaluation of concentration of heat generation in space. *Lithos* **1**, 268–74.

—— 1969. Temperature with depth resulting from frictionally generated heat during metamorphism. *Geol. Soc. Am. Mem.* **115**, 495–502.

REUBER, I., WHITECHURCH, H. & CARON, J. M. 1982. Setting of gabbroic dykelets in an ophiolitic complex by hydraulic fracturing. *Nature, Lond.* **296**, 141–3.

SAXENA, S K. 1979. Garnet-clinopyroxene geothermometry. *Contrib. Mineral. Petrol.* **70**, 229–35.

SCHOLZ, C. H. 1980. Shear heating and the state of stress on faults. *J. geophys. Res.* **85**, 6174–84.

——, BEAVAN, J. & HANKS, T. C. 1979. Frictional metamorphism, argon depletion and tectonic stress on the Alpine Fault, New Zealand. *J. geophys. Res.* **84**, 6770–82.

SEARLE, M. P. 1980. The metamorphic sheet and underlying volcanic rocks beneath the Semail ophiolite in the northern Oman mountains of

Arabia. Thesis, PhD, The Open University (unpubl.).
—— & MALPAS, J. 1980. Structure and metamorphism of rocks beneath the Semail ophiolite of Oman and their significance in ophiolite obduction. *Trans. R. Soc. Edinburgh* **71**, 247–62.
—— & —— 1982. Petrochemistry and origin of sub-ophiolite metamorphic and related rocks in the Oman Mountains. *J. geol. Soc. Lond.* **139**, 235–48.
SIBSON, R. H. 1977. Fault rocks and fault mechanisms. *J. geol. Soc. Lond.* **133**, 191–213.
SMITH, A. G. & SPRAY, J. G. (in press). A half-ridge transform model for the Hellenic–Dinaric ophiolites. *In*: DIXON, J. E. & ROBERTSON, A. H. F. (eds) *The Geological Evolution of the Eastern Mediterranean.* Special Publication of the Geological Society of London.
SPRAY, J. G. 1980. Some ophiolite-related metamorphic rocks from the Eastern Mediterranean and Britain. Thesis, PhD, Univ. Cambridge (unpubl.).
—— 1982. Mafic segregations in ophiolite mantle sequences. *Nature, Lond.* **299**, 524–8.
—— & RODDICK, J. C. 1980. Petrology and $^{40}Ar/^{39}Ar$ geochronology of some Hellenic sub-ophiolite metamorphic rocks. *Contrib. Mineral. Petrol.* **72**, 43–55.
—— & —— 1981. Evidence for Upper Cretaceous transform fault metamorphism in West Cyprus. *Earth planet. Sci. Lett.* **55**, 273–91.
—— & WILLIAMS, G. D. 1980. The sub-ophiolite metamorphic rocks of the Ballantrae Igneous Complex, SW Scotland. *J. geol. Soc. Lond.* **137**, 359–68.
——, BEBIEN, J., REX, D. C. & RODDICK, J. C. (in press). Age constraints on the igneous and metamorphic evolution of the Hellenic–Dinaric ophiolites. *In*: DIXON, J. E. & ROBERTSON, A. H. F. (eds) *The Geological Evolution of the Eastern Mediterranean.* Special Publication of the Geological Society of London.
THUIZAT, R., WHITECHURCH, H., MONTIGNY, R. & JUTEAU, T. 1981. K-Ar dating of some infra-ophiolite metamorphic soles from the Eastern Mediterranean: new evidence for oceanic thrustings before obduction. *Earth planet. Sci. Lett.* **53**, 302–10.
TRELOAR, P. J., BLUCK, B. J., BOWES, D. R. & DUDEK, A . 1980. Hornblende-garnet metapyroxenite beneath serpentinite in the Ballantrae ophiolite of SW Scotland and its bearing on the depth provenance of obducted oceanic lithosphere. *Trans. R. Soc. Edinburgh* **71**, 201–12.
WHITECHURCH, H. & PARROT, J. F. 1974. Les écailles métamorphiques infra-péridotitiques du Baër-Bassit (nord-ouest Syrie). *ORSTOM* **6**, 173–84.
WILLIAMS, H. & SMYTH, W. R. 1973. Metamorphic aureoles beneath ophiolite suites and Alpine peridotites: tectonic implications with west Newfoundland examples. *Am. J. Sci.* **273**, 594–621.
YUEN, D. A., FLEITOUT, L. SCHUBERT, G. & FROIDEVAUX, C. 1978. Shear deformation zones along major transform faults and subducting slabs. *Geophys. J.R. astr. Soc* **54**, 93–119.

J. G. SPRAY, Department of Earth Sciences, University of Cambridge, Cambridge CB2 3EQ.

Initiation of subduction zones along transform and accreting plate boundaries, triple-junction evolution, and forearc spreading centres—implications for ophiolitic geology and obduction

J. F. Casey & J. F. Dewey

SUMMARY: Most large ophiolite complexes have been interpreted as representing the obducted remnants of oceanic basement of forearc regions. A knowledge of the tectonic setting of ophiolite formation as well as the mechanism of their entrapment within forearc regions is essential to achieve a more complete understanding of the often complicated geology of ophiolite complexes and the tectonic processes that result in their evolution and ultimate obduction. The mechanism of entrapment of an ophiolite complex within a forearc region is intimately related to the mechanisms involved in trench initiation. The most easily documented and apparently likely mechanisms of trench initiation and isolation of oceanic basement within forearcs are: (i) polarity flips along weakened back-arc/island-arc interfaces, and (ii) plate boundary evolution involving the conversion of active transform and accreting plate boundaries into subduction zones. These latter mechanisms of trench initiation are predicted by plate theory to occur by the evolution of transform and accreting plate boundaries upon (usually) gradual and (less often) abrupt changes in their position with respect to the instantaneous pole of relative motion, and by the evolution of certain types of triple junctions that leads to rapid conversion of transform and accreting plate boundaries to subduction zones. These plate boundary evolutionary schemes involve spatially and temporally complex overlap of subduction and plate accretion processes in forearc regions. The overlapping processes occurring during these conversions may result in 'atypical' oceanic-crustal magmas that are produced during or soon after sea-floor spreading within the forearc region. These magmas may not resemble typical mid-ocean-ridge basaltic (MORB) compositions. Such 'abnormalities' in the petrology of some ophiolite complexes need not necessarily imply that a process other than sea-floor spreading was the dominant plate-boundary process responsible for the generation of these ophiolite complexes. The term 'island-arc ophiolite' attached to several ophiolite complexes on the basis of their petrology may not appropriately reflect the dominant plate-boundary process (i.e. sea-floor spreading) and should be used cautiously and only if supported by non-petrologic data (e.g. field observations and structural relationships).

Although ophiolite obduction is a discrete major event occasionally punctuating the geologic record and usually marking the destruction of a previously stable Atlantic-type continental margin (Dewey 1976; Moores 1982), it is the necessary, often capricious, result of normal plate-boundary evolution in a wide variety of tectonic settings. In a more than two-plate mosaic, plate boundaries continuously change their positions in the reference frame of the instantaneous rotation poles that describe their relative motion (Dewey 1975a, b). This means that slip rates and direction across plate boundaries change through time, and that plate boundaries may evolve from one fundamental type to another.

Many large ophiolite complexes have been interpreted as the remnants of oceanic basement of forearc regions (e.g. Gealey 1980). These 'forearc' ophiolites may be obducted as the overriding plate during attempted subduction of previously stable continental margins (Dewey 1976; Casey 1980; Moores 1982). In our opinion, the conversion process by which a segment of oceanic crust, formed during sea-floor spreading, is suddenly or gradually trapped within a forearc region upon initiation of a subduction zone, is intimately related to the complicated geology of most ophiolite complexes. Such complexities have led to the conclusion that some ophiolite complexes have not formed at accreting plate margins and cannot therefore be compared to contemporary oceanic crust.

We outline in this paper a proposal that the evolution of transform and accreting plate boundaries into trenches, and the development of certain types of triple junctions, affords an adequate explanation of many common features and complexities of obducted ophiolites.

These include: (i) the commonly short time period between accretion and detachment of the ophiolite; (ii) the high temperature, transported, subjacent, dynamothermal aureoles; (iii) the relative orientations of the trends of ophiolite belts with respect to their sheeted dyke trends; (iv) the occurrence of highly deformed transform fault-generated ophiolitic rocks along the frontal edges of some ophiolite nappes; (v) the well-developed stratiform nature of the ophiolite suite; (vi) the occurrence of sheeted dykes with well-defined preferred orientations; and (vii) the occurrence of MORBs and/or boninitic volcanics as well as other REE and incompatible-element-depleted basaltic and silicic volcanics within some ophiolite complexes.

We suggest that many primitive island arcs have been initiated along oceanic fracture zones or along spreading centres, and that others are propagated or lengthened during the evolution of certain types of triple junctions. Because magmatic fronts within ensimatic arcs usually lie at least 100 km from the active trench, once the subduction is initiated, entrapment and preservation of fairly pristine oceanic crust will tend to occur within the forearc region (Matsuda & Uyeda 1971; Dickinson & Seely 1979). This ophiolitic basement of the forearc will be largely unaffected by a protracted history of arc magmatism. If significant tectonic erosion of the ophiolitic forearc basement does not occur after its entrapment (i.e. during the course of subduction), attempted subduction of a passive (Atlantic-type) continental margin will result in obduction of the leading edge of the overriding plate which includes the trapped ophiolitic forearc basement. In this paper we attempt to briefly review the various methods proposed to distinguish the tectonic environment of ophiolite generation as well as various obduction models. We also attempt to show from observations of present plate configurations and from plate reconstructions derived from the study of ophiolite-bearing orogenic belts that, upon initial contraction of an ocean basin, subduction zones have a strong tendency to initiate within, and not at the stable margin of, oceanic basins. This takes place by the conversion of previously existing plate boundaries to subduction zones.

Subduction initiation and ophiolite obduction models

The Wilson Cycle requires perpetual rupture and oceanic opening followed by closure, culminating in the collision of its bordering land masses. The initial closing of an Atlantic-type ocean basin requires initiation of a subduction zone within the basin or near a continental margin. The site of initiation of a subduction zone upon contraction of an ocean is a subject seldom discussed using data from orogenic belts. Dewey & Bird (1970) suggested that the direct transformation of an Atlantic-type continental margin to an Andean margin was likely when a trench originates at or very close to a continental margin. The direction of underthrusting in this case would be beneath the continent and would initiate due to the gravitational instability of old, thick oceanic lithosphere. Similarly, the majority of obduction models have been influenced by the fact that most present-day subduction zones dip toward the continents, not away from them. These models include obduction of an ophiolite just prior to subduction of a spreading ridge beneath an Andean-type margin (Christensen & Salisbury 1975; Dewey 1976), and obduction of a small, young back arc basin opened between the continental margin and an island arc (Dewey & Bird 1971; Armstrong & Dick 1974; Dewey 1976; Smith & Woodcock 1976).

Transform motion preceding ophiolite obduction has recently been incorporated into a number of obduction models. Karson & Dewey (1978) suggested that the obduction site of the Bay of Islands Ophiolite was a transform–fracture zone system lying proximal to and parallel with the ancient continental margin of North America. This fracture zone is now represented by a narrow belt of highly deformed ophiolitic rocks along the Gulf of St Lawrence, known as the Coastal Complex (Fig. 1). This deformed belt has an autochthonous relationship with the Bay of Islands Ophiolite, and appears to have been the leading edge of the once-intact ophiolitic nappe (Karson 1977, this volume; Karson & Dewey 1978). Similarly, Saleeby (1977) proposed that the emplacement of the Kings-Kaweah ophiolitic belt of the Sierra Nevada in California occurred upon a change in relative motion after the ophiolite was juxtaposed with the raw edge of a continent along a transform fault. Brookfield (1976) suggested this same scenario as a general model for the emplacement of many large ophiolite complexes throughout the world, and furthermore contended that these ophiolites were not associated with subduction zones or island arcs. He also proposed that the final phase of emplacement involved gravity sliding.

We distinguish two types of passive margins; the first are Atlantic-type margins bordering large cratonic areas and typically large ocean basins, the second are back-arc passive margins created at the trailing edges of smaller back-arc spreading basins or at the edge of trapped back-arc basins. It has often been suggested that passive Atlantic-type continental margins are likely sites for the initiation of subduction zones (e.g. Dewey & Bird 1970; Karig 1982). The rationale for this seems to be the fact that the simplest way to picture the development of an Andean-type margin is to initiate subduction at a stable Atlantic-type continental margin. However, what is overlooked is that present-day and recently active Andean orogenic belts have often undergone long and very complex geologic histories prior to the development of the Andean-type margin. Part of that history has involved accretion of older island-arc terrains to the

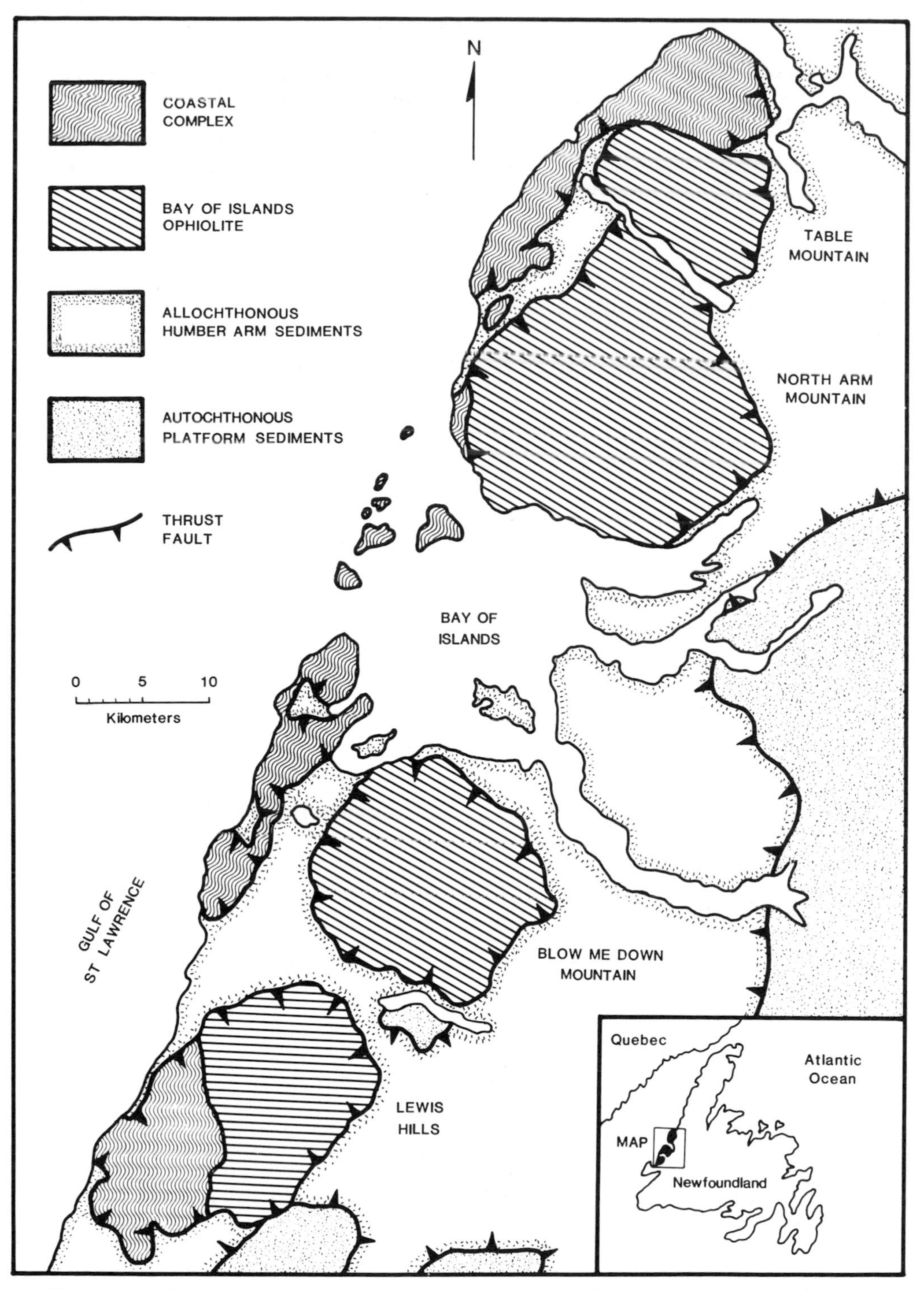

FIG. 1. Schematic representation of the internal geometric features of the Bay of Islands Ophiolite. The Coastal Complex represents a NE–SW trending belt of deformed ophiolitic rocks interpreted to have been generated along an oceanic fracture zone (Karson & Dewey 1978) adjacent to the Bay of Islands Complex which consists of a relatively undeformed ophiolitic sequence interpreted to have formed at a spreading centre outside the transform deformation zone (Casey *et al.* 1983). The mean trend of dykes in the BOIC is shown by solid black lines. The southernmost massif (the Lewis Hills) has suffered a 45° anticlockwise rotation during obduction. The regional trend of the Coastal Complex (NE–SW) and sheeted dykes (NW–SE) are respectively parallel to and perpendicular to the leading edge of the ophiolitic nappe. The transform-affected ophiolitic sequence is situated at the front of the nappe.

continental edge, attendant destruction of previously stable margins, and arc polarity flips following collision (e.g. Ben-Avraham *et al.* 1981). Part of the rationale for predicting that stable Atlantic-type margins are likely places for trench initiation is that the oceanic lithosphere is denser and thicker there and more capable of sinking into the asthenosphere, driven by gravitational forces. The subsidence-driving mechanisms observed at these margins are however adequately explained by thermal subsidence, continental-crustal stretching and sediment loading (Steckler & Watts 1981). Flexure of the oceanic lithosphere along older margins usually has a large wavelength and involves dips of much less than 1° on the oceanic basement/sedimentary rock interface (Lehner & de Ruiter 1977; Grow *et al.* 1979; Watts & Steckler 1979). The main problem in initiation of a subduction zone is concentrating the stress necessary to flex and rupture a thick lithosphere. We suggest that along older, Atlantic-type margins the interface between continental and oceanic lithosphere is probably one of the least likely places to rupture the lithosphere and initiate subduction. This is because the continental edge is usually characterized by the oldest and thickest oceanic lithosphere within the basin. It is flexurally and torsionally the strongest and would require very high stress levels to initiate rupture and cause subduction. We, therefore, do not find it peculiar that within the geologic record *direct* transformation from Atlantic- to Andean-type continental margins is seldom recorded or demonstrated. Stable Atlantic-type margins are more likely destroyed and softened by attempted subduction beneath colliding arcs having polarities away from the margin. Exceptions to such direct transformations may occur within embryonic (Red-Sea type) basins if ocean closure is attempted soon after rifting. This may have occurred along the northern Spanish margin in the late Cretaceous as initiation of subduction occurred in response to partial closure of the Bay of Biscay (Malod *et al.* 1982).

On the other hand, a freqent mechanism of subduction initiation is by polarity flip (e.g. New Hebrides) in which the island-arc/oceanic-lithosphere interface along a back-arc margin becomes a locus for trench initiation (Karig 1982). We regard this as a site in which the flexural and torsional rigidity of the lithosphere is extremely low. The thickness of the elastic or short-term lithosphere is diminished here because of the unusually high geothermal gradient in the arc region (Matsuda & Uyeda 1971). The stress levels required to initiate rupture at these types of margins are probably smaller and may initially lead to rapid thermal resorption of the lithosphere within the hot underlying mantle. The mantle drag forces resisting subduction would therefore be initially minimal. Segments of trapped or spreading back-arc basins may be isolated in the forearc during polarity flips and may be obducted upon island-arc collision.

Spreading centres and segments of transform plate boundaries represent zones where the lithosphere also tends to be hot and thin as well as flexurally and torsionally weak. Thus, they represent pre-established zones of weakness in the lithosphere along which rupture has already occurred which combined with the kinematics of plate-boundary evolution, appears to make these likely sites of trench initiation.

Therefore, we propose that the most easily documented and apparently likely mechanisms of trench initiation are polarity flips along weak back-arc/island-arc interfaces, and plate-boundary evolution involving the change of transform and accreting plate boundaries into subduction boundaries. In an Atlantic-type basin where island arcs are not generally present, mechanisms involving polarity flips cannot operate initially. We believe that, upon initiation of closure of an Atlantic-type basin, certain existing plate boundaries will evolve and lead to trench formation as the result of changes in relative plate motion.

Basal metamorphic aureoles

A frequently imposed constraint in interpreting the tectonic environment of ophiolite generation and the mechanism of ophiolite obduction, results from the common occurrence of dynamothermal metamorphic 'aureoles' attached to the base of mantle tectonite sections of ophiolite complexes. These aureoles are interpreted as the basal mantle detachment zone along which ophiolite obduction is initiated (Williams & Smyth 1973). They usually consist of 50–500 m of polyphase-deformed metamorphic rocks displaying an inverted metamorphic gradient from granulite facies near the top, to low greenschist facies near the base. The protoliths of these aureoles are inferred to have been predominantly mafic plutonic and volcanic rocks and sedimentary rocks of oceanic affinity (Jamieson 1980; Malpas 1979; Searle & Malpas 1980).

These aureoles have been generally interpreted as formed by obduction of young, hot oceanic lithosphere with the heat source for metamorphism contained as residual heat within an overriding thrust wedge of young lithosphere. Various models also include a component of frictional heat (Church & Stevens 1971; Dewey &

Bird 1971; Graham & England 1976; Malpas 1977, 1979; Jamieson 1980). This has led to a variety of obduction models which hold that the site of formation (i.e. accretion) of the ophiolite must be proximal to the continental margin over which obduction occurs (Malpas 1977; Jamieson 1980).

In addition, these obduction models have been influenced by the fact that most present-day subduction zones dip towards, and not away from, the closest continent. One such model is the obduction of an ophiolite from the underthrust plate just prior to subduction of a spreading ridge beneath an Andean-type margin (Christensen & Salisbury 1975; Dewey 1976). This model, is however, inconsistent with the fact that most continental margins over which large ophiolite sheets have been obducted record a passive margin (i.e. Atlantic-type) history until their destruction at the time of obduction (Dewey 1976; Gealey 1977; Williams 1979). Obduction, therefore, usually punctuates the stratigraphic record of the previously stable margin with a major orogenic episode. Another commonly proposed model involves obduction of oceanic lithosphere from a small and young back-arc basin opened between the continental margin and an island arc (Dewey & Bird 1971; Armstrong & Dick 1974; Dewey 1976; Smith & Woodcock 1976). A similar model involving obduction of lithosphere from an embryonic (Red Sea-type) basin has also been proposed (Malpas 1977; Karson & Dewey 1978). All these models help to satisfy the proposed constraint that the lithosphere must be young and hot at the time of obduction in order to create the basal dynamothermal metamorphic aureole.

The stratigraphic and structural events surrounding the obduction of the Bay of Islands Ophiolite Complex of western Newfoundland are summarized in Fig. 2 and illustrate some of the problems associated with the above models that attempt to explain the heat source for metamorphic aureoles and incorporate subduction towards the continental margin. Many of the relationships about to be discussed are not unique to the Bay of Islands Complex, but illustrate common problems along margins where large ophiolites have been obducted. Many obduction models thus far proposed for the Bay of Islands Ophiolite emphasize that the site of ophiolite formation must be proximal to the site of obduction—i.e. proximal to the east-facing early-Palaeozoic continental margin of North America (Dewey & Bird, 1971; Malpas 1977, 1979; Williams *et al.* 1977; Karson & Dewey 1978; Jamieson 1980). Of these, the two most-quoted mechanisms for producing this proximity are: (i) continent–continent rifting and the creation of a small embryonic basin (Red Sea or Gulf of California type) (Malpas 1977; Karson & Dewey 1978), and (ii) continent/detached-island-arc rifting and opening of a small back-arc basin separating the east-facing North American continental margin from a more easterly island arc with continental basement (Dewey & Bird 1971; Dewey 1976). These models require obduction soon after ophiolite formation to satisfy proposed heat requirements for the metamorphic auerole.

There are some seemingly difficult problems with models that predict formation of the Bay of Islands Ophiolite proximal to the North American continent and obduction shortly after formation. The geology of western Newfoundland records late Precambrian ocean opening and the development of a stable continental margin until the time of obduction during mid-Ordovician (Williams & Stevens 1969; Stevens 1970, 1976; Williams 1979). The age of rifting and the birth of the proto-Atlantic ocean is given by the ~600 Ma ages on rift-related mafic dykes and volcanics along the margin (Stukas & Reynolds 1974), and by the earliest sedimentological and morphological development of the margin in the Late Proterozoic and early-Cambrian time (Williams & Stevens 1969; Stevens 1970, 1976). The margin appears to have reached maturity by the middle Cambrian (Stevens 1970), and there is little evidence of a prolonged rifting stage. If the Bay of Islands Complex formed as an oceanic tract lying proximal to the rifted margin, the age of ophiolite formation should be only slightly younger than that of the initial rifting of the margin. On the contrary, there is a large disparity (~90 m.y.) between the age of rifting (600 Ma) and the age of ophiolite formation (508 Ma) (Fig. 2). If the duration of spreading was 90 m.y., a slow half-spreading rate of 1 cm yr^{-1} would have generated an ocean 1800 km wide, while a maximum rate of 10 cm yr^{-1} would have produced a basin 18,000 km wide. Whatever the actual spreading rate, the implication of this age disparity is that a large ocean would have been created after initial rifting, and a segment of oceanic lithosphere having the age of the Bay of Islands Complex would likely lie a substantial distance from the margin. Thus, the Bay of Islands oceanic lithosphere would have been transported a large distance before obduction onto the continental margin of early Palaeozoic North America. This transport may have occurred by incorporation of the Bay of Islands oceanic lithosphere into the forearc of an overriding plate, above a subduction zone that dipped to the east, away from the stable continental margin of eastern North America, and which consumed a progressively older

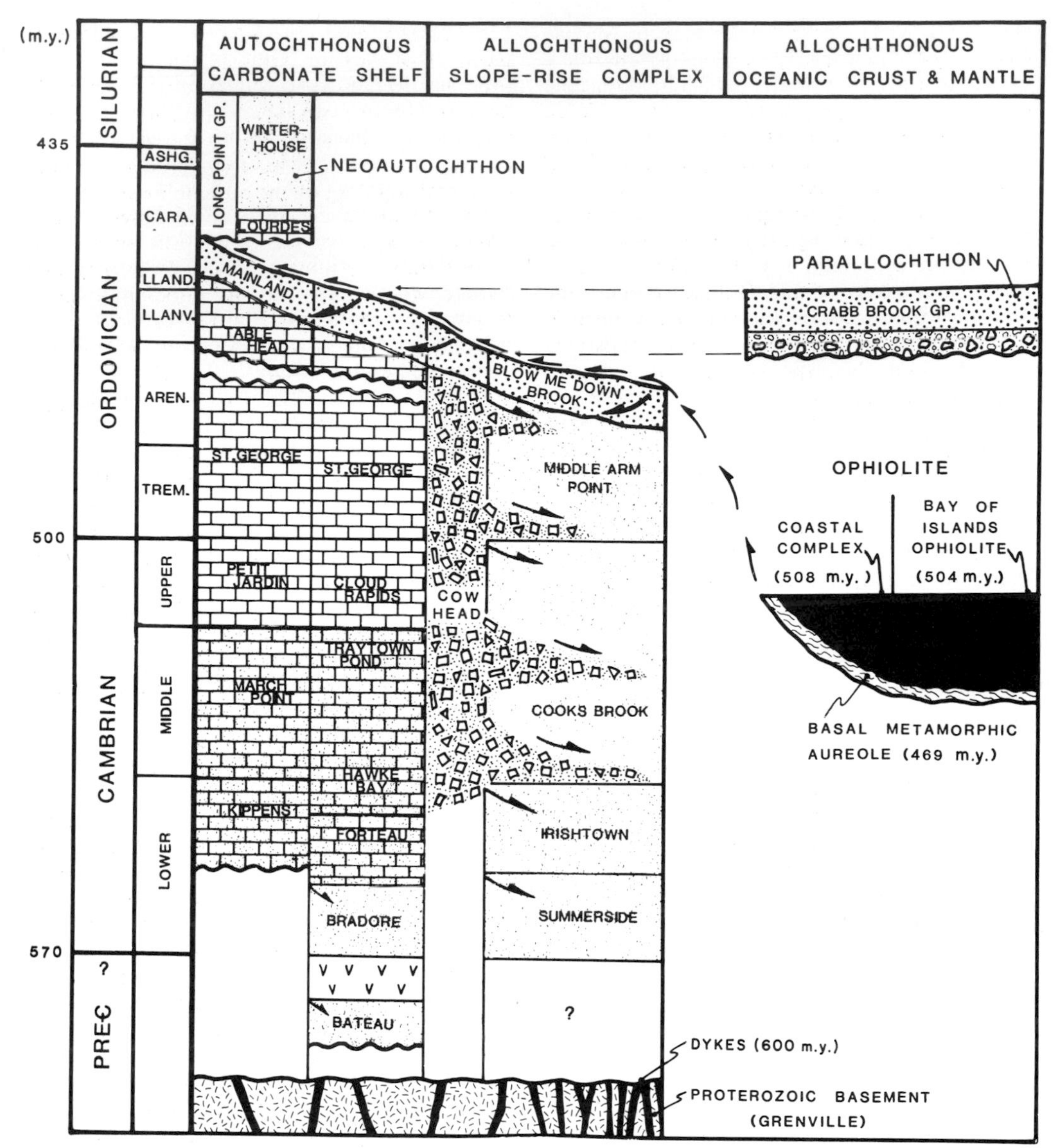

FIG. 2. Reconstructed stratigraphic relationships of western Newfoundland, modified after Williams (1971). Random dashes = crystalline Grenville basement; fine stipple = pre- or postorogenic clastics; bricks = carbonates; shaded bricks = interbedded carbonates and clastics; angular clasts = bank-edge carbonate breccias; coarse and very coarse stipple = synorogenic clastics; V = flood basalts; steep thick lines = basalt dykes; arrows = sedimentary provenance direction; half arrows = tectonic transport; thick lines = unconformities; ophiolite complex shown as mafic and ultramafic layer (no difference in age implied) over schematic aureole (of younger age than position shown); fossil control (Stevens 1976; Casey & Kidd 1981).

oceanic tract adjacent to the margin. Other possibilities that may explain the age disparity result when ophiolite formation occurs within the forearc region near certain types of triple junction (discussed in the following sections).

Another important age relationship is the disparity (i.e. 40 m.y.) between the age of ophiolite formation and the age of ophiolite obduction (Fig. 2). The age of formation of the Bay of Islands Ophiolite is given by U-Pb dates of 504 ± 10 Ma and 508 ± 5 Ma (Mattinson 1975, 1976), and by Sm-Nd internal isochrons that yield ages of 508 ± 6 Ma and 501 ± 13 Ma (Jacobsen & Wasserburg 1979). Amphibolites from the basal dynamothermal aureole give an $^{40}Ar/^{39}Ar$ age of 469 ± Ma (Dallmeyer & Williams 1975). This cooling age on the aureole has been interpreted as the approximate time of initiation of obduction onto the previously stable continental margin of North America, and this age is consistent with its stratigraphic development (Stevens 1970). The Bay of Islands Ophiolite is 10–12 km thick. If oceanic lithosphere of this thickness had undergone a normal oceanic thermal decay history prior to obduction (Sclater *et al.* 1971), or even a back-arc cooling history (Watanabe *et al.* 1977), the temperature at 12 km depth would have been less than 300°C. It would therefore appear that the Bay of Islands lithosphere was thermally mature at the time of obduction, yet argon retention at the level of the aureole was apparently not achieved until near the time of obduction. There would not seem to be an adequate source of residual heat at the time of obduction to produce the highest parts of the basal metamorphic aureole which require temperatures of 800–950°C (Malpas 1979; Jamieson 1980). In addition, thermodynamic calculations suggest that only a minor temperature increase (~200°C) will be generated by friction during underthrusting (Mercier 1976; Malpas 1979).

Malpas (1977) estimated that temperatures of 800°C were attained at the contact between the mafic parts of the aureole and the overlying sheared lherzolites in the Bay of Islands. Jamieson (1980) calculated temperatures of 900–950°C at the top of the aureole beneath the White Hills peridotite in northwestern Newfoundland. Temperatures decrease progressively with depth until near its base, 800 m below the peridotite–aureole contact in the case of Hare Bay, a temperature of 360°C is estimated. Pressure estimates of 7–10 kb at the top of the aureole at Hare Bay (Jamieson 1980) are difficult to reconcile in terms of the present structural thickness of the ophiolite in the Bay of Islands Complex (10–12 km). A continuous pressure gradient from 7–10 kb to an estimated value of 2 kb at the base of the aureole was obtained by Jamieson (1980). Two kilobars more closely coincides with the present structural thickness estimates of the overlying ophiolite. The steep, inverted, pressure–temperature gradient through the aureole would suggest that the aureole was formed sequentially from top to bottom and represents a composite of rocks accreted to the base of the ophiolite over a long period under progressively cooler temperature conditions (Jamieson 1980) and shallower depths (or confining pressures). Jamieson (1980) has suggested that this accretion to the base of the ophiolite takes place during the displacement of the ophiolitic section onto the continental margin and explains the discrepancy between pressure differences displayed through the aureole (~5 kb) and its structural thickness (~800 m) as a result of structural thinning of the aureole itself.

An alternative explanation proposed here is that, during subduction, the peridotite member of the overriding ophiolitic forearc was structurally thinned as the dip of the subduction zone progressively shallowed (Fig. 3) as the arc became compressional in character prior to collision (see Dewey 1980). The aureole may begin to form in a completely intra-oceanic environment along a subduction zone at the time of initiation of a trench. The protoliths for the aureole rocks are dominantly mafic volcanic and plutonic rocks and minor sediments (Jamieson 1980; Malpas 1979), all of which may represent the top part of the downgoing oceanic lithosphere, deformed, metamorphosed and accreted to the base of the overriding ophiolite complex. The detachment zone (i.e. the subduction zone), which is marked by these accreted oceanic rocks, presumably shallowed progressively with time as the overlying mantle wedge was expelled and the overriding plate was in turn structurally thinned. The metamorphic auerole would be the product of a form of burial metamorphism brought about by the tectonic insertion of crustal rocks to deep levels in the mantle and accretion of slivers of these rocks to the hanging wall of the subduction zone. During initiation of subduction, the lithosphere may have been young and thermally immature and the mantle wedge overlying the subduction zone may have been initially thick and at an elevated temperature at the level of detachment. Relatively high-temperature/high-pressure metamorphic equivalents of the downgoing oceanic crust and mantle would be accreted to the base of the overlying mantle wedge. This wedge would be structurally thinned and cooled with time as the dip of the subduction zone shallowed, and the isotherms became depressed. Therefore, with time, progressively lower-temperature and lower-pressure metamorphosed oceanic-crustal rocks would be accreted to the base of the overlying mantle wedge along the subduction plane. The length of time over which the auerole formed, therefore, may be exceptionally long if this scenario is correct. The aureole may not actually represent a form of short-term dynamothermal contact metamorphism with the heat supplied by the residual heat contained within a hot, thin (10–12 km) ophiolite sheet as proposed by most authors. The upper portion of the aureole may form in a completely intraoceanic

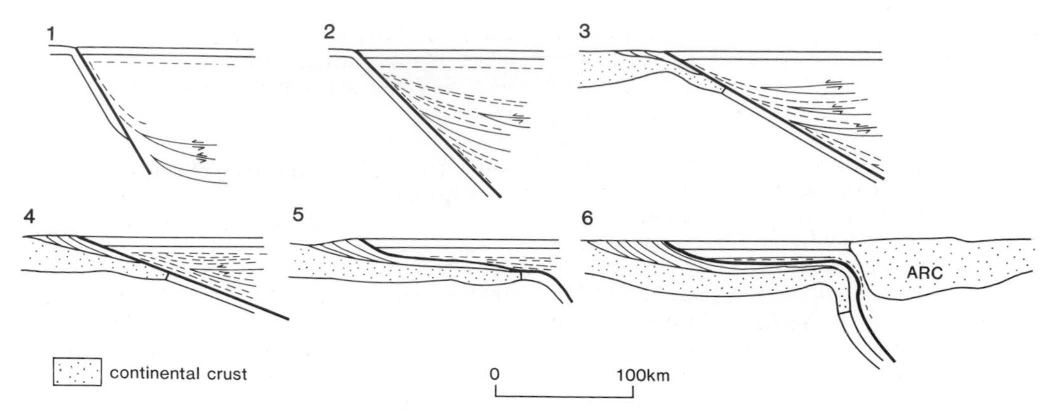

FIG. 3. Schematic model depicting 'flattening' of the subduction plane by progressive shallowing of dip and consequent expulsion or structural thinning of the mantle wedge above the subduction zone. Removal of the thick mantle wedge is accomplished by initially high-temperature solid-state flow under low-stress conditions and later by more localized lower-temperature solid-state flow under high-stress conditions. Thick black line represents the basal metamorphic 'aureole' which consists of material derived from the downgoing oceanic plate and accreted to the hanging wall of the subduction zone under progressively lower temperature and lower-pressure conditions. The aureole is largely formed prior to the time of obduction and final emplacement.

environment at deep levels in the mantle, while the lower part may involve continental-margin sedimentary and rift-related volcanic rocks accreted at shallow depths (Jamieson 1980). The difference between the structural thinning model proposed here and previous models is basically one of how quickly the heat source is dissipated, how large it is, and over how long a time period the aureole develops. We regard the radiometric cooling age of the aureole as representing a lower limit for the time of obduction, but it probably does not reflect the time of initiation of subduction or the time in which the bulk of the aureole is forming.

Ophiolite petrology, sedimentary rocks and structure as discriminants of tectonic setting

Although the similarities of ophiolites with contemporary oceanic crust and mantle are, in general, well acknowledged in the geological community, detailed petrological studies of many ophiolites have shown that there are important geochemical discrepancies within some ophiolites. Miyashiro (1973) suggested, mainly on the basis of major-element chemistry of lavas and dyke rocks from the Troodos Complex in Cyprus, that the complex formed within a magmatic arc. Although Miyashiro's proposal was heavily criticised beause hydrothermal alteration may have severely obscured the original major-element chemistry, recent analysis of fresh glasses from the Troodos Massif suggest some of the volcanics within the massif do have island-arc affinities (Robinson pers. comm.). In addition, analysis of geochemical data using plots of relatively immobile major and minor elements to estimate the tectonic setting of ophiolite formation suggests that lavas from many other ophiolites do have island-arc affinities, while others show typical MORB compositions (Pearce & Cann 1973; Pearce & Norry 1979; Pearce 1980).

There are, on the other hand, several important geological constraints that argue against a magmatic island-arc origin and support a sea-floor spreading origin for many of the ophiolites which show geochemical dissimilarities with MORBS or BABBs (back-arc basin basalts). If some ophiolites form within magmatic arcs, we would expect that ophiolite stratigraphies would show evidence of a long and complicated history of igneous and metamorphic events, and a variety of fine- to coarse-grained, petrologically distinct igneous rocks that are chaotically intermixed within the crust. As pointed out by Moores (1982), ophiolitic crust should be much thicker than normal oceanic crust. In general, however, there is a tendency for ophiolitic crust to be somewhat thinner on average than the oceanic average (Coleman 1977). In addition, most ophiolite stratigraphies display evidence that a discrete, short-lived igneous event was responsible for its formation. This is indicated by the orderly arrangement of lithologic units, the regionally sharp and non-random subhorizontal contacts between coarse-grained gabbroic and fine-grained hypabyssal dyke rocks, and the general lack of abundant sedimentary rocks interstratified with pillow lavas. Perhaps the most convincing argument against the formation of ophiolites within magmatic island arcs is the mere existence of the sheeted-dyke unit (Coleman 1977). Some ophiolite complexes extend for hundreds of kilometres perpendicular to the strike of dyke trends indicating spreading (i.e. 100% extension) of at least these distances (Moores 1982). Spreading of such large distances without developing some sort of spreading centre is difficult to reconcile with observed tectonic activity within the magmatic-arc environments.

If one accepts that the structure of ophiolites conclusively indicates that their formation is the result of sea-floor spreading despite petrologic discrepancies, one is left with two possible tectonic settings for ophiolite generation; either along a mid-ocean spreading centre or back-arc spreading centre. There appears however, to be few, if any, definite petrological distinction that can discriminate between contemporary MORBs and BABBs (Hawkins 1977, 1980; Pearce & Norry 1979). Moores (1982) has suggested that the type of sedimentary sequences overlying ophiolites may allow back-arc-spreading ophiolites to be distinguished from mid-ocean-ridge-generated ophiolites. While we accept the notion that a thick sequence of pelagic sediments may tend to indicate an open ocean environment of spreading, the occurrence of volcaniclastic sequences overlying ophiolitic lavas may not necessarily indicate a back-arc origin. For example, it would be difficult to distinguish forearc volcaniclastic sedimentary aprons from those of back arcs.

One recent development in ophiolite petrology is the recognition of boninitic lavas in many complexes (e.g. Cameron *et al.* 1979). These lavas, which resemble high magnesium andesites and basaltic komatiites, are found in only two tectonic settings, i.e. in igneous basement complexes of forearcs and as parts of extrusive units within some ophiolites (Cameron *et al.* 1979). This is a striking observation and supports the contention that ophiolites, whatever their origin, are incorporated within a forearc tectonic setting prior to their obduction.

Briefly stated, the ophiolite dilemma seems to be that, while the structure of ophiolites suggests that sea-floor spreading is the dominant plate-boundary process responsible for ophiolite formation, the petrology of some ophiolites indicates that magmas typically generated along subduction zones are often incorporated within ophiolite volcanic sequences. It would seem, therefore, that it might be advantageous to propose and test models for the tectonic setting of ophiolite formation and obduction that are consistent with the range of both geochemical and geological observations within ophiolites rather than ignoring some facts and giving others exaggerated importance.

Initiation of subduction along transform faults: implications for ophiolite geology and obduction

If initiation of subduction zones and ensimatic arcs occurs at times along transform–fracture-zone systems, then the leading edge of the overriding forearc oceanic basement might be expected to have suffered an earlier transform history. Magnetic lineations would be roughly orthogonal to the magmatic arc and trench line, a similar relationship to that of the trapped marginal-basin anomalies. If obduction of this forearc oceanic basement occurs, the leading edge of the ophiolitic nappe, if preserved, would be expected to have undergone a previous transform history, and the orientation of sheeted dykes (which statistically define a former ridge axial surface and lie parallel to magnetic lineations) should be roughly perpendicular to the frontal portions of the ophiolite nappe and the orientation of the ophiolite belt.

Depending on how soon after the formation of the forearc oceanic-crust subduction is initiated, and on the duration of subduction after its initiation, the sedimentary sequences above these types of forearcs may be varied. If initiation of subduction occurs shortly after formation (as must have been the case in places along the Izu-Bonin-Mariana arc, where an extinct spreading ridge, only slightly older than the arc itself, is preserved in the West Philippine Sea), arc volcanic detritus may be the first sediments deposited on the newly formed oceanic forearc basin. If initiation of subduction is much later than formation of the oceanic crust underlying the forearc, the first sediments may be typical abyssal sediments followed later by volcaniclastic sequences. If subduction occurs only for a short time prior to collision (e.g. Oman), which is probably a common situation (Dewey 1976), and a well-developed arc has not been produced, volcaniclastic sequences overlying the forearc oceanic basement may be scarce or completely absent. The contention of some workers (e.g. Moores 1982) that arc-derived volcaniclastic sediments overlying ophiolites necessarily indicate a back-arc spreading origin for ophiolites may be unwarranted, since it would be difficult to preclude a forearc setting simply on the basis of a volcaniclastic cover.

The possibility of arc volcanics being superimposed on the oceanic basement of the forearc is likely upon initiation of subduction, especially if the overriding and recently hydrated underriding lithosphere is young and thermally immature. Boninitic volcanics reported from a number of ophiolite complexes have usually been interpreted as off-axis volcanics, but may represent the early stages of island arc magmatic activity within forearc settings. This is supported by the fact that no analogous rock types have been documented within deep ocean basins. Boninites have been interpreted as the products of partial melting of a previously depleted mantle under hydrous conditions and at shallow depths (Cameron *et al.* 1979). Such arc-related volcanics may therefore be associated with the forearc region and may become particularly abundant as the magmatic front is approached. In addition, sea-floor spreading may not cease immediately within the forearc upon initiation of subduction (Lee & Hilde 1982). The ophiolite produced by such spreading within the forearc may not have a typical MORB character. This may explain the petrologic character of some ophiolite complexes (Pearce 1980; Cameron *et al.* 1980).

Present-day subduction zones with possible early transform histories

There are present-day examples of subduction zones that have, at least in part, been initiated within main ocean basins far from any continental margin. This results in the formation of ensimatic island arcs on the overriding plate. These types of subduction zones can be identified from the geologic and geophysical nature of the oceanic basins in the rear of island arcs. Back-arc basins originate in two ways: by back-arc spreading (Karig 1971; Packham & Falvey 1971; Sleep & Toksoz 1971; Scholl *et al.* 1975; Watts & Wiessel 1975) and by entrapment of old oceanic crust upon initiation of a subduction zone (Uyeda & Ben-Avraham 1972; Cooper *et al.* 1976; Loudon 1976; Watts *et al.* 1976; DeWit 1977; Ghosh *et al.* in press).

The West Philippine Sea (Uyeda & Ben-Avraham 1972; Loudon 1976; Watts *et al.* 1976) and the Bering Sea Basin (Cooper *et al.* 1976) are two well-documented examples of trapped marginal basins. Features common to both include relatively low heat flow, a thick sedimentary cover, layered oceanic seismic struture and magnetic lineations that strike at a high angle to the adjacent magmatic arc or remnant arc. The youngest of these anomalies are older than the oldest rocks contained within the adjacent arc or remnant arc. In addition, the West Philippine Sea contains an extinct spreading centre, identified on the basis of elevation *v.* age relationships (Sclater *et al.* 1971), the symmetry of magnetic anomalies (Watts *et al.* 1976), heat flow data, and sediment thicknesses (Mogi 1970; Uyeda & Ben-Avraham 1972). A striking feature of both the West Philippine and Bering Sea trapped basins is that magnetic lineations in each have an orthogonal relationship with the adjacent arc or remnant arc (Figs 4 & 5). The trend of the arc or remnant arc is, therefore, in a former transform–fracture-zone orientation. This relationship was noted by Uyeda & Ben-Avraham (1972) in the West Philippine Sea, and they suggested that it was originally part of a large ocean trapped by initiation of subduction along a former transform–fracture-zone system. The Bering Sea Basin also has the orthogonal relationship between magnetic lineations and the magmatic arc which led Dewey & Casey (1979) to suggest that ensimatic subduction zones and the resulting magmatic arcs may commonly nucleate along a transform–fracture-zone system during changes in relative plate motion. Another example of a trapped back-arc basin which apparently preserves a fossil spreading centre is the Venezuelan basin (Ghosh *et al.* in press).

Because magmatic fronts of island arcs usually lie at least 100 km and commonly at greater distances from the active trench, initiation of subduction zones within ocean basins leads to the entrapment of oceanic crust within the forearc region as well as in the back-arc region. The oceanic foundation beneath the magmatic arc will be considerably modified by arc-related igneous and tectonic activity. Although few geophysical studies directed at determining the basement structure of modern forearcs are available, geophysical results (Curray *et al.* 1977; Purdy *et al.* 1977; Dickinson & Seely 1979) support the existence of an oceanic foundation for some forearcs. Dredging along the inner walls of the Tonga and Puerto Rican trenches has yielded oceanic suites that included ultramafic tectonites and basalts (Fisher & Engel 1969). Ophiolitic rocks have been reported as basement to structural highs bordering the seaward edges of some modern forearc basins (Seely 1979) and as older basement to modern and ancient arc terrains (Shiraki 1971; Williams 1979).

Ophiolite examples

Most continental margins over which large allochthonous ophiolite nappes (e.g. the Semail in Oman, the Papuan Ophiolite of New Guinea, the Antalya Ophiolite of Turkey, the Villa de Cura of northern Venezuela, the Northern Appenine Ophiolites of Italy) have been obducted, record histories similar to the early-Palaeozoic

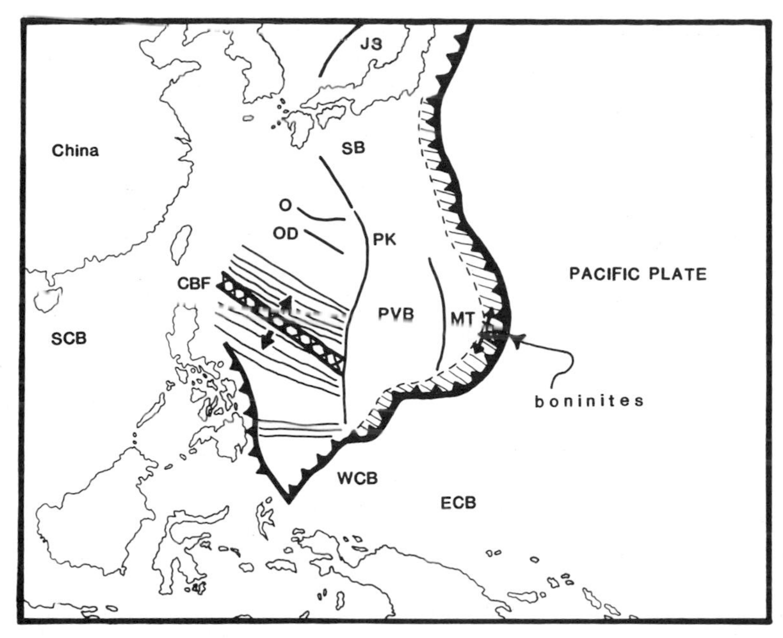

FIG. 4. Schematic map of magnetic lineations (18–21) in the 'trapped' West Philippine Sea symmetrically distributed about the Central Basin Fault (CBF) (Watts *et al.* 1976) interpreted to represent an extinct spreading centre (Ben-Avraham *et al.* 1972). Note that the anomalies are approximately orthogonal to the Palau-Kyushu Ridge (PK) remnant arc suggesting that the trench line formed in an approximate transform orientation. The Central Basin Fault is oblique with respect to other magnetic lineations in the basin and appears to indicate a change in the pole of relative motion at the time of ridge shut off. Anomalies are projected into Mariana forearc. Note that during subsequent opening of Parece Vela Basin (PVB) and Mariana trough (MT) back-arc basins, the trench line and active arc progressively underwent oroclinal bending, but that anomalies and sheeted-dyke orientations in the trapped forearc should maintain an orientation perpendicular to the trench line. JS = Japan Sea; SB = Shikoku Basin; OD = Oki-Daito Ridge; SCB = South China Basin; WCB = West Caroline Basin; ECB = East Caroline Basin.

continental margin of eastern North America, over which the Bay of Islands Ophiolite was obducted (Fig. 2). These margins record a rifting phase, no direct evidence of magmatic-arc activity prior to obduction, and a stable development until the time of ophiolite obduction. There is no subduction-related arc magmatism recorded prior to obduction, and the ages of the obducted ophiolites, where known, are much younger than the ages of initiation (i.e. rifting) of the stable margins. These similarities suggest that, in most cases, the stable margins did not come into existence due to intra-arc splitting caused by subduction beneath former Andean-type margins, and that these ophiolite complexes did not originate in marginal basins of the Japan Sea type between a stable margin and a magmatic arc. In the majority of cases, the polarity of underthrusting (i.e. subduction) seems to be directed away from the continental margin. The large disparity between the ages of initiation of the stable margins and the ages of the obducted ophiolite complexes indicates that these ophiolites probably did not form proximal to the stable continental margins, but were transported to the margins by subduction processes. Thus, models that place the site of formation adjacent to the continental margin appear suspect.

Evidence presented by Nelson & Casey (1979) and Casey (1980) indicates that the Bay of Islands Ophiolite Complex and the once-continuous arc or composite arc terrain east of the Baie Verte Lineament are highly allochthonous with respect to the ancient continental margin of eastern North America. The coincidence of stratigraphic and tectonic events in the early-Palaeozoic stable platform regions of western Newfoundland and the adjacent arc terrain of central Newfoundland at the time of ophiolite obduction suggests that these terrains were juxtaposed during obduction, and support the contention that arc-continent collision was the mechanism of emplacement of the Bay of Islands Ophiolite Complex (Church & Stevens 1971; Strong 1974; Stevens 1976; Malpas & Stevens 1977; Nelson & Casey 1979). The Bay of Islands-Hare Bay Ophiolite would represent

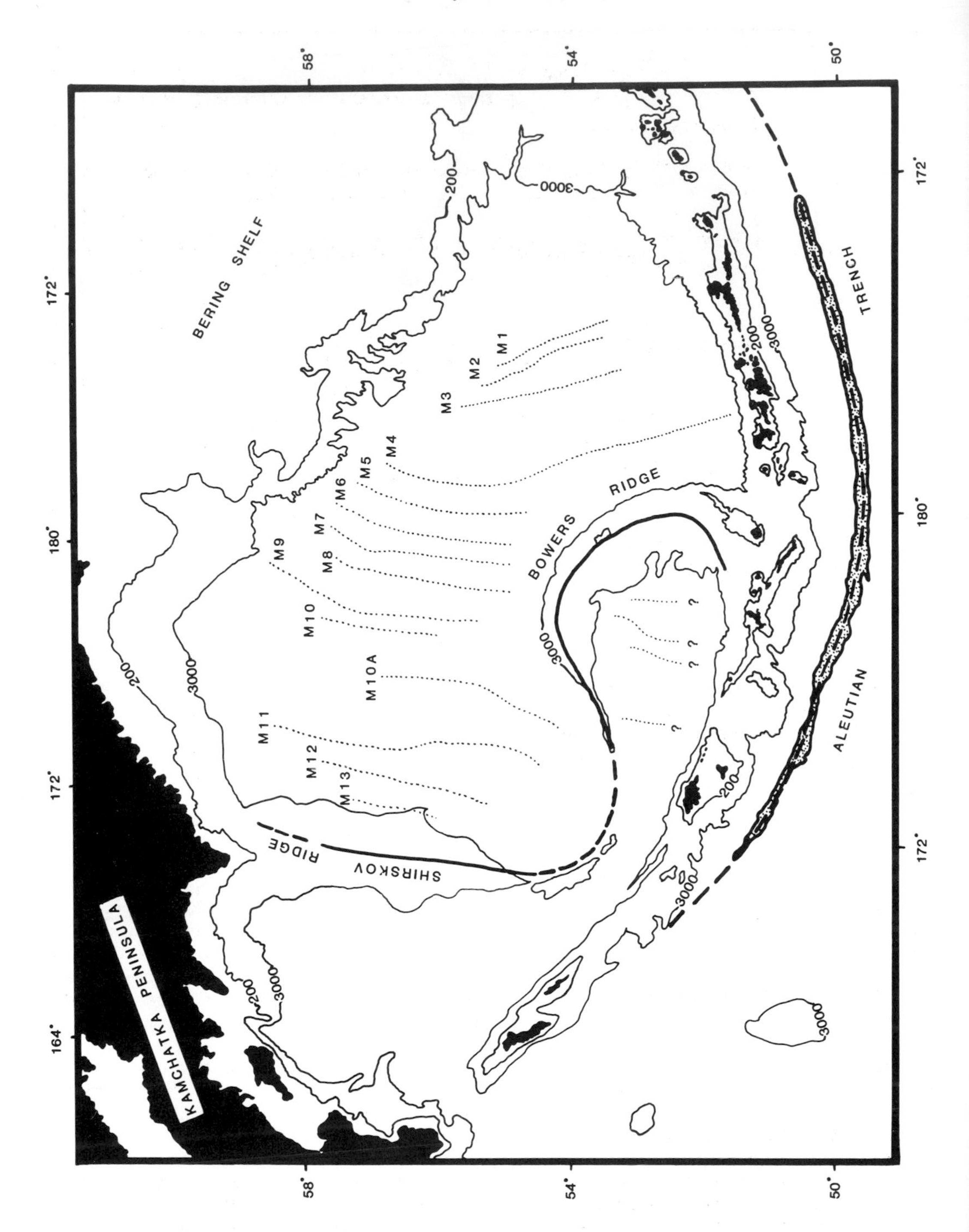

FIG. 5. Map of the Bering Sea 'trapped' back-arc showing magnetic lineations M1–M13 (dotted lines) documented by Cooper *et al.* (1976). Dashed lines represent extension of observed anomaly trends into the forearc region. Note that the anomalies maintain a nearly orthogonal relationship with respect to the trench line suggesting that the trench has initiated in a transform–fracture-zone orientation. Sheeted-dyke trends within the trapped forearc ophiolite would presumably parallel the magnetic lineations.

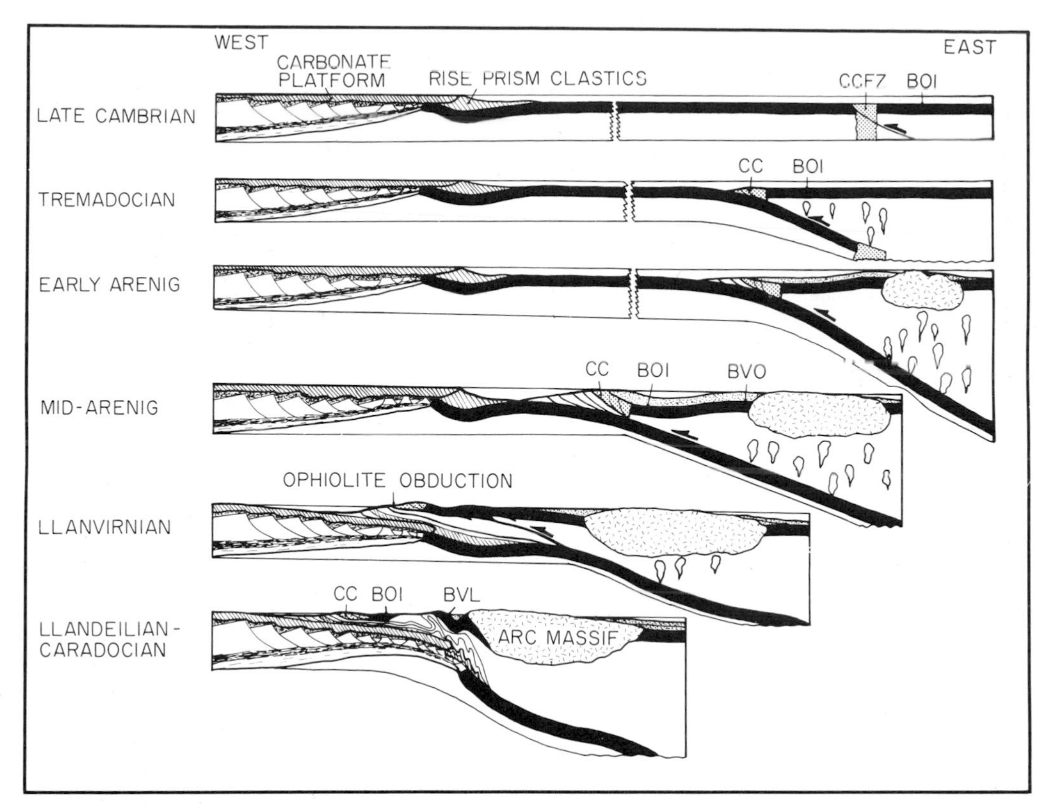

FIG. 6. Schematic plate tectonic reconstruction for the evolution of the eastern margin of North America in Western Newfoundland from late Cambrian to mid-Ordovician. Vertical exaggeration is minimal. Trench formation and the initiation of arc activity occurs shortly after ophiolite formation (late Cambrian). Site of trench initiation is along the trend of the Coastal Complex transform. The Bay of Islands Complex and Coastal Complex are trapped as the ophiolitic basement of the forearc. Upon attempted subduction of the stable east-facing continental margin of North America, this forearc basement is obducted and finally emplaced in the Llandeilo-early Caradoc. CCFZ = Coastal Complex Transform/Fracture Zone; BOI = Bay of Islands Complex; BVL = Baie Verte Lineament.

the seaward edge of the basement to the forearc, while the Baie Verte Ophiolites would represent the arcward edge of the forearc lying above an east-dipping subduction zone (Fig. 6).

Obduction of ophiolitic forearc basements was depicted in some of the early models of ophiolite obduction (Church & Stevens 1971; Dewey & Bird 1971), and this obduction mechanism has been supported by later syntheses (Zimmerman 1972; Dewey 1976; Gealey 1977; Nelson & Casey 1979). Gealey (1980) has suggested that obduction of forearcs during arc–continent collisions is the dominant mechanism of ophiolite obduction, although each mountain belt must be investigated thoroughly before definite conclusions can be reached.

However, the model of Karson & Dewey (1978), whereby the localization of the site of initiation of obduction of the Bay of Islands occurs along a transform-fracture zone adjacent to the proto-North American stable margin, can easily be incorporated into the arc-continent collision model if the transform–fracture-zone system is regarded as the site of initiation of subduction within, and not at, the edge of a fairly large proto-Atlantic Ocean (Dewey & Casey 1979).

The leading edge of the obducted Bay of Islands Ophiolite consists of highly deformed ophiolitic rocks of the Coastal Complex interpreted to have evolved within an oceanic transform–fracture-zone system (Karson 1977; Karson & Dewey 1978; Casey *et al.* 1983; Karson, this volume). The sheeted-dyke orientations within the undeformed portion of the ophiolite adjacent to the Coastal Complex are statistically orthogonal to the trend of the preserved fracture zone and the regional structural trends of the orogen; they are also sub-parallel to the inferred direction of transport of the allochthons. If the

Bay of Islands represents an obducted forearc as contended above, the internal geometric configuration of the complex (Fig. 1) is consistent with the nucleation of a subduction zone along an oceanic transform–fracture-zone system within the proto-Atlantic Ocean.

These same geometrical conditions are satisfied in the Troodos Ophiolite of Cyprus where the Arakapas Fault Zone (Moores & Vine 1971; Simonian & Gass 1978) affects the southern part of the Troodos Massif. The trend of the fault zone is approximately E–W, parallel to the long dimension of the massif and perpendicular to the statistical orientation of sheeted dykes within the relatively undeformed portion of the massif. The Arakapas Fault Zone has been interpreted as a fossil oceanic transform fault by Moores & Vine (1971) and Simonian & Gass (1978). The fault zone is situated at the leading edge of the ophiolite nappe and its orientation is perpendicular to the inferred structural front of the ophiolite nappe (Biju-Duval *et al.* 1976; Lapierre & Rocci, 1976).

Conversion of transform–fracture-zone systems to subduction zones

Dewey & Bird (1970) suggested a direct transformation of an Atlantic-type continental margin to an Andean-type margin by the initiation of a subduction zone near a previously stable Atlantic-type margin. Information gathered from orogenic belts where large ophiolite nappes have been emplaced suggests, however, that initial orogenic activity at the continental margin is initiated by overthrusting and obduction towards the previously stable margin, and that magmatic-arc activity is absent from the stable margin prior to obduction. This apparently indicates that subduction polarity was away from the continental margin at the beginning of orogenic activity. Also, the remnants of ensimatic island arcs have been identified as suture-bound fragments in many of the world's orogenic belts. Their development appears to have been synchronous with the stable development of the continental margins with which they have been juxtaposed. It would seem, therefore, that many former subduction zones were initiated within an oceanic basin, not along stable margins. Present-day arc or remnant-arc trends have been observed parallel to the trend of former transform–fracture-zone trends deduced by the orientation of magnetic anomalies in trapped back-arc basins. Transform-generated features are also found along the leading edges of some ophiolite nappes. It is therefore also suggested that many ensimatic arcs have a strong tendency to nucleate along transform–fracture-zone systems.

If such a model is applicable to the initiation of trenches, it must satisfy constraints imposed by plate theory, and there must be something special about transform–fracture-zone systems which makes them preferential sites for the initiation of subduction zones. In a sense transform boundaries are indeed special in that their trends necessarily describe small circles about the pole of relative motion demanded by the geometry of plate motion. Similarly, the constraints imposed on transform evolution are largely geometrical. There are no such orientation constraints placed on convergent and divergent plate boundaries by plate theory, although an intrinsic feature of accretionary plate boundaries is that they lie along great circles passing through the poles of relative motion between the two plates. Divergent spreading segments are, therefore, commonly orthogonal to the direction of spreading and to the transforms that offset them.

Because the Earth's lithosphere comprises more than a dozen major and minor plates, their relative motion is necessarily described by instantaneous rotation poles. Plate boundaries move relative to these instantaneous rotation poles such that the slip direction and rate across them changes with time (Dewey 1975a, b). Changes in the direction of relative motion caused by their shift require that transform plate boundaries change character or adjust their trends to remain parallel to the changing direction of relative plate motion.

Adjustments of transform orientations to changing relative motion poles are well documented in the ocean basins (Menard & Atwater 1968, 1969; Fox *et al.* 1969; Olivet *et al.* 1974). In response to large and abrupt changes in relative plate motion during periods of major plate reorganization, a transform boundary will be placed under tension or compression, depending on the sense of rotation of great circles and on the sense of offset of the transform (i.e. the plate configuration) following the shift in pole of rotation. There are many ways in which adjustments to the new relative motion vector may be facilitated. For extensional transforms, the nucleation of small ridge segments offset by shorter transforms, adjustment fractures, extension of ridge segments, asymmetric spreading, and ridge jumps may all be involved in the adjustment (Menard & Atwater 1968, 1969). For compressional transforms, unless the transform boundary adjusts its trend, a subductive component will develop along it. Once initiated, the convergence across the boundary might be short-lived if it is alleviated by the development of a new adjustment fracture in an appropriate orientation for purely transform motion. Asymmetric spreading and ridge jumping, leading to a reduction in the transform offset, will also help facilitate adjustment. There are undoubtedly numerous other ways to accomplish adjustment so that the nearly pure transform motion is maintained. Alternatively, the convergent motion might become permanent across the transform trend, especially if the transform offset is large and the change in relative motion is large and abrupt.

The high frequency of transform-fault–fracture-zone systems offsetting accreting plate margins and subduction zones in the world's ocean basins introduces inhomogeneity and permanent zones of weakness into the oceanic lithosphere that represent likely places to initiate subduction zones. They represent sites in which the rheology of the lithosphere changes along their length, and more abruptly, across them (Karson & Dewey 1978). Transform and fracture-zone lithosphere also appears to contain a thinner crustal portion (2–3 km) than adjacent lithosphere produced outside them (Fox *et al.* 1979). Long transforms seem to be likely places to initiate subduction. If the length of the transform is large, the length of any new adjustment fracture upon a change in relative motion must also be large. Because of the initial strength of the lithosphere away from fracture zones and the frictional resistance to the initial formation of a new adjustment fracture which would increase with its length, continued motion along the previous plate boundary may be favoured and result in convergence. This convergence may eventually require that the relative motion resulting from the change in direction be taken up along the older transform boundary without the creation of a new adjustment fracture. This may especially be the case near ridge segments, where hotter and more easily deformable material lies at a shallow level. Nucleation of a subduction zone may occur along the active transform and create instabilities that allow extension of the subduction zone along old lines of weakness (i.e. fracture zones), eventually creating a new plate boundary within an intra-oceanic domain.

Figure 7 depicts a possible scenario for the initiation of a subduction zone along a transform–fracture-zone system for the case of a transform offsetting two ridge segments. This represents only one plate configuration among many involving transforms, which may be applicable to the initiation of a subduction zone. This case may be particularly applicable to the Palau-Kyushu remnant arc fronting the West Philippine Sea Basin (Fig. 4), and to the original palaeogeography of the Bay of Islands Ophiolite Complex as previously outlined. Figure 7a depicts the original ridge-transform ridge plate configuration and spreading direction prior to a change in the direction of relative motion. Figure 7b depicts the situation shortly after the change in relative motion has produced a compressive transform. Assuming that a new adjustment fracture does not develop, there must be convergence across the original transform domain and the inception of lithosphere overlap. Figure 7c depicts a more advanced stage of lithospheric overlap at which time the downgoing slab may begin to restrict upward mantle flow behind the now convergent transform. After a short period of spreading within the forearc, the spreading centre may become extinct (Fig. 7d), as appears to have been the case in the West Philippine Sea just after change in the spreading direction and rate (Loudon 1976; Watts *et al.* 1976; Lee & Hilde 1982). At this point, a three-plate system must be produced, and the subduction zone already established within the old transform domain will extend laterally along the older fracture-zone systems to accommodate the convergent motion. At the same time, magmatic-arc activity may be initiated on the oceanic basement of the overriding plate, approximately 100 km from the trench line, preserving a forearc basin underlain by pristine oceanic crust unaffected by arc magmatism. Closer to the former ridge axis, where the underlying mantle is young and thermally immature, partial melting at shallow levels of a previously depleted mantle under hydrous conditions along a subduction zone could produce arc-related volcanic rocks, such as boninites, even in the forearc areas. This forearc activity would dissipate with time as the lithosphere cooled, eventually leading to a well-defined magmatic front and an island arc developed to the rear of the forearc region (~100 km from the trench).

If spreading in the overriding forearc does not cease soon after subduction initiates, as appears to be the case in the West Philippine Sea (Lee & Hilde 1982), accretion of parts of the forearc oceanic basement may impart a distinctive boninitic character to the entire ophiolite stratigraphy because of the unusual subduction-related source of these magmas that rise to ultimately crystallize along an accretionary boundary. We emphasize that the sea-floor spreading process is probably nearly identical to that at mid-ocean ridges, but that basalt chemistries may be significantly different.

Initiation of subduction along ridge segments

Because ridge segments also change their position continuously with respect to the instantaneous rotation pole, spreading rate and ridge orientations will change with time. In addition, abrupt changes in the position of the instantaneous rotation poles during episodes of major changes

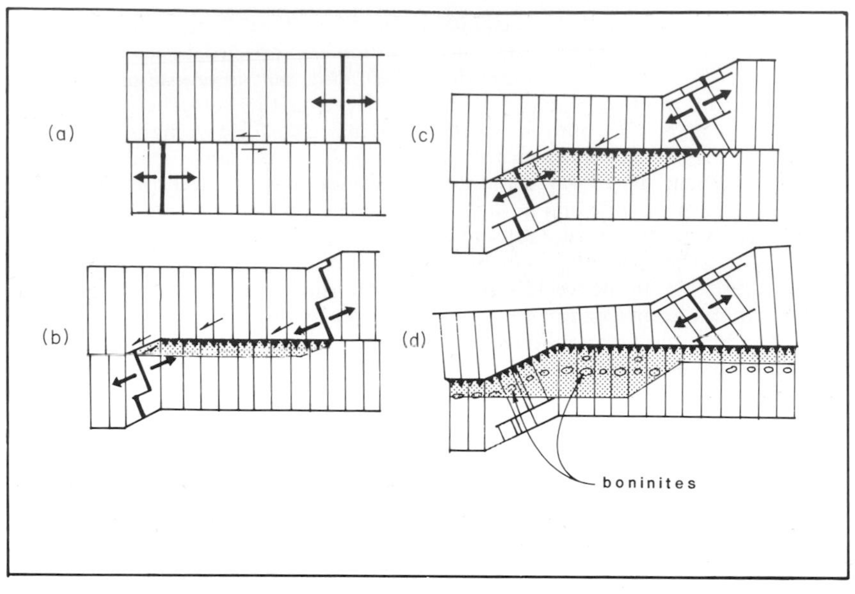

FIG. 7. Schematic plate model for the initiation of a subduction zone along a transform–fracture-zone system in the case of a transform offsetting two ridge segments. For discussion refer to text. Thin lines represent magnetic anomalies; thick black lines are spreading ridges; thin double lines mark extinct ridge segments; medium straight lines represent transform-fault/fracture-zone systems; black triangles indicate a subduction zone; open triangles indicate a former subduction zone; triangles point in the direction of underthrusting; shaded areas represent areas of lithospheric overlap (i.e. the extent of subduction lithosphere) where arc-related magmatism may occur. Note that ridge areas can be within this zone of lithospheric overlap and that continual spreading may lead to ophiolites with a petrologic character linked to subduction processes. Areas away from spreading ridges may preserve off-axis island arc-related volcanic centres (ellipses on overriding plate) within the forearc region which are formed during the early stages of arc evolution.

in plate motion may result in compression across the former spreading plate boundary. In either case, accommodation of plate motion may be achieved by the development of a subduction zone along part of the fomer ridge segment, and this may occur diachronously along the ridge segment over a long period due to normal plate boundary evolution or alternatively, abruptly along a large part of the ridge if there has been a sudden major change in plate motion. In the case of continuous evolution, if an accretionary plate boundary passes through a particular great circle containing the instantaneous motion pole, it must be continuous with a subduction zone. If migration of the ridge is such that it moves into a zone where the vectors describing the motion of each plate with respect to some fixed reference frame are no longer divergent but convergent, the boundary will change from accreting to subducting.

In Fig. 8 (after Dewey 1975a), an example of the transformation of an accretionary plate boundary to a subduction zone is depicted. In this case the ridge extends to the instantaneous rotation pole that describes the relative motion for plates *b* and *c*. Beyond this point the motion between plates *b* and *c* is convergent and a subduction zone must exist. Points 1, 2 and 3 represent oceanic crust and mantle generated successively at the *bc* ridge which shortly after accretion become part of the leading edge of plate *b* (the overriding plate) at a subduction zone. If a continent on plate *c* were to collide with this trench line, the ophiolitic leading edge of plate *b* (ophiolite complexes 1, 2, 3) would be obducted onto the continent. The forearc ophiolite overlying the subducted plate would be extremely young and the trend of the trench-line would be parallel to the former ridge axis. Sheeted dykes would be oriented parallel with the trench-line and the long axis of the forearc basin. This situation may be appropriate for the ocean closure event that resulted in the obduction of the Semail Ophiolite in Oman as dykes there are oriented subparallel to the ophiolite belt (Pallister 1981).

The initiation of subduction in this area of abnormally high geothermal gradient would

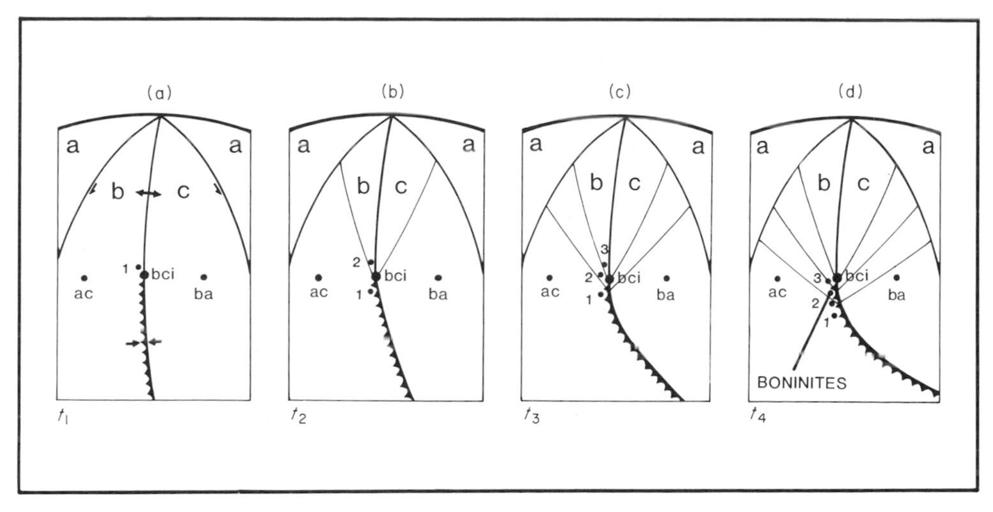

FIG. 8. Lower-hemisphere equal-angle stereographic (Wulff net) projections of the evolution of the boundary between two plates *b* and *c* from an initial (A) to a final (D) configuration. The motion of *b* and *c*, with respect to *a* (taken as fixed) are prescribed by finite rotations around fixed poles *ba* and *ac*. The increments of rotation about *ba* and *ac* between successive frames (t_1, t_2, t_3, t_4) are 20°. Plate boundaries shown in (a) are designated as follows: solid line with diverging arrows = accreting boundary; solid lines with half arrows parallel to lines = transforms (arrows indicate slip directions); toothed line = subduction zone. (Discussion in text.)

produce arc-related volcanic activity within the forearc region due to the shallow depth of melting along the subduction zone. Boninitic and other arc-related magmas may be superimposed on the ophiolite stratigraphy a short time after its accretion, as seems to have been the case in the Semail Ophiolite (Alabaster *et al.* 1980). Discordant plutons, thought to be the equivalent of these upper lavas in the Semail, crosscut the ophiolite stratigraphy, demonstrating that these magmas post-date ophiolite formation at the ridge axis. Volcaniclastic sequences may develop above these ophiolites if arc development is completed prior to obduction (i.e. collision). If the arc develops over a long period, the ophiolitic basement may be overlain by a thick section of volcaniclastics. Therefore, volcaniclastic sequences directly overlying ophiolites may again not necessarily be an appropriate criteria for distinguishing between back-arc and normal mid-ocean ridge oceanic crust and upper mantle, especially if subduction is initiated near a ridge axis.

Evolution of triple junctions and forearc spreading centres

Karig (1982) has argued recently that, under certain circumstances, large coherent ophiolite sheets may be produced within forearc settings rather than representing oceanic lithosphere trapped as the basement to a forearc upon initiation of subduction. Such ophiolites might evolve when a back-arc basin spreading centre intersects the forearc and trench (e.g. Marianas) to form a trench–trench–ridge triple junction (Fig. 9a). This model is attractive for certain ophiolite complexes in that it provides another mechanism of ophiolite formation on the overriding plate in a forearc environment. This model could lead to the development of ophiolite magmas that are not typical MORBs. Other models proposed by Karig (1982) involve ridge–transform–trench triple-junction evolution (e.g. Tonga-New Hebrides; Fig. 9b) or conversion of trench–ridge or arc–arc transforms into subduction zones. These latter cases are consistent with similar transform models previously proposed (Dewey & Casey 1979; Casey 1980). Karig (1982) has also argued that ophiolite complexes tend not to be trapped within the forearc region behind trenches because he believes that trenches have been initiated predominantly by polarity flips along back-arc oceanic-lithosphere/arc-lithosphere interfaces or at the continental-lithosphere/oceanic-lithosphere interface along Atlantic-type continental margins. We have already noted the lack of evidence in the geologic record indicating direct conversion of an Atlantic-type to Andean-type margin. Polarity flipping does seem to be an important process, and may allow entrapment of back-arc oceanic basement within the newly developed forearc. We believe,

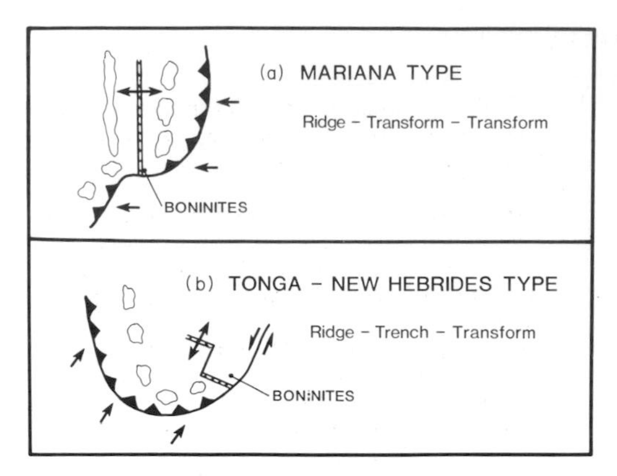

FIG. 9. Examples of forearc spreading centres in the western Pacific (modified after Karig 1982). In the case of (a) (the Mariana ridge–transform–transform triple junction) cessation of spreading will result in conversion of transform boundaries to convergent boundaries. In case (b) (the Tonga-New Hebrides ridge–transform–trench triple junction) migration of the triple junction will extend the forearc region.

however, that subduction zones must at times initiate within oceanic basins. Several possible examples of modern-day ensimatic arcs can be identified where trapped back-arc basins are older than their frontal arcs. Possible examples of these basins include the Aleutian basin (Cooper *et al.* 1976), the Venezuelan basin (Ghosh *et al.* in press), the West Philippine Sea (Uyeda & Ben-Avraham 1972; Watts *et al.* 1976), the South Fiji basin and the Central Scotia Sea (De Wit 1977). We regard these types of arcs as likely places for the entrapment of a forearc ophiolite, and do not see the need to restrict ophiolite formation to special triple-junction localities. The fact that, with the exception of its western part, the Pacific basin is largely devoid of ensimatic arcs may simply be the result of rapid spreading rates within and rapid migration of these ensimatic arcs to its margins where they ultimately collide with continents or other island arcs (e.g. Ben-Avraham *et al.* 1981).

Summary and conclusions

The internal geology and ordered stratigraphy of ophiolite complexes, as well as their tectonic position and mode of emplacement within mountain belts, indicate that they are formed at accretionary plate boundaries and then become part of the leading edge of an overriding plate above a subduction zone. Commonly, this transition appears to occur shortly after ophiolite formation. The conversion from a divergent to a convergent tectonic regime appears to have occurred along previously developed active transform and accretionary plate boundaries when subduction was initiated upon changes in poles of relative motion. In other cases, plate accretion near or within forearc regions may occur where back-arc spreading centres intersect forearc areas at ridge–trench–trench, ridge–transform–trench, or ridge–transform–transform triple junctions (Karig 1982). Both tectonic regimes may result in subduction and accretion processes overlapping in space and time. This may account for the petrologic character of 'off-axis' volcanics at the top of, as well as throughout, the ophiolite stratigraphy where, accretionary and subduction processes occur synchronously in the same area. Therefore, whilst some ophiolites may have 'island-arc' petrologic affinities, they should not necessarily be discounted as the end products of the sea-floor spreading process. Ophiolite formation by sea-floor spreading, whether at a normal mid-ocean ridge or at a back-arc spreading centre that traverses a usually non-volcanic forearc region, would seem to best explain the geologic character of many obducted ophiolite complexes.

In most cases, ophiolite obduction can be viewed as the culminating event in the closure of an ocean basin as a stable continental margin is subducted beneath a trapped ophiolitic basement within the arc–trench gap. Metamorphic aureoles attached to the base of ophiolites, and exhibiting inverted metamorphic gradients, can be viewed as ultramafic, mafic and sedimentary rocks that have been polyphasally deformed, metamorphosed, and accreted over a range of progressively lower temperatures and pressures to the hanging wall of a subduction zone in which the ophiolite is the overriding plate. These changing temperature and pressure conditions may be caused by the progressive structural thinning and expulsion of the mantle wedge above the subduction plane. The normal thermal decay of young, hot oceanic lithosphere will also contribute to the decaying temperature conditions, but cannot explain the observed pressure gradients within the aureoles.

ACKNOWLEDGMENTS: The authors appreciate the comments, suggestions and thoughtful review of the anonymous reviewer. They have helped a great deal in revising the manuscript. Amanda Graham and Mike Dix aided in the preparation of this manuscript. This work has been supported by NSF Grants EAR80-26445 and EAR7403246A01.

References

ALABASTER, T., PEARCE, J. A., MALLICK, D. I. J. & ELBOUSHI, I. 1980. The volcanic stratigraphy and location of massive sulphide deposits in the Oman ophiolite. *In*: PANAYIOTOU, A. (ed.) *Ophiolites. Proc. Int. Ophiolite Symp. Cyprus 1979*. Geol. Surv. Dept., Cyprus, pp. 751–8.

ARMSTRONG, R. L. & DICK, H. J. B. 1974. A model for the development of thin overthrust sheets of crystalline rock. Geology **2**, 35–40.

BAROZ, F., DESMET, A. & LAPIERRE, H. 1976. Les traits dominants de la geologie de Chypre. *Bull. Soc. géol Fr.* **18**, 429–37.

——, —— & —— 1975. Trois famillies volcaniques preorogeniques à Chypre: Comparison et discussion geotectonique. *3e Reun. Ann. Sci. Terre, Montpellier*, p. 126.

BEN-AVRAHAM, Z., BOWEN, C. & SEGAWA, J. 1972. An extinct spreading center in the Philippine Sea. *Nature, Lond.* **240**, 453–5.

——, NUV, A., JONES, D. & COX, A. 1981. Continental accretion: from oceanic plateaus to allochthonous terrains. *Science* **213**, 4503, 31–40.

BIJU-DUVAL, B., LAPIERRE, H. & LETOUZEY, J. 1976. Is the Troodos Massif (Cyprus) allochthonous? *Bull. Soc. geol. Fr.* **18**, 1347–56.

BROOKFIELD, M. E. 1976. The emplacement of giant ophiolite nappes. I. Mesozoic–Cenozoic examples. *Tectonophysics* **37**, 247–303.

BRUNN, J. H., ARGRIADIS, I., MARCOUX, J. & RICOU, L. E. 1977. Commentaires sur la note: 'Is the Troodos Massif allochthonous?' présentee' par B. Biju-Duval, H. Lapierre et J. Letouzey. Discussion d une origine nord ou seidtarique. *C. R. Somm. Soc. géol Fr.* **6**, 344–5.

CAMERON, W. E., NISBET, E. G. & DIETRICH, V. J. 1979. Boninites, komatiites and ophiolitic basalts. *Nature, Lond.* **280**, 550–53.

——, —— & —— 1980. Petrographic dissimilarities between ophiolities and ocean floor basalts. *In*: PANAYIOTOU, A. (ed.) *Proceedings International Ophiolite Symposium*. Cyprus Geological Survey Dept., Cyprus, pp. 182–92.

CASEY, J. F. 1980. *The geology of the southern part of the North Arm Mountain Massif, Bay of Islands Ophiolite Complex, Western Newfoundland with application to ophiolite obduction and the genesis of the plutonic portions of oceanic crust and upper mantle*. Ph.D. dissert., SUNY, Albany, 620 p.

—— & KIDD,, W. S. F. 1981. A parallochthonous group of sedimentary rocks unconformably overlying the Bay of Islands Ophiolite Complex, North Arm Mountain, Newfoundland. *Can. J. Earth Sci.* **18**, 1035–50.

——, KARSON, J. A., ELTHON, D., ROSENCRANTZ, E. & TITUS, M. 1983. Reconstruction of the geometry of accretion during formation of the Bay of Islands Ophiolite Complex. *Tectonics (in press)*.

CHRISTENSEN, N. S. & SALISBURY, M. H. 1975. Structure and constitution of the lower oceanic crust. *Rev. Geophys. space Phys.* **13**, 57–86.

CHURCH, W. R. & STEVENS, R. K. 1971. Early Paleozoic ophiolite complexes of the Newfoundland Appalachians as mantle-ocean crust sequences. J. Geophys. Res., **76**, 1461–1466.

COLEMAN, R. G. 1977. *Ophiolites–Ancient Oceanic Lithosphere?* Springer-Verlag, New York, 229 pp.

COOPER, A. K., SCHOLL, D. W. & MARLOW, M. L. 1976. Plate tectonic model for the evolution of the eastern Bering Sea Basin. *Bull. Geol. Soc. Amer.* **87**, 1119–26.

CURRAY, J. R., SHOR, G. G., RAITT, R. W. & HENRY, M. 1977. Seismic refraction and reflection studies of crustal structure of the Eastern Sunda and Western Banda Arcs. *J. Geophys. Res.* **82**, 2479–89.

DALLMEYER, R. D. & WILLIAMS, H. 1975. $^{40}Ar/^{39}Ar$ release spectra of hornblende from the Bay of Islands metamorphic aureole, western Newfoundland: their bearing on the timing of ophiolite obduction at the Ordovician continental margin of Eastern North America. *Can. J. Earth Sci.* **12**, 1685–90.

DEWEY, J. F. 1975a. Finite plate evolution, some implications for the evolution of rock masses on plate margins. *Am. J. Sci.* **275A**, 260–84.

—— 1975b. Plate tectonics. *Rev. Geophys. space Phys.* **13**, 3.

—— 1976. Ophiolite obduction. *Tectonophysics* **31**, 93–120.

—— 1980. Episodicity, sequence and style at convergent plate boundaries. *In*: STRANGEWAY, D. W. (ed.) *The Continental Crust and its Mineral Deposits*. Spec. Pap. Geol. Assoc. Can. **20**, 533–76.

—— & BIRD, J. M. 1970. Mountain belts and the new global tectonics. *J. Geophys. Res.* **75**, 2625–47.

—— & —— 1971. Origin and emplacement of the ophiolite suite: Appalachian ophiolites in Newfoundland. *J. Geophys. Res.* **76**, 3179–206.

—— & CASEY, J. F. 1979. Nucleation of subduction zones along transform faults and ophiolite emplacement. Abstracts with Programs, *Geol. Soc. Amer.* **11**, 413.

DE WIT, M. J. 1977. The evolution of the Scotia Arc as a key to the reconstruction of Southwestern Gondwanaland. *Tectonophysics* **37**, 53–81.

DICKINSON, W. R. & SEELY, D. R. 1979. Structure and stratigraphy of forearc regions. *AAPG Bull.* **63**, 2–31.

FISHER, R. L. & ENGEL, C. G. 1969. Ultramafic and basaltic rocks dredged from the nearshore flank of the Tonga trench. *Bull. Geol. Soc. Am.* **80**, 1373–8.

FOX, P. J., DETRICH, R. & PURDY, M. 1979. Evidence for crustal thinning near fracture zones: implications for ophiolites. Abstracts of Papers Submitted, *International Ophiolite Symposium, Cyprus*, Cyprus Geological Survey Dept., Nicosia, p. 114.

——, PITMAN III, W. C. & SHEPARD, F. 1969. Crustal plates in the central Atlantic: evidence for at least two poles of rotation. *Science* **165**, 487–9.

GEALEY, W. K. 1977. Ophiolite obduction and geologic evolution of the Oman Mountains and adjacent areas. *Bull. Geol. Soc. Amer.* **88**, 1183–91.

extremity of the deformed Appalachian miogeosynclinal belt. *Can. J. Earth Sci.* **6**, 1145–57.
WILLIAMS, H., HIBBARD, J. P. & BURSNALL, J. T. 1977 Geologic setting of asbestos-bearing ultramafic rocks along the Baie Verte Lineament Newfoundland. *Geol. Survey Canada,* **77-1 A**, 351–60.
ZIMMERMAN, J., JR 1972. Emplacement of the Vourinos ophiolitic complex, Northern Greece. *Geol. Soc. Am. Mem.* **132**, 255–39.

J. F. CASEY, Department of Geosciences, University of Houston, Central Campus, Houston, TX 77004, USA.
J. F DEWEY, Department of Geological Sciences, Science Laboratories, Smith Road, Durham DH1 3LE, UK.

Emplacement of ophiolitic rocks in forearc areas: Examples from central Japan and Izu-Mariana-Yap island arc system

Y. Ogawa & J. Naka

SUMMARY: Recent results of ocean dredging of ophiolitic rocks in forearc areas in the west Pacific region are summarized and discussed together with the field data in the ophiolitic belts in the Setogawa and Mineoka forearc belts in central Japan. Metamorphism, deformation and sedimentation of the dismembered ophiolites in these regions indicate that they were emplaced at first in transform-fault areas or oceanic fracture zones, and they became the subsequent zones of initiation of subduction. Sometimes the mass of dismembered ophiolite was tectonically and sedimentarily mixed with island-arc materials and incorporated into the forearc belts during subsequent forearc tectonics.

The idea of obduction of oceanic lithosphere, for the emplacement of ophiolitic rocks onto land usually occurs at the time of two-plate collision or closure of oceanic basin (Dewey 1976). In some cases, oceanic plates may have been trapped and accreted to the continental side at the time of initiation of a subduction zone (Karig 1982). Saleeby (1981) discussed ophiolite emplacement along a continental margin by transcurrent or transform tectonics, and Suppe *et al.* (1981) explained ophiolite transfer across a subduction zone.

Several examples of dismembered ophiolites are known from dredging and drilling in the present forearc areas, as mentioned later, whilst few examples have hitherto been reported in open oceans, except from fracture zones. This suggests that there is still another possibility to explain the initial emplacement of dismembered ophiolites in forearc areas. Simple obduction or trapping or off-scraping of oceanic lithosphere does not satisfactorily explain the occurrence of dismembered ophiolite in forearc areas.

This paper introduces several recent examples of forearc dismembered ophiolites, both from the dredge results of the Izu-Mariana-Yap arc system and from the land geology of the young accretionary prisms in central Japan. We propose a process whereby the dismembered ophiolite was first formed as ophiolite mélange in a transform or fracture-zone area and then the zone became the site for the initiation of subduction zone within the oceanic area, as has already been mentioned by Karson & Dewey (1978) and Karig (1982).

Examples of the dredged ophiolite in the Izu-Mariana-Yap arc system

Figure 1 and Table 1 summarize the dredge-haul points and results briefly in the Izu-Mariana-Yap arc system, and Fig. 2 schematically shows the general profiles of several examples of forearc areas of Mariana, Tonga and Guatemala. The trench landward slope areas are at least partly, and in some cases largely, underlain by dismembered ophiolite. In many cases, ophiolitic assemblages are obtained from the lower trench slopes or from the flanks of small knolls ('seamounts') in the middle trench slope or near structural high areas. The dredged material from all areas shows many similarities. The ophiolitic rocks are dismembered and all the rock types, originally situated at different levels, are now mixed together. The rocks range from upper-mantle materials to ocean-floor materials; from peridotite to basalt or pelagic sediments through gabbro, dolerite and their metamorphic derivatives. Peridotites are mostly harzburgite with minor amounts of wehrlite, lherzolite or dunite; they are more or less serpentinized to lizardite and chrysotile. Basaltic rocks are usually tholeiitic and rarely alkalic. Island-arc materials may be involved, but, if present, they are in minor amounts. Gabbros and dolerite are usually metamorphosed or altered under high geothermal gradients, from amphibole to zeolite facies through intermediate greenschist or pumpellyite/prehnite facies. Retrogressive metamorphism is common and veins of prehnite or zeolite are overprinted in the metamorphic rocks. Above all, cataclastic deformation usually occurred in these rocks, and many varieties of deformation, from ductile to brittle and from mylonite to cataclastic breccia, are all mixed. These occurrences of dredged ophiolites show many similarities to those from the oceanic fracture zones in the Mid-Atlantic Ridge (Bonatti *et al.* 1971, 1974; Fox *et al.* 1976; Bonatti & Hamlyn 1981), and in fracture zones identified in onland ophiolites such as in Cyprus (Simonian & Gass 1978) or in the Bay of Islands, Newfoundland (Karson & Dewey 1978). Another interesting point is that the dredged samples include many sedimentary rocks made only of ophiolitic clasts (IGCP Project Working Group 'Ophiolite' 1977, Naka & Uehara 1983). In the samples from KH82-4 in the Bonin area

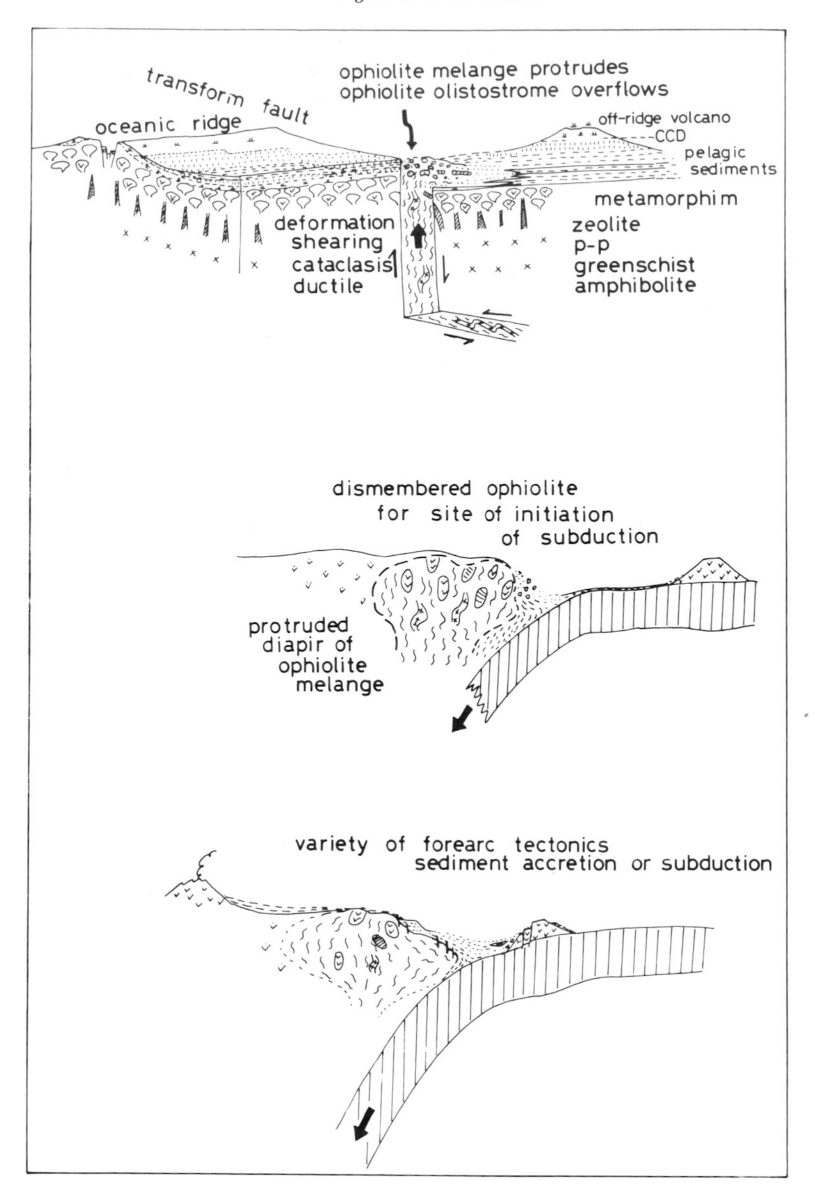

FIG. 5. Schematic model of the several stages of emplacement of dismembered ophiolite formed in a transform-fault zone to a forearc area. (a) First, ophiolite mélange is formed along a transform fault where it is metamorphosed and deformed in a high geothermal gradient and under lateral shearing. It is then protruded by normal faulting in a non-transform segment of the fracture zone (Karson & Dewey 1978). Next the overflowed dismembered ophiolite makes an ophiolite olistostrome. The large mass of dismembered ophiolite provides the site for the initiation of subsequent subduction (b). Upheaval of the diapir may cause tension tectonics in the forearc region, where trench sediments are mostly subducted, but the tectonic style may vary with other factors (c). In some cases, the ophiolite mass may be minor and an accretionary prism may be formed by compressional tectonics as in the Setogawa belt.

present Philippine Sea plate and western Pacific island-arc system is as follows. From 50 Ma until 40 Ma, the Pacific side of the Philippine Sea plate was subducted southward in the southern hemisphere (Fig. 4a). An island-arc system developed to the north and a large transform fault developed to the east. Several ridges were formed within the plate or along the transform fault. When the plate began to rotate clockwise and move northward (Fig. 4b), it approached the area of Japan. At that stage, a transform boundary may have existed along the western part of the island arc (indicated by '?' in Fig. 4b). Between 25 and 17 Ma, the eastern transform–island-arc portion (now a part of the Kyushu-Palau ridge and Izu-Mariana arc) was separated from west to east by the opening of the Shikoku basin (Fig. 4c). This spreading is proved by several palaeomagnetic studies (Watts & Weissel 1975; Kobayashi & Nakada 1978; Klein & Kobayashi 1980), though there are several discrepancies between the age and opening styles proposed. During the opening of the Shikoku basin the area moved northward (Klein & Kobayashi 1980), and consequently some parts of the island-arc, ridge or transform areas might have accreted to SW Japan. The oceanic and ophiolitic rocks now distributed within the Tertiary sections of the Shimanto accretionary belt may be correlated to these.

To the south of the Shikoku basin, on Yap Island, amphibolite and greenschist are unconformably overlain by a middle Tertiary conglomeratic formation which contains pebbles and breccias of only ophiolitic origin such as amphibolite, plagiogranite and diorite (Shiraki 1971). Shiraki (1971) concludes that the rocks indicate a fully oceanic affinity and neither an island-arc nor continental origin. The IGCP Project Working Group 'Ophiolite' (1977) also reported similar rock assemblages in several localities around the Yap Islands.

The geology and tectonics in the island-arc system in the west Pacific region must be further studied, but many island arcs have ophiolite on their lower trench slope areas, as mentioned previously. These may be parts of the fossil fracture zones or transform-fault zones formed at an early stage in the development of the Philippine Sea plate.

Plausible emplacement of ophiolitic rocks into forearc areas

It is concluded from the above that many forearc ophiolites are the product of transform-fault or fracture-zone processes, where a large amount of serpentinized peridotite was formed and intermingled with various blocks of ophiolitic rocks which include metamorphic rocks showing high geothermal gradients and cataclastic deformation (Fig. 5). The rock assemblages should be called serpentinite or ophiolitic mélange, and the mass is liable to protrude as a diapir along the fault zone. The sheared serpentinite mass is essentially rather light and ductile (the density of serpentinized peridotite ranges from 2.1 to 2.7 g cm^{-3}, Bonatti & Hamlyn 1981). Thus along the transform fault or fracture zone a line of serpentinite mélange is formed. As discussed by Karson & Dewey (1978), the tectonics of such fault zones is explained by strike-slip movement in the transform segment and by normal fault movement in the non-transform segment. During this shearing a completely dismembered ophiolite zone may become a buoyant ridge which acts as a barrier to later subduction (Fig. 5).

There is, however, another possible explanation for the emplacement of oceanic or ophiolitic materials into the forearc areas; that is, that such blocks may be transferred either by a trapping or off-scraping mechanism (Karig 1982). But this model does not explain the huge volumes of dismembered ophiolite including both the deeper and shallower levels of the oceanic lithosphere. Also it fails to explain the presence of various metamorphic rocks and various styles of deformation ranging from ductile to brittle cataclastic deformation. Rather, the dismembered ophiolitic rocks discovered in the present forearc areas resemble those found in transform-fault zones, particularly in their metamorphic and deformational characteristics. Furthermore, some of the recent discussions come from the trench-slope areas, whose tectonics are largely explained by normal faulting without evident accretionary phenomena (e.g. Aubouin *et al.* 1982; von Huene & Arthur 1982). Off-scraping of the oceanic crust cannot be denied, but is not favoured for the large-scale emplacement of forearc ophiolitic rocks. In the Setogawa belt, the seamount masses may represent the down-sliding blocks of a seamount to the trench, and the ophiolitic mass may represent the fossil transform or fracture zone. These rock assemblages are complicated by later events, but this model essentially explains how the oceanic and ophiolitic materials are emplaced into this forearc area.

ACKNOWLEDGMENTS: The idea was chiefly developed during the cruise of the Ocean Research Institute of University of Tokyo, KH82-4 in 1982. We thank the shipboard scientists, especially Profs K. Kobayashi and K. Konishi. Thanks are extended to Messrs T. Furuta, T. Tanaka, Drs K. Fujioka, S. Tonouchi, T. Matsuda, T. Ishii and J. K. Leggett for various discussions during the cruise or in the field. Special thanks are due to Profs K. Kanmera and N. Nasu for valuable discussions. The early version of the manuscript was improved by Dr S. Lippard. This study was partly supported by the Grant-in-Aid from the Ministry of Education, Japan (Sogo A-43401) and by the fund of the Cooperative Programme of the Ocean Research Institute of University of Tokyo (No. 82108).

Obduction processes in ancient, modern and future ophiolites

M. P. Searle & R. K. Stevens

SUMMARY: Geologic and tectonic processes of ophiolite obduction deduced from structural mapping, timing of deformation and metamorphism and palinspastic reconstruction of ancient ophiolite complexes compare closely with modern processes operative in the western Pacific. Many ongoing arguments about the origin of ophiolite complexes (oceanic, island arc or marginal basin) and their emplacement (subduction-related or not) are largely semantic and depend on exactly what individuals regard as island arcs, marginal basins or subduction zones. A review of tectonic processes currently active in the western Pacific region shows a wide range in size, shape and duration of arc-basin complexes and some of these show uniformitarian models similar to those proposed for emplacement of ancient ophiolites. The Tethyan ophiolites of the Semail (Oman Mountains) and Spontang (Western Himalaya of Ladakh-Zanskar) and the Ordovician western Newfoundland ophiolites are compared with recent examples of the Mariana basin, Yap trench, Banda arc, Luzon arc and Papua New Guinea. It is concluded that they are all connected in some way with marginal basins and island arcs in varying stages of development and maturity. Emplacement of these ophiolites is related to continental underthrusting of a trapped marginal basin and accretionary prism comparable to present-day examples of the N Australian margin underthrusting Timor and the Banda Sea and the SE China margin underthrusting the Luzon arc.

Although ophiolite complexes are widely regarded as fragments of oceanic crust and upper mantle obducted onto continental margins, controversy has remained over their origin (e.g. oceanic crust, island arc or marginal basin) and emplacement mechanisms. Numerous models for ophiolite emplacement have been proposed, which, on the basis of tectonic processes, may be divided into:

(i) Collision–subduction–accretion processes (e.g. Church & Stevens 1971; Dewey & Bird 1970, 1971; Dewey 1976; Smith & Woodcock 1976; Malpas & Stevens 1977; Welland & Mitchell 1977; Gealey 1977; Searle & Malpas 1980) with subduction polarity dipping away from the continent (Tethyan-type) or towards the continent (Cordilleran-type).

(ii) Gravity sliding processes (e.g. Williams & Smyth 1973; Glennie *et al.* 1973; Stoneley 1975; Coleman 1977).

(iii) Gravity spreading processes (e.g. Elliott 1976; Searle & Malpas 1980).

(iv) Transform-fault processes (e.g. Brookfield 1977; Karson & Dewey 1978).

No single model is likely to account for the emplacement of all obducted ophiolites, and a combination of different processes may have acted to account for their present position. The origin of ophiolite complexes as remnants of oceanic (Atlantic-type) basins or small marginal basins associated with island arcs is vital in determining the emplacement mechanism. For example, many ophiolites appear to be formed only a very short time prior to obduction when they retain a high heat content. This suggests a marginal basin rather than the edge of an Atlantic-type basin where the crust will be old, cool and far removed from an active spreading centre.

Primary requirements for palinspastic reconstruction and study of processes leading up to ophiolite obduction are accurate and detailed maps, good biostratigraphic control of sediments beneath and above the ophiolite, the construction of balanced cross-sections, dating of ophiolite formation and other volcanic and metamorphic rocks in the allochthon and seismic, gravity and magnetic data for sub-surface control.

Three ophiolite complexes that have been studied in detail show examples of processes operating during emplacement. The structural and tectonic evolution of the Semail ophiolite (Oman), the Spontang ophiolite in the West Himalaya and the West Newfoundland ophiolites are briefly described. A review of the complex tectonic processes operating in the western Pacific shows several analogous regions and comparison can be made with these ancient ophiolite complexes and their palinspastic reconstructions.

Tectonic processes in ophiolite obduction

Semail ophiolite (Oman)

All the above data requirements are available for the Semail ophiolite in SE Arabia which forms the largest and best exposed ancient ophiolite complex in the world (Glennie *et al.* 1973, Open University Oman Ophiolite map sheets 1–4, Bailey 1981). Recent detailed mapping in the Dibba zone and Musandam Mountains in the UAE has provided tighter structural control on a complete cross-section from Mesozoic shelf carbonates to the base of the Semail ophiolite (Searle *et al.* 1983). For fuller details of stratigraphy and regional setting the reader is referred to Glennie *et al.* (1973), Hopson *et al.* (1981) and Smewing (1980).

The structural evolution of the area can be broadly related to the destruction of a passive Mesozoic continental margin beginning in Cenomanian time, shortly after the formation of the Semail ophiolite complex (Fig. 2). Coleman (1981) and Hopson *et al.* (1981), working in the SE Oman Mountains, believed the ophiolite crustal sequence was formed at a mid-oceanic spreading centre, while Pearce *et al.* (1981), working in the N Oman Mountains, believe that it formed by back-arc spreading accompanied by eruption of basalts and andesites from discrete volcanic centres above a short-lived subduction zone. Sub-ophiolite metamorphic rocks of greenschist and amphibolite facies were formed during the Cenomanian–Campanian thrusting of Tethyan basin sediments and volcanics beneath the ophiolite. This zone of underthrusting must have penetrated to a depth of at least 12 km below the petrologic Moho into the mantle to account for this thickness of peridotite now present in the Oman Mountains (Hopson *et al.* 1981).

The strongly imbricated metamorphic rocks outcrop in a wide area of the Dibba zone as well as in tectonic windows through the ophiolite in Oman (Searle & Malpas 1980, 1982). Such narrow zones of high-grade dynamothermal metamorphic rocks, formed at 750–900°C at 3–5 kb pressure, seem to have been produced by residual heat and shearing along a mantle-tapping thrust system that Searle & Malpas (1980, 1982) argue must have been a short-lived subduction zone. True blueschist-facies rocks do not occur in the metamorphic sheet, but do crop out east of Muscat in the pre-Permian basement rocks. Garnet, glaucophane, phengite schists formed at 400–450°C and 4–8 kb pressure have late Cretaceous K-Ar ages (Lippard 1983) and appear to support an ophiolite emplacement model of attempted continental-margin subduction.

U-Pb ages on plagiogranite zircons from the ophiolite give a mean age of 95 Ma (Tilton *et al.* 1981). $^{40}Ar/^{39}Ar$ dates on hornblendes from sub-ophiolite amphibolites give a mean total fusion age of 90±3 Ma (Lanphere 1981), indicating that subduction beneath the ophiolite occurred immediately after, or possibly even during, formation of the ophiolite (Fig. 1). A late Cretaceous reconstruction of the Arabian continental margin in the N Oman Mountains is shown in Fig. 2. The ticked lines show the positions of the basal Hawasina, Haybi and Semail thrusts. Different styles of deformation exist either side of the Semail thrust. A relatively undeformed ophiolite slice (450 × 50 × 1–14 km) overlies highly imbricated, folded and thrust Haybi and Hawasina rocks. The dominant process operating during emplacement appears to be a foreland (downward) progradation of thrusting in which successive duplexes are accreted to the hanging-wall of the uppermost Semail-ophiolite thrust sheet. This is an analogous process to that responsible for accretionary prisms above modern subduction zones (Seely *et al.* 1974; Karig & Sharman 1975).

Upper (Lasail Unit) lavas of the Semail ophiolite and some dykes have 'arc'-related geochemistry—high relative abundances of K, Rb, Sr, Ba and Th and depletion (relative to MORB) in elements of high ionic potential including Ti, Y, Zr, Hf and Cr (Pearce *et al.* 1981). Overlying these lavas is a late-stage basalt and rhyolite unit (Alley Unit) erupted in a submarine graben (Pearce *et al.* 1981). There is evidence to suggest that tensional stresses existed in the hanging wall of the late-Cretaceous subduction zone whilst compressive stresses were operating in the footwall forming the Hawasina and Haybi duplexes. The Cenomanian–Turonian subduction zone, thought by Searle & Malpas (1980, 1982) to be responsible for obduction of the Oman ophiolite, would have been a separate and earlier event, unrelated to the late Tertiary–Recent subduction zone in the Gulf of Oman (Farhoudi & Karig 1977; Coleman 1981).

Failure and sudden collapse of a previously stable continental margin immediately prior to ophiolite emplacement, is recorded by spectacular carbonate slope deposits along the N margin of the Dibba zone (Searle *et al.* 1983). The fact that only a narrow zone of tectonically repeated Sumeini (shelf edge), Hawasina (basin) and Haybi rocks separate the shelf edge from the ophiolite suggests that the Cretaceous shelf edge collided with the NE-dipping subduction zone that was responsible for the original displacement of the ophiolite (Fig. 11a).

Spontang ophiolite, Western Himalaya

The Spontang ophiolite is a late-Cretaceous slab of Tethyan oceanic crust and mantle obducted onto the Mesozoic Zanskar shelf carbonates of the N Indian continental margin (Fig. 3) some time after the Lower Maastrichtian (Fuchs 1979; Searle 1983). A rapid deepening event is recorded by the Kangi-la Fm. flysch sediments which contain Campanian-L. Maastrichtian foraminifera (Fuchs 1979). This is interpreted to coincide with the emplacement of the Spontang ophiolite southwards which was

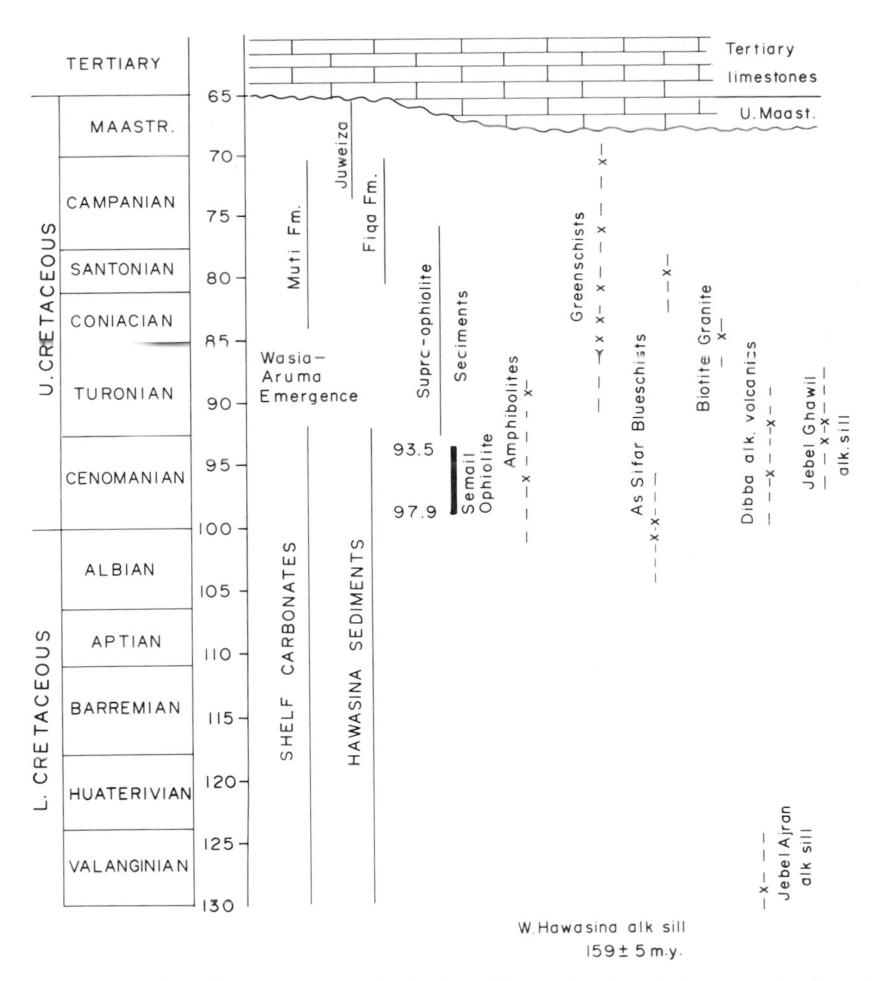

FIG. 1. Age ranges of late Cretaceous rocks in the Oman Mountains thrust belt (see text for discussion and sources of data).

completed before the Eocene when shallow-marine limestones transgressed over the whole Zanskar zone. Collision of the continental margin with the late-Cretaceous Dras island arc, closing the forearc basin, occurred in the late Cretaceous (Brookfield & Reynolds 1981; Searle 1983). The Dras arc is a thick sequence of hornblende andesites and basalts erupted in the Tethyan basin during Cretaceous time. It is continuous westward with the Kohistan arc of N Pakistan (Coward *et al.* 1982) and forms a major component of the western part of the Indus suture zone. The Spontang ophiolite must therefore be a relict part of the forearc basin and must have been obducted by northward subduction, southward overthrusting when the continental margin collided with the Dras-Kohistan arc (Fig. 11b). The structure of the Tibetan-Tethys zone in the Zanskar Mountain Range is now dominated by intense isoclinal-tight folding and thrusting (N-directed back thrusting) caused by continued convergence of the Indian and Tibetan plates after ophiolite obduction and overturning of the whole Indus suture zone during the climax of the Himalayan orogeny in the Oligocene–Miocene (Fig. 4).

West Newfoundland ophiolites

In west Newfoundland, large masses of ophiolites and associated rocks form the highest thrust sheets in an imbricated assemblage of basinal, slope and shelf rocks. The whole pile now rests on shelf sediments and has been interpreted as rocks formed at a passive continental margin during latest Proterozoic–Llanvirnian (Ordovician) time. The ophiolitic rocks have been the subject of many papers (see Coleman 1977; Karson &

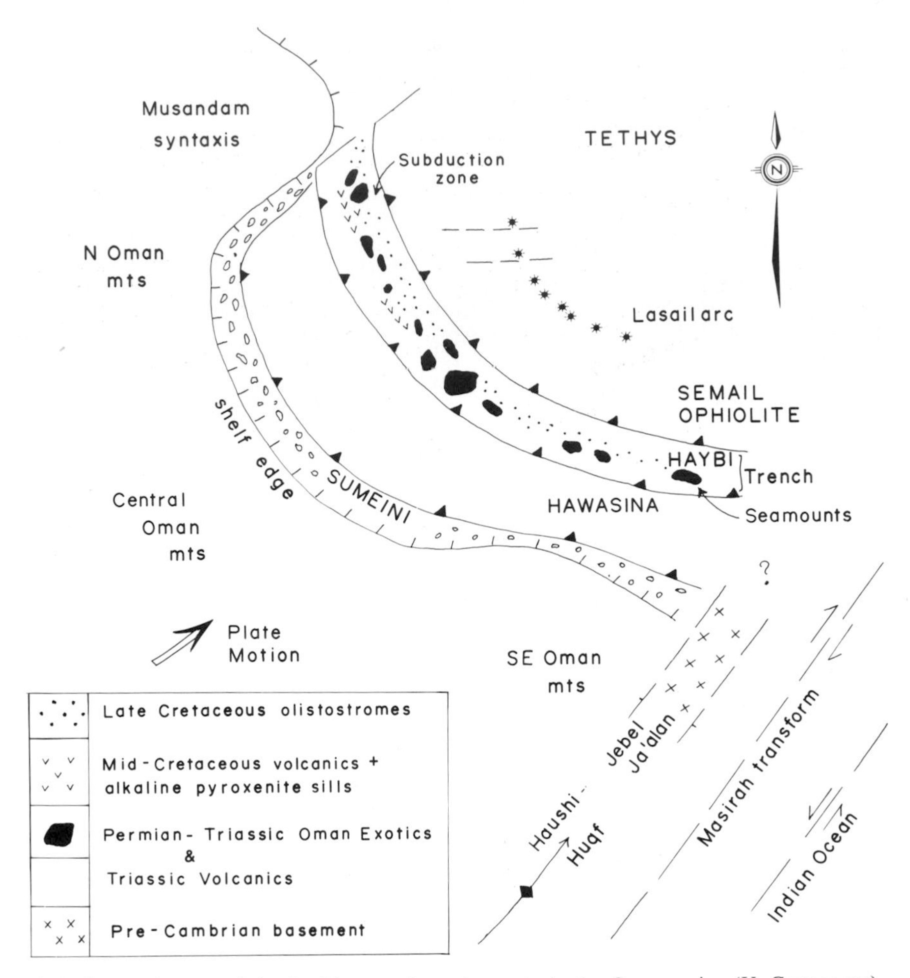

FIG. 2. Palinspastic map of the Arabian continental margin in the Cenomanian (U. Cretaceous).

Dewey 1978; Malpas & Talkington 1979; for references). Geophysically, petrologically and chemically the 10 km thick ophiolites strongly resemble oceanic crust and mantle and all recent workers interpret them as such.

An assortment of oceanic rocks underlie the W Newfoundland ophiolites. The metamorphic sheet (basal 'aureole' of previous workers), formed during the early transport of the ophiolite whilst it was still hot, occurs as a narrow zone of inverted metamorphic rocks from pyroxene and garnet amphibolites down to greenschists and phyllites immediately below the ophiolites (Malpas 1979; Jamieson 1980). The protoliths are best studied in Hare Bay (Williams & Smyth 1973; Jamieson & Talkington 1980) where alkali pillow lavas and intrusions as well as tholeiitic igneous rocks, shale and minor chert, limestone and greywacke can be recognized.

In the Bay of Islands area the next lowest slices are made up of the Coastal Complex (Williams 1975; Karson & Dewey 1978). Four assemblages have been recognized, but essentially the Complex comprises a structurally lower assemblage of ankaramitic lavas and agglomerates (Skinner Cove sequence) of presumed Tremadoc age (Strong 1974) and a complex association of schists and gneisses of mafic to ultramafic composition with minor sediments cut by mafic dykes and large trondhjemite masses. The nature of the contact between the Coastal Complex and the ophiolite is ambiguous. Locally red shale and greywacke separate the two. Elsewhere the contact is a mélange, often with large masses of volcanic rock. In the Lewis Hills, Karson & Dewey (1978) describe a contact marked by a screen of ultramafic rocks with dykes they suggest are related to the ophiolite cutting the deformed rocks of the Complex. In the Bonne Bay area there is evidence suggesting that the Coastal

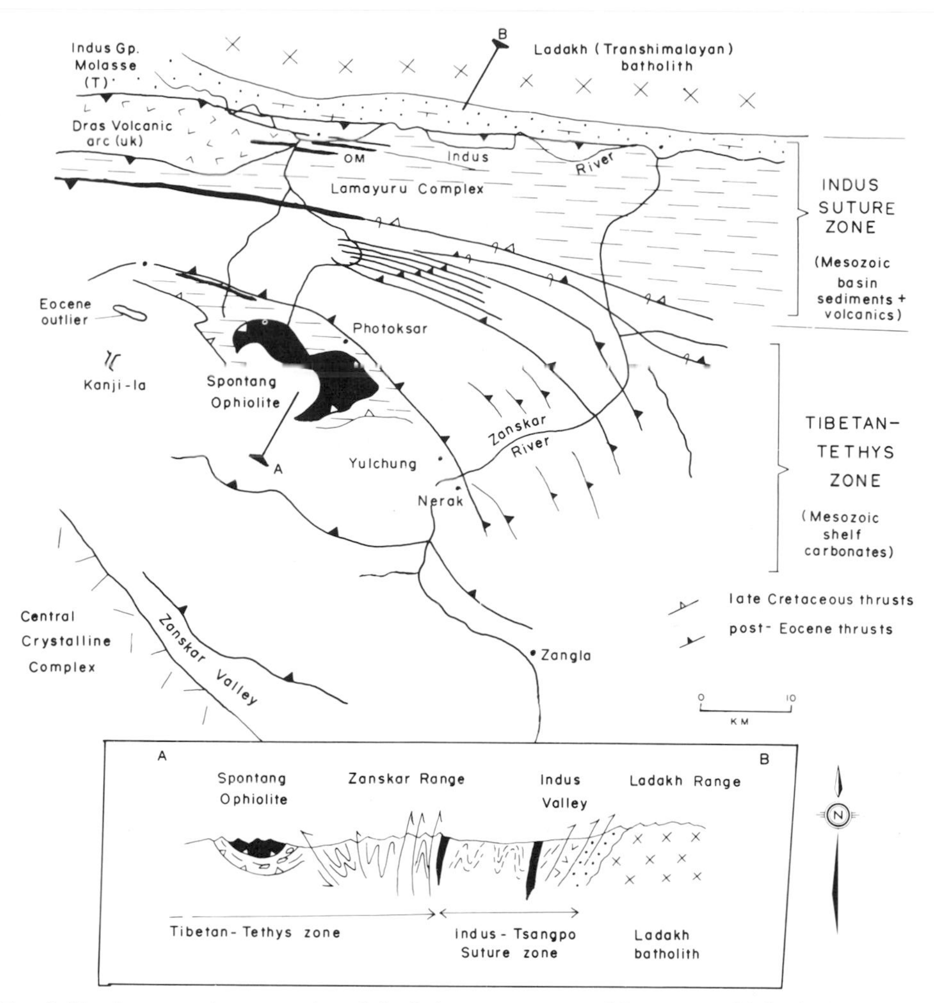

FIG. 3. Sketch map and cross-section of the Indus suture zone and Spontang ophiolite in the Ladakh Himalaya (NW India).

Complex has been tectonically reworked into the overlying basal metamorphic sheet of the ophiolite.

Two main hypotheses have been proposed for the origin of the Coastal Complex. Both agree that the alkali volcanic rocks represent within-plate ocean-island basalts, but the deformed rocks and trondhjemites have been interpreted as a remnant arc (Malpas *et al.* 1973) or as products of an early Ordovician fracture zone (Karson & Dewey 1978). The fracture-zone hypothesis has been adequately explored but several outstanding problems remain. For example, the origin of large volumes of acid intrusions which are rarely known from modern transform-fault environments or fossil transforms in ophiolites (e.g. in Oman and Troodos) is difficult to explain. Furthermore, new precise zircon dates from the Bay of islands ophiolite yield an age of 485(+1.9, −1.2) Ma (Dunning & Krogh 1982) significantly younger than the Coastal Complex (508 ± 5 Ma—Mattinson 1975) and younger than previous age determinations for the ophiolites. On the Van Eysinga (1978) age scale there is only about 10 m.y. before the break-up of the shelf and the start of ophiolite transport. On the Ross *et al.* (1982) scale the two events are almost synchronous (Fig. 5).

Three conclusions can be drawn from a study of the time chart (Fig. 5). First, the Coastal Complex 'arc' trondhjemite is significantly older than the Bay of Islands ophiolite. Secondly the Bay of Islands ophiolite was formed just prior to obduction, probably by back-arc spreading (see Fig. 11c). Thirdly, the 'suture-zone' ophiolites along the Baie Verte Lineament and Dunnage zone span U. Tremadoc to U. Arenig-L. Llanvirn

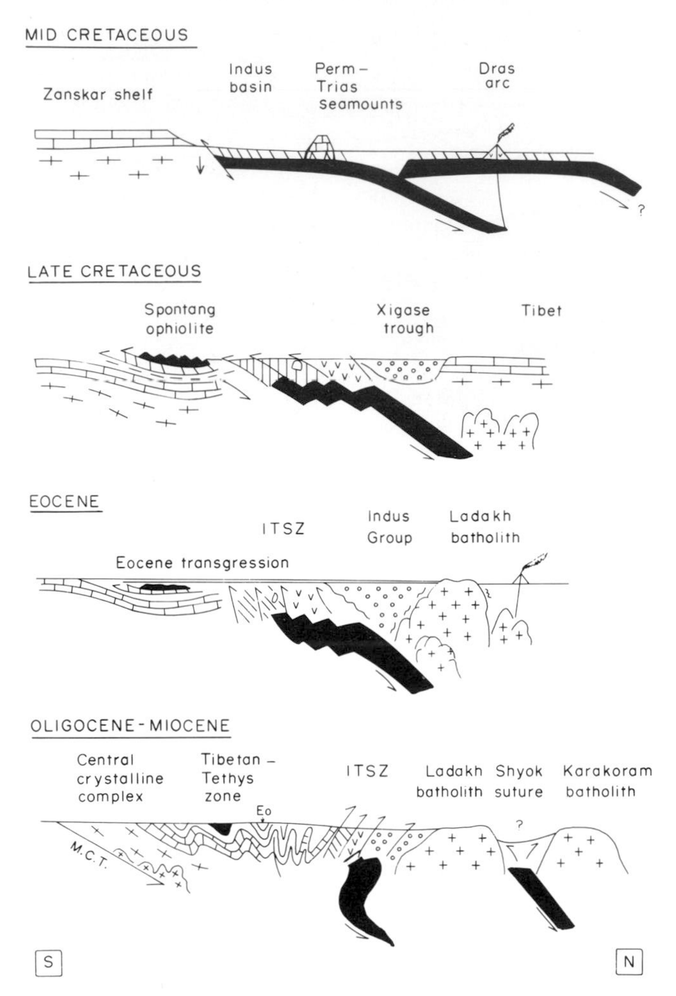

FIG. 4. Geotectonic evolution of the Indus suture zone and Zanskar (Indian) continental margin in the late Cretaceous–Tertiary (after Searle, 1983).

time and form the basement on which the central Newfoundland island-arc volcanics were subsequently erupted, above a newly formed E-dipping subduction zone.

Current data suggest the following scheme for the evolution of west Newfoundland. Break-up of a continental mass with the establishment of a shelf, slope, basin topography by early Cambrian times; the open marine faunas suggest that an ocean basin was in existence at this time, but its extent is not known. At some time in the late Cambrian or Tremadoc, E-directed subduction was initiated in this basin leading to development of an arc (Coastal Complex) built on oceanic crust. This event had little effect on the continental margin. During the Arenig, spreading took place behind, to the east of the arc, now a remnant arc, and gave rise to oceanic crust that is now represented by all the ophiolites preserved in west Newfoundland (Fig. 11c). Spreading was at an angle to the margin as indicated by a regional NW–SE orientation of sheeted dykes. Collison resulted in attempted subduction of the continental margin and obduction of the ophiolites. The Coastal Complex may be the remnant arc, while the 10 km thick pile of calc-alkaline rocks of Notre Dame Bay represents a new arc built on the ophiolite as a result of a newly initiated E-dipping subduction zone beneath central Newfoundland. Notre Dame Bay shows an almost continuous record of igneous events from Tremadocian to Caradocian times when emplacement of the ophiolites was completed. Later igneous events can be interpreted as the buoyant upwelling of the subducted marginal continental crust of North America.

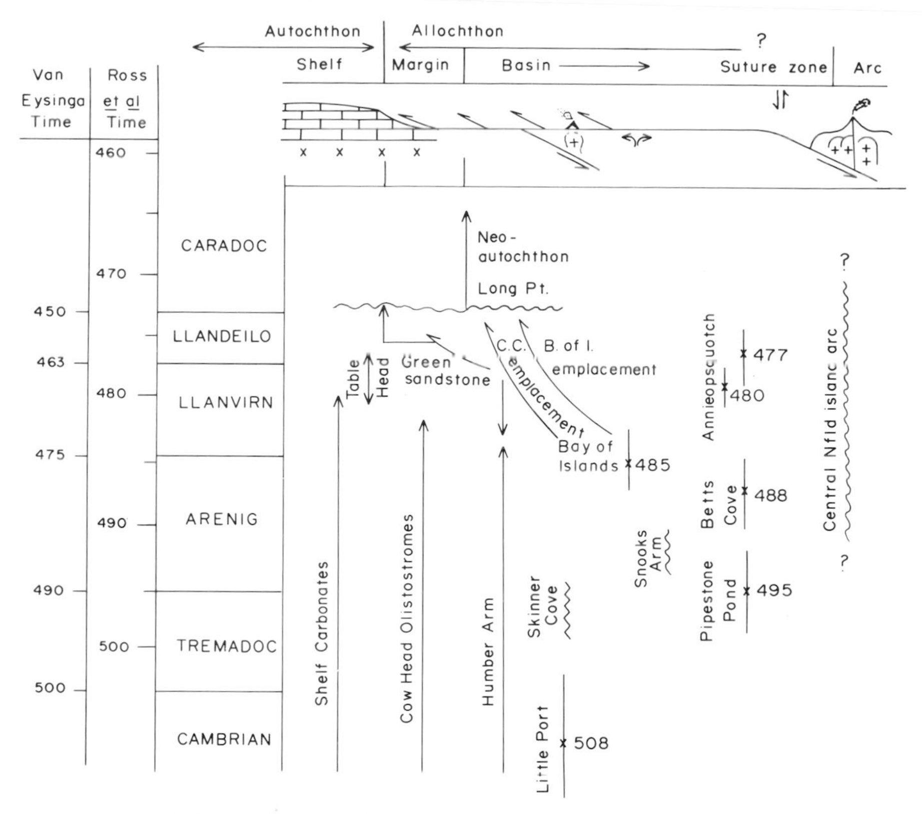

FIG. 5. Age range of L. Ordovician rocks in the West Newfoundland Appalachians (see text for discussion and sources of data). Crosses are radiometric ages with errors; zigzag lines are fossil ages.

Tectonic processes occurring in the western Pacific

Mariana arc basin

One of the most active arc systems in the western Pacific is the Bonin–Mariana ridge system (Karig 1971) separating the Philippine and Pacific oceanic plates (Fig. 6). Karig (1971) suggested that the Mariana island arc that formed above a subduction zone split longitudinally and new oceanic crust was generated by spreading in the back-arc basin, a hypothesis that recent geologic, geophysical and drilling work has largely confirmed (e.g. Sharaskin *et al.* 1981; Hussong & Uyeda 1981).

DSDP drilling in the *West Mariana (back-arc) Basin* shows that the upper crust consists of high alumina olivine tholeiites with flat REE patterns and low trace-element contents (Dietrich *et al.* 1978) typical of mid-ocean ridge basalts (MORBs). Parallel magnetic anomalies and high heat-flow values indicate the existence of newly formed oceanic crust. Submarine ridges (e.g. Kyushu-Palau Ridge) are interpreted as remnant arcs (Karig 1972). Short-lived subduction zones have been operative, with spreading in the West Philippine Basin from 45–37 Ma, the Parece-Vela Basin 25–18 Ma and the Mariana trough 2–3 Ma (Dietrich *et al.* 1978). This shows that the zone of underthrusting has migrated oceanward (E) with time and that back-arc spreading and arc development has been short-lived and immature.

In intra-oceanic arcs, such as the *Mariana arc*, magmatism is dominated by island-arc tholeiites, high-Mg andesites and rare boninites (Tarney *et al.* 1981; Sharaskin *et al.* 1981). Whereas basalts erupted at back-arc spreading centres appear to be geochemically indistinguishable from MORBs, those from island arcs are distinct. Island-arc lavas show an enrichment in elements of low ionic potential (K, Rb, Sr, Ba, Th, Ce) relative to elements of high ionic potential (Ta, Nb, Hf, Zr, Ti, Y and Yb), and exhibit an increase in their $^{87}Sr/^{86}Sr$ ratio relative to $^{143}Nd/^{144}Nd$ (Hawkesworth *et al.* 1977; Pearce *et al.* 1981). Both these features are thought to be caused by interaction of seawater-derived fluids driven off the subduction zone. Some ophiolite complexes such as the Semail and Troodos show similar geochemical

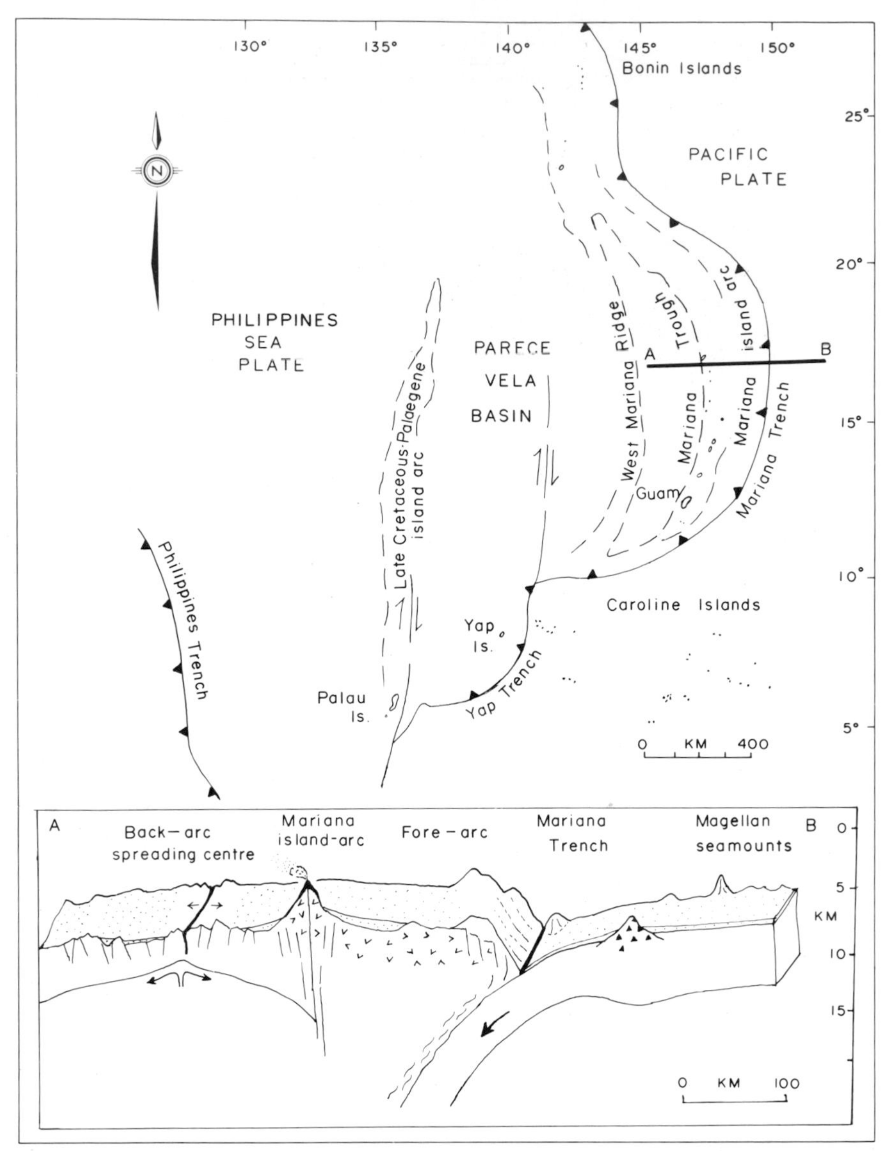

FIG. 6. Map and cross-section of the Mariana arc–trench–basin system (after Karig 1971; Hamilton 1979 and Hussong & Uyeda 1981).

features in their upper volcanic stratigraphy and are interpreted as immature island arcs (Pearce 1975; Pearce *et al.* 1981).

The *Mariana Trench* is 8–10 km deep and reaches 11.5 km S of Guam (Hamilton 1979). The Pacific plate to the east is dotted with numerous seamounts which are composed of alkalic 'within-plate' volcanics sometimes associated with atoll-reef carbonate cappings and are similar to the Triassic Haybi volcanics and 'Oman Exotics' beneath the Semail ophiolite (Searle *et al.* 1980; Searle & Graham 1982). Within-plate, ocean-island alkali basalts also occur in allochthonous slices beneath the west Newfoundland ophiolites (Skinner Cove volcanics, Strong 1974; and the Cape Onion–Ireland Point volcanics, Jamieson 1980).

Yap arc-trench system

The Yap arc-trench system is the southward extension of the Mariana trench but the Yap 'arc' is formed of metamorphic rocks (Hawkins & Batiza 1977). Dredge samples from the trench

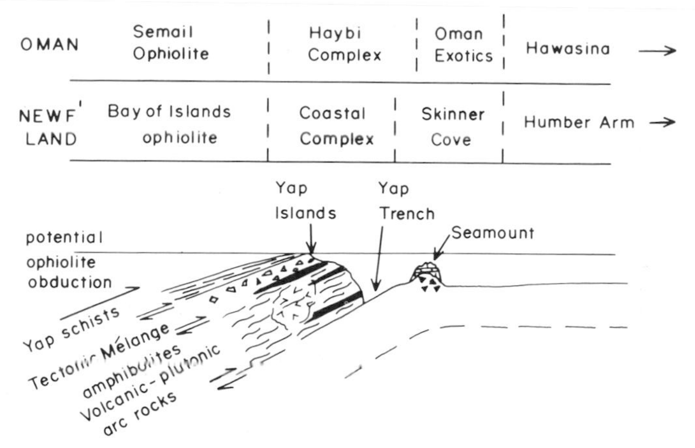

FIG. 7. Cross-section of the Yap trench and Yap Islands (after Hawkins & Batiza 1977) with appropriate equivalents of the Oman and Newfoundland rocks for comparison (see text for discussion).

wall are mostly foliated amphibolites and also, calcite-diopside-grossularite (+ hydrogrossular) marbles (metamorphosed impure limestones) similar to rocks presently found in the sub-ophiolite metamorphic sheet in Oman (Searle & Malpas 1980, 1982). These metamorphic rocks are tectonically overlain by the Yap ultramafic schists along a mélange zone (Fig. 7). This provides a present-day analogue to the formation of the metamorphic sheet beneath the Bay of Islands (Malpas *et al.* 1973) and Oman (Searle & Malpas 1980, 1982) ophiolites. Seismic reflection profiles across the Yap trench show swarms of irregular seamounts and atolls, some of which are plastered against the Yap 'arc' (Hawkins & Batiza 1977). It is possible that termination of subduction was caused by blocking of subduction zones by these seamounts, a theory that may also apply to Oman where isolated, up to 1000 m thick, Exotic carbonates immediately underlie the ophiolite (Searle & Graham 1982). Palinspastic reconstruction places the Exotics along the zone of underthrusting beneath the Semail ophiolite (Fig. 11a) and Exotic carbonate and volcanic-island basement also provides the protoliths for some of the metamorphic rocks beneath the ophiolite (Searle & Malpas 1980, 1982).

Continental margin-arc collision

Nowhere in the world today is a continental margin underthrusting or being overthrust by a slab of oceanic crust from an Atlantic-type mid-ocean ridge. Two examples in the Western Pacific, the collision of the Banda arc with the N Australian continental margin and the Luzon arc with Taiwan and the SE China margin, appear to be the only modern examples of a continental margin underthrusting an island arc and trapping a marginal oceanic basin apparently in the early stages of obduction.

Banda Sea–Timor area

Detailed geologic studies in the Indonesian region by Van Bemmelen (1949), together with plate tectonic interpretations by Hamilton (1973, 1977, 1979) and a recent marine geophysical study in the Banda Sea (Bowin *et al.* 1980), allow an understanding of processes occurring today at a modern arc–continent collision zone. Seismic reflection profiling shows the Australian shelf sequence dipping beneath the accretionary prisms of the outer Banda arc at the Timor and Seram troughs (Fig. 8). The 35–40 km thick continental crust appears to be underthrusting the Banda arc along active subduction zones at these 3000 m deep trench systems (Fig 11d). Oceanic crust underlies the Banda Sea (back-arc) and Weber Trough (forearc) suggesting that the Banda Sea could be trapped oceanic crust in the early stages of obduction onto the N Australian margin (Bowin *et al.* 1980).

The *outer 'arc'* passes through the island of Timor and circles the Banda Sea to Seram Island south of Irian Jaya, New Guinea. It is composed mainly of olistostromes, mélanges, imbricated sedimentary rocks (Mesozoic radiolarites, calcilutites and marls) and metamorphic rocks. The sedimentary rocks are interpreted as largely Australian continental-margin sequences now stacked up in an accretionary prism (Audley-Charles *et al.* 1972; Barber *et al.* 1977; Carter *et al.* 1976). These authors interpret Australian continental-shelf rocks as underlying the whole of Timor and being overthrust by continental-margin rocks, bathyal sediments, volcanics,

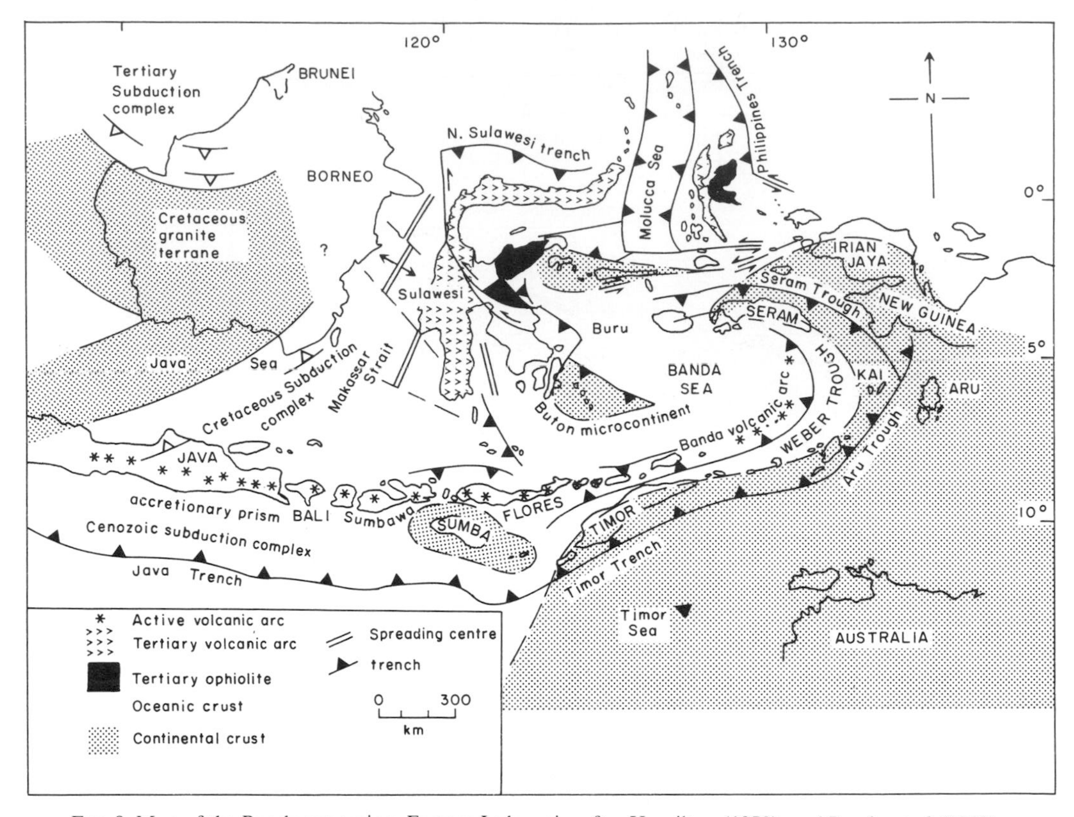

FIG. 8. Map of the Banda arc region, Eastern Indonesia, after Hamilton (1979), and Bowin *et al.* (1980).

serpentinite and metamorphic rocks. Metamorphism apparently increases towards the trench (Weber Trough) from phyllites and greenschists, to highest-grade granulites and migmatites on the islands of Kur and Fadol on the brink of the Weber Deep. The metamorphic rocks have been described by Van Bemmelen (1970) and Barber & Audley-Charles (1976) and are interpreted as metamorphosed continental-margin sediments.

The *Weber Trough* is a 7 km deep trench that separates the inner volcanic arc from the outer accretionary prism island chain. The crust is similar in thickness (8–11 km) and velocity to that beneath the central Banda Sea and is interpreted by Bowin *et al.* (1980) to be trapped forearc oceanic crust. The crust in the Weber Deep is however overlain by a 3.2 km thick sediment sequence. The development of subduction-related faulting between the inner and outer Banda arcs in the Flores–Timor region involved vertical displacements of 3000 m in the late Pleistocene–Holocene (Audley-Charles & Hooijer 1973).

The volcanically active *inner arc* extends from Sumatra through Java, Bali, Lombok, Sumbawa, Flores to Pantar. A presently inactive segment continues east, N of the Timor Trough and active volcanoes occur in the segment from Damar Island to Banda Islands, W of the Weber Trough. Seismic-refraction data show that oceanic crust exists either side of the inner arc and Bouguer gravity anomalies suggest that the islands are underlain by oceanic crust (Bowin *et al.* 1980). The late Pliocene–Pleistocene volcanics are composed mostly of andesites except for the Banda Islands where basalts outcrop.

Eastern Papua-New Guinea

The Papuan ophiolite outcrops at the NE corner of Papua-New Guinea (PNG) (Davies & Smith 1971). The 16 km thick sequence of Cretaceous oceanic upper-mantle-crust includes 8 km of peridotite and has been obducted onto or underthrust by the northern margin of the Australian continental plate (Fig. 9). The ophiolite has been interpreted as the basement beneath an Eocene island arc that collided with PNG in Miocene time (Hamilton 1979). Emplacement of the ophiolite occurred along the Owen Stanley thrust fault which may have been an early Tertiary subduction zone (Davies & Smith 1971). Metamorphism of the Mesozoic sialic sediments

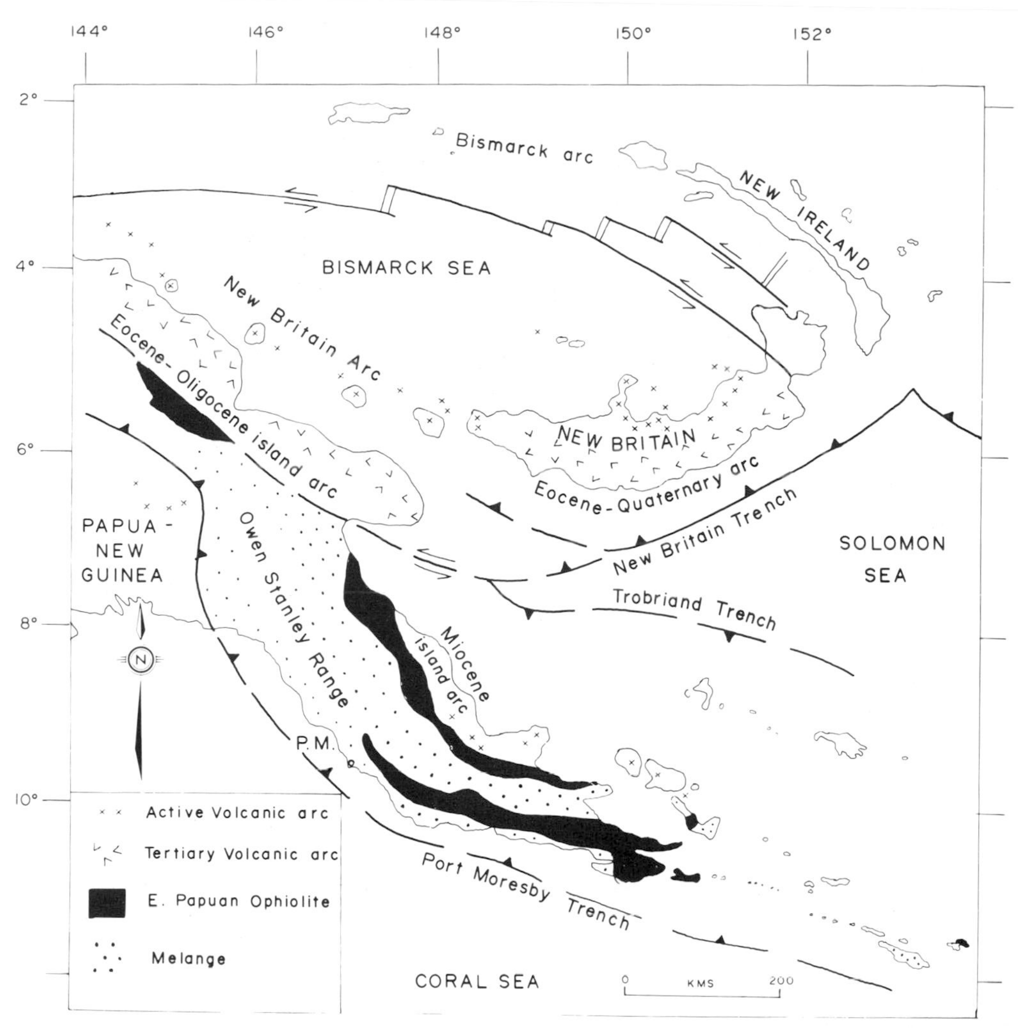

FIG. 9. Map of Eastern Papua-New Guinea, after Davies & Smith (1971) and Hamilton (1979).

below the Owen Stanley fault (Eocene) is directly related to emplacement of the ophiolite and includes some high-temperature hornblende and pyroxene granulite and high-pressure glaucophane-lawsonite blueschists as well as widespread greenschists. Arrival of the Australian continental crust (and marginal sediments) at the zone of subduction caused subduction to cease and become active in the New Britain–Solomon Island trench to the north. A broad belt of subduction-related mélanges, strongly imbricated, crops out beneath and SW of the ophiolite thrust sheet, with metamorphism increasing towards the north, i.e. towards the base of the Papuan ophiolite. Pliocene granodiorite plutons intrude the metamorphic rocks in the D'Entrecasteaux islands off the north coast (Davies & Smith 1971), possibly resulting from partial melting of Australian continental-crust rocks dragged down to depth beneath PNG.

Taiwan–Luzon area

A marine geophysical survey involving bathymetry, magnetics, gravity and seismic reflection was conducted in 1975 and, together with on-land geological surveys in Taiwan, Luzon and adjoining islands, has allowed a comprehensive understanding of the tectonics of this complex region (Bowin *et al.* 1978). Rapid depression of the SE China continental shelf occurred during the middle and late Miocene as it approached the subduction zone of the Luzon arc (Figs 10 & 11e). The *South China Sea* was largely formed by sea-

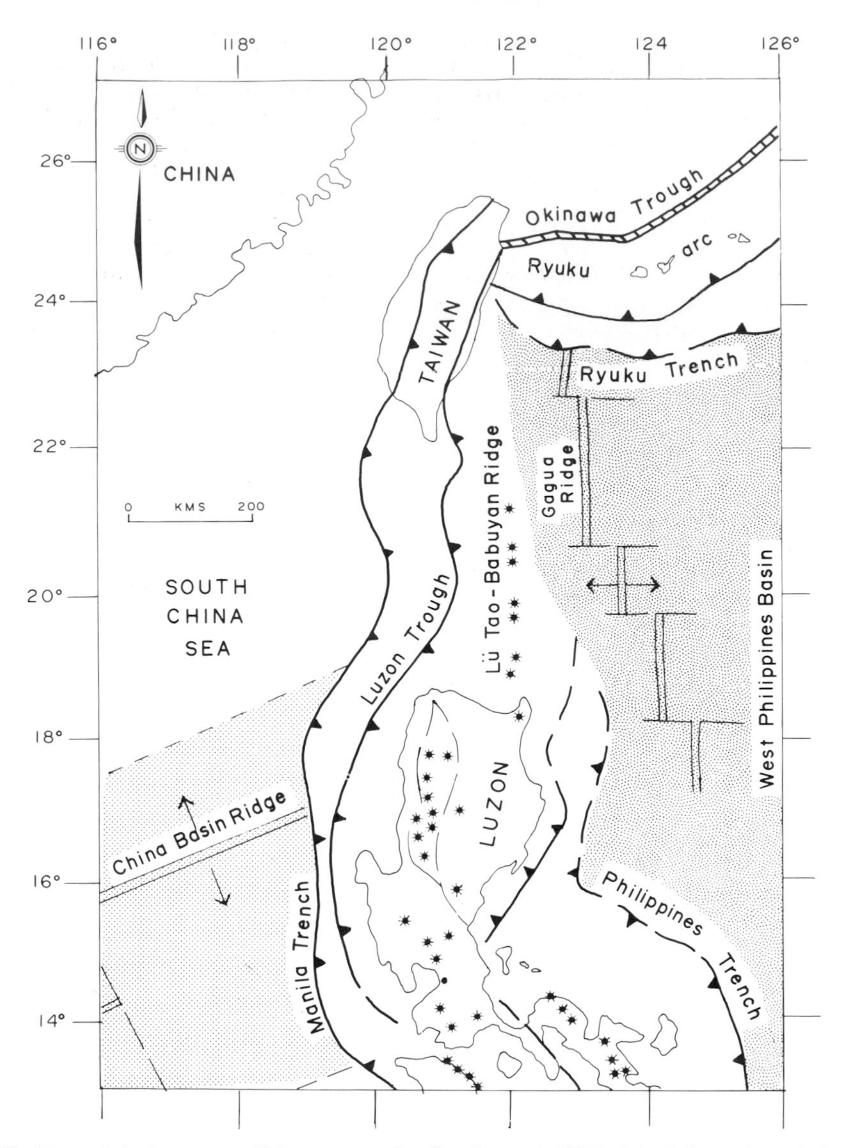

FIG. 10. Map of the Luzon arc-Taiwan area, after Bowin *et al.* (1978). Stippled area is underlain by oceanic crust, spiked circles are active volcanoes.

floor spreading in late Jurassic–early Cretaceous time. The contact between the crust of the Philippine and Asian plates now occurs beneath the deformed accretionary wedge of the *Manila trench.* Geophysical profiles show that this deformed belt associated with the Manila Trench and Luzon trough continues northward and is exposed in the Central Range of Taiwan (Bowin *et al.* 1978).

The *Central Range of Taiwan* is composed mainly of metamorphosed clastic sedimentary rocks and limestone marbles of Palaeozoic age and unconformably overlying Eocene–Miocene metamorphosed clastic continental-margin deposits. Deformation is intense and late Tertiary metamorphism increases from W to E. Along the eastern margin glaucophane schists and ophiolitic fragments have been reported (Bowin *et al.* 1978). The western foothills are composed of thrust sheets of Eocene–Pleistocene sedimentary rocks.

The *Luzon arc* comprises subaerial andesitic volcanoes in Luzon Island and its northward continuation in a submarine ridge termed the *Lü Tao-Babuyan Ridge*. The deformed accretionary wedge of the Manila trench occupies the west flank of the ridge whilst the ridge volcanoes are composed of hornblende and augite andesites and olivine basalts. East of the Luzon arc the *Gagua Ridge* is thought to be an extinct spreading centre from which MORB-type volcanics may have been extruded in the West Philippine Basin (Fig. 10).

Thus, taking a section from W to E and eliminating complexities such as the flipped subduction of Luzon Island (Karig 1973; Roeder 1977), a broad comparison of structural positions of major units can be made with Oman and Newfoundland. A continental margin is depressed beneath the deformed accretionary wedge of continental-margin clastic sediments (of Central Taiwan, Humber Arm-Fleur de Lys in Newfoundland, Hawasina in Oman). Going oceanward, metamorphism increases towards the suture zone where olistostromes, arc volcanics and high-grade metamorphics (± blueschists) occur (Luzon trough, Baie Verte zone in Newfoundland, Haybi complex in Oman). Finally, above the hanging wall of the trench, the ophiolite complexes of Oman and Newfoundland form the structurally highest unit with associated supra-subduction-zone volcanics.

Summary

A combination of geological and geophysical evidence from recent surveys in the Pacific region reveal that the following tectonic processes are currently operating:

1 Initiation of short-lived intra-oceanic subduction zones, producing:
2 Small, immature island arcs composed of depleted island-arc tholeiites, hornblende and augite andesites, olivine basalts, gabbros and diorites.
3 Rapid plate motions—up to 10 cm yr^{-1} in West Mariana basin (Karig 1973).
4 Sealing of active subduction zones soon after initiation and propagating oceanward (e.g. Manila subduction zone and West Mariana zones).
5 Generation of new ocean crust from back-arc spreading as shown by bathymetry, parallel magnetic anomalies and drilling (e.g. West Mariana basin, Ryuku back-arc basin).
6 Extensional rifting tectonics and vertical faulting in forearc and back-arc basins above the hanging wall of the subduction zone.
7 Formation of amphibolites, calc-silicate marbles, greenschists, serpentinite and tectonic mélanges in hanging wall of an arc–trench gap subduction complex beneath an overthrust marginal basin (e.g. Yap Island).
8 Metamorphism of continental-margin clastics in an accretionary prism with metamorphic grade increasing oceanwards (e.g. Taiwan, E Papua). High-pressure glaucophane schists occur along the former trench.
9 Rapid development of deep trenches (e.g. 8 km in Ryuku Trench, 10 km in Mariana trench).
10 Sudden collapse of a previously stable continental margin as documented in N Oman and W Newfoundland producing olistostrome debris flows. Downwarping of the continental margin and crust beneath an arc–trench complex and marginal basin (e.g. N Australian margin beneath Timor and the SE China margin beneath the S China Sea and Taiwan).
11 Incorporation of swarms of seamounts (within-plate alkali basalts and carbonate cappings) on inner trench wall as seen in the Yap trench, Mariana trench and Manila trench, and proposed for the Oman Exotics.
12 Imbricated sediment wedge (accretionary prism) thrust over the continental margin (e.g. Outer Banda arc over N Australian shelf, Taiwan thrust sheets over SE China margin).
13 Continent-arc collision can result in trapped ocean crust in forearc region (e.g. Weber Trough, east of Banda Islands) or back-arc (Banda Sea).
14 Flipping of subduction-zone polarity with time—e.g. Philippine plate beneath Luzon, Roeder (1977), comparable to the flipped subduction zone becoming inactive in the Indus suture zone, Himalaya, in the late Tertiary.

Conclusions

All these tectonic processes listed above have also occurred during the formation and emplacement of ancient ophiolite complexes. Conditions required for emplacement of large ophiolite thrust sheets onto continental margins are:

1 Collapse of the continental margin on a massive scale immediately prior to emplacement.
2 A rapidly deepening foredeep on the continental margin accumulating a great thickness of syn-orogenic flysch.
3 A thrust fault that cuts the mantle (a subduction zone).
4 Underthrusting of oceanic sediments and volcanics beneath the ophiolite producing a dynamo-thermal metamorphic sheet.
5 Regional uplift of the ophiolite and sinking of a foreland-migrating foredeep causing a regional surface slope dipping towards the continent. Gravity spreading can then thrust

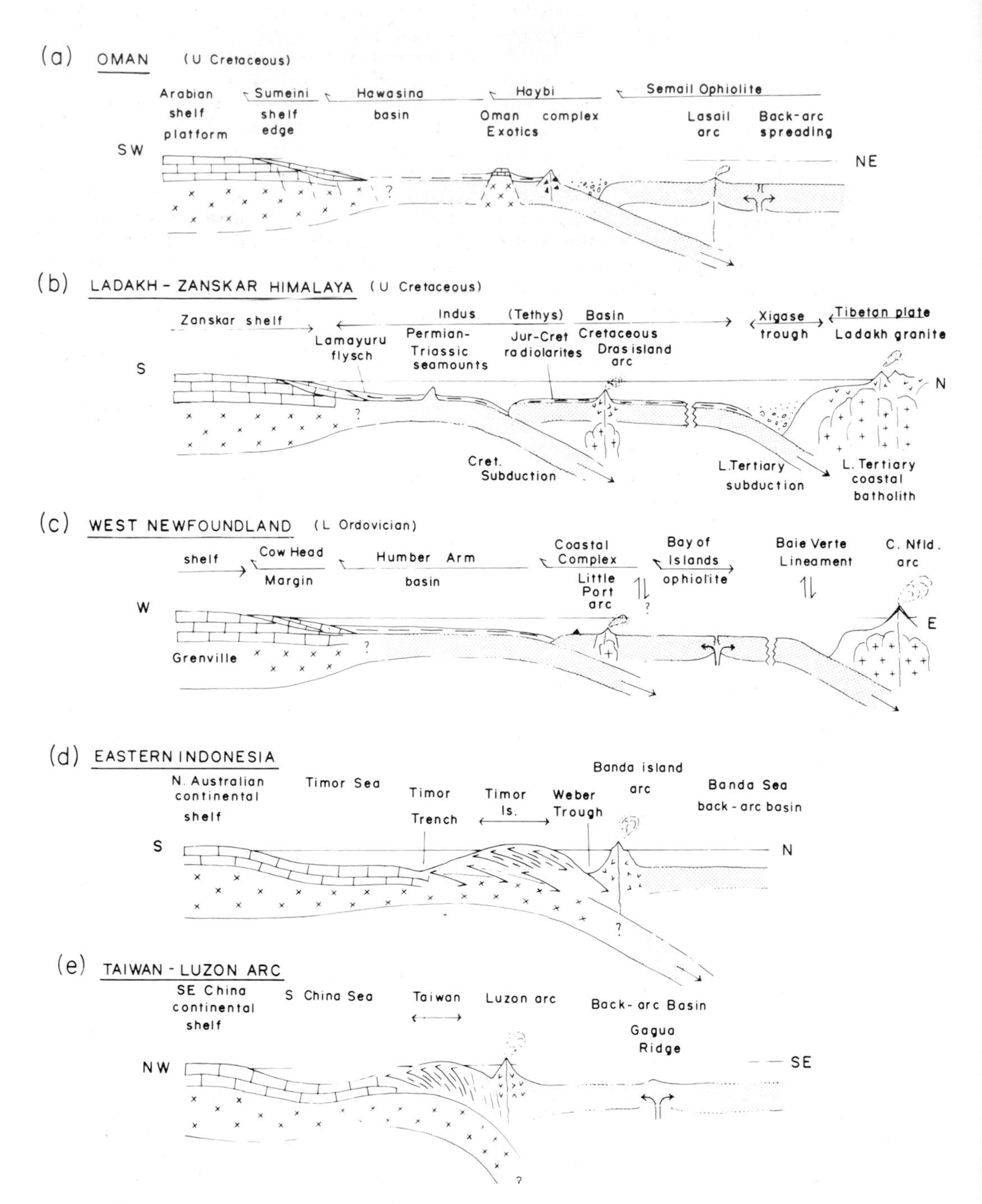

FIG. 11. Palinspastic cross-sections through (a) the Oman continental margin in the Cenomanian (see Fig. 3), (b) the Zanskar (Indian) continental margin in the late Cretaceous, and (c) the N American continental margin in W Newfoundland during the L. Ordovician. Present-day cross-sections (d) across the N Australian continental margin and Banda arc after (Carter *et al.* 1976, Barber & Audley-Charles 1976), and (e) the Luzon arc–S China Sea margin (after Bowin *et al.* 1978). (See text for discussion.)

telescoped oceanic sediments and ophiolite 'uphill' onto the continental margin (Elliott 1976).

Present-day cross-sections through arc–continent collision zones such as the Banda arc-Australia and Luzon arc-SE China compare closely with palinspastically restored cross-sections through the ophiolite belts of Oman, W Himalaya and W Newfoundland (Fig. 11). Bathymetric and geophysical profiles across the Timor Sea and S China Sea (Bowin *et al.* 1978, 1980) indicate a downwarping of the continental margin (to a depth of 25 km beneath Timor Island) compatible with the proposed models of partial subduction of a continental margin beneath the ophiolite sequences of Oman (Searle & Malpas 1980, 1982) and W Newfoundland (Malpas & Stevens 1977).

Seismic sections across the Banda arc do not define a single subduction surface as occurs along the Java Trench or the Andean Pacific margin—rather it is diffusely scattered across the width of the forearc region comparable to deformation beneath obducted ophiolites. Changing stress distribution is brought about by the presence at the site of subduction of a continental margin with a corresponding thick sequence of continental-margin sediments.

Lavas produced by back-arc spreading are chemically indistinguishable from typical MORB; lavas erupted in immature island arcs are distinct showing depletion in elements of high ionic potential and high $^{87}Sr/^{86}Sr$ ratios. Boninites, the volcanic products of partial melting of peridotite in hydrous conditions, were erupted in the Bonin arc and the arc rocks above the Papuan ophiolite belt above subduction zones. They are also found in the upper lavas of the Troodos, Othris (Greece) and Betts Cove (Newfoundland) ophiolites.

Metamorphic soles to obducted ophiolites were formed during the early stages of displacement, soon after ophiolite formation while it still retained a high heat content. Thermodynamic calculations for Oman (Searle & Malpas 1980, 1982; Ghent & Stout 1981) and Newfoundland (Malpas 1979; Jamieson 1980) suggest that temperatures of 750–900°C and pressures of 3–5 kb were attained. Such temperatures and pressures could only have existed at a spreading centre or in a marginal basin where high heat flow exists above an active subduction zone.

No single model can explain the obduction of all ophiolite complexes. However detailed mapping and structural analysis of ancient ophiolite complexes, particularly of rocks beneath them, show that comparisons to present-day processes at arc–continent boundaries are the closest uniformitarian analogues.

ACKNOWLEDGMENTS: Field work upon which these ideas were founded was supported in Oman (MPS) from 1977–80 by an Open University grant and an Overseas Development grant (UK) to Prof. I. G. Gass, and in 1981–82 by Amoco International Oil Company; in the Western Himalaya (MPS) in 1981–82 and in W Newfoundland (RKS) in 1968–82 by National Science & Engineering Research Council Canada grants. We are grateful to all these sources.

References

AUDLEY-CHARLES, M. G. & HOOIJER, D. A. 1973. Relation of Pleistocene migrations of pygmy stegodonts to island arc tectonics in eastern Indonesia. *Nature, Lond.* **241**, 197–8.

——, CARTER, D. J. & MILSOM, J. S. 1972. Tectonic development of eastern Indonesia in relation to Gondwanaland dispersal. *Nature, Lond.* **239**, 35–9.

BAILEY, E. H. 1981. Geologic map of Muscat-Ibra area, Sultanate of Oman. *J. Geophys. Res.* **86**. pocket map.

BARBER, A. J. & AUDLEY-CHARLES, M. G. 1976. The significance of the metamorphic rocks of Timor in the development of the Banda arc, eastern Indonesia. *Tectonophysics* **30**, 119–28.

——, —— & CARTER, D. J. 1977. Thrust tectonics in Timor. *Geol. Soc. Australia Jour.* **24**, 51–62.

BOWIN, C., LU, R. S., CHAO-SHING LEE & SCHOUTEN, H. 1978. Plate convergence and accretion in Taiwan-Luzon Region. *Bull. Am. Assoc. Petrol. Geol.* **62**, 1645–72.

——, PURDY, G. M., JOHNSTON, C., SHOR, G., LAWVER, L., HARTONO, H. M. S. & JEZEK, P. 1980. Arc-continent collision in Banda Sea Region. *Bull. Am. Assoc. Petrol. Geol.* **64**, 868–914.

BROOKFIELD, M. E. 1977. The emplacement of giant ophiolite nappes. I. Mesozoic–Cenozoic examples. *Tectonophysics* **37**, 247–303.

—— & REYNOLDS, P. H. 1981. Late Cretaceous emplacement of the Indus suture zone ophiolitic melanges and an Eocene–Oligocene magmatic arc on the northern edge of the Indian plate. *Earth Planet. Sci. Lett.* **55**, 157–62.

CARTER, D. J., AUDLEY-CHARLES, M. G. & BARBER, A. J. 1976. Stratigraphical analysis of island arc–continental margin collision in eastern Indonesia. *J. Geol. Soc. Lond.* **132**, 179–98.

CHURCH, W. R. & STEVENS, R. K. 1971. Early Palaeozoic ophiolite complexes of the Newfoundland Appalachians as mantle-oceanic crust sequences. *J. Geophys. Res.* **76**, 1460–6.

COLEMAN, R. G. 1977. *Ophiolites, Ancient Oceanic Lithosphere.* Springer-Verlag, New York, 229 pp.

—— 1981. Tectonic setting for ophiolite obduction in Oman. *J. Geophys. Res.* **86**, 2497–508.

COWARD, M. P., JAN, M. Q., REX, D., TARNEY, J., THIRLWALL, M. & WINDLEY, B. F. 1982. Geotectonic framework of the Himalaya of N. Pakistan. *J. geol. Soc. Lond.* **139**, 299–308.

DAVIES, H. L. & SMITH, I. E. 1971. Geology of Eastern Papua. *Bull. geol. Soc. Am.* **82**, 3299–312.
DEWEY, J. F. 1976. Ophiolite obduction. *Tectonophysics* **31**, 93–120.
—— & BIRD, J. M. 1970. Mountain belts and the new global tectonics. *J. Geophys. Res.* **75**, 2625–47.
—— & —— 1971. Origin and Emplacement of the Ophiolite Suite: Appalachian Ophiolites in Newfoundland. *J. Geophys. Res.* **76**, 3179–206.
DIETRICH, V., EMMERMANN, R., OBERHANSLI, R. & PUCHELT, H. 1978. Geochemistry of basaltic and gabbroic rocks from the West Mariana Basin and Mariana Trench. *Earth Planet. Sci. Lett.* **39**, 127–44.
DUNNING, G. & KROGH, T. 1982. Tightly clustered, precise U-Pb (zircon) ages of ophiolites from the Newfoundland Appalachians. *Geol. Soc. Am. N.E. Section Meeting* (Abstract).
ELLIOTT, D. 1976. The motion of thrust sheets. *J. Geophys. Res.* **81**, 949–63.
FARHOUDI, G. & KARIG, D. E. 1977. Makran of Iran and Pakistan as an active arc system. *Geology* **5**, 664–8.
FUCHS, G. 1979. On the Geology of Western Ladakh. *Jahrb Geol.* **B-A**, 513–40.
GEALEY, W. K. 1977. Ophiolite obduction and geologic evolution of the Oman Mountains and adjacent areas. *Bull. geol. Soc. Am.* **88**, 1183–91.
GHENT, E. D. & STOUT, M. Z. 1981. Metamorphism at the base of the Samail ophiolite, southeastern Oman Mountains. *J. Geophys. Res.* **86**, 2557–72.
GLENNIE, K. W., BOUEF, M. G. A., HUGHES-CLARK, M. W., MOODY-STUART, M., PILAAR, W. F. H. & REINHARDT, B. M. 1973. Late Cretaceous nappes in the Oman Mountains and their geologic evolution. *Bull. Am. Assoc. Pet. Geol.* **57**, 5–27.
HAMILTON, W. 1973. Tectonics of the Indonesian region *Bull. Geol. Soc. Malaysia* **6**, 3–10.
—— 1977. Subduction in the Indonesian region. *In*: TALWANI, M. (ed.) *Island Arcs, Deep Sea Trenches and Back-arc Basins*. Am. Geophys. Union Maurice Ewing Series, **1**, 15–31.
—— 1979. Tectonics of the Indonesian Region. *U.S. Geol. Survey Professional Paper 1078*, 345 pp.
HAWKESWORTH, C. J., O'NIONS, R. K., PANKHURST, R. J., HAMILTON, P. J. & EVENSON, N. M. 1977. A geochemical study of island arc and back-arc tholeiites from the Scotia Sea. *Earth Planet. Sci. Lett.* **36**, 253–62.
HAWKINS, J. & BATIZA, R. 1977. Metamorphic rocks of the Yap arc-trench system. *Earth Planet. Sci. Lett.* **37**, 216–29.
HOPSON, C. A., COLEMAN, R. G., GREGORY, R. T., PALLISTER, J. S. & BAILEY, E. H. 1981. Geologic section through the Samail Ophiolite and Associated Rocks along a Muscat-Ibra transect, southeastern Oman Mountain. *J. Geophys. Res.* **86**, 2527–44.
HUSSONG, D. M. & UYEDA, S. 1981. Tectonics in the Mariana Arc: results of recent studies including DSDP Leg 60. *Oceanol. Acta* 203–12.
JAMIESON, R. A. 1980. Formation of metamorphic aureoles beneath ophiolites—Evidence from the St Anthony Complex Newfoundland. *Geology* **8**, 150–4.
—— & TALKINGTON, R. 1980. A Jacupirangite-syenite assemblage beneath the White Hills Peridotite, northwestern Newfoundland. *Am. J. Sci.* **280**, 459–77.
KARIG, D. E. 1971. Origin and development of marginal basins in the western Pacific. *J. Geophys. Res.* **76**, 2542–61.
—— 1972. Remnant Arcs. *Bull. geol. Soc. Am.* **83**, 1057–68.
—— 1973. Plate convergence between the Philippines and Ryuku Islands. *Marine Geology* **14**, 153–68.
—— & SHARMAN, G. F. 1975. Subduction and accretion in trenches. *Bull. geol. Soc. Am.* **86**, 377–89.
KARSON, J. & DEWEY, J. F. 1978. Coastal complex, western Newfoundland, an early Ordovician fracture zone. *Bull. geol. Soc. Am.* **89**, 1037–49.
LANPHERE, M. A. 1981. K-Ar ages of metamorphic rocks at the base of the Semail Ophiolite. *J. Geophys. Res.* **86**, 2777–82.
LIPPARD, S. J. 1983. Cretaceous high pressure metamorphism in NE Oman and its relationship to subduction and ophiolite nappe emplacement. *J. geol. Soc. Lond.* **140**, 97–104.
MALPAS, J. 1979. Dynamothermal aureole beneath the Bay of Islands ophiolite in western Newfoundland. *Can. J. Earth Sci.* **16**, 2086–101.
—— & STEVENS, R. K. 1977. The origin and emplacement of the Ophiolite Suite with examples from Western Newfoundland. *Geotectonics (USSR)* **11**, 453–66.
—— & TALKINGTON, R. (eds) 1979. *Ophiolites of the Canadian Appalachians and Soviet Urals*. Memorial University of Newfoundland, 165 pp.
——, STEVENS, R. K. & STRONG, D. F. 1973. Amphibolite associated with the Newfoundland ophiolite—its classification and tectonic significance. *Geology* **1**, 45–7.
MATTINSON, J. M. 1975. Early Paleozoic ophiolite complexes of Newfoundland: Isotopic ages of zircons. *Geology* **3**, 181–3.
PEARCE, J. A. 1975. Basalt geochemistry used to investigate past tectonic environments on Cyprus. *Tectonophysics* **25**, 41–67.
——, ALABASTER, T., SHELTON, A. W. & SEARLE, M. P. 1981. The Oman Ophiolite as a Cretaceous arc-basin complex: evidence and implications. *Philos. Trans. R. Soc. London Series A*, **300**, 299–317.
ROEDER, D. 1977. Philippine arc system—collision or flipped subduction zones? *Geology* **5**, 203–6.
ROSS, R. J. *et al.* 1982. *The Ordovician System in the United States*. International Union of Geological Sciences, publication 12.
SEARLE, M. P. 1983. Stratigraphy, structure and evolution of the Tibetan-Tethys zone in Zanskar and the Indus suture zone in the Ladakh Himalaya. *Philos. Trans. R. Soc. Edinburgh* **73**, 203–217.
—— & GRAHAM, G. M. 1982. 'Oman Exotics'—Oceanic carbonate build-ups associated with the early stages of continental rifting. *Geology* **10**, 43–9.
—— & MALPAS, J. 1980. Structure and metamorphism of rocks beneath the Samail ophiolite of Oman and their significance in ophiolite obduction. *Philos. Trans. R. Soc. Edinburgh* **71**, 247–62.
—— & —— 1982. Petrochemistry and origin of sub-ophiolitic metamorphic and related rocks in the

Oman Mountains. *J. geol. Soc. Lond.* **139**, 235–48.
——, James, N. P., Calon, T. J. & Smewing, J. D. 1983. Sedimentological and structural evolution of the Arabian continental margin in the Musandam Mountains and Dibba zone, United Arab Emirates. *Bull. geol. Soc. Am.* (in press).
——, Lippard, S. J., Smewing, J. D. & Rex, D. C. 1980. Volcanic rocks beneath the Semail ophiolite nappe in the northern Oman Mountains and their tectonic significance in the Mesozoic evolution of Tethys. *J. geol. Soc. Lond.* **137**, 589–604.
Seely, D. R., Vail, P. R. & Walton, G. C. 1974. Trench-slope model. *In*: Burk, C. A. & Drake, C. L. (eds) *The Geology of Continental Margins*. Springer-Verlag, New York, pp. 249–60.
Sharaskin, A. Y., Bogdanov, N. A. & Zakariadze, G. A. 1981. Geochemistry and timing of the marginal basin and arc magmatism in the Philippine Sea. *Philos. Trans. R. Soc. London Series A*, **300**, 287–97.
Smewing, J. D. 1980. Regional setting and petrologic characteristics of the Oman Ophiolite in North Oman. *Ofioliti special issue Tethyan Ophiolites, Vol. 2, Eastern Area*, 335–78.
Smith, A. G. & Woodcock, N. H. 1976. Emplacement model for some Tethyan ophiolites. *Geology* **4**, 653–6.
Stoneley, R. 1975. On the origin of ophiolite complexes in the southern Tethys region. *Tectonophysics* **25**, 303–22.
Strong, D. F. 1974. An off-axis alkali volcanic suite associated with the Bay of Islands ophiolites, Newfoundland. *Earth Planet. Sci. Lett.* **21**, 301–9.
Tarney, J., Saunders, S. D., Mattey, D. P., Wood, D. A. & Marsh, N. G. 1981. Geochemical aspects of back-arc spreading in the Scotia Sea and western Pacific. *Philos. Trans. R. Soc. London Series A*, **300**, 263–85.
Tilton, G. R., Wright, J. E. & Hopson, C. A. 1981. Uranium-lead isotopic ages of Samail ophiolite, Oman, with application to Tethyan ocean ridge tectonics. *J. Geophys. Res.* **86**, 2763–76.
Van Bemmelen, R. W. 1949. *The Geology of Indonesia*, 2nd edition (reprinted 1970) Govt. Printing Office, The Hague, 732 pp.
Van Eysinga, F. W. B. 1978. *Geological Time Table*. Elsevier Publishing Co. Amsterdam.
Welland, M. J. P. & Mitchell, A. H. G. 1977. Emplacement of the Oman Ophiolite: A mechanism related to subduction and collision. *Bull. geol. Soc. Am.* **88**, 1081–8.
Williams, H. 1975. Structural succession, nomenclature and interpretations of transported rocks in western Newfoundland. *Can. J. Earth Sci.* **12**, 1874–94.
—— & Smyth, W. R. 1973. Metamorphic aureoles beneath ophiolite suites and Alpine peridotites: tectonic implications with west Newfoundland examples. *Am. J. Sci* **273**, 594–621.

M. P. Searle & R. K. Stevens, Department of Earth Sciences, Memorial University of Newfoundland, St Johns, Newfoundland A1B 3X5, Canada.

The structural variety in Tethyan ophiolite terrains

N. H. Woodcock & A. H. F. Robertson

SUMMARY: We examine the structural arrangement of some important ophiolitic terrains from Greece, through Turkey, Cyprus and Iran, to Oman. No two areas have an identical geometry or kinematic history and gross comparisons cannot support any single model of ophiolite emplacement. However, more detailed comparisons reveal similarities between individual structural units in the different areas. These units include, for example, the ophiolite sheet itself, slivers of metamorphic rocks, imbricated platform margin sequences, deformed syn-orogenic clastics or mélange, and allochthonous supra-ophiolite sheets. Each of these units may reflect a distinct process operating during emplacement. We describe and interpret these units in the framework of two stacking geometries, one produced by dip-slip-dominated emplacement and the other by strike-slip. These geometries highlight how the presence or absence of particular units may closely constrain possible ophiolite emplacement models. This point is illustrated by the difficulties in applying the 'emplaced forearc' model to Oman and Cyprus.

In this paper we show that the major differences in structural organization of Tethyan ophiolite terrains are incompatible with a single emplacement mechanism. There are, however, instructive similarities between some individual structural elements of these terrains. We try here to identify these distinctive elements, to summarize them briefly, and to indicate examples of each. We treat only relatively complete ophiolites, not ophiolitic mélanges. Many of the examples come from our own published work, though comparable examples are found elsewhere in the literature.

The relative spatial relationships of the various structural elements are generalized in Figs 1 and 2. We emphasize that in any one ophiolite terrain some of the elements may be missing. Figure 1 shows relationships where an ophiolite slab has been emplaced by mainly dip-slip displacements. This includes pure thrust tectonics or gravity sliding/spreading. All ophiolite emplacement must involve a component of dip-slip, simply to allow the uplift of the dense ophiolitic crust to the same structural level as less dense continental or marginal crust. However some regions show a strong strike-slip emplacement component. Figure 2 shows geometrical relationships in this case. These dip-slip and strike-slip models can be regarded as idealized end members of a spectrum of possible emplacement histories. Many areas show a complex polyphase history in which the dominance of dip-slip or strike-slip varies through time.

Dip-slip emplacement

Dip-slip emplacement commonly results in a moderate to low dipping ophiolite slab separating sub-ophiolite from supra-ophiolite units. The units are referred to by their number on Fig. 1.

Unit 1

The ophiolites may range from a relatively undeformed sheet to discrete blocks and ultimately to ophiolite mélange. The major sheets are commonly affected by ductile shear zones near the base, and are cut by faults throughout. The shear zones may post-date or be synchronous with serpentinization of the lower levels of the ophiolite. In such cases they are usually emplacement-related, but higher-grade shear zones may be difficult to distinguish from earlier spreading- or transform-related peridotite fabrics (e.g. Oman: Smewing 1980; Boudier & Coleman 1981). Brittle faults may shorten or extend the sheet and may often coexist. The Semail ophiolite, Oman shows NW-striking listric thrusts (e.g. in the Wuqbah Block) and steep, variably trending, extension faults often bounding wide corridors between intact ophiolite blocks (maps of Glennie *et al.* 1974 and Open University 1980–81). The Pindos ophiolite, Greece (Brunn 1956) shows a similar coexistence of thrust and extension faults. The extension faults have been taken as evidence for gravity-driven emplacement (e.g. Elliott 1976), although they can equally well be explained as compression-parallel tension cracks in a laterally unconstrained ophiolite sheet being pushed from one end. In an extreme case only ophiolite olistostrome mélange is preserved on the foreland (e.g. Kastel Formation, SE Turkey: Rigo de Righi & Cortesini 1964).

Unit 2

Discontinuous slices of metamorphic rocks commonly underlie Tethyan ophiolite sheets (Woodcock & Robertson 1977; Karamata 1980). They are dominated by ophiolitic rocks metamorphosed to greenschist and amphibolite facies, though minor eclogites and glaucophane schists

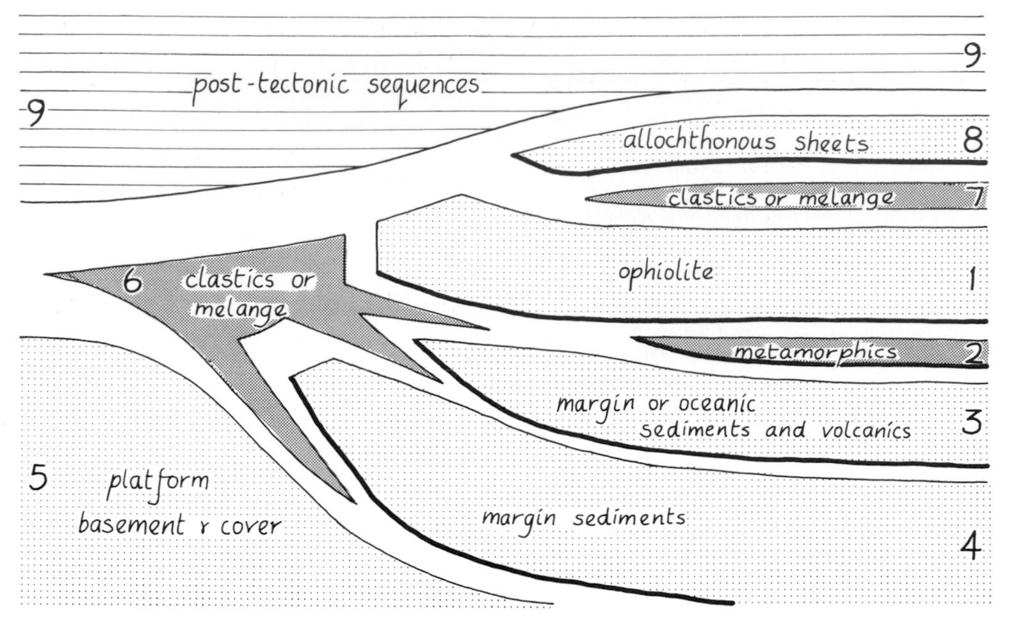

FIG. 1. Diagrammatic exploded section across an ophiolite terrain dominated by dip-slip-emplacement tectonics (pure thrust or gravity regimes). Emplacement is from right to left. Pre-emplacement sequences are in light stipple, syn-emplacement sequences in deep stipple and post-emplacement units are horizontally ruled. Thick lines denote tectonic basal contacts to units. (Full discussion in the text.)

may occur (e.g. Oman: Lippard 1983). The grade generally increases towards the ophiolite. The metamorphics may show two or three phases of deformation each at progressively lower grades. The most intact examples are in Oman (Searle & Malpas 1980) and in Yugoslavia (Karamata 1974). During thrusting the metamorphic sheet may be tectonically thickened, thinned or cut out altogether. Dismembered metamorphic sheets are more usual, and late semi-ductile and brittle deformation may reduce them to a tectonic mélange. A good example is the Semail mélange in Oman where blocks of ophiolitic and margin rocks lie in a serpentinite matrix (e.g. Searle & Malpas 1980). The sub-ophiolite metamorphic rocks are thought to be accreted to the hanging wall of a major reverse shear zone that initially detaches the ophiolite sheet from the ocean floor. As the displacement increases, progressively higher levels of the footwall ophiolite are accreted at progressively lower P–T conditions. Spray (this volume) reviews the process in detail, stressing that initial oceanic slicing commonly occurs millions of years before final emplacement on to the margin.

Unit 3

This unit is characterized by sedimentary and volcanic sequences originally deposited on mafic or oceanic crust or on the outer edge of a platform margin, now irregularly deformed into fault-bounded slices or dismembered to mélange. Basinal sediments, cherts or pelagic limestones, are common and margin sequences may be floored by alkalic or transitional mafic extrusives. Carbonate build-ups, originally deposited on continental slivers or lava highs, may have complicated the margin palaeogeography (e.g. Oman Exotics: Graham 1980). Only modest amounts of deformation of these sequences may thus produce a mixed tectonic/sedimentary mélange texture, with 'exotic' limestone and lava blocks floating in a sedimentary matrix. Examples of this type of mélange occur in Oman (Hawasina mélange: Graham 1980) and in Othris, Greece (Smith *et al.* 1979). This component predates overriding by the ophiolite sheet and can be a precursor olistostrome to the ophiolite or derived by local disruption of the outer margin. Another possible structural style in unit 3 is the repeated thrust imbrication of ocean-floor igneous and sedimentary rocks. Where this can be proved to be diachronous, unit 3 might represent an accretionary complex at a subduction zone. Each oceanic slice may be depositionally overlain by and then tectonically imbricated with syn-emplacement flysch or olistostrome, representing trench fill and discussed as unit 6. Possibly because the Tethyan ocean basin was relatively narrow (hundreds rather than thousands of kilometres) and because the ocean floor and margin successions are commonly condensed, few examples of definite subduction complexes have yet been recorded. In a general way Sengor & Yilmaz (1981) interpret much of East Anatolia as a late Cretaceous–Paleogene accretionary terrain. Specifically the Upper Cretaceous oceanic lavas and sediments of the Maden Complex, SE Turkey, show sequential imbrication suggesting an accretionary origin as a forearc complex above a north-dipping subduction zone (Aktas & Robertson, in press).

Unit 4

Mélanges become less common structurally downwards and forwards and are underlain by coherent thrust sheets. Sedimentary sequences predominate. They were usually deposited closer to the platform than those in unit 3, and often include thick redeposited facies. Examples are the Hawasina allochthon, Oman (Glennie *et al.* 1974; Graham 1980), the Meterizia Formation, Othris (Smith *et al.* 1979) and equivalents in Baër-Bassit, Syria (Delaune Mayere *et al.* 1977). Even the most distal sequences commonly show passive margin successions which cover much of the time between initial rifting and final emplacement of the ophiolite (e.g. Halfa Formation, Oman). Although in some areas, for instance the Hawasina window, Oman, unit 4 has been overridden by the ophiolite sheet, in other areas, such as Othris or the SW foothills of the Oman Mountains, this has not happened. Graham (1980) deduced that the Hawasina units were assembled by thrusting ahead of, rather than beneath, the ophiolite sheet.

Unit 5

The platforms onto which Tethyan ophiolites were emplaced are often only weakly deformed. They are usually autochthonous on a continental basement and typically comprise thick sequences of shallow-water carbonate. These sequences may be gently folded or block-faulted. Major detachment of the carbonates from their basement seems uncommon during dip-slip ophiolite emplacement and is only evident after more complex compressional tectonics (e.g. southern Turkey: Monod 1976; Kyrenia Range, Cyprus: Aubert & Baroz 1974; Dibba Zone, UAE). In platform areas not directly overthrust by the ophiolite sheet there may be deformation due to flexural downwarp ahead of the advancing allochthons. For example, in the Dibba Zone (UAE) the platform and Hawasina allochthon are separated by a thin, nearly autochthonous clastic and pelagic unit (Aruma Group: Allemann & Peters 1972) reflecting platform subsidence prior to emplacement. Typical 'sub-ophiolite' platforms are those of Oman (Glennie *et al.* 1974), Othris (Smith *et al.* 1979), the western Hellenides (Smith & Moores 1974), the western Bay Daglari adjacent to the Lycian Nappes (Poisson 1978) and the Arabian margin in Eastern Turkey (Rigo de Righi & Cortesini 1964).

Unit 6

The sequences in this unit are sediments broadly contemporaneous with ophiolite emplacement, rather than the deformed pre-emplacement sequences that characterize units 1–5. They may range from olistostromes and fan turbidites to alluvial deposits, often shed forwards from the advancing nappe pile. In some areas (e.g. Othris: Smith *et al.* 1979) this debris supply began early in the emplacement history and was then overthrust. There it now interdigitates as mélange between units 3, 4, and 5. In Oman, small volumes of margin-derived sediment (Aruma Group) were shed on to the platform then overthrust by the Hawasina allochthon. Ophiolitic detritus was only shed off the front of the allochthons as emplacement was ending, and the resulting Juweiza Formation (Glennie *et al.* 1974) now remains as a weakly folded blanket overlapping the platform.

Sediments of unit 6 may form an accretionary subduction complex if they are shed onto subducting oceanic crust and then diachronously imbricated by thrusting. In this case they would now appear to interdigitate with oceanic sediments of unit 3. The Pindos Zone, central Greece, shows some evidence for diachronous stacking of flysch and underlying pelagic sediments (Smith 1976, Lorsong 1979), though the gross geometry in this region is complicated by the two-stage emplacement history of the Pindos/Othris ophiolite belt (Smith 1976; Smith *et al.* 1979). Another case is the Eocene Maden Complex in SE Turkey in which olistostrome mélange and oceanic crust is sequentially sliced immediately prior to tectonic emplacement over the platform (Aktas & Robertson, in press). Norman (in press) also describes the Ankara mélange of Northern Turkey as an accretionary complex generated from early Jurassic into Tertiary time.

Unit 7

We now consider supra- and intra-ophiolite units. Unit 7 has only been recognized so far in Oman and comprises material derived entirely from the ophiolite and the sub-ophiolite units. The lower part is the Zabyat Formation which lies with depositional contact on the upper

Poisson, A. 1978. *Recherches géologiques dans las Taurides occidentales (Turquie).* Thèse de Docteur des Sciences, Universite de Paris-Sud.

Reuber, I. 1982. Mylonitic shear zones of millimetric to kilometric scale related to a transform fault in an ophiolitic complex (Antalya, Turkey). *Int. conf. on Planar and Linear Fabrics of Deformed Rocks, Mitt. ETH. Zurich,* **239a**, 233–6.

Ricou, L.-E., Argyriadis, I. & Marcoux, J. 1975. L'axe calcaire du Taurus, un alignement de fenêtres arabo-africaines sous des nappes radiolaritiques, ophiolitiques et metamorphiques. *Bull. Soc. geol. Fr.* **17**, 1024–44.

Rigo de Righi, M. & Cortesini, A. 1964. Gravity tectonics in foothills structure belt of south-east Turkey. *Bull. Am. Ass. Petrol. Geol.* **48**, 1911–37.

Robertson, A. H. F. 1977. The Moni Melange, Cyprus: an olistostrome formed at a destructive plate margin. *J. geol. Soc. London* **133**, 447–66.

—— & Hudson, J. D. 1974. Pelagic sediments in the Cretaceous and Tertiary history of the Troodos Massif, Cyprus. *Spec. Publ. Int. Assoc. Sedimentol.* **1**, 403–36.

—— & Woodcock, N. H. 1979. The Mamonia Complex, southwest Cyprus; the evolution and emplacement of a Mesozoic continental margin. *Bull. geol. Soc. Am.* **90**, 651–65.

—— & —— 1980a. Strike-slip related sedimentation in the Antalya Complex, SW Turkey. *Spec. Publ. Int. Assoc. Sedimentol.* **4**, 127–45.

—— & —— 1980b. Tectonic setting of the Troodos massif in the east Mediterranean. *In*: Panayiotou, A. (ed.) *Ophiolite—Proc. Int. Ophiolite Symp. Cyprus, 1979*, pp. 36–46.

—— & —— 1981a. Bilelyeri Group, Antalya Complex: deposition on a Mesozoic continental margin in SW Turkey. *Sedimentology*, **28**, 381–99.

—— & —— 1981b. Alakir Çay Group, Antalya Complex, SW Turkey: A deformed Mesozoic carbonate margin. *Sediment. Geol.* **30**, 95–131.

—— & —— 1981c. Gödene Zone, Antalya Complex, SW Turkey: volcanism and sedimentation on Mesozoic marginal ocean crust. *Geol. Rdsch.* **70**, 1177–214.

—— & —— 1983. Zabyat Formation, Semail Nappe, Oman: sedimentation on to an ophiolite during emplacement. *Sedimentology* **30**, 105–16.

—— & —— 1983. Genesis of the Batinah mélange above the Semail ophiolite, Oman. *J. Struct. Geol.* **5**, 1–17.

Searle, M. P. & Malpas, J. 1980. Structure and metamorphism of rocks beneath the Semail ophiolite of Oman and their significance in ophiolite obduction. *Trans. R. Soc. Edinburgh*, **71**, 247–62.

Sengör, A. M. C. & Yilmaz, Y. 1981. Tethyan evolution of Turkey; a plate tectonic approach. *Tectonophysics*, **75**, 181–241.

Shelton, A. W. & Gass, I. G. 1980. Rotation of the Cyprus microplate. *In*: Panayiotou, A. (ed.) *Ophiolites—Proc. Int. Ophiolite Sym. Cyprus, 1979*, pp. 61–65.

Simonian, K. O. & Gass, I. G. 1978. Arakapas fault belt Cyprus: A fossil transform belt. *Bull. geol. Soc. Am.* **89**, 1220–30.

Smewing, J. D. 1980. An Upper Cretaceous ridge-transform intersection in the Oman ophiolite *In*: Panayiotou, A. (ed.) *Ophiolites—Proc. Int. Ophiolite Symp. Cyprus, 1979*, pp. 407–13.

Smith, A. G. 1976. Plate tectonics and orogeny: A review. *Tectonophysics* **33**, 215–85.

—— & Moores, E. M. 1974. Hellenides. *In*: Spencer, A. M. (ed.) '*Mesozoic and Cenozoic Orogenic Belts*', *Spec. Publ. geol. Soc. Lond.* **4**, 159–85.

—— & Woodcock, N. H. 1976. Emplacement model for some 'Tethyan' ophiolites. Geology **4**, 653–6.

—— & —— & Naylor, M. A. 1979. The structural evolution of a Mesozoic continental margin, Othris Mountains, Greece. *J. geol. Soc., London* **136**, 589–602.

Spray, J. G. & Roddick, J. C. 1981. Evidence for Upper Cretaceous transform fault metamorphism in West Cyprus. *Earth planet. Sci. Lett.* **55**, 273–91.

Swarbrick, R. E. 1980. The Mamonia Complex of SW Cyprus: A Mesozoic continental margin and its relationship with the Troodos Complex. *In*: Panayiotou, A. (ed.) *Ophiolites—Proc. Int. Ophiolite Sym. Cyprus, 1979*, pp. 86–92.

—— & Naylor, M. A. 1980. The Kathikas Mélange, south-west Cyprus: late Cretaceous, submarine debris flows. *Sedimentology* **27**, 63–78.

Waldron, J. W. F. in press. Structural history of the Antalya Complex in the 'Isparta Angle', SW Turkey. *In*: Dixon, J. E. & Robertson, A. H. F. (eds) *The Geological Evolution of the Eastern Mediterranean*, Special Publication of the Geological Society of London, 13.

Welland, M. J. P. & Mitchell, A. H. G. 1977. Emplacement of the Oman ophiolite: a mechanism related to subduction and collision. *Bull. geol. Soc. Am.* **88**, 1081–8.

Woodcock, N. H. & Robertson, A. H. F. 1977. Origin of some ophiolite-related metamorphic rocks in the 'Tethyan' belt. *Geology* **5**, 373–6.

—— & —— 1981. Wrench-related thrusting along a Mesozoic–Cenozoic continental margin: Antalya Complex, SW Turkey. *In*: McClay, K. R. & Price, N. J. (eds) *Thrust and Nappe Tectonics*. Special Publication of the Geological Society of London, pp. 359–62.

—— & —— 1982a. Stratigraphy of the Mesozoic rocks above the Semail ophiolite, Oman. *Geol. Mag.* **119**, 67–76.

—— & —— 1982b. Wrench and thrust tectonics along a Mesozoic–Cenozoic continental margin: Antalya Complex, SW Turkey. *J. geol. Soc. London* **139**, 147–63.

—— & —— 1982c. The upper Batinah Complex, Oman: allochthonous sediment sheets above the Semail ophiolite. *Can. J. Earth Sci.* **19**, 1635–56.

Yilmaz, P. O. & Maxwell, J. C., 1981. K-Ar investigations from the Antalya Complex ophiolites, SW Turkey. Abstracts, Ophiolites and Actualism, Florence 1981, *Ofioliti*, **6**, 49.

N. H. Woodcock, Department of Earth Sciences, Downing Street, Cambridge CB2 3EQ.
A. H. F. Robertson, Grant Institute of Geology, West Mains Road, Edinburgh EH9 3JW.

REGIONAL STUDIES

An ophiolite suite in Fiji?

H. Colley

SUMMARY: In SW Viti Levu, Fiji rocks formerly described as part of an island-arc succession are regarded as possibly representing the upper part of an ophiolite suite. Foraminiferal oozes, cherts, red clays, Fe-Mn metalliferous sediments, fine-grained volcanic turbidites, and reworked polymict lapillistones can be equated with Layer 1 of the oceanic lithosphere. Underlying these sediments are pillow basalt sequences and deeper structural levels in the suite may be represented by a tonalite body intruded by a dyke swarm. The chemistry of the igneous rock is ambiguous showing affinities with both arc tholeiites and oceanic tholeiites.

Emplacement of the ophiolite suite against arc rocks has occurred along arcuate thrusts with the ophiolitic rocks, principally Layer-1 sediments, being folded along the leading edge of the thrusted block.

The Oligocene–Lower Miocene age of the ophiolite suite suggests it is an obducted portion of the South Fiji Basin, however, emplacement from this direction requires a complex tectonic model involving up to 180° anticlockwise rotation of Viti Levu. This is supported to some degree by palaeomagnetic data.

Melanesia is a region of considerable tectonic complexity which on a geographical basis can be divided into an inner arc system (Papua New Guinea–New Caledonia) and an outer arc system (Solomon Islands–Vanuatu–Fiji–Tonga). A number of marginal basins are associated with these arc systems and these include the Solomon Sea, Woodlark Basin, and Coral Sea in the Papua New Guinea–Solomons region and the South Fiji Basin (Minerva Plain), Fiji Plateau, and Lau Basin in the Fiji area (see Fig 5c). Reviews of the tectonic evolution of Melanesia are provided by Packham & Andrews (1975), Coleman & Packham (1976), Weissel (1981), and Malahoff *et al.* (1982a).

Most tectonic models for the region involve one or two subduction flips along the Pacific–India plates' boundary (e.g. Mitchell & Warden 1971, Karig & Mammerickx 1972; Gill & Gorton 1973; Colley & Warden 1974; Parrot & Dugas 1980) thus there is considerable scope within the models for emplacement of ophiolite suites into the Melanesian arcs. Such models explain the occurrence of ophiolite suites described in Papua New Guinea (Davies & Smith 1971), New Caledonia (Guillon & Routhier 1971), the Solomon Islands (Hughes & Turner 1977; Neef & Plimer 1979), Vanuatu (Colley & Warden 1974) and Tonga (Ewart & Bryan 1972). Parrot & Dugas (1980) regard these ophiolite suites as part of a disrupted ophiolite belt related to an Eocene subduction zone.

Within the Melanesian arcs, Fiji is regarded as anomalous in that ophiolite rocks appear to be absent. This absence cannot be attributed to non-geological factors for the larger islands are eroded to relatively deep structural levels and reconnaissance mapping (1:50,000 scale) was virtually complete by the mid-1970s.

Geology of Fiji

The geological history of Fiji is apparently restricted to the Cenozoic, the oldest known rocks being reef limestones of Eocene age (Cole 1960) and the youngest being volcanic ashes erupted on Taveuni in historic time (Rodda 1974). The geology of Fiji can be described in three natural stages each of which reflects major changes in the geological evolution of the island group.

Upper Eocene to Middle Miocene stage

This stage is dominated by fragmental volcanic rocks and derived sediments; flow units and reef limestones are relatively minor. The rocks are of arc and marginal basin origin and it is the latter that form the main topic of this paper.

Middle to Upper Miocene stage

During this stage, plate boundary changes in Melanesia resulted in a period of relatively strong deformation in Fiji known locally as the Tholo Orogeny. Plutonism accompanied the orogeny with the intrusion of tonalite and minor gabbro bodies of arc tholeiite composition (Gill & Stork 1979) which collectively approach batholithic proportions. Another consequence of the orogeny, as suggested in this paper, was the emplacement of ophiolitic rocks in the Fiji arc.

Upper Miocene to Recent stage

In Fiji, the Upper Miocene to Recent was a period of very voluminous and widespread volcanism which produced rocks of diverse chemistry. Arc tholeiites, calc-alkaline andesites, shoshonitic rocks, and ocean-island basalts have

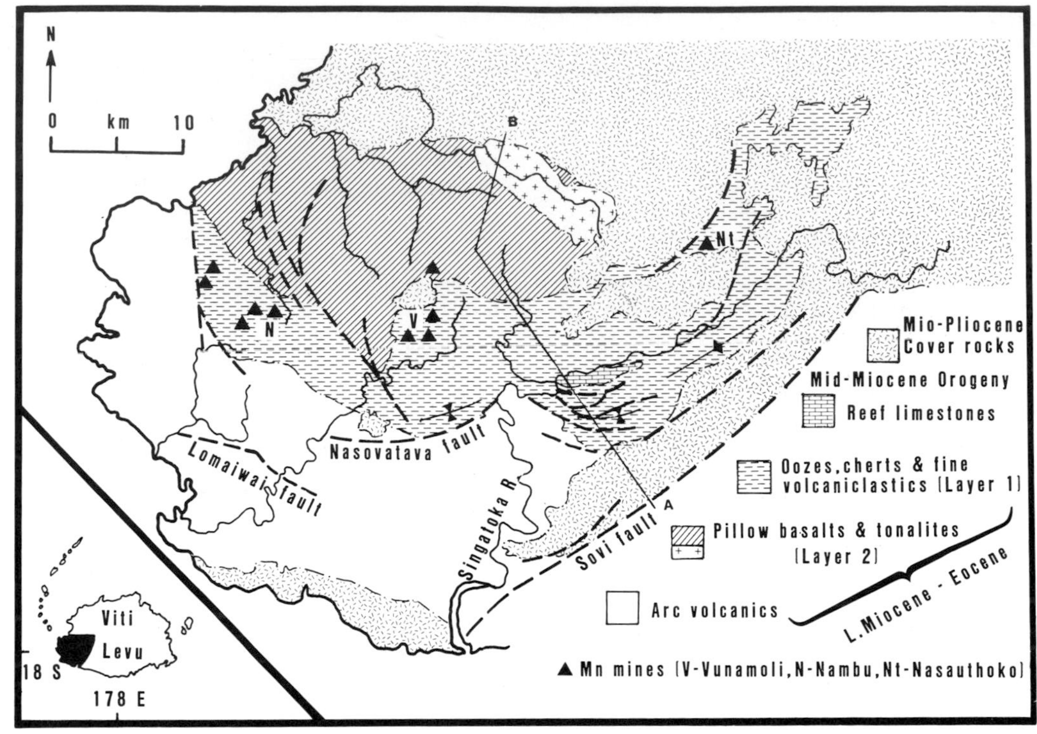

FIG. 1. Geological sketch map of SW Viti Levu showing the distribution of ophiolitic rocks. The Layer-1 sediments shown by dashed ornamentation have been mapped as the Singatoka Sedimentary Group (Houtz 1959; 1960; 1963) and the Layer-2 pillow basalts shown by cross-hatch ornamentation have been mapped as Wainimala Group (Bartholomew 1960). The arc volcanics (no ornamentation) are part of the Wainimala Group.

been reported (Gill 1970, 1976; Hindle & Colley 1981). The increasing oceanic character of the later volcanics has been related to the progressive isolation of Fiji from the Tonga–Kermadec subduction zone (Gill 1976) or to the influence of an extensional tectonic regime in Fiji during the openings of adjacent marginal basins (Colley & Hindle 1984).

The ophiolite suite

The proposed ophiolitic rocks in Fiji have been recognized in Upper Eocene to Lower Miocene sequences in SW Viti Levu (Fig. 1). Previous mapping in this part of the island has distinguished two lithostratigraphic units of pre-Middle Miocene age (Houtz 1959, 1960, 1963).

The *Wainimala Group* consists of flows, volcanic breccio-conglomerates, fine-grained volcaniclastic sediments and reef limestones which are representative of a primitive island-arc succession. Foraminifera from limestones close to the assumed base of this unit are Tertiary *b–c* (Cole 1960) whilst those in limestones close to the top of the unit are Tertiary *e–f* (Cole in Houtz 1960; Hirst 1965). Generally, the volcanics have an arc-tholeiite chemistry (Gill 1970); however, there is widespread alteration to spilitic and keratophyric compositions with the development of zeolite and greenschist-facies mineral assemblages. Crook (1963) relates the alteration to burial metamorphism.

The *Singatoka Sedimentary Group* has been described as consisting mainly of greywackes, sandstones and argillites (Bartholomew 1960; Houtz 1960; Rodda 1967). Carbonate beds, principally reef limestones, have generally been considered as relatively minor components (e.g. Houtz 1960), although some workers (e.g. Skiba 1964) have commented on the widespread distribution of well-bedded, non-reefal, green limestones. Following Houtz (1960) it was assumed that the Singatoka Sedimentary Group represented essentially reworked Wainimala Group debris deposited in a basin produced by subsidence within the arc. Thus the Singatoka sediments were regarded as an integral part of the arc succession in Viti Levu.

In this paper, the whole of the Singatoka Sedimentary Group is recognized as part of an ophiolite suite corresponding to Layer 1 of oceanic lithosphere (dashed ornamentation in Fig. 1). In addition pillow basalts occurring to the north of the Singatoka sediments (cross-hatch ornamentation in Fig. 1), formerly mapped as part of the Wainimala Group (Bartholomew 1960), are thought to represent Layer 2 of oceanic lithosphere.

Evidence for this re-interpretation of the geology of SW Viti Levu comes from various sources. Recent fieldwork by the author has shown that a large proportion of the Singatoka Sedimentary Group consists of pelagic sediments with formainiferal oozes being particularly widespread. Carter (in Skiba 1964) has studied the fauna of some of the oozes and concludes that the sediments were deposited in deep-water, open-sea conditions. Support for this has been obtained from gravity measurements and utilization of the Airy hydrostatic equilibrium equation to indicate that water depths in the Singatoka sediment basin were of the order of 3000 m (Worthington 1974). Red clays and cherts are associated with the oozes and there are also minor interbeds of fine-grained volcanic turbidites and reworked, polymict lapillistones. All these facies types are similar to those reported by Klein (1975) in cores from DSDP sites 285 and 286 in the South Fiji Basin.

Reef limestones of the Tertiary *e* stage (Whipple 1934) in the Singatoka Sedimentary Group are concentrated along the southern faulted contact of the ophiolite suite with Wainimala Group arc rocks and younger cover rocks (Fig. 1). The abundance of reef limestones, which probably represent a barrier-reef complex, suggests that the Singatoka basin was shallowing towards the arc.

A significant feature of the deeper-water sequences of the Singatoka sediment basin is the widespread occurrence of Fe-Mn metalliferous sediments; 'nests' of managanese deposits are found in the Nambu, Vunamoli, and Nasauthoko areas (Fig. 1). The mound-like form of these deposits, the occurrence of syngenetic, stratiform Fe and Mn cherts flanking the mounds, the association of Fe and Mn minerals with nontronite and red clays, and the location of most deposits close to, or at the pillow lava–pelagic sediment interface are characteristics shared by hydrothermal mounds described from oceanic spreading centres such as the Galapagos Ridge (Honnerez *et al.* 1981).

North of the Layer-1 sediments of the Singatoka Sedimentary Group there is a transition into pillow-basalt sequences corresponding to Layer 2 of oceanic lithosphere. Continuous sequences approaching 1 km in thickness are exposed in road cuttings and along creeks. Initial chemical studies show that the pillow basalts (Table 1; No. 1) are distinct from arc tholeiites of the Wainimala Group (Table 1; No. 2). The former have higher TiO_2, MgO, Rb, and Sr and lower SiO_2, Al_2O_3, and Na_2O. This distinction is demonstrated more clearly by the REE patterns (Fig. 2) with the pillow basalts having the concave downward pattern typical of ocean-ridge basalts and the arc tholeiites showing a fractionated pattern with LREE enrichment.

Along the northern margin of the pillow-basalt outcrop small pockets of gabbro would seem to suggest deeper levels within Layer 2. Also in this northern area there is an outcrop of tonalite, the Yavuna pluton (Fig. 1), covering some 40 km^2. Previously this body has been assigned to the *Tholo Plutonic Suite*, the lithostratigraphic unit consisting of synorogenic tonalite-gabbro intrusions of generally Upper Miocene age (11–7 Ma).

Within this unit the Yavuna pluton was markedly anomalous in having a much older age, 34 Ma (McDougall 1963), and being conspicouously different in the field in that it is cut by a swarm of doleritic dykes. The dykes (Table 1; No. 3) have a similar chemistry to the Layer-2 pillow basalts and show a concave downward REE pattern (Fig. 2). It is therefore tempting to class the Yavuna pluton as an oceanic plagio-

TABLE 1. *Representative analyses of Upper Eocene to Middle Miocene igneous rocks*

	1	2	3	4
SiO_2	45.65	53.50	50.88	72.18
TiO_2	1.34	1.09	1.22	0.32
Al_2O_3	15.17	17.18	15.07	13.95
Fe_2O_3	2.15	3.82	3.79	1.07
FeO	7.10	4.95	7.50	1.61
MnO	0.15	0.25	0.19	0.09
MgO	12.58	4.05	4.75	0.90
CaO	9.79	7.33	8.51	2.98
Na_2O	2.60	3.77	3.28	4.50
K_2O	0.45	0.60	0.35	1.21
P_2O_5	0.12	0.24	0.10	0.06
L.O.1	2.95	2.85	3.55	1.28
Total	100.05	99.60	99.19	100.15
Rb	17	5.2	4.9	13
Sr	285	186	164	193
Ba	—	98	44	340
Zr	—	76	72	118
Ni	—	6	16	—
Cr	—	7	12	8
Y	—	29	26	22

1. Olivine basalt, Nandi River, Viti Levu 2. Arc tholeiite, Viti Levu (average of 6) 3. Dolerite dyke, Yavuna, Viti Levu 4. Tonalite, Yavuna, Viti Levu. Data from Gill (1970); Gill & Stork (1979); Colley (unpublished).

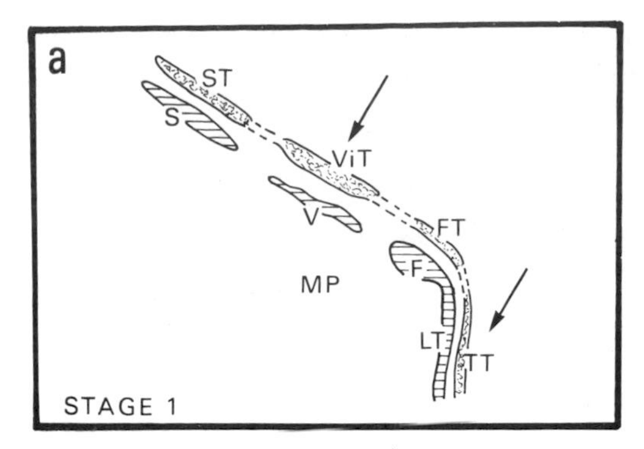

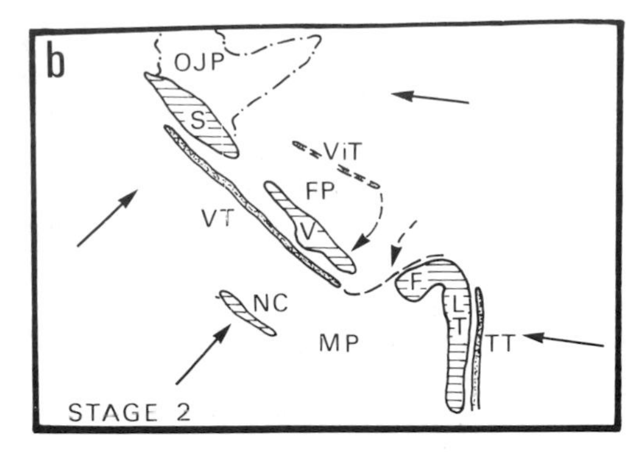

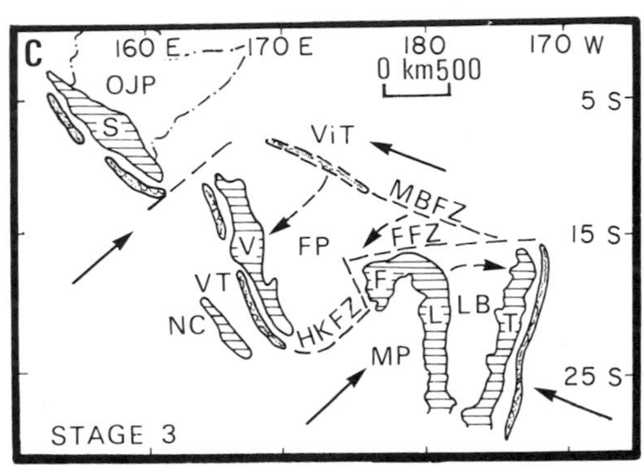

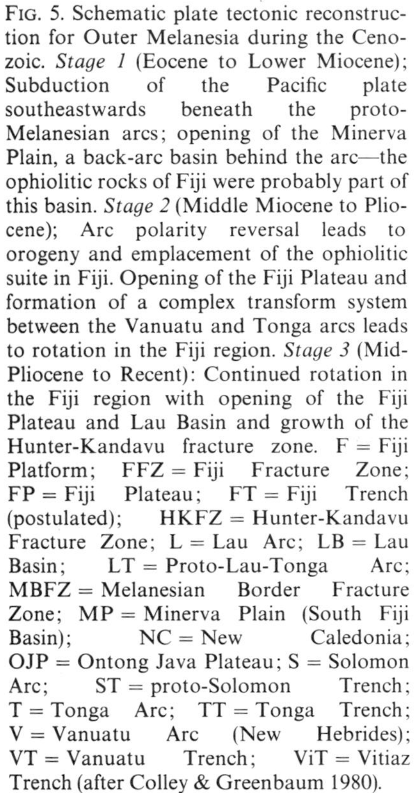
FIG. 5. Schematic plate tectonic reconstruction for Outer Melanesia during the Cenozoic. *Stage 1* (Eocene to Lower Miocene); Subduction of the Pacific plate southeastwards beneath the proto-Melanesian arcs; opening of the Minerva Plain, a back-arc basin behind the arc—the ophiolitic rocks of Fiji were probably part of this basin. *Stage 2* (Middle Miocene to Pliocene); Arc polarity reversal leads to orogeny and emplacement of the ophiolitic suite in Fiji. Opening of the Fiji Plateau and formation of a complex transform system between the Vanuatu and Tonga arcs leads to rotation in the Fiji region. *Stage 3* (Mid-Pliocene to Recent): Continued rotation in the Fiji region with opening of the Fiji Plateau and Lau Basin and growth of the Hunter-Kandavu fracture zone. F = Fiji Platform; FFZ = Fiji Fracture Zone; FP = Fiji Plateau; FT = Fiji Trench (postulated); HKFZ = Hunter-Kandavu Fracture Zone; L = Lau Arc; LB = Lau Basin; LT = Proto-Lau-Tonga Arc; MBFZ = Melanesian Border Fracture Zone; MP = Minerva Plain (South Fiji Basin); NC = New Caledonia; OJP = Ontong Java Plateau; S = Solomon Arc; ST = proto-Solomon Trench; T = Tonga Arc; TT = Tonga Trench; V = Vanuatu Arc (New Hebrides); VT = Vanuatu Trench; ViT = Vitiaz Trench (after Colley & Greenbaum 1980).

(Fig. 5c) the openings of the Fiji Plateau and Lau Basin, and the development of the Hunter-Kandavu transform fracture zone would lead to anticlockwise rotation in the Fiji region.

Support for rotation has been provided by a number of recent palaeomagnetic studies in Fiji (Cassie 1978; James & Falvey 1978; Malahoff *et al.* 1982b). These studies have been confined to rock sequences on Viti Levu of Upper Miocene to Pliocene age (i.e. cover rocks to the ophiolite suite). James & Falvey (1978) report 21° of anticlockwise rotation in the last 4 Ma; Cassie (1978) has 31° of anticlockwise rotation in the last 6 Ma; and Malahoff *et al.* (1982b) have 90° of anticlockwise rotation in the last 7 Ma. Such values do not approach the 180° of rotation that is required, however, it is possible that emplacement of the ophiolite suite occurred at 14 Ma and rotation could have commenced soon after. To speculate by taking the highest rate of rotation ($12° \mathrm{Ma}^{-1}$—Malahoff *et al.* 1982b) and to suggest that rotation began at 14 Ma, then values approaching 180° are obtained. It can be added that Worthington (1974) on the basis of alignment of aeromagnetic lineations indicated the possibility of up to 180° of rotation for Viti Levu.

Conclusions

In part of the succession in SW Viti Levu formerly regarded as representing an arc sequence, an ophiolite suite indicative of Layers 1 and 2 of oceanic lithosphere has been proposed. Evidence for this contention is both varied and compelling and includes the recognition of pelagic, turbiditic, and volcaniclastic sediments characteristic of Layer-1 oceanic lithosphere; a fauna in the pelagic sediments which indicates open sea–deep water conditions; and the thrusted nature of the contact between the ophiolitic rocks and arc rocks (Figs 1 & 3).

The chemistry of the rocks is somewhat ambiguous with pillow lavas and dykes showing

REE patterns typical of oceanic tholeiites. Overall though the chemistry would seem to suggest a closer affinity with arc tholeiite; a condition that might prove to be not unusual in rocks formed during the early stages of opening of a marginal basin. The Yavuna tonalite may represent melting of arc crustal material during a high-temperature episode at the beginning of marginal basin extension, subsequent intrusion by dykes indicates some transition towards oceanic-tholeiite chemistry.

Emplacement is problematical in that the age of the Fiji ophiolite suggests it to be obducted South Fiji Basin; however, to achieve this a complex tectonic model involving up to 180° anticlockwise rotation of Viti Levu is required. Palaeomagnetic data support anticlockwise rotation but the maximum degree of rotation so far documented is only 90° (Malahoff *et al.* 1982b).

The alternative model—that the Fiji ophiolite represents obducted Pacific Ocean floor—allows a much simpler tectonic model but appears invalid on age criteria; such ocean floor should be of Cretaceous age whereas the Fiji ophiolite is no older than the Oligocene.

Further palaeomagnetic work in Fiji, particularly on rocks older than 7 Ma, is needed to resolve the emplacement problem. Such a solution would have important implications for regional tectonic models of the SW Pacific.

ACKNOWLEDGMENTS: The author would like to thank the Director of the Mineral Resources Department, Fiji for logistic support during fieldwork in Fiji. Dr D. Greenbaum, Mr B. Rao and Mr W. H. Hindle provided much useful information and discussion. Dr P. Harvey (Nottingham University) and Dr. N. Walsh (King's College, London) provided chemical analyses. This work formed part of a research project in Fiji funded by the NERC.

References

BARTHOLOMEW, R. W. 1960. Geology of the Nandi area, western Viti Levu. *Bull. geol. Surv. Fiji* **7**, 27 pp.

CASSIE, R. A. 1978. *Palaeomagnetic Studies of the Suva Marl*. BSc (Hons.) Thesis University of Sydney, 72 pp.

CROOK, K. A. W. 1963. Burial metamorphic rocks from Fiji. *N.Z.Jl Geol. Geophys.* **6**, 681–704.

COLE, W. S. 1960. Upper Eocene and Oligocene larger foraminifera from Viti Levu, Fiji. *Prof. Pap. U.S. geol. Surv.* **374-A**, 7 pp.

COLEMAN, P. J. & PACKHAM, G. H. 1976. The Melanesian borderlands and India–Pacific plates' boundary. *Earth Sci. Rev.* **12**, 197–233.

COLEMAN, R. G. 1977. *Ophiolites*. Springer-Verlag, Berlin, 229 pp.

COLLEY, H. & GREENBAUM, D. 1980. The mineral deposits and metallogenesis of the Fiji Platform. *Econ. Geol.* **75**, 807–29.

—— & HINDLE, W. H. 1984. The volcano-tectonic evolution of Fiji and adjoining marginal basins. *In*: KOKELAAR, B. P. & HOWELLS, M. F. (eds) *Marginal Basin Geology: Volcanic and Associated Sedimentary and Tectonic Processes in Modern and Ancient Marginal Basins*. Special Publication of the Geological Society, London.

—— & WARDEN, A. J. 1974. Petrology of the New Hebrides. *Bull. geol. Soc. Am.* **85**, 1635–46.

DAVIES, H. L. & SMITH, I.E. 1971. Geology of eastern Papua. *Bull. geol. Soc. Am.* **82**, 3299–312.

EWART, A. & BRYAN, W. B. 1972. Petrography and geochemistry of the igneous rocks from Eua, Tongan Islands. *Bull. geol. Soc. Am* **83**, 3281–98.

FALVEY, D. A. 1975. Arc reversals and a tectonic model for the North Fiji Basin. *Bull. Aust. Soc. Explor. Geophys.* **6**, 47–9.

GILL, J. B. 1970. Geochemistry of Viti Levu, Fiji and its evolution as an island arc. *Contrib. Mineral. Petrol.* **27**, 179–203.

—— 1976. From island arc to oceanic islands; Fiji, southwestern pacific. *Geology* **4**, 123–6.

—— & GORTON, M. P. 1973. A proposed geological and geochemical history of eastern Melanesia. *In*: COLEMAN, P. J. (ed.) *The Western Pacific: Island Arcs, Marginal Seas, and Geochemistry*. Univ. West. Aust. Press, Perth, 459–67.

—— & STORK, A. L. 1979. Miocene low-K dacites and trondhjemites of Fiji. *In*: BARKER, F. (ed.) *Trondhjemites, Dacites and Related Rocks*. Elsevier, Amsterdam, 629–49.

GUILLON, J. H. & ROUTHIER, P. 1971. Les stades d'évolution et de mise en place des massifs ultramafiques de Nouvelle Calédonie. *Bull. BRGM.* **4**, 5–37.

HEEZEN, B. C. & FORNARI, D. J. 1976. *Pacific Ocean: Sheet 20 Geological World Atlas*. UNESCO, Paris.

HINDLE, W. H. & COLLEY, H. 1981. An oceanic volcano in an island arc setting—Seatura volcano, Fiji. *Geol. Mag.* **118**, 1–12.

HIRST, J. A. 1965. Geology of east and north-east Viti Levu. *Bull. geol. Surv. Fiji* **12**, 51 pp.

HONNEREZ, J., VON HERZEN, R. P., BARRETT, T. J., BECKER, K., BENDER, M. L., BORELLA, P. E., HUBBERTEN, H.-G., JONES, S. C., KARATO, S., LAVERNE, C., LEVI, S., MIGDIZOV, A. A., MOORBY, S. A. & SCHRADER, E. L. 1981. Hydrothermal mounds and young ocean crust of the Galapagos: preliminary deep sea drilling results, leg 70, *Bull. geol. Soc. Am.* **92**, 457–72.

HOUTZ, R. E. 1959. Regional geology of Lomaiwai-Momi, Nandronga, Viti Levu, *Bull. geol. Surv. Fiji* **3**, 20 pp.

—— 1960. Geology of Singatoka area, Viti Levu. *Bull. geol. Surv. Fiji* **6**, 19 pp.

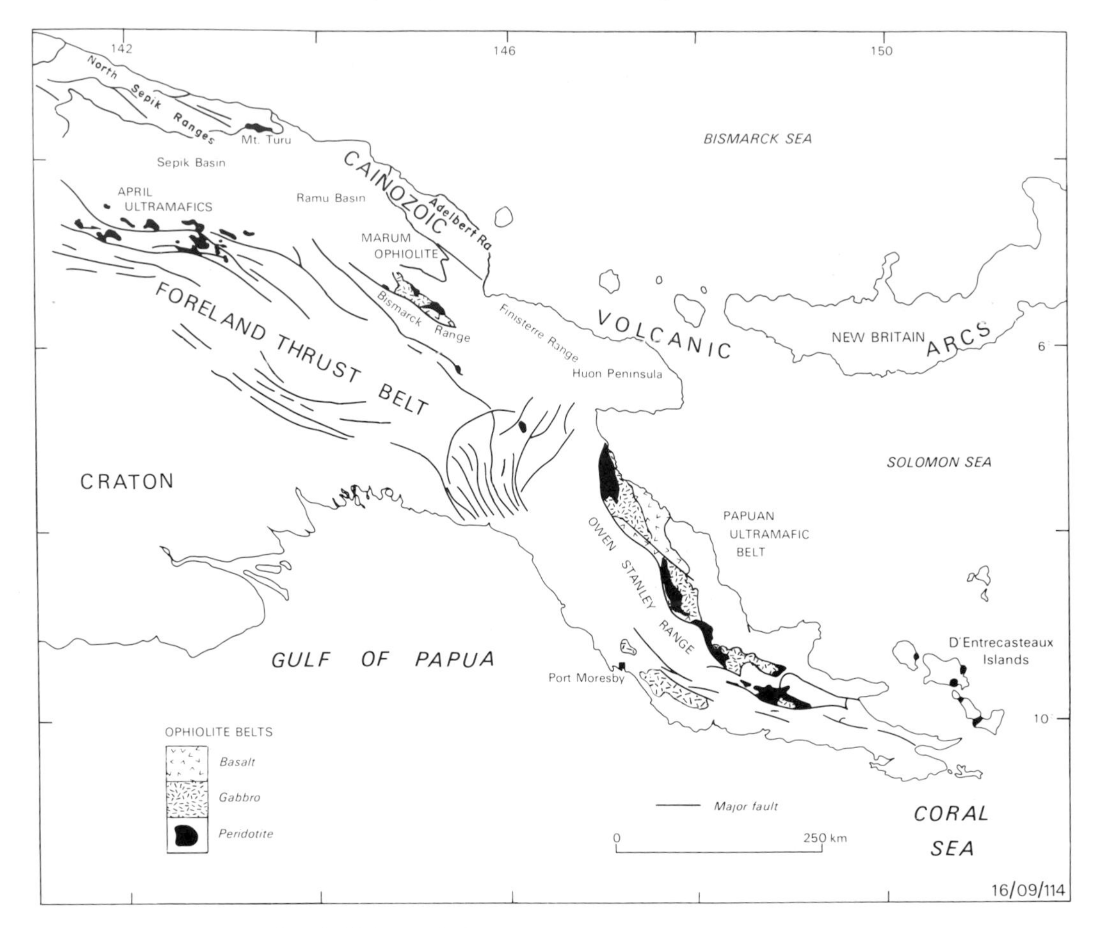

FIG. 1. Major ophiolite complexes of Papua New Guinea.

volcanics and sediments, and is underlain by metamorphic rocks with faulted contact. The metamorphic rocks consist of a contact metamorphic aureole of greenschist to granulite facies at the ophiolite contact; an underlying mass of mafic metamorphics; and, below this, a large volume of felsic schist and gneiss which forms the greater part of the Owen Stanley Range.

The felsic metamorphics (Kagi metamorphics: Pieters 1978) grade from garnet-bearing gneiss, closer to the ophiolite, through biotite and chlorite schist to unmetamorphosed Cretaceous, Paleocene and Eocene sediments further from the ophiolite (Fig. 3; Pieters 1978). Metamorphic foliation generally has steep dip.

The mafic metamorphics (Emo metamorphics: Pieters 1978) consist of metamorphosed basalt, dolerite and gabbro of tholeiitic character, and have the form of a shallow-dipping sheet, about 130 km long, 30 km exposed width, and more than 1 km thick, which is interposed between the underlying felsic metamorphics and the overlying ophiolite complex (Fig. 3). Metamorphic grade is generally greenschist facies with some development of lawsonite and glaucophane. Metamorphic foliation dips consistently at 10–30° to the E and NE, but with some local variations in dip—see map in Pieters (1978). Pieters concluded that the mafic metamorphics are a tectonic slice, bounded above and below by thrust faults. The significance of the mafic thrust sheet is discussed further below.

The contact metamorphic aureole also is mafic, but differs from the Emo metamorphics in having mineral assemblages which indicate high rather than low T/P conditions (Davies 1971, p. 23; 1980a).

The igneous rocks of the PUB ophiolite crystallized in the Jurassic or Cretaceous, and the complex was emplaced in the Eocene (K-Ar ages on gabbro and basalt, and on granulite from the metamorphic aureole; Davies 1980a).

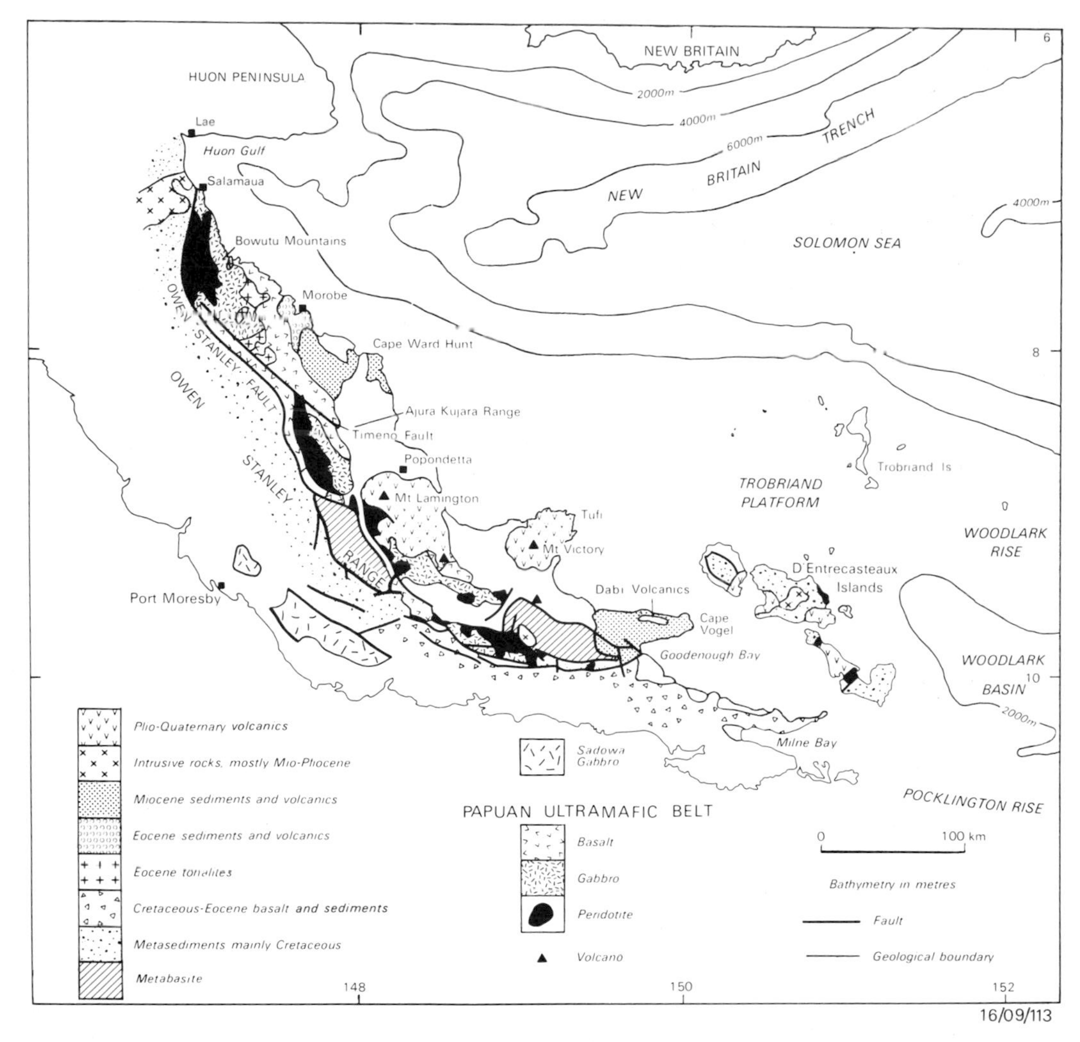

FIG. 2. Papuan ultramafic belt and Sadowa Gabbro.

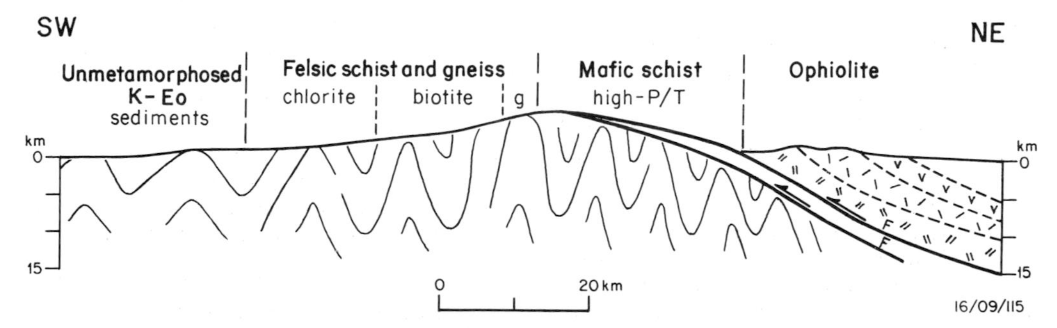

FIG. 3. Cross-section through the Owen Stanley Range shows distribution of mafic (Emo) and felsic (Kagi) metamorphics, and zonation in felsic metamorphics. Folds are diagrammatic; g : garnet (after Davies 1980b).

Marum ophiolite complex

The Marum complex (Fig. 4) is an incomplete ophiolite sequence which is exposed for 90 km along the northern flank of the Bismarck Range, immediately south of the Ramu-Markham valley, in the north-central part of the Papua New Guinea mainland (Jaques 1981). The complex consists of tectonite peridotite overlain by cumulus peridotite and gabbro with a total thickness of 3–4 km.

The tectonite peridotite consists mostly of harzburgite, commonly with thin dunite and orthopyroxenite layers and minor dunite lenses, cut by enstatite pyroxenite dykes. The harzburgite has a uniformly refractory mineralogy—olivine Fo_{92}, orthopyroxene En_{93} and chrome-rich spinel (53–63% Cr_2O_3)—and, like the PUB tectonites, is highly depleted in lithophile elements, containing less than 0.3% CaO and Al_2O_3.

The cumulus peridotite and gabbro sequence is grossly layered from dunite and peridotite at the base, through troctolite, plagioclase pyroxenite and norite/gabbro to ferrogabbro and anorthositic gabbro at the top. The mappable major units (Fig. 4) are dunite, peridotite, pyroxenite and norite/gabbro. Superimposed on the gross layering are at least three major cyclic units and an undetermined number of smaller cyclic units; the cyclic units are from 10 m to several hundreds of metres thick, and record initial deposition of olivine and chrome spinel, followed by pyroxene, then plagioclase. Cryptic variation is exhibited by the major cumulus phases: olivine Fo_{93-79}, clinopyroxene Wo_{44-34} En_{51-52} Fs_{4-14}, orthopyroxene Wo_{3-4} En_{87-74} Fs_{10-23}, chrome spinel (63–32% Cr_2O_3), and plagioclase An_{92-63}, but is limited and irregular compared with layered stratiform intrusions (Jaques 1981). Magnetite crystallized late in the sequence and is commonly associated with interstitial quartz. Although igneous layering and structures and cumulus textures are abundant, most rocks show textural and mineralogical evidence of subsolidus re-equilibration. The layered cumulates show simple differentiation trends from Mg-rich (dunite) to CaO-rich (wehrlite) to Fe-rich and Al_2O_3-rich gabbroic cumulates. REE abundances are similar to those of the PUB gabbros and imply a parent magma strongly depleted in LREE, and with lower abundances of REE, Ti, Zr and Y than MORB (Jaques *et al.* 1983). The entire sequence is intruded by gabbro pegmatite dykes, and the uppermost gabbros are intruded by rare micro-gabbro and quartz dolerite.

The plutonic rocks of the Marum complex form a broadly folded thrust sheet, more than 4 km thick, which rests on a subsidiary thrust sheet of spilitic pillow lava, lava breccia, hyaloclastite and argillite up to 1 km thick (Section CD in Fig. 4; Tumu River Basalt). The pillow lavas range from magnesian tholeiites to highly differentiated ferrobasalts, and are enriched in LREE, Ti, Zr, and other 'incompatible' elements similar to 'transitional' MORB (Jaques *et al.* 1978). Although originally mapped as part of the Marum ophiolite, marked differences in mineralogy and chemistry argue against any genetic relationship between the Tumu River basalt and the plutonic rocks.

The thrust sheets overlie low-grade metasediments, mostly dark calcareous shale, siltstone and limestone (Asai Shale), of late Cretaceous to Eocene age. In the southeast and east, both the ophiolite and the Asai Shale are faulted against graphitic slate and phyllite of the Mesozoic Goroka Formation.

The plutonic rocks crystallized in the late Mesozoic or, possibly, earliest Tertiary, and the Tumu River basalt probably in the Eocene (K-Ar age and age of associated microfauna, respectively; Jaques in press). The complex was emplaced after the middle Eocene and before the middle Miocene, as indicated by relationships with country rock; emplacement in the late Oligocene–early Miocene was favoured by Jaques & Robinson (1977).

April ultramafics

The April ultramafics include a number of bodies of ultramafic rock which lie within the foothills and mountains south of the Sepik plains in western Papua New Guinea, (Fig. 1; Dow *et al.* 1972; Davies 1982a; Davies & Hutchinson 1982). The bodies range in area up to 100 km^2 and some appear to be sheet-like and shallow-dipping; maximum thickness may be 2–3 km.

The composition of the smaller bodies is masked by serpentinization. The larger bodies are typically unaltered, strongly deformed lherzolite, with some textural evidence of igneous cumulate origin. Minor layered gabbro is associated with some of the ultramafics, but there are no complete sequences which include major units of gabbro and basalt. Nor is there any recognizable refractory tectonite harzburgite or dunite.

The mineralogy of the lherzolites compares with that of the ultramafic cumulates of the Marum complex. Pyroxenes and olivines have 100 Mg/(Mg + Fe) ratios in the range 85–88, and pyroxenes generally contain between 1.5 and 3.0% Al_2O_3 (H.L. Davies, in progress).

The ultramafic bodies are thought to have originated as thrust sheets, and to have been modified by later tectonism, including E–W left-

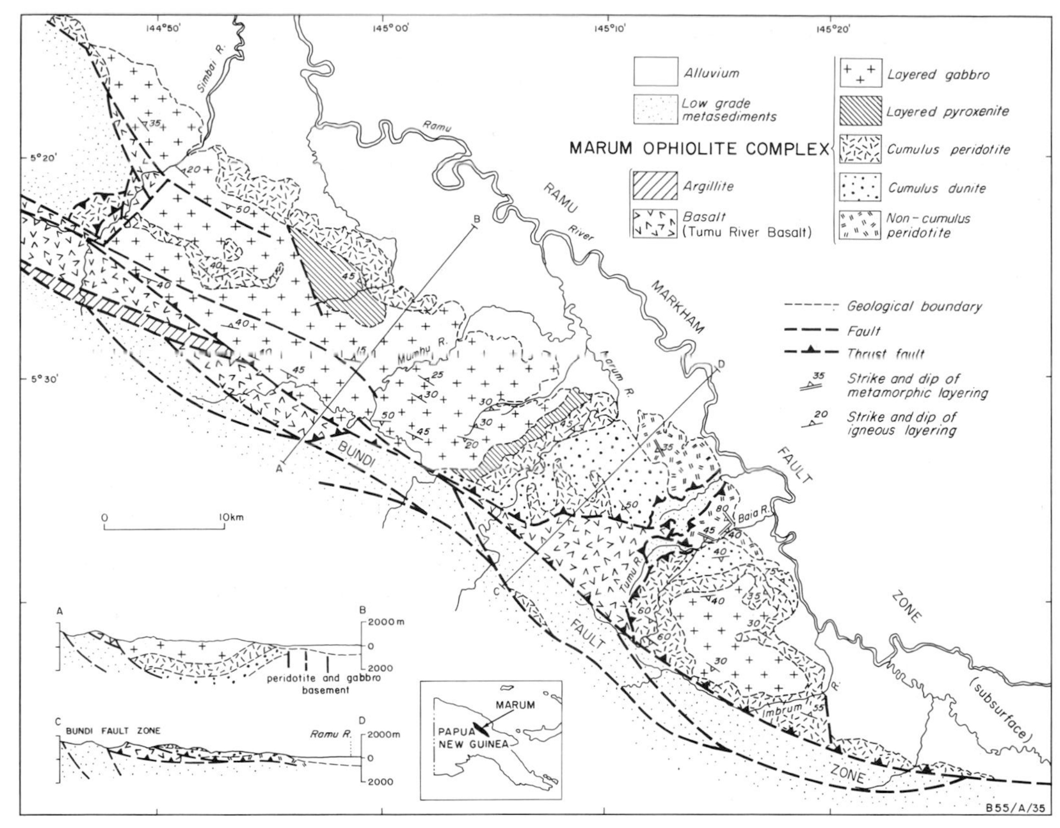

FIG. 4. Marum ophiolite (after Jacques 1981).

lateral strike-slip faulting. They abut against, overlie and underlie metamorphosed and unmetamorphosed volcanics and sediments of diverse lithologic character in an association which is here termed the Sepik complex (Davies 1982b). The metamorphics include discrete developments of both high-*P/T* (eclogite and glaucophane-epidote rocks; Ryburn 1976), and normal-*P/T* (amphibolite and greenschist) facies. The youngest recognized protolith is Eocene and the isotopic age of metamorphic minerals is Oligocene (Page 1976).

Emplacement

The ophiolite complexes lie on the deformed outer margin of the Australian craton, or of a rifted segment of Australian craton in the case of the PUB (Davies & Smith 1971, Fig. 3; Drummond *et al.* 1979), and are flanked externally by Palaeogene volcanic arcs. This common configuration suggests that all of the ophiolites were emplaced by the one process and in similar tectonic environment. We conclude that they were emplaced by arc–continent collisions. In each case the subduction system is thought to have dipped away from the Australian craton, because the ophiolites structurally overlie high *P/T* metamorphics (inferred subduction complex) at the craton margin, and dip towards Palaeogene island-arc sequences. The ophiolites are believed to represent segments of oceanic forearc basement. Details of collision-emplacement models are given in Davies (1976, 1980a) and Jaques & Robinson (1977).

The structural setting is clearest in the case of the PUB: surface geological mapping and seismic refraction and gravity data indicate that the ophiolite dips away from the craton, is more or less continuous with oceanic crust and mantle of the Solomon Sea, and is partly underlain by low-velocity material (subducted low-density crust?; Finlayson *et al.* 1976a, 1976b, 1977). Subduction-related Eocene volcanic arc andesites and tonalite intrusives (Davies 1971; Jaques & Chappell 1980) are exposed in the northern part of the complex, and possibly in the extreme southeast (high-Mg andesites: Dallwitz *et al.* 1966; Jenner 1981; Walker & McDougall 1982), and may extend offshore (Davies 1980a).

The structural setting of the Marum complex is similar. The ophiolite dips northeastward, away from the craton and towards the Oligocene–early Miocene island-arc-type volcanics of the Adelbert and Finisterre Ranges (Jaques & Robinson 1977; Milson 1981; this volume). The lack of any pronounced metamorphic aureole beneath the ophiolite indicates that the complex may have been displaced by later fault movements, possibly from an initial position above the Goroka metamorphics.

The structural setting of the April ultramafics is more complex: ultramafic thrust sheets are interleaved with weakly and moderately to strongly metamorphosed sediments and volcanics in what appears to be a kilometres-thick composite thrust sheet, here termed the Sepik complex. The complex probably formed by tectonic stacking in an Eocene subduction system, followed by further tectonism during Oligocene arc–continent collision, and continued convergence after collision (Davies 1982a). Eocene arc volcanics which presumably once lay to the north of the subduction system now form part of the complex. Oligo-Miocene volcanics which now lie to the north of the complex are thought to be part of the later Finisterre volcanic arc, which was accreted by collision in the earliest Miocene (Davies 1982b); this was presumably the same event which emplaced the Marum ophiolite.

Since emplacement of the Sepik complex, continuing NE–SW convergence has been accommodated partly by movement on E–W left-lateral strike-slip faults within the complex, and partly by compressive fracturing of the crust, and folding and thrust-faulting of superjacent sediments, to the south of the complex, on the southern slopes of the main cordillera (foreland thrust belt; Fig. 1; Davies 1982b, in prep.: cf. Jenkins 1974). Compressive fracturing of the crust is indicated by fault-bounded basement uplift and by shallow seismicity with NE–SW horizontal compressive character (Ripper & McCue, 1983).

Emo metabasite thrust sheet

We now discuss the significance of the metamorphosed mafic rocks which underlie the PUB ophiolite, the Emo metamorphics of Pieters (1978). These rocks appear to form a gently dipping slab, at least 1 km thick, 130 km long and at least 30 km across, and to be interposed between ophiolite above and felsic Kagi metamorphics below. Metamorphic grade is greenschist with lawsonite and some glaucophane (Pieters 1978). We endorse Pieters' interpretation that the mafic metamorphics are probably a thrust sheet, and here seek to develop a model for emplacement of the ophiolite which takes this into account.

One of us had previously mapped variably metamorphosed mafic rocks within the faulted zone which separates the ophiolite from the main mass of Emo and Kagi metamorphics (the area between the Timeno and Owen Stanley fault systems; Davies 1971), and had concluded that these were part contact metamorphic aureole, and part basalt from the top layer of the ophiolite which had been drawn down into the imbricate zone during thrusting (Davies 1971, p. 23). We do not think that these processes are adequate to explain the development of a thrust sheet of the size and character of the Emo metabasite sheet.

We propose an alternative model, which is illustrated in Fig. 5. The model is constrained by the following:

(i) The present-day configuration of crust and mantle are defined, within broad limits, by surface geological mapping, seismic refraction and gravity data (Finlayson *et al.* 1977, Fig. 9; Davies 1980a, Fig. 5).

(ii) The Emo thrust sheet, and much or most of the metamorphic foliation within the sheet, has shallow to moderate easterly and northeasterly dip.

(iii) Metamorphic mineral assemblages indicate that the Emo rocks have been subjected to pressures in the range 4–8 kb, and temperatures of 300–450°C (based on estimates of P and T for lower greenschist facies, and on stability field of lawsonite).

In the model illustrated in Fig. 5, we assume that the Emo metabasite thrust sheet had an initial NE–SW dimension much greater than 30 km, and that Pieters' (1978) estimates of thickness, at 0.8–1.2 km, are conservative. We also assume that the shallow to moderate dipping metamorphic foliation, and the associated metamorphic minerals, were produced by underthrusting in a subduction system, to depths of 15–25 km below surface.

The illustration (Fig. 5) shows NE-dipping subduction of the Australian plate in the late Cretaceous and early Palaeogene. At some time before the Australian craton reached the subduction zone, and possibly as it reached the outer swell, a secondary rupture developed on the approximate line of the continent–ocean crust boundary. As subduction proceeded, in the Palaeogene, the oceanic crust (Emo) was subducted and metamorphosed, and was in turn underthrust by continental crust. The low-density continental crust was subducted at a shallow angle, and eventually caused the subduction system to fail. At this time convergence was taken

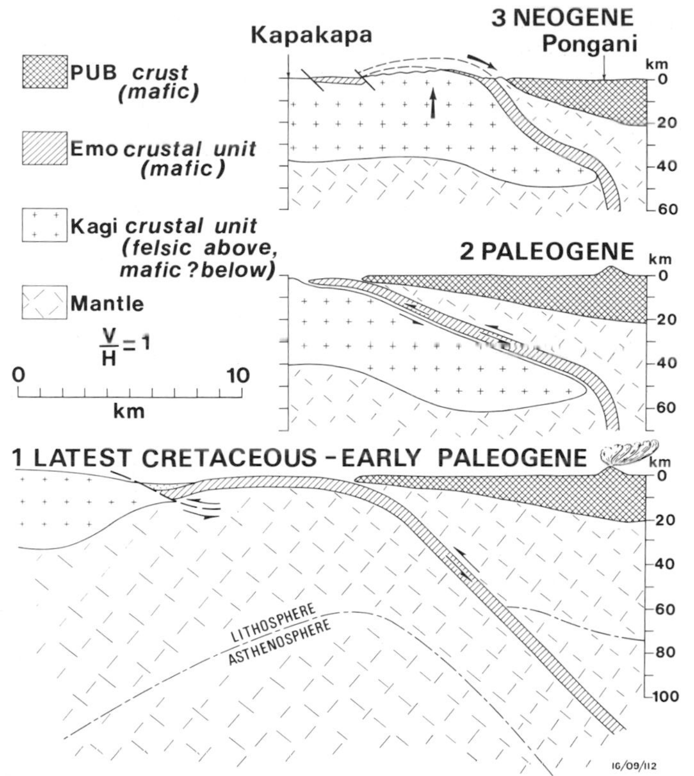

FIG. 5. Cross-sections: emplacement of Emo metabasite thrust sheet and PUB ophiolite.

up elsewhere, possibly in the New Britain arc (Fig. 1) and the Papuan area became passive or extensional. Relaxation of compressive stress permitted the buoyant emergence of the subducted continental crust. Emergence was accompanied by reversed movement of the PUB thrust sheet through a distance of 30–60 km, presumably driven by gravity. (Reversed movement is predicated, rather than erosion, because no significant record of PUB erosion products has been found in nearby Neogene sediments.)

Discussion

The essence of the above model is that a secondary and sympathetic rupture of the crust must have developed in the downgoing plate at some distance from the subduction zone, and that limited shallow subduction of continental lithosphere ensued, starting from this rupture. The rupture may have started at the boundary between continental and oceanic crust, as the continent approached the outer swell of the subduction system.

Nicolas & Le Pichon (1980) have reviewed the stress field within subducting oceanic crust, as indicated by theoretical considerations and observed seismicity, and have deduced that a secondary rupture could develop, in response to bending and compression of elastic upper lithosphere. They predict that compressive stress at the base of elastic upper lithosphere, immediately seaward of the trench, will generate shears which, in the right circumstances, might propagate upward and seaward to reach the surface in the region of shallow compressional stress seaward of the outer swell. They go on to note that multiple thrusting of oceanic crust, by this process, could be a first step in the eventual emplacement of ophiolite by arc–continent collision.

Examples of metabasite thrust sheets beneath ophiolite are known from other parts of Papua New Guinea. One example is the Tumu River Basalt thrust sheet beneath the Marum complex. Another is the metabasalt thrust sheet beneath the southeasternmost part of the PUB ophiolite (Suckling-Dayman massif: Davies 1980b). And another is the mafic amphibolite gneiss which underlies ultramafic rocks in the D'Entrecasteaux Islands (Figs 1 & 6; Mebulibuli Peninsula: Davies & Ives 1965; Davies 1973). The Suckling-Dayman massif is also a possible example of secondary

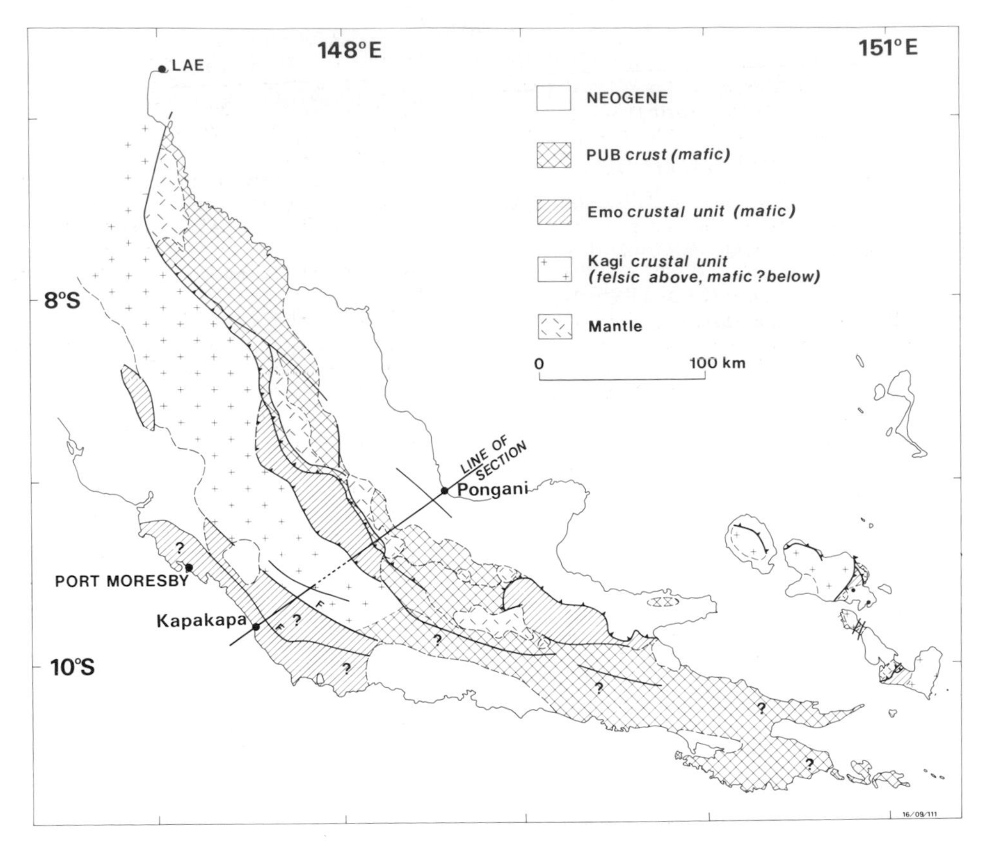

FIG. 6. Interpretative map of Papuan peninsula shows possible extent of Emo and PUB allochthons.

rupture, or 'double subduction', associated with ophiolite emplacement (Fig. 6 in Davies 1980b).

Conclusions

Some of our conclusions regarding emplacement of ophiolites in Papua New Guinea may have more general application:

(i) The emplacement of the PUB ophiolite appears to have required the development of a secondary rupture or sympathetic thrust fault in the downgoing plate, parallel to and at some distance from the subduction system, before continental crust had entered the subduction system. The secondary thrust led to emplacement of a thrust sheet of (metamorphosed) ocean crust between the ophiolite and the underlying continental crust.

(ii) After emplacement, the PUB ophiolite moved through 30–60 km in reverse direction, to expose the subducted substratum. This movement may have been driven by buoyant emergence of the subducted substratum and presumably required a tensional environment. Emergence of the folded thrust plane in the Suckling-Dayman massif, similarly, required reversed movement of the PUB ophiolite (Davies 1980b). Other domal or anticlinal 'metamorphic core complexes' (Davis & Coney 1979) may have similar origin.

(iii) Tension after collision permits development of simple structure, with ophiolite dipping off formerly subducted metamorphics—e.g. the Papuan peninsula. Continued convergence after collision leads to complex structure of interleaved thrust sheets of ophiolite and metamorphosed and unmetamorphosed volcanics and sediments—e.g. western PNG.

ACKNOWLEDGMENTS: This paper is published by permission of the Director, Bureau of Mineral Resources.

References

AUBOUIN, J. 1976. Pacific and Mediterranean tectonics: An attempt at comparison. *25th Internat. Geol. Congress, Sydney* (verbal presentation).

D'ADDARIO, G. W., DOW, D. B. & SWOBODA, R. 1976. Geology of Papua New Guinea 1:2 500 000. *Bur. Miner. Resour. Aust.*

DALLWITZ, W. B., GREEN, D. H. & THOMPSON, J. E. 1966. Clinoenstatite in volcanic rock from the Cape Vogel area. Papua. *J. Petrol.* **7**, 375–403.

DAVIES, H. L. 1968. Papuan Ultramafic Belt. *Proc. 23rd Internat. Geol. Congress, Prague* **1**, 209–20.

—— 1971. Peridotite-gabbro-basalt complex in eastern Papua: an overthrust plate of oceanic mantle and crust. *Bur. Miner. Resour. Aust. Bull.* **128**, 48 pp.

—— 1973. Fergusson Island, Papua New Guinea—1:250 000 Geological Series. *Bur. Miner. Resour. Aust. Explan. Notes*, 25 pp.

—— 1976. Papua New Guinea ophiolites. *25th Internat. Geol. Congress, Sydney, Excursion Guide 52A*, 13 pp.

——1982a. Mianmin—1:250 000 geological series. *Geol. Surv. Papua New Guinea Explan. Notes*, 44 pp.

—— 1980a. Crustal structure and emplacement of ophiolite in southeastern Papua New Guinea. Association mafiques ultra-mafiques dans les orogenes. *Colloques Internationaux du CNRS* **272**, 17–33.

—— 1980b. Folded thrust fault and associated metamorphics in the Suckling-Dayman massif, Papua New Guinea. *Amer. J. Sci.* **280-A**, 171–91.

—— 1982b. The Papua New Guinea thrust belt, longitude 141°–144° East (abstr.). *In*: *Abstracts 11th BMR Symposium. Bur. Miner. Resour. Aust. Rec.* 1982/3, 24.

—— & HUTCHINSON, D. S. 1982. Ambunti—1:250 000 geological series. *Geol. Sur. Papua New Guinea Explan. Notes*, 48 pp.

—— & IVES, D. J. 1965. Geology of Fergusson and Goodenough Islands, Papua. *Bur. Miner. Resour. Aust. Rep.* **82**, 75 pp.

—— & SMITH, I. E. 1971. Geology of eastern Papua. *Bull. Geol. Soc. Am.* **82**, 8299–312.

DAVIS, G. H. & CONEY, P. J. 1979. Geologic development of the Cordilleran metamorphic core complexes. *Geology* **7**, 120–4.

DRUMMOND, B. J., COLLINS, C. D. N. & GIBSON, G. 1979. The crustal structure of the Gulf of Papua and northwest Coral Sea. *BMR J. Aust. Geol. Geophys.* **4**, 341–51.

DOW, D. B., SMIT, J. A. J., BAIN, J. H. C. & RYBURN, R. J. 1972. Geology of the South Sepik region. *Bur. Miner. Resour. Aust. Bull.* **133**.

ENGLAND, R. N. & DAVIES, H. L. 1973. Mineralogy of ultramafic cumulates and tectonites from eastern Papua. *Earth planet. Sci. Lett.* **17**, 416–25.

FINLAYSON, D. M., DRUMMOND, B. J., COLLINS, C. D. N. & CONNELLY, J. B. 1976a. Crustal structure under the Mount Lamington region of Papua New Guinea. *In*: JOHNSON, R. W. (ed.) *Volcanism in Australasia*. Elsevier, Amsterdam, pp. 259–74.

——, DRUMMOND, B. J., COLLINS, C. D. N. & CONNELLY, J. B. 1977. Crustal structure in the region of the Papuan Ultramafic Belt. *Phys. Earth Planet. Inter.* **14**, 13–29.

——, MUIRHEAD, K. J., WEBB, J. P., GIBSON, G., FURUMOTO, A. S., COOKE, R. J. S. & RUSSELL, A. J. 1976b. Seismic investigation of the Papuan Ultramafic Belt. *Geophys. J. Roy. Astron. Soc.* **44**, 45–60.

JAQUES, A. L. 1981. Petrology and petrogenesis of cumulate peridotites and gabbros from the Marum ophiolite complex, northern papua New Guinea. *J. Petrol.* **22**, 1–40.

——, in press. Ophiolites of Papua New Guinea. In: BOGDANOV, N. (ed.) *IGCP Project 39 Ophiolites.*

—— & CHAPPELL, B. W. 1980. Petrology and trace element geochemistry of the Papuan Ultramafic Belt. *Contrib. Miner. Petrol.* **75**, 55–70.

—— & ROBINSON, G. P. 1977. The continent/island-arc collision in northern Papua New Guinea. *BMR J. Aust. Geol. Geophys.* **2**, 289–303.

——, —— & TAYLOR, S. R. 1978. Geochemistry of LIL-element enriched tholeiites from the Marum ophiolite complex, northern Papua New Guinea. *BMR J. Aust. Geol. Geophys.* **3**, 297–310.

——, —— & —— 1983. Geochemistry of cumulus peridotites and gabbros from the Marum ophiolite complex, northern Papua New Guinea. *Contrib. Miner. Petrol.* **82**, 154–64.

JENKINS, D. A. L. 1974. Detachment tectonics in Western Papua New Guinea. *Bull. Geol. Soc. Am.* **85**, 533–48.

JENNER, G. A. 1981. Geochemistry of high-Mg andesites from Cape Vogel, Papua New Guinea. *Chem. Geol.* **33**, 307–32.

MILSOM, J. S. 1981. Neogene thrust emplacement of a frontal arc in New Guinea. *In*: MCCLAY, K. R. & PRICE, N. J. (eds) *Thrust and Nappe Tectonics*. Spec. Publ. Geol. Soc. London **9**, pp. 417–24.

NICOLAS, A. & LE PICHON, X. 1980. Thrusting of young lithosphere in subduction zones with special reference to structures in ophiolitic peridotites. *Earth planet. Sci. Lett.* **46**, 397–406.

PAGE, R. W. 1976. Geochronology of igneous and metamorphic rocks in the New Guinea highlands. *Bur. Miner. Resour. Aust. Bull.* **162**, 117 pp.

PIETERS, P. E. 1978. Port Moresby-Kalo-Aroa, Papua New Guinea—1:250 000 Geological Series. *Bur. Miner. Resour. Aust. Explan. Notes*, 55 pp.

RIPPER, I. D. & MCCUE, K. F., 1983. The seismic zone of the Papuan Fold Belt. *BMR J. Aust. Geol. Geophys.* **8**, 147–56.

RYBURN, R. J. 1976. Median tectonic line in Papua New Guinea: a continent–island arc collision suture (Abstr.). *Bur. Miner. Resour. Aust. Rec.* 1976/35.

SMITH, I. E. & DAVIES, H. L. 1976. Geology of the southeast Papuan mainland. *Bur. Miner. Resour. Aust. Bull.* **165**, 1–72.

WALKER, D. A. & MCDOUGALL, I. 1982. $^{40}Ar/^{39}Ar$ and K-Ar dating of altered glassy volcanic rocks: the Dabi Volcanics, P.N.G. *Geochim. Cosmochim. Acta* **46**, 2181–90.

H. L. DAVIES & A. L. JAQUES, Bureau of Mineral Resources, PO Box 378, Canberra City, ACT 2601, Australia.

The gravity field of the Marum ophiolite complex, Papua New Guinea

J. Milsom

SUMMARY: The ophiolitic Marum complex crops out in northern New Guinea and to the south is in fault contact with sialic rocks of the continental core of the island. It is overlain to the north by thick sediments of the Ramu basin which give rise to a major gravity low.

Gravity surveys over the complex and its margins have defined a Bouguer anomaly high offset towards the northern edge of the outcrops of basic rock. The anomaly is similar in form to, but much smaller than, the anomaly associated with the Papuan Ultramafic Belt to the east. Also, it is superimposed on a very much lower background field, so that positive Bouguer anomaly values occur only at the very peak of the high. A geological model of the area consisting of only three major units; continental crust, Neogene sediments and a heavily faulted island-arc frontal zone of mainly ultramafic rocks, is compatible with the gravity data. Other interpretations which are gravitationally acceptable are less easily accomodated to the geological constraints. A problem still exists in explaining the 'thin-skin' overthrust in the southeastern part of the complex.

Since the Australian continental mass broke away from Antarctica and began its long drift north in the late Cretaceous, the interactions between the northern continental margin and the westward-moving Pacific plate have been absorbed in the New Guinea region. At the present time, these interactions are causing the fragmentation of the borderland into a number of small ocean basins and slices of continental crust, but during and even before the Tertiary a number of collisions occurred between island-arc systems and the continental margin. One consequence of these collisions has been that large masses of ultramafic and associated mafic rocks have been emplaced along the northern and eastern flanks of the main orogenic belt. These complexes have been best mapped and described in eastern New Guinea (Dow *et al.* 1972); they also occur in the western part of the island (Visser & Hermes 1962) but have not been studied there in any detail.

The most easterly body, the Papuan Ultramafic Belt, is also the largest (Fig. 1). The full sequence of ophiolite rock types is developed, from basal peridotites through gabbros to basalts with intercalated deep-sea limestones and cherts (Davies 1971). Sheeted-dyke complexes are not well exposed but have been seen in some places (Davies 1980). The quite extensive geological and geophysical information now available all points to the belt having been emplaced by thrusting from the north or east (cf. Milsom 1973; Finlayson *et al.* 1977), although the form of the overthrust has been modified by later normal and transcurrent faulting (cf. Davies & Smith 1971).

In central New Guinea, ultramafic thrust sheets (April Ultramafics; Dow *et al.* 1972) are found scattered along the northern slopes of the central mountain range. Outcrops are generally elongated parallel to the WNW regional trends and vary in length from a few metres to more than 50 km. The smaller masses have been intensely sheared and serpentinized whereas the larger ones have less regular margins and consist almost entirely of unaltered peridotite and dunite. Mafic rocks do occur, commonly as small blocks of gabbro at the margins of the main ultramafic outcrops. They form too small a component of the total mass to allow the term ophiolite, which implies an association, to be used.

Ryburn (1976) has mapped blueschists to the north of the April Ultramafics which, he notes, are very similar to those associated with the large ultramafic overthrust on New Caledonia; he considers them to be the metamorphosed equivalents of the gabbroic and basaltic rocks that outcrop to the south. The blueschists are separated from low-pressure, high-temperature metamorphic rocks further north by a major suture, the Frieda Fault. If the April Ultramafics do represent one part of an ophiolite, then much of the missing mafic component must presumably have suffered metamorphism and fault displacement.

Marum complex

The Marum complex lies both geographically and geologically midway between the coherent Papuan Ultramafic Belt and the much-dismembered April Ultramafics (Fig. 1). Both mafic and ultramafic plutonic components are present and have been further subdivided into a number of well-defined cumulus and non-cumulus types (Jaques 1981). However, the only basaltic rocks present (Tumu River Basalts) have been overthrust by the plutonic rocks and are chemically quite distinct from them, having more in common with tholeiites from oceanic islands and aseismic ridges (Jaques *et al.* 1978). Tuffaceous argillites

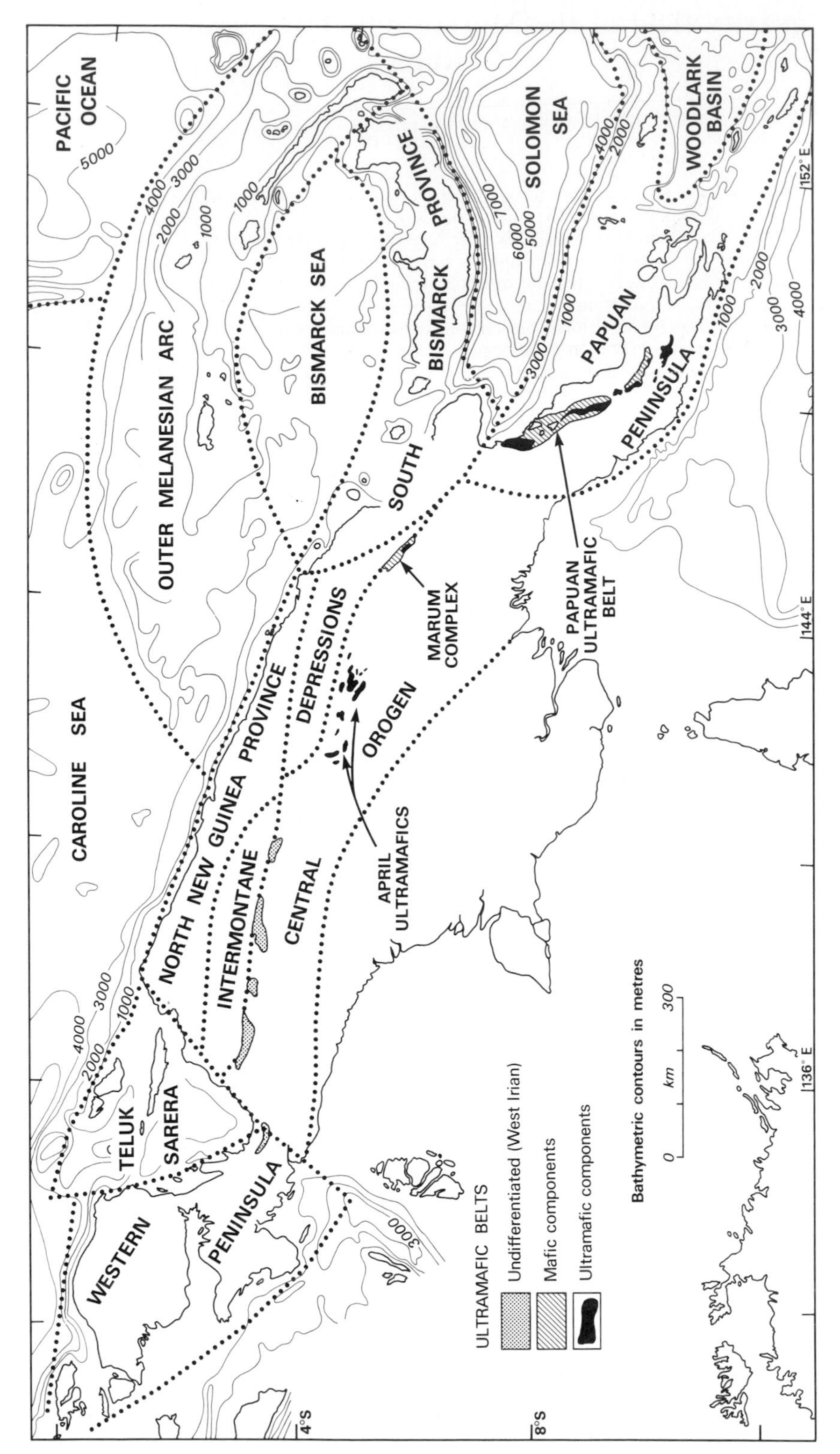

FIG. 1. Main geological subdivisions of the New Guinea mobile belt, showing the locations of the major ophiolites.

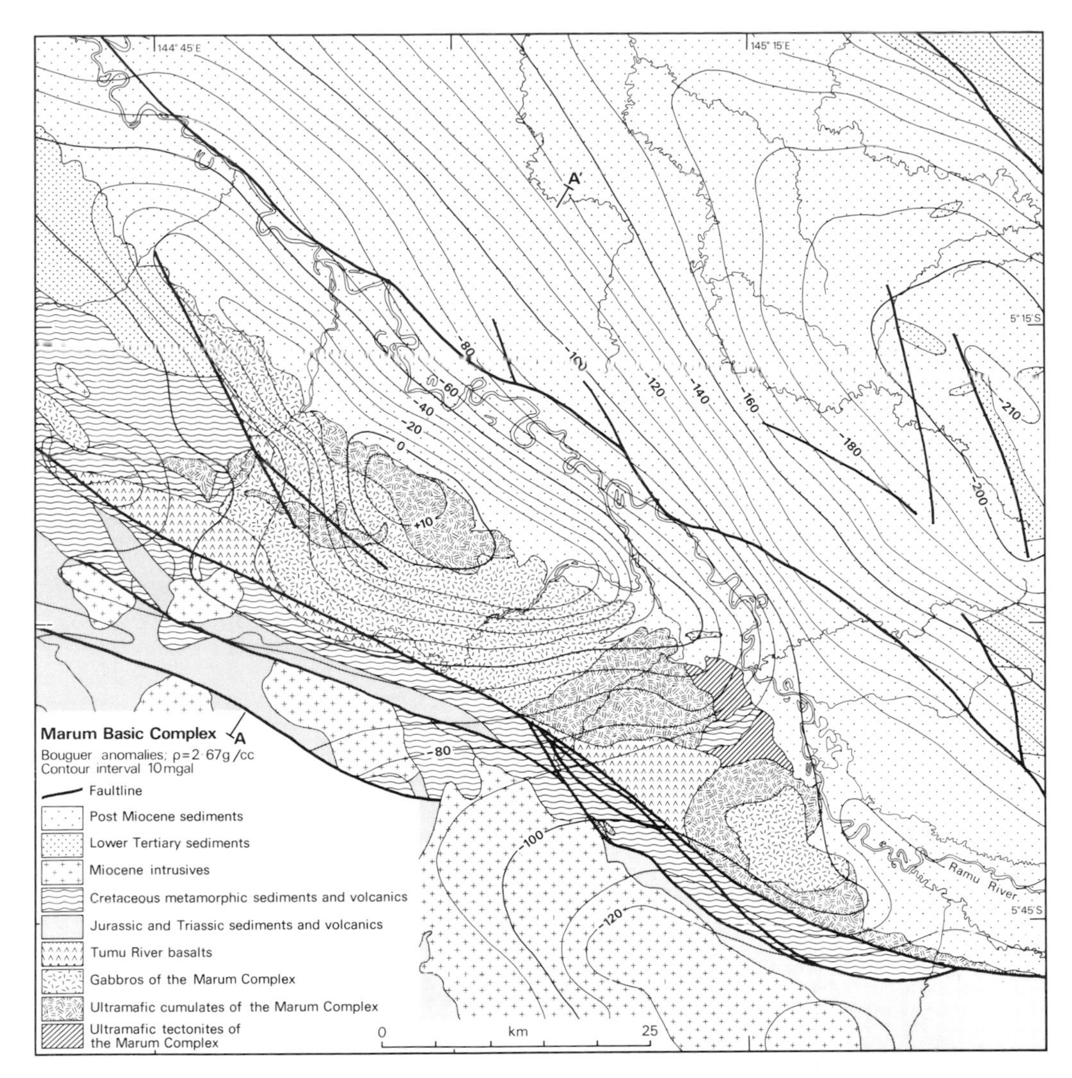

FIG. 2. Geology and simple Bouguer anomalies in the Marum area. Geology after Jaques *et al.* (1978).

and greywackes that overlie the basalts contain abundant detritus derived from basaltic to andesitic volcanics (A.L. Jaques, pers. comm. 1982).

Direct evidence for arc-related rocks in the Marum area is thus not strong, and such as it is refers mainly to volcanogenic sediments which seem to form part of the underthrust plate. Nevertheless, it is generally assumed that all the various ultramafic bodies along the northern flank of the central ranges in New Guinea have a common origin, and the most generally favoured hypothesis is that they represent segments of a former island arc. In no case is the time of emplacement well controlled, but at the Marum and in most other areas it must lie somewhere between the late Eocene and the early Middle Miocene (Dow 1977). Information from the Marum complex should assist in understanding the transition, which seems to be common, between coherent ophiolites and fragmentary ultramafic sheets.

Gravity surveys

Interpretations of gravity surveys carried out by the Australian Bureau of Mineral Resources in the vicinity of the Papuan Ultramafic Belt were made both before (Milsom 1973) and after (Finlayson *et al.* 1977) seismic-refraction information became available. Very large gravity anomalies are associated with the belt, offset towards the Solomon Sea to give a pattern almost universally interpreted as indicating thrusting from that direction. Neither the April Ultramafics nor the ultramafics in western New Guinea have been systematically covered by gravity surveys, but there are indications that large positive anomalies are associated with some of the major outcrops (Visser & Hermes 1962).

In 1968, the Bureau of Mineral Resources carried out a gravity survey of much of the Papua New Guinea intermontane depression (Fig. 1). There were only a few readings in the Marum Complex outcrop area, but a large gravity high was defined, extending out across the Neogene sediments of the Ramu valley (Watts 1969). In 1974 the Geological Survey of Papua New Guinea added stations in the mountainous and heavily forested areas further south (Milsom 1975). Combination of the results has allowed the Marum gravity high to be more completely defined.

In so far as was possible, gravity station sites were selected so that local terrain corrections were minimized, but even so it is probable that up to 10 mgal of correction should be added to stations away from the Ramu valley. Uncertainties in the station heights, which were determined barometrically, may lead to additional errors of 2 or 3 mgal. Regional terrain effects may also be large but would produce only a relatively slowly varying background to the more local gravity anomalies. The available topographic maps do not allow accurate computation of the corrections but fortunately the gravity features of interest are sufficiently large for useful information to be obtained from the simple Bouguer anomalies.

Bouguer anomaly map

The main features of the Bouguer anomaly map (Fig. 2) are the Marum Complex 'high', the 'low' to the south associated with the thickening crustal root beneath the central ranges and the 'low' to the north due to the deep Neogene sedimentary basin in the Ramu valley. The combined effect of the two mass deficits is a regional low of such a size that positive Bouguer anomaly levels are reached only at the extreme peak of the Marum high.

Gravitationally, the Marum complex is divided into two distinct zones. Rather more than half, in the NW and centre, can be associated with the Bouguer anomaly high which has a peak-to-base amplitude of about 80 mgal and which is offset to the NE of the axis of the outcrop of the complex. A similar offset has been noted in the case of the Papuan Ultramafic Belt (Milsom 1973). Anomaly levels over the SE part of the complex, however, seem to be barely affected by the dense ultramafic rocks seen there in outcrop. The gravitational difference clearly reflects a considerable difference in subsurface geology and this is also implied by the surface mapping. Outcrop in the NW is mainly of gabbro, with some cumulus peridotite exposed near the Bouguer anomaly maximum. A much greater variety of rock types has been mapped in the SE where the basal thrust is exposed in places, enclosing windows of Tumu River Basalts and weakly metamorphosed sediments (Jaques 1981). The quite steep gravity gradient which cuts across the outcrop of the complex correlates reasonably well with a cross-cutting contact between mafic and ultramafic rocks, along which are parallel exposures of narrow belts of peridotite and pyroxenite cumulates. Although no major faults with such a trend have been mapped in this region, their existence is strongly suggested by the gravity data.

Similarly, contrasting gravity patterns are seen associated with the Papuan Ultramafic Belt, where strong Bouguer anomaly highs are found over outcrops in places where the complete ultramafic and mafic sequence is exposed but not where the thrust sheets consist almost entirely of ultramafic rock. In eastern Papua, however, the change between the two environments occurs gradually over a distance of more than 100 km, whereas in the Marum complex it occupies 10 km at the most.

A rather unusual feature of the Marum outcrop pattern is that the rocks closest to the surface trace of the contact with the underlying sediments belong to the higher, mafic, layers, whereas some of the deeper-level ultramafic components are exposed at the NE edge of the complex. Southerly dips have been reported by Jaques (1981) from the SW margin and can explain this pattern, as can complex imbrication. If imbrication is not invoked, and the large gravity anomaly and relatively low degree of serpentinization make this rather the less likely solution, the degree of tilt which must be supposed to have occurred depends on the nature of the original emplacement process.

Figure 3a is a schematic representation of a simple oceanic overthrust model in which, after erosion has occurred, the deepest levels of the ophiolite are found outcropping closest to the thrust. A large subsequent tilt would be needed to produce the outcrop pattern actually seen in the NW part of the Marum area and the source body for any large gravity effect would be removed in the process. In the island-arc model (Fig. 3b), only the mafic rocks would be expected at the

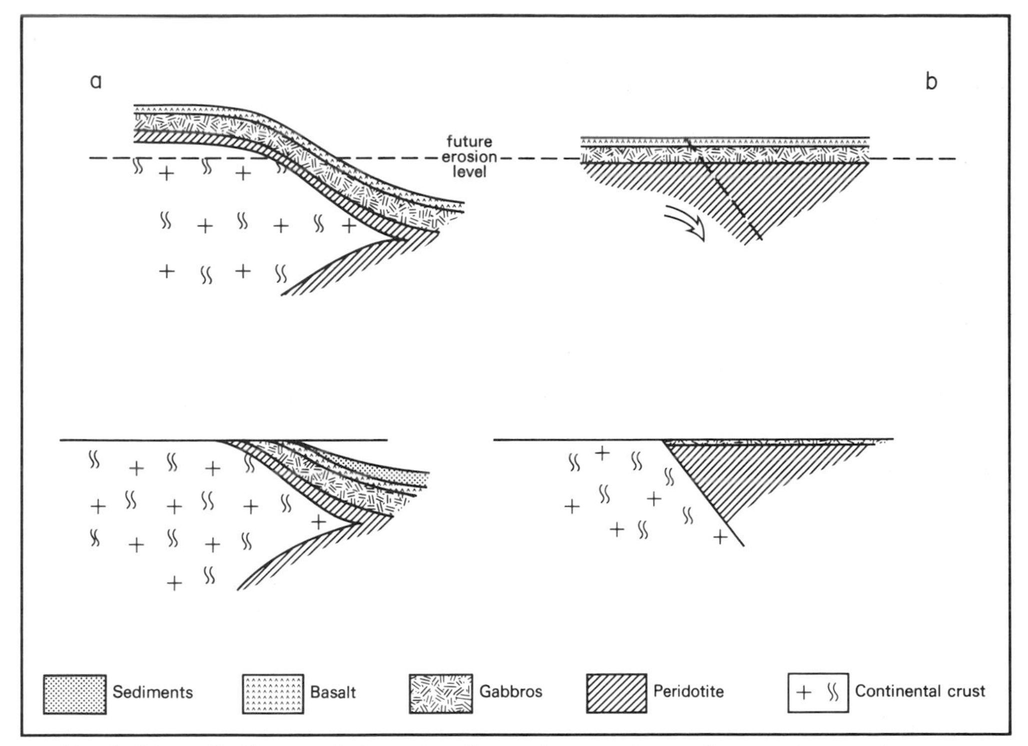

FIG. 3. Schematic diagram of alternatives for emplacement (upper diagram of each pair) and later modification by uplift and erosion (lower diagrams) of ophiolites. Sequence (a) represents the obduction model of Coleman (1971), sequence (b) represents an arc-continent collision.

surface immediately after collision but a relatively small amount of later tilting and erosion would expose the ultramafic layer in places. In the Marum area, such tilting could be a consequence of listric normal faulting and rotation of basement slices at the margin of the tensional Ramu basin.

Bouguer anomaly profile

A profile has been constructed from the Bouguer anomaly contours in the area of the gravity maximum and has been interpreted in terms of strike-limited bodies of polygonal cross-section (Fig. 4). Strike lengths of the order of 20–30 km have been assumed for bodies which form part of the Marum Complex itself and of several hundred kilometres for the Ramu valley sediments. Only a very simple model has been used since the uncertainties in Bouguer anomaly values—due to terrain effects and likely elevation errors—do not allow for any very great precision in interpretation. It is evident that the solution can be found in terms of only two major features, these being a high-density, mainly ultramafic mass which is the subsurface continuation of the outcropping Marum complex, and a deep basin to the NE filled with low-density sediments. In addition, a background level of –60 mgal has been assumed. This corresponds to a crustal thickness some 4 km greater than standard, which seems reasonable for an area at the margin of a major mountain range. Choice of a different background would not greatly alter the density contrasts and rock volumes required in the near-surface parts of the model. The sediments have been assumed to have bulk density contrasts with standard crust of –0.2 and –0.5 g cm^{-3}; the near-surface mafic rocks of the complex have been assigned a contrast of +0.2 g cm^{-3} (again with respect to normal crust), the Tumu River basalt wedge a contrast of +0.05 g cm^{-3} and the upper portion of the ultramafic mass a contrast of +0.35 g cm^{-3}. Downfaulted blocks of the ophiolite have been allocated progressively lower bulk density contrasts, both because they are likely to preserve more of of the lighter upper layers of the complex and also because the density of 'standard' crust may be assumed to increase with depth. The model does not otherwise allow for the possible change in composition of the thrust sheet from oceanic to quasi-continental island-arc, but this could be accommodated without difficulty by reducing the rather implausible thickness of the Neogene sedimentary pile.

Ophiolites and the tectonic evolution of the Arabian Peninsula

R.G. Coleman

SUMMARY: Arabian ophiolites mark periods of important tectonic activity. Their petrotectonic evolution gives insights on movements of the Arabian plate through Proterozoic and Phanerozoic times. The upper Proterozoic ophiolite belts in Saudi Arabia and NE Africa indicate that oceanic crust of this age was incorporated within the continental crust during its Precambrian evolution. Recent studies on the Semail ophiolite in Oman reveal that it was emplaced onto the Arabian plate in the Campanian during the closure of Neo-Tethys. An igneous complex of sheeted dykes, layered gabbro, granophyre, and basaltic-rhyolitic lava on the SE margin of the Red Sea represents new oceanic crust formed some 20–24 m.y. ago during the initial opening of the Red Sea when the continental crust became attenuated.

The distribution of ophiolites in space and time identifies major tectonic events in the development of the Arabian Peninsula over the last 1000 Ma (Fig. 1). The Precambrian ophiolites are related to the late Precambrian continental accretion within the Pan-African mobile belt (Bakor *et al.* 1976). The Oman Mesozoic ophiolites record the history of the Tethyan Ocean and its disappearance (Glennie *et al.* 1974), and the Tertiary ophiolites along the Arabian Red Sea coastal plain near Jizan reveal details of the formation of new oceanic crust during the initial opening of the Red Sea as the Arabian plate separated from Africa in the Oligocene (Coleman *et al.* 1979). Thus, within this relatively small area, ophiolites are part of at least three of the major late Proterozoic and Phanerozoic tectonic events.

Proterozoic ophiolites

The Precambrian Shield of western Arabia consists of a series of island arcs and small ocean basins that developed during the upper Proterozoic. These generally N–S-trending belts have been telescoped and invaded by syntectonic and post-tectonic diorite to granite plutons that effectively cratonized this eastern side of the African plate (Greenwood *et al.* 1976; Delfour 1979; Schmidt *et al.* 1979; Shackleton 1979; Gass 1982).

Recent studies in the past 10 years on the Arabian Shield have confirmed the presence of ophiolites and have attempted to integrate their presence into the evolution of the shield (Al-Shanti & Mitchell 1976; Bakor *et al.* 1976; Garson & Shalaby 1976; Frisch & Al-Shanti 1977; Gass 1977; Delfour 1979; Church 1980; Engel *et al.* 1980; Al-Shanti & Hakim 1981).

The Arabian Shield is generally considered to be a northward extension of the Mozambique belt, and the Proterozoic igneous and tectonic events are usually referred to as the Pan-African event (Kennedy 1964; Shackleton 1979; Gass 1982). The distribution of these ophiolites in north-eastern Africa and Saudi Arabia as compiled by Shackleton (1979) suggests a series of N–S-trending belts that may represent sutures that developed during the Proterozoic cratonization (Fig. 1). The presence of the various ophiolite belts along prominent sutures in Saudi Arabia strongly suggests that during late Proterozoic cratonization small amounts of oceanic crust were incorporated along plate boundaries. The actual nature of the oceanic crust is not clear, but at Jabal Al Wask, Jabal Ess, and Bir Umq the sequences are thought to be more complete by some authors and may be related to back-arc basins or small rifted intracontinental seaways (Bakor *et al.* 1976; Al-Shanti & Roobol 1979; Al-Rehaili & Warden 1980). The extent of the ocean crust actually consumed cannot be established by the presently exposed relationships. It could be that the oceans may have been small and some of the intrusive relationships reported by Kemp (1980) and Kemp *et al.* (1980) represent transitional material along attenuated continental margins. Where I have seen these occurrences the allochthonous relationships are not always clear. The irregularity of the layered gabbros and sheeted-dyke occurrences, along with the association of calc-alkaline volcanics within the upper sections of these mafic-ultramafic complexes, requires proximity to island-arc activity or adjacent mobilized underlying crust. The presence of pelagic sedimentary units interlayered with the mafic volcanics on the other hand, points to an oceanic environment during the development of some ophiolites such as Jabal Ess (Al-Shanti & Roobol 1979). Reconstruction of the Nabitah suture by Schmidt *et al.* (1979) reveals that this suture has important tectonic implications. The ophiolite material must mark a zone of imbrication that formed during late Proterozoic accretion of oceanic crust during Halaban time (700–800 Ma). A U-Pb lead age of 822 ± 12 Ma (Kemp *et al.* 1980) on zircon from a co-genetic plagiogranite from Jabal Wask suggests that these ophiolites formed shortly before they were accreted onto the continental margin during

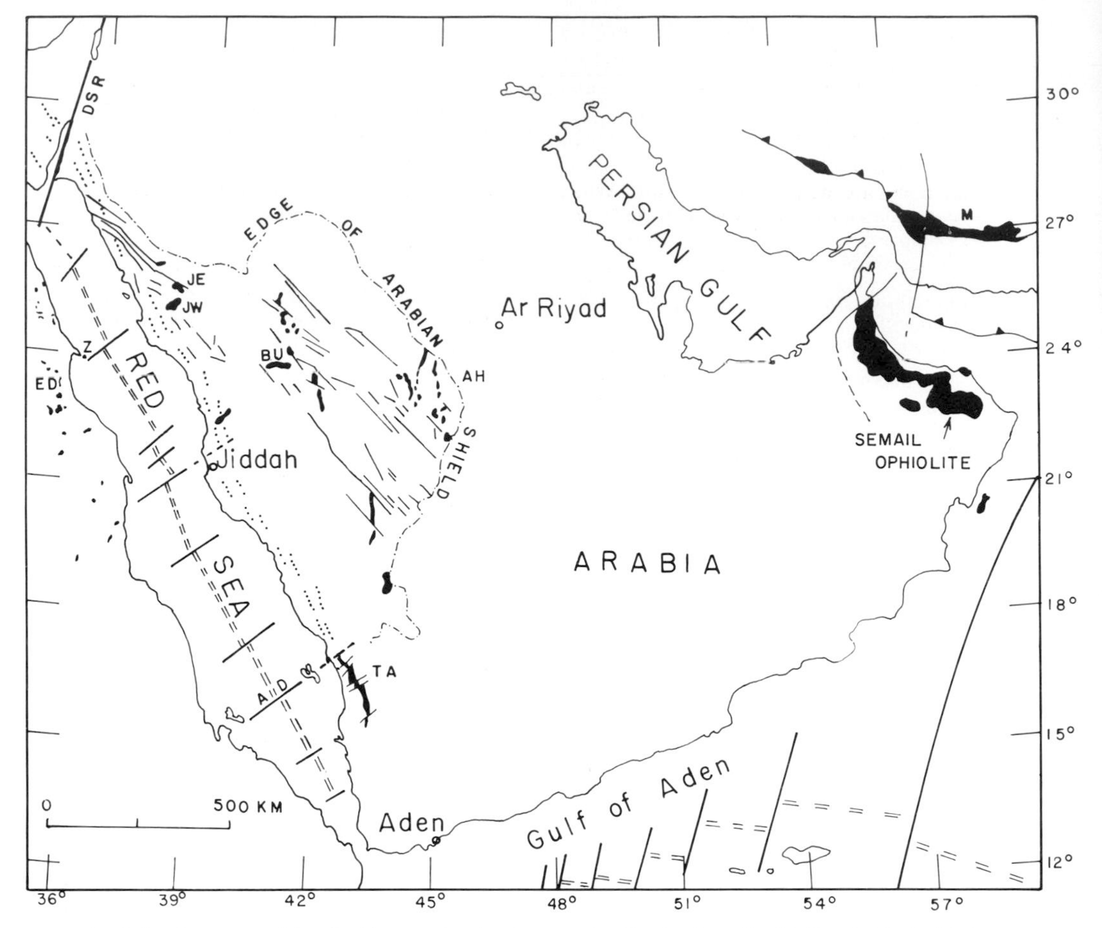

FIG. 1. Outline map of the Arabian peninsula showing the distribution of ophiolites in black, Makran Iran (M), Eastern Desert, Egypt (ED), Jabal Ess (JE), Jabal Wask (JW), Bir Umq (BU), Al-Amar-Idsas and Halaban-Itithal (AH), Zabargad Island (Z), and Tihama Asir Complex (TA). Major faults shown as dark line, Dead Sea Rift (DSR) and Ad Darb transform (AD). Active spreading centres in the Red Sea and Gulf of Aden are shown as double bars between transform faults (solid lines). Subduction trace in the Gulf of Oman shown as a solid line with barbs on the overriding plate, this same ornamentation is shown for the Zagros thrust.

Halaban time (700–800 Ma) as part of west-directed subduction against a small marginal basin. Continued collapse and imbrication of the oceanic crust may have transported the Bir Umq ophiolite further west and inboard of the Nabitah suture. The extensive metamorphism, increasing southward, of the dismembered ophiolites in the Nabatah suture, may be due to a deeper level of erosion and their incorporation into a still active island-arc system.

The Al-Amar-Idsas suture and Halaban-Itithal ophiolite belt represent zones of convergence eastward of the Nabitah suture and may have similar origins. Stacey & Stoeser (1983) report a lead–lead isotopic age of 694 ± 5 Ma on plagioclase from the Urd gabbro in the Halaban-Itithal ophiolite belt. Subparallel arcuate suture zones produced by continental accretion of island arc-back arc basins in the Klamath Mountains of western North America bear a striking resemblance. The ophiolites in the Klamath often lack sheeted dykes and have been metamorphosed by calc-alkaline intrusions developed during subduction (Irwin 1979).

Oman mesozoic ophiolite

The Semail ophiolite is part of an elongate belt in the Middle East that forms an integral part of the Alpine mountain chains marking the northern boundary of the Arabian–African plate (Fig. 1). The mountainous part of Oman extending from the Musandam Peninsula to Ra's Al Hadd forms a mountain chain geologically distinct from the rest of the Arabian peninsula. The Oman Mountains form the eastern margin of the Arabian continental platform and contain great thicknesses of autochthonous shelf carbonate rocks that correlate with those of Saudi Arabia and those of the Zagros foldbelt of Iran.

In the central part of the Oman Mountains, updomed areas have been eroded to expose pre-Permian crystalline basement. Metasedimentary and metavolcanic rocks within the northern and eastern parts of these domes reach greenschist facies, and radiometric data (K/Ar) reveal a Hercynian age for this metamorphic event (327 ± 16 Ma) (Glennie *et al.* 1974). Recent work by Lippard (1983) reveals mid-late Cretaceous blueschist-facies metamorphism overprinting these same basement rocks, and he has suggested that these high-pressure, low-temperature rocks may have formed as a result of NE- directed subduction prior to the Semail ophiolite emplacement.

Overlying the pre-Permian basement rocks are the Middle Permian to Middle Senonian shallow marine shelf carbonate rocks. These autochthonous sedimentary rocks are collectively referred to as the Hajar Supergroup (Glennie *et al.* 1974). A late Cretaceous unconformity is marked by deposition of the Conacian to Campanian Muti formation upon the Hajar Supergroup. To the west, in the northern part of Oman, the Muti grades laterally into the Campanian and Maastrichtian Fiqa and Simsima formations. In some areas, these sedimentary rocks rest unconformably on the Semail ophiolite and the Hawasina nappes. The Maastrichtian Juweiza formation, which forms a unit within the same basin, contains clastic debris derived from both the Semail and Hawasina nappes. These sedimentary rocks all occupy a foredeep on the continental margin (Brown 1972), and their stratigraphic relations suggest that the Semail ophiolite was emplaced during the late Cretaceous (Glennie *et al.* 1974; Graham 1980).

The Hawasina Group is an imbricated assemblage of thin nappes consisting chiefly of pelagic marine sedimentary rocks (Glennie *et al.* 1974). The Hawasina nappes are tectonically sandwiched between the Semail ophiolite nappe above and the autochthonous shelf carbonates of the Hajar Supergroup below. In northern Oman, volcanic rocks (Haybi complex) up to 700 m thick occupy an imbricate zone above the Hawasina nappes (Searle *et al.* 1980). These volcanic rocks are also part of a mélange that consists of cherts and limestones surrounding huge exotic blocks of Permian and Triassic limestones. The lower volcanics (200–230 Ma) are alkaline basalts and their pyroclastic derivatives; whereas, the upper volcanics are tholeiitic pillows interbedded with chert and limestone (Searle *et al.* 1980). The lower Haybi Complex volcanics are interpreted as products of Triassic rifting marking the early development of the Tethys and upper tholeiitic volcanics may represent off-axis volcanic activity of Tethyan spreading (Searle *et al.* 1980).

The Semail ophiolite, now forming the highest nappe of the allochthonous sequence in Oman, is considered to represent Tethyan oceanic crust and upper mantle. Reconstruction of the ophiolite stratigraphy in the Muscat-Ibra area shows a basal peridotite tectonite (9–12 km thick) consisting primarily of harzburgite with abundant dunite (Boudier & Coleman 1981). Discontinuous narrow zones of metamorphic rock form screens between the base of the peridotite and the underlying mélange belt. Right at the basal contact narrow zones ($\leq$2 m) of garnet amphibolite form discontinuous lenses that grade downward into low-grade amphibolite with interlayered quartzites that, in turn, grade into greenschist facies; and finally into unmetamorphosed Hawasina units (Ghent & Stout 1981).

The peridotite tectonite is overlain by cumulus ultramafic rocks and gabbro and exhibit some tectonic fabric in the northern sector (Smewing 1980). The cumulus section (3–5 km) has a narrow basal zone ($<$0.5 km) of olivine + chromite cumulates overlain by several kilometers of three-phase olivine + clinopyroxene + plagioclase cumulates (Pallister & Hopson 1981; Smewing 1981). A gradual transition into high-level gabbro is marked by the virtual disappearance of igneous layering. The high-level gabbro is a thin ($<$1 km) unit of isotropic non-cumulus gabbro with extremely variable textures and proportions of phases. These irregular features are related in part to fractionation of the magma giving rise to the layered gabbro and to the development of discrete bodies of plagiogranite (Pallister & Hopson 1981). These leucocratic bodies may have developed *in situ* either by partial melting of the roof rocks or by extreme differentiation of a hydrous melt (Gregory & Taylor 1981).

A sheeted-dyke complex (1–1.6 km thick) overlies the high-level gabbro and penetrates upward into and locally feeds pillow lava accumulations (Pallister 1981). The dykes are mostly medium- to fine-grained aphyric diabase that has been pervasively altered to zeolite and upper-greenschist-facies assemblages in a downward-increasing intensity of alteration. Most of the sheeted-dyke complex apparently predates the crystallization of the high-level gabbro and plagiogranite; however, at least some of the dykes are intrusive along a horizontal direction and project downwards into the upper part of the gabbro sequence.

Overlying the dyke sequence is a volcanic section consisting of pillowed submarine lavas and sheet flows. These lavas are chiefly microphyric basalts with phenocrystic assemblages that can be related to the same stage of magma fractionation that produce the Semail cumulus gabbro. In northern Oman, three divisions are recognized (Alabaster *et al.* 1980) in the volcanics: (i) Geotimes unit (1–0.5 km thick), the lowest unit, consists of interlayered aphyric pillow basalts; (ii) Lasail unit fractionated volcanics (1 km thick) consisting of basalts, andesites, and rhyolites; (iii) upper Alley unit (0.5 km thick), a fractionated basalt-rhyolite sequence considered to be developed from 'off-axis' eruptions. Copper-bearing massive sulphide deposits are localized between the Geotimes and Lasail units and umbers are present between all of the units and represent hydrothermal products.

Uranium-lead ages on zircons from co-magmatic plagiogranites give a range in age of 93.5–97.9 Ma for the igneous development of the Semail ophiolite (Tilton *et al.* 1981). Radiolarian from chert sequences in the Geotimes or axis unit are Cenomanian, and those from off-axis lavas (Alley and Lasail units) are Cenomanian to earliest Turonian (Tippit *et al.* 1981). K/Ar ages on hornblende from amphibolites at the base of the Semail ophiolite are 90 ± 3.0 Ma and mark the beginning of detachment in the oceanic realm (Lanphere 1981; Ghent & Stout 1981). Plate configurations and palinspastic reconstructions during the late Cretaceous demonstrate that northward subduction and island-arc volcanism in Iran did not commence until after the Semail ophiolite was emplaced during the Campanian (Lanphere 1981). Pearce *et al.* (1981), based on geochemical evidence, suggest that the Semail ophiolite represents a submarine arc-basin complex developed above a short-lived subduction zone dipping northeast. The ocean crust of either back-arc basin or ocean-ridge origin was detached when the African plate began to collide with Eurasia. Still-hot parts of the oceanic crust near or within the late Cretaceous magma chamber and underlying mantle provided a weak zone for detachment of the still-hot oceanic lithosphere (Boudier & Coleman 1981). The detached nappe of oceanic crust was deformed and eroded as it rose above sea level. Later, gravity sliding was initiated southward and the Semail nappe was emplaced onto the now-subsiding Arabian passive margin (Graham 1980), perhaps accompanied by tectonic convergence and underthrusting of the continental margin (Coleman 1981). Formation of blueschists in the Saih Hatat basement and autochthonous sediments may be the result of this ocean–continental crust convergence.

Tertiary ophiolites

Along the eastern coastal plain of the Red Sea near Jizan, Saudi Arabia, the Tertiary Tihama-Asir complex consisting of subvolcanic mafic rocks crops out between Precambrian cratonic rocks of the Arabian shield and younger Tertiary sediments of the coastal plain (Coleman *et al.* 1979). East of Jizan, within the Arabian plate, dykes intrude Mesozoic–Paleozoic sediments and Precambrian metamorphic rocks along a NW trend that parallels the present-day axial trough of the Red Sea. West of the Precambrian crust, new oceanic crust consists of a narrow zone of subparallel dykes sometimes chilled against earlier dykes or screens of volcanic agglomerate, granophyre, and basaltic lavas. Layered gabbros, granophyric masses, and leucocratic dykes may represent products of magmatic differentiation or assimilation developed during the initial stages of rifting. Late Miocene tilting of the sediments and the dyke swarm toward the axial trough and later erosion have exposed this initial continental–oceanic-crust boundary along a narrow zone of the coastal plain. The dyke swarm is exposed discontinuously for about 200 km from the Yemen border northward to Ad Darb where it is terminated by a NE-trending fault that may be the landward extension of a transform fault that offsets the present Red Sea spreading axis. North of the Ad Darb fault, the sheeted dyke swarm gives way to generally isolated dykes of hypabyssal quartz gabbro, quartz monzogabbro, quartz monzonite, and quartz syenite. These dykes are coarse to fine grained and may be aphyric to porphyritic with chilled margins against the surrounding Precambrian country rock; individual dykes extend uninterrupted for more than 50 km.

Blank (1977) has shown that these dykes can be traced by their prominent aeromagnetic lineaments from the Yemen border to the Gulf of Aqabah. These same dykes are present in the Sinai Peninsula where they have been offset by approximately 107 km of sinistral movement along the Gulf of Elat–Dead Sea Rift (Bartov *et al.* 1980). Even though these dykes have the same compositional range as the Tihama-Asir dykes (Blank 1977; Coleman *et al.* 1979; Bartov *et al.* 1980), they rarely form conjugate systems and represent a maximum of only 20% distension of the Precambrian crust, whereas to the south, the Tihama-Asir complex represents almost entirely new oceanic crust. North of Ad Darb and south of Jiddah, there is no evidence for tilting toward the Red Sea axial trough, here the Paleozoic Wajid sandstone is horizontal and the continental dykes are vertical. Thus two distinct igneous manifestations of rifting can be distinguished on the eastern side of the Red Sea: (i) sheeted diabase and rhyolitic dykes with associated centres of layered gabbro and granophyre formed by nearly continuous spreading, (ii) swarms of continental hypabyssal dykes of quartz gabbro to quartz syenite separated by extensive screens of Precambrian rock or volcanic rocks formed penecontemporaneously. Gradation between these two types may have occurred at various stages of the opening of the Red Sea.

The Tihama-Asir complex does not have a classic ophiolite statigraphy. The sheeted dykes form the bulk of the complex and invade Precambrian basement and the overlying Paleozoic–Mesozoic sediments along a NW trend paralleling the axial trough of the Red Sea. Exposures along streams cutting the Tihama-Asir complex show a transition over about 100 m from dykes separated by screens of Precambrian schist and Wajid sandstone to dykes with only screens of basalt, agglomerate, and granophyre. Schmidt *et al.* (1982) have shown that rift-valley volcanism preceded these dyke sequences and that the volcanic products were bimodal associated with freshwater sediments. The basal complex of Fuerteventura in the Canary Islands displays a similar dyke swarm cutting Cretaceous marine sediments and mafic plutonic rocks (Stillman *et al.* 1975). Layered gabbro bodies within the dyke swarm, also invaded by dykes, appear to be contemporaneous and at the same structural level, but not below the dyke swarm. The largest layered gabbro in the Tihama-Asir is 8 km long and from 1.5 to 2.5 km wide. This gabbro body has an irregular saucer shape and is surrounded by dykes except for its roof which consists of a granophyric sheet.

Rhythmic layering results from the accumulation of plagioclase, clinopyroxene, and olivine crystals in layers up to several metres thick. Neither variation in An content nor modal abundance of the ferromagnesian minerals is systematically related to stratigraphy.

Granophyre intrudes the dykes and gabbros as sheet-like and stock-like masses and in some instances is older than the dyke complex, as it forms screens between the dykes. Many leucocratic dykes within the dyke swarm have the same mineral assemblages as the granophyres. Judging from the proportions of granophyre and gabbro, the granophyre cannot represent a differentiation product of the gabbros at the present level of exposure. The actual source of the granophyre may represent melted fractions of the Precambrian crust mobilized during crustal attenuation and rise of the mantle. The high initial Sr^{87}/Sr^{86} ratios of these granophyres tend to support this idea (Coleman *et al.* 1979).

Nearly all of the rocks making up the Tihama-Asir complex have undergone low-temperature hydrothermal alteration. The resulting secondary mineral assemblage developed under conditions of zeolite or greenschist-facies metamorphism, although original igneous textures are preserved. The hydrothermal alteration has apparently resulted from circulating hot meteoric water within the dyke swarm beginning shortly after its formation and perhaps continuing up to the present (Taylor 1980). The chemical parameters, as determined by Coleman *et al.* (1979), for the Tihama-Asir complex and associated rocks indicate that the primary magma emplaced into Precambrian crust became contaminated, and at the same time some of the crust was melted by ascending basaltic magmas to produce the granophyres. K/Ar ages on gabbro, granophyre, and hornfels from the Tihama-Asir complex range from 20 to 24 Ma.

The formation of the Tihama-Asir complex is related to crustal attenuation accompanied by the formation of sheeted dykes, layered gabbros, and granophyres. A complete Bouguer gravity-anomaly map by Gettings (1977) covering the Tihama-Asir complex and areas westward nearly to the Red Sea Axial trough shows that the complex is part of the Miocene oceanic crust that extends westward under the sedimentary rocks of the coastal plain to the axial trough. This same gravity map shows that the contact between Precambrian crust and the Tihama-Asir complex marks a fundamental boundary where new oceanic crust produced by Red Sea rifting first accreted to the Arabian plate.

The setting of the Tihama-Asir complex at the edge of the rifted Arabian plate invites comparison with current models of ocean-crust growth and also with fragments of more ancient allochthonous oceanic crust (ophiolites). The complex is tilted up to 30–40° towards the Red Sea and is intruded into the Jizan Group consisting of freshwater lacustrine deposits and bimodal volcanic rocks (Schmidt *et al.* 1982). There are no pillow lavas forming the upper parts of the complex, and there is no evidence that initial rifting was accomplished under a cover of seawater. The on-land rifting such as is now taking place in the Asal rift near Djibouti is perhaps similar. The bimodal nature of the Aden and Little Aden volcanoes of South Arabia, similar to the hypabyssal Tihama-Asir complex, may represent a similar spreading event in an

E–W direction of the southern Arabian Coast (Cox *et al.* 1969). The dyke complex is similar to that described in Cyprus (Moores & Vine 1971) but contains screens of granophyre, vesiculated basalt, and agglomerate of the Jizan Group which suggest a more complicated volcanic evolution than is visualized for mid-ocean spreading ridges. Even though the layered gabbros of the complex are structurally situated within the dyke complex and formed in shallow magma chambers, their lithologies and mineral assemblages are similar to those proposed for present-day spreading centres. Seismic refraction and gravity studies indicate crustal sections consisting of 9 km of mafic material and 6 km of sediments under the coastal plain (Blank *et al.* 1979). Harzburgite inclusions in the Holocene Al Birk volcanics resting above the tilted Tihama-Asir complex may come from underlying ultramafic rocks that represent depleted upper mantle under the new oceanic crust (Ghent *et al.* 1980). Similar harzburgites on Zabargad Island in the northern part of the Red Sea have been interpreted as oceanic mantle exposed along a transform fault indicating that upper-mantle rocks occupy shallow depths in the Red Sea (Bonatti & Otonella 1981). Even though a complete 'ideal' ophiolite is not exposed in the Tihama-Asir complex, these rocks obviously represent new oceanic crust formed in the initial opening of the Red Sea.

Conclusions

These brief accounts of Arabian ophiolites demonstrate that in each case the rock assemblage was involved in the development of ocean-like new crust and now occupies significant plate boundaries. Proterozoic ophiolites of the Arabian Shield may have developed within an intra-ocean system related to island arcs and marginal basins. The actual documentation of each ophiolite occurrence in the Proterozoic remains incomplete and further work needs to be carried out on the age and the actual tectonics of imbrication. It is not clear if these ophiolite belts are remnants of back-arc basin crust, mid-ocean-ridge crust, or fragments of a small ocean produced by intra-continental rifting and attenuation. The ophiolite belts do clearly mark boundaries of convergence between intra-oceanic areas that formed during the progressive cratonization of the Arabian–Nubian Shield (Gass 1982).

The Oman ophiolite represents oceanic crust formed at a spreading centre in the Neo-Tethys during the late Cretaceous that was detached and emplaced nearly synchronously onto the African Plate with those ophiolites of the eastern Mediterranean, Syria and Iran. The emplacement marks the zone of closure between African and Eurasian plates and the disappearance of the Neo-Tethys ocean.

The Tihama-Asir ophiolite represents an autochthonous portion of the proto-Red Sea crust where MORB-type dykes with associated gabbros intrude continental crust and bi-modal volcanics associated with freshwater sediments. Thick sections of continental clastic material overlie the igneous complex providing the tie to its formation along a passive continental margin.

Formation of ocean crust related to plate movements can be identified as far back as late Proterozoic on the Arabian Peninsula. The actual nature of the oceans during this time is obscure but the younger Mesozoic and Tertiary Arabian ophiolites can be clearly related to important plate movements.

References

ALABASTER, T., PEARCE, J. A., MALLICK, D. I. J. & ELBOUSHI, I. M. 1980. The volcanic stratigraphy and location of massive sulphide deposits in the Oman ophiolite. *In*: PANAYIOTOU, A. (ed.) *Ophiolites, Proceedings International Symposium, Cyprus, Geological Survey Department*, pp. 751–7.

AL-REHAILI, M. H. & WARDEN, A. J. 1980. Comparison of the Bir Umq and Haundah ultrabasic complexes, Saudi Arabia. Evolution and Mineralization of the Arabian–Nubian Shield. *Inst. Applied Geology, King Abdulaziz University Bulletin No. 3*, **4**, 143–56.

AL-SHANTI, A. M. & HAKIM, H. 1981. Mafic-ultramafic complexes of the Arabian Shield and their mineral occurrences. *UNESCO Int. Symp. Metallogeny of Mafic and Ultramafic Complexes*, **2**, 1–20.

—— & MITCHELL, A. H. G. 1976. Late Precambrian subduction and collision in the Al Amar-Idsas region, Arabian Shield, Kingdom of Saudi Arabia. *Tectonophysics* **30**, T41–7.

—— & ROOBOL, M. J. 1979. A late Proterozoic ophiolite complex at Jabal Ess in northern Saudi Arabia. *Nature, Lond.* **279**, 488–94.

BAKOR, A. R., GASS, I. G. & NEARY, C. R. 1976. Jabal Al Wask, northwest Saudia Arabia, an Eocambrian back-arc ophiolite. *Earth planet. Sci. Lett.* **30**, 1–9.

BARTOV, Y., STEINITZ, G., EYAL, M. & EYAL, Y. 1980. The sinistral movement along the Gulf of Elat—its age and relationship to the first phase of movement in the Red Sea. *Nature, Lond.* **285**, 220–1.

BLANK, H. R. JR 1977. Aeromagnetic and geologic study of tertiary dikes and related structures on the Arabian margin of the Red Sea. *In*: *Red Sea*

Research 1970–1975 Mineral Resources Bulletin 22, Dir. Gen. Mineral Resources, Saudia Arabia, G-1–G-18.

——, HEALY, J. H., ROLLER, J., LAMSON, R., FISHER, F., MCCLEARN, R. & ALLEN, S. 1979. Seismic refraction profile Kingdom of Saudi Arabia. *U.S. Geological Survey Saudi Arabian Mission, Project Report* **259**, 1–49.

BONATTI, E. P. R. & OTONELLO, G. 1981. The upper mantle beneath a young oceanic rift: peridotites from the island of Zabargad (Red Sea). *Geology* **9**, 474–9.

BOUDIER, F. & COLEMAN, R. G. 1981. Cross section through the peridotite in the Samail Ophiolite, Southeastern Oman. *J. Geophys. Res* **86**, 2573–92.

BROWN, G. F. 1972. Tectonic map of the Arabian Peninsula, Saudi Arabian *Dir. Gen. Mineral Resources, Arabian Peninsula Map AP-2* 1:400,000).

CHURCH, W. R. 1980. Late Proteozoic ophiolites. *In*: *Association mafiques ultra-mafiques dans les orogenes, Colloques internationaux du CNRS*, **272**, 105–117.

COLEMAN, R. G. 1981. Tectonic setting for ophiolite obduction in Oman, *J. Geophys. Res.* **86**, 2497–508.

——, HADLEY, D. G., FLECK, R. G., HEDGE, C. T. & DONATO, M. M. 1979. The Miocene Tihama Asir ophiolite and its bearing on the opening of the Red Sea. *In*: AL-SHANTI, A. M. S. (ed.) *Evolution and Mineralization of the Arabian–Nubian Shield*, Vol. 1, Pergamon Press, New York, pp. 173–86.

COX, K. G., GASS, I. G. & MALLICK, D. I. J. 1969. The evolution of the volcanoes of Aden and Little Aden, South Arabia. *Q. Jl. geol. Soc. Lond.* **124**, 283–308.

DELFOUR, J. 1979. Upper Proterozoic volcanic activity in the northern Arabian Shield, Kingdom of Saudi Arabia. *Evolution and Mineralization of the Arabian -Nubian Shield*, Institute of Applied Geology, King Abdulaziz University Bulletin No. 3, **2**, 59–76.

ENGLE, A. E. J., DIXON, T. H. & STEIN, D. J. 1980. Late Precambrian evolution of Afro-Arabian crust from ocean arc to craton. *Bull. geol. Soc. Amer.* **81**, 699–706.

FRISCH, W. & AL-SHANTI, A. 1977. Ophiolite belts and the collision of island arcs in the Arabian shield. *Tectonophysics* **43**, 293–306.

GARSON, S. & SHALABY, I. M. 1976. Precambrian-lower Paleozoic plate tectonics and metallogenesis in the Red Sea region. *Geological Association of Canada, Special Paper* **14**, 574–96.

GASS, I. G. 1977. The evolution of the Pan African crystalline basement in NE Africa and Arabia. *J. Geol. Soc. Lond.* **134**, 129–38.

—— 1982. Upper Proterozoic (Pan African) calc-alkaline magmatism in north-eastern Africa and Arabia. *In*: THORPE, R. S. (ed.) *Andesites*. John Wiley & Sons, pp. 591–609.

GETTINGS, M. E. 1977. Delineation of the continental margin in the southern Red Sea region from new gravity evidence. *Red Sea Research 1970–1975, Mineral Resources Bulletin 22, Director General Mineral Resources Saudi Arabia*, K1–11.

GHENT, E. D., COLEMAN, R. G. & HADLEY, D. G. 1980. Ultramafic inclusions and host alkali olivine basalts of the southern coastal plain of the Red Sea, Saudi Arabia. *A. J. Sci.* **280-A**, 499–527.

—— & STOUT, M. Z. 1981. Metamorphism at the base of the samail ophiolite, southeastern Oman Mountains. *J. Geophys. Res.* **86**, 2557–72.

GLENNIE, K. W., BOEUF, M. G. A., HUGHES-CLARKE, M. W., MOODY-STUART, M., PILAAR, W. F. H. & REINHARDT, B. M. 1974. Geology of the Oman Mountains. *Vehr. K. Ned. geol. Mijnbouwkd. Genoot.* **31**, 423 pp.

GRAHAM, G. 1980. Evolution of a passive margin and nappe emplacement in the Oman mountains. *In*: PANAYIOTOU, A. (ed.) *Ophiolites Proc. Int. Ophiolite Symposium, Cyprus, Mins. of Agriculture & Natural Resources Geological Survey Dept.* pp. 414–423.

GREENWOOD, W. R., HADLEY, D. G., ANDERSON, R. E., FLECK, R. J. & SCHMIDT, D. L. 1976. Late Proterozoic cratonization in south-western Saudi Arabia. *Phil. Trans. Roy. Soc. Lond.* **A280**, 517–727.

GREGORY, R. T. & TAYLOR, H. P. 1981. An oxygen isotope profile in a section of Cretaceous Oceanic Crust, Samail Ophiolite, Oman: Evidence for $\delta^{18}O$ buffering of the Oceans by deep (>5 km) seawater-hydrothermal circulation at mid-ocean ridges. *J. Geophys. Res.* **86**, 2737–55.

HILDRETH, W. 1981. Gradients in silicic magma chambers. Implications for lithospheric magmatism. *J. Geophys. Res.* **86**, 10153–92.

IRWIN, P. 1979. Ophiolitic terranes of part of the western United States. *In*: *International Atlas of Ophiolites.* Geologic Society of America MC-33.

KEMP, J. 1980. Geology of the Wadi Al Ays quadrangle, Sheet 25c, Kingdom of Saudi Arabia. *BRGM Open-File Report* BRGM-OF-01-2.

——, PELLATON, C. & CALVEZ, J. V. 1980. Geochronological investigations and geological history in the Precambrian of northwestern Saudi Arabia. *BRGM Open-file Report* BRGM-OF-01-1.

KENNEDY, W. Q. 1964. The structural differentiation of Africa in the Pan-African (± 500 m.y.) tectonic episode. *Res. Inst. Afr. Geol., Univ. Leeds 8th Ann. Rep.* p. 48–9.

LANPHERE, M. A. 1981. K-Ar ages of metamorphic rocks at the base of the Samail ophiolite, Oman. *J. Geophys. Res.* **86**, 2777–82.

LIPPARD, S. J. 1983. Evidence for Cretaceous high-pressure metamorphism in NE Oman and its possible relationship to subduction and ophiolite nappe emplacement. *J. geol. Soc. Lond.* **140**, 97–104.

MOORES, E. M. & VINE, F. J. 1971. Troodos Massif, Cyprus and other ophiolites and oceanic crust: Evolution and implications, *Phil. Trans. Roy. Soc. London*, **A286**, 443–66.

PALLISTER, J. S. 1981. Sheeted dike complex of the Samail ophiolite near Ibra, Oman. *J. Geophys. Res.* **86**, 2661–72.

—— & HOPSON, C. A. 1981. Samail ophiolite plutonic suite: Field relations, phase variations, cryptic variation and layering; a model of a spreading-ridge magma chamber. *J. Geophys. Res.* **86**, 2593–644.

PEARCE, J. A., ALABASTER, T., SHELTON, A. W. & SEARLE, M. P. 1981. The Oman ophiolite as a Cretaceous arc basin complex: evidence and implications. *Phil. Trans. Roy. Soc. London,* **A300**, 299–317.

SCHMIDT, D. L., HADLEY, D. G. & BROWN, G. F. 1982. Middle Tertiary continental rift and evolution of the Red Sea in southwestern Saudi Arabia: *Saudi Arabian Deputy Ministry for Mineral Resources Open File Report USGS*-OF-03-6, 56 pp.

——, —— & STOESSER, D. B. 1979. Late Proterozoic crustal history of the Arabian Shield, southern Najd Province, Kingdom of Saudi Arabia, *Evolution and Mineralization of the Arabian-Nubian Shield, Institute of Applied Geology, King Abdulaziz University Bulletin No. 3,* **2**, 41–58.

SEARLE, M. P., LIPPARD, S. J., SMEWING, J. D. & REX, D. C. 1980. Volcanic rocks beneath the Semail ophiolite nappe in the northern Oman mountains and their significance in the Mesozoic evolution of Tethys. *J. geol. Soc. Lond.* **137**, 589–604.

SHACKLETON, R. M. 1979. Precambrian tectonics of north-east Africa. *Evolution and Mineralization of the Arabian-Nubian Shield. Institute of Applied Geology, King Abdulaziz University Bulletin No. 3,* **4**, 1–6.

SMEWING, J. D. 1980. Regional setting and petrological characteristics of the Oman ophiolite in North Oman. *In*: ROCCI, G. (ed.) *Ofioliti, Special Issue Tethyan Ophiolites. Vol. 2. Eastern Area,* pp. 55–78.

—— 1981. Mixing characteristics and compositional differences in mantle-derived melts beneath spreading axes; evidence from cyclically layered rocks in the ophiolite of north Oman. *J. Geophys. Res.* **86**, 2645–60.

STACEY, J. S. & STOESER, D. B. 1983. Distribution of oceanic and continental leads in the Arabian-Nubian Shield. *Saudi Arabian Deputy Ministry for Mineral Resources Open File Report USGS-OF-03-55*, 36 pp.

STILLMAN, C. J., FUSTER, J. M., BENNELL-BAKER, M. J., MUNOZ, M., SMEWING, J. D. & SAGREDO, J. 1975. Basal complex of Fuerteventura (Canary Islands) is an oceanic intrusive complex with rift-system affinities, *Nature, Lond.* **257**, 469–71.

TAYLOR, H. P. JR 1980. Stable isotope studies of spreading centers and their bearing on the origin of granophyres and plagiogranites, *In*: ALLEGRE, C. & AUBOUIN, J. (eds) *Orogenic Mafic and Ultramafic Association.* Cen. National Research Science, No. 272, pp. 149–66.

TILTON, G. R., HOPSON, C. A. & WRIGHT, J. E. 1981. Uranium-lead isotopic ages of the Samail ophiolite, Oman, with applications to Tethyan Ocean Ridge tectonics. *J. Geophys. Res.* **86**, 2763–76.

TIPPIT, P. R., PESSAGNO, E. A., JR & SMEWING, J. D. 1981. Age of the Samail ophiolite based on radiolarian biostratigraphy. *J. Geophys. Res.* **86**, 2756–2.

R. G. COLEMAN, Geology Department, Stanford University, California, USA.

Ophiolites of the northern Caribbean: A reappraisal of their roles in the evolution of the Caribbean plate boundary

G. Wadge, G. Draper & J.F. Lewis

SUMMARY: The ophiolites of the northern Caribbean represent latest Jurassic and Cretaceous oceanic crust which was obducted from the mid-Cretaceous to the Paleogene. The majority of these ophiolites are early Cretaceous in age and were emplaced during the late Cretaceous. The late Cretaceous geological history can be successfully interpreted in terms of a northward-migrating island arc which subducted Atlantic crust to the south beneath Pacific crust. This arc collided with southern Yucatan in Campanian times to form the Guatemalan ophiolites, and in Paleogene times with the Bahamas platform to form the central and western Cuban ophiolites. Both these collisions involved underthrusting by carbonate-mantled continental forelands. However, the eastern extension of this subduction zone, the North Coast belt of Hispaniola, has not been thrust over the Bahamas platform. The provenance of these arc-related ophiolites should be Atlantic ocean crust. Back-arc spreading to the south of this arc during the late Cretaceous – Paleocene may have produced the source material for the Jamaican and Oriente ophiolites and also for the Dumisseau Fm. of the Southern Peninsula of Haiti. The oldest ophiolites in central Hispaniola and Puerto Rico are possible candidates for crust of Pacific provenance, perhaps isolated during an early Cretaceous change of subduction polarity. However, our preferred model for the Cordillera Central of Hispaniola is as the base of an accretionary prism with southward-dipping subduction. All but the Cuban ophiolites have suffered varying degrees of strike-slip dismemberment during the Cenozoic.

Ophiolitic rock associations form belts along three of the four boundaries of the Caribbean plate. These rocks show great variety and differ from the classical Alpine–Tethyan types. There has been considerable interest in the origin of these rocks and their significance with respect to the tectonic evolution of the Caribbean (see Dengo 1972; Khudoley & Meyerhoff 1971; and Case 1980 for previous reviews). The aim of this paper is to present a concise account of the present state of knowledge of the ophiolitic associations of the northern Caribbean. After a brief discussion of the general setting we present two tables summarizing the composition and emplacement histories of each ophiolite. The present-day crustal structure across the ophiolite belts is briefly described and the final section of the paper discusses the ophiolites in terms of current plate tectonic models of the area.

Present-day motion along the northern boundary of the Caribbean plate is mainly strike-slip. Left-lateral displacement between the North American and Caribbean plates has been complicated by the construction of a narrow strip of oceanic crust in the Cayman Trough and minor plate consumption in Hispaniola, but essentially throughout the Cenozoic the boundary has been conservative. During the Cretaceous and part of the Paleogene, however, the boundary was convergent. The ophiolites discussed here were emplaced during that period of convergence. The plates involved were oceanic to transitional in structure except in Guatemala, where continental crust was involved. We are not concerned with the ophiolite belt of the Pacific coast of Central America nor with the Cretaceous ophiolites of Venezuela (Villa de Cura). We accept that these ophiolites are relevant to Caribbean ophiolites generally but we do not have the space to discuss them here and they are also omitted from Fig. 3.

For the sake of descriptive expediency we recognize thirteen ophiolite provinces within the northern Caribbean and Fig. 1 shows the location of these ophiolites. There are, of course, many more individual outcrops of ophiolitic rock types than thirteen, and similarly we have not applied strict stratigraphic criteria in our choice of what to call ophiolites. Only one of these northern Caribbean ophiolites appears to contain the full sequence from metamorphic peridotite to pelagic sediments that characterize the classic ophiolite sequence (Coleman 1977). In Guatemala, two regions are recognized; a southern belt close to the Motagua fault zone and a northern belt close to the Polochic fault zone. Cuba contains the most extensive ophiolite outcrops (>6500 km^2; Khudoley & Meyerhoff 1971). A western belt in the three westernmost provinces contains many small (<100 km^2) bodies. Further east, large continuous outcrops are found: Las Villas, Camaguey, Gibara/Holguin, Nipe/Cristal and Purial. Jamaica has only a small area of ophiolite in the southern Blue Mountains. Hispaniola has three main subparallel belts. From north to south these are the E–W-trending North Coast belt, the NW–SE-trending Cordillera Central belt and the Southern Peninsula belt. Finally, the ophiolitic rocks of Puerto Rico are confined to outcrops in the SW corner of the island. There are a few minor, isolated outcrops of ophiolitic rocks that do not fit easily into these thirteen provinces and these are not shown in Fig. 1.

Ophiolite sequences

Table 1 indicates the presence or absence of the classic subdivisions of ophiolite stratigraphy within each region. The lithology or mineralogy are shown in order of observed frequency of occurrence. A dash indicates that this unit is absent or has not yet been recognized.

The Sierra de Santa Cruz in the northern Guatemala belt contains a full ophiolite sequence (Rosenfeld 1981) from 3 to 7 km thick, which is probably the best preserved in the northern Caribbean. The western part of the Sierra is notable for the massive sulphide mineralization in the hydrothermally altered pillow basalts (Petersen & Zantop 1980). In contrast to the northern belt, the southern belt is generally more dismembered and the ultramafics show greater degrees of serpentinization. Serpentinite mélanges with a wide variety of blueschist- and eclogite-facies inclusions are common (Lawrence 1976; Bertrand & Vuagnat 1980).

In Cuba, the lack of recent mapping by geologists using modern concepts of ophiolite stratigraphy hampers comparisons. Undoubtedly the surface outcrops generally contain a high percentage of ultramafic rocks relative to other components (e.g. 85% ultramafics, 10% gabbros in Camaguey, Flint *et al.* 1948). Borehole data, however, which show at least a 5 km section in Las Villas, suggest that this balance is not maintained at depth. Spilitized pillow lavas and diabase are much more common in boreholes, particularly on the southern side of the island (e.g. Soto 1978a). The ophiolites of the Sierras, Nipe/Cristal and Purial in easternmost Cuba are notable for having an exposed basement of mainly metatholeiitic rocks (partly blueschist facies) (Boiteau *et al.* 1972), but they do not have stratigraphically continuous extrusive rocks associated with the overlying serpentinite thrust sheets. Sheeted-dyke complexes have not been described from any of the Cuban ophiolites. In eastern Jamaica, pelagic sediments and pillow basalts are found in association with isotropic gabbros, but the presumed ultramafic sections of the ophiolite are found separately in small, fault slivers and blocks along strike of a major Cenozoic fault zone (Wadge *et al.* 1982).

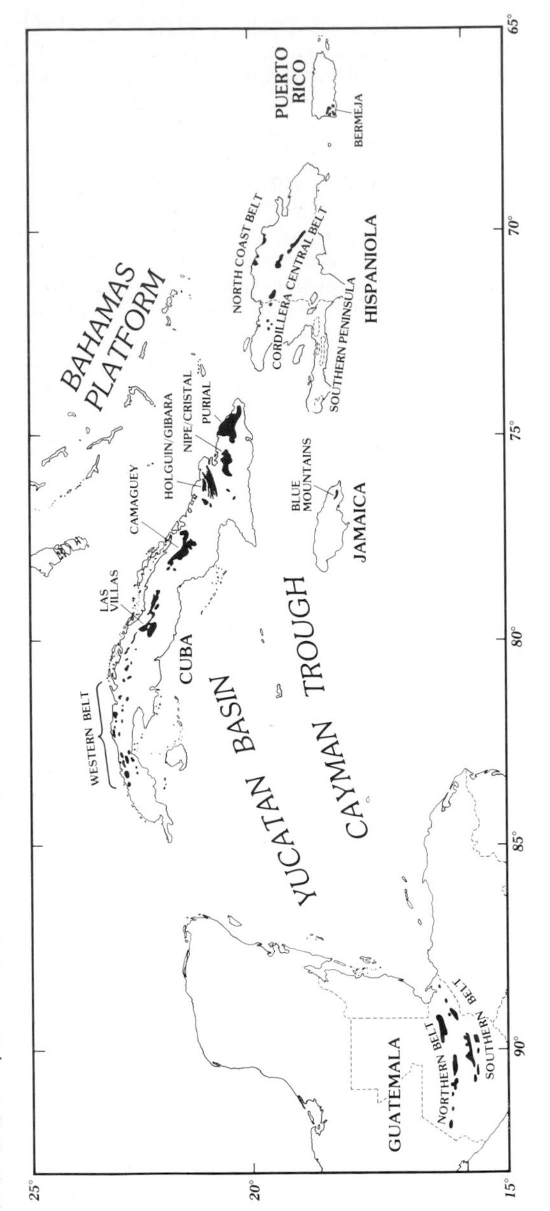

FIG. 1. The location of the thirteen ophiolite provinces of the northern Caribbean discussed in the text are shown in solid black. The outcrop of mafic rocks in the Southern Peninsula of Haiti is shown in dashed outline.

Two major subparallel belts occur in northern Hispaniola. In the North Coast belt the rock associations are dominated by serpentinite, and contain tectonic inclusions and fault blocks of high-pressure metamorphic rocks (Nagle 1974). To the east, this belt has been found to extend along the southern margin of the Puerto Rico trench (Perfit *et al.* 1980). The Rio San Juan Complex (Eberle *et al.* 1980) is the most extensive outcrop comprising ultramafic rocks along the coast, a large gabbro complex in the centre and a belt of amphibolite (metagabbro) in the south.

The central gabbro complex is separated from the ultramafic rocks by a wide zone of breccia containing high-pressure metamorphic rock. Farther west in the Puerto Plata complex the gabbroic bodies are tectonite blocks.

In the Cordillera Central of Hispaniola, an essentially linear outcrop of peridotite lies NE of a belt of mafic metamorphic rocks, the Duarte Complex, and a parallel belt of unmetamorphosed volcanic rocks, the Siete Cabezas Formation (Bowin 1966). Tectonic inclusions display either low-pressure/high-temperature metamorphism or no metamorphism. No unequivocal pillow lavas or sheeted dykes have been identified. The main belt in the Dominican Republic appears to extend westward into the complex Douvray/Terrier Rouge area of Haiti (Nicolini 1977). Here there are small ultramafic-mafic intrusive bodies, two of which were emplaced along a major fault zone, apparently a thrust. As in the east, basalts with associated radiolarian cherts are present in the complex, but dyke swarms and pillow basalts are absent. The gabbroic rock types are mainly norites, and hornblende-bearing gabbros are common.

The Dumisseau Formation of the Southern Peninsula of Haiti is an uplifted Upper Cretaceous sequence of pelagic sediments with interlayered basalts and dolerites. Although this sequence has been called an ophiolite in the literature (Maurasse *et al.* 1979), the lower ultramafic parts are not exposed. In southwestern Puerto Rico serpentinite, cherts, pillow basalts and tholeiitic amphibolite are the principal rock types (Mattson 1960, 1974), but the whole sequence is strongly tectonized and the original relations between these rocks are unclear.

Emplacement histories

The minimum ages of formation of the ophiolites shown in Table 2 are the oldest paleontologically determined ages of rocks apparently deposited on the oceanic crust. Where paleontological data are unavailable we have included isotopic ages. By 'emplacement' we mean the process of transfer from the ocean crust to a subaerial or continental basement. The age range of emplacement is determined either from the paleontological age of sedimentary rocks considered to have been deposited on the ophiolite in shallow water or on land, or from the age of sedimentary rocks containing ophiolite-derived fragments.

The Guatemalan, Cuban, Puerto Rican and Cordillera Central ophiolites appear to be at least as old as early Cretaceous, whereas the Jamaican and North Coast Hispaniola ophiolites may be of late Cretaceous age. Emplacement of many of the ophiolites took place at some stage during the period 80–50 Ma. Incremental emplacement over a long period in an environment such as an accretionary prism might produce a wide range of apparent emplacement ages. In parts of western Cuba, serpentinite-bearing clastic rocks of Campanian–Maestrichtian age testify to surface exposure of serpentinite at this time (Khudoley & Meyerhoff 1971), though the last period of thrusting was not until Eocene time (Mossakovskiy & Albear 1978).

Each of the ophiolite belts of Hispaniola has undergone different tectonic histories. The age of formation of the Cordillera Central belt is based on a K-Ar age of 127 Ma for a hornblendite intrusion cutting rocks of the Duarte Complex (Bowin 1975), but the relationship of the Duarte Complex to the ultramafic rocks, and hence the age of the latter, is not clear. Bowin (1966) suggested that emplacement of the ultramafic belt occurred before middle Albian times because no rocks younger than this are found in contact with the peridotite. However, the present fault boundaries of the main peridotite body are of Paleogene age (Haldemann *et al.* 1980). The age of formation of the North Coast belt is also uncertain and is based on the isotopic ages (90–70 Ma) of metamorphic inclusions in serpentinite. Pre-Paleocene emplacement is clear because the serpentinites are unconformably overlain by Paleocene volcanics (Nagle 1966). Maurasse *et al.* (1979) present paleontological evidence for a middle-to-late Cretaceous age for the rocks of the Southern Peninsula. The emplacement of these rocks involved progressive vertical uplift of fault blocks such that by the late Eocene shallow-water facies sediments were deposited on earlier deeper-water sediments whilst nearby deep-water sedimentation continued until the Miocene. However, this was superposed on a series of southward-dipping thrusts, possibly of Maestrichtian age (Mercier de Lepinay *et al.* 1979), whose significance is not fully understood.

The North Coast belt appears to belong to a subduction-zone complex in which oceanic crust was subducted to the south. Eberle *et al.* (1980) suggested that the gabbroic rocks of the Rio San Juan complex separated from the underlying serpentinite and moved southwards whilst the periodotite moved vertically upwards after emplacement. In contrast to the North Coast belt of Hispaniola, the direction of emplacement is from south to north in Guatemala and Cuba. However, in both these areas underthrusting by a carbonate foreland was involved and the original subduction-zone structures have been largely destroyed. However, in Holguin/Gibara the thin imbricate ultramafic/carbonate sheets described

TABLE 1. *Compositions of northern Caribbean ophiolites*

		Peridotites	Serpentinization	Gabbro	Leucocratic rocks	Diabase	Lavas	Sedimentary	Selected refs
Guatemala	North	Harzburgite* Lherzolite dykes Websterite*	Moderate (Lizardite-Chrysotile)	Layered (Pl-cpx-opx)	Qz-diorite Hb-diorite	Dyke swarm (Pl-cpx-opx) (Rodingites)	50% pillowed	Limestone Cherts	Rosenfeld (1981) Peterson & Zantop (1980) Bertrand & Vuagnat (1976, 1980)
Guatemala	South	Harzburgite* Lherzolite	Extensive (Antigorite in N, Lizardite-Chrysotile in S)	Greenschist facies (Pl-cpx-opx)	—	Rodingite	Prehnite-Pumpellyite facies	Cherts	Muller (1979) McBirney (1963) Bertrand & Vuagnat (1980, 1975, 1976) Lawrence (1976)
Cuba	West	Harzburgite* Lherzolite, Dunite, Wehrlite	Decreases downwards	Yes	Granodiorite dykes	Yes	Yes	Si-Lmst., Chert, Tuffs	Mossakovskiy & Albear (1978) Soto (1978b) Fonseca *et al.* (1980)
Cuba	Las Villas	Harzburgite*	Extensive	Ol. gabbro Hb. gabbro	—	Hb-dolerite (Rodingite)	Pillows, breccias, mafic tuffs. (Pl-cpx-Ol)	Cherts, Radiolarites	Mossakovskiy & Albear (1978), Pardo (1975), Ducloz & Vuagnat (1962) Meyerhoff & Hatten (1968)
Cuba	Camaguey	Harzburgite* Dunite*	Extensive	Ol. gabbro Troctolite (Chromite deposits below gabbro)	Anorthosite	Few sheets, dykes	—	—	Flint *et al.* (1948)

Cuba	Holguin-Gibara	Harzburgite Lherzolite, Dunite, Pyroxenite	Inversely proportional to body size	Yes (Uralitized dykes in serpentinite)	Minor Anorthosite	—	—	—	Keijzer (1945–7) Kozary (1968)
Cuba	Nipe-Cristal	Harzburgite* Pyroxenite, Dunite	Extensive	Troctolite Gabbro	Diorites	Uralitized dykes in serpentinites	—	—	Keijzer (1945–7) Lewis & Straczek (1955)
Cuba	Purial	Harzburgite* Dunite, Chromite	Variable (Antigorite-Chrysotile)	Olivine gabbro Troctolite Hb. gabbro	Anorthosite	Non-sheeted (Pl-cpx)	—	Meta-cherts Limestones	Guild (1947) Boiteau *et al.* (1972)
Jamaica	Blue Mts.	Lherzolite* Harzburgite*	Extensive (Antigorite)	Isotropic (Pl-cpx)	Tonalite	Non-sheeted	80% massive 20% pillowed (Pl-cpx)	Limestones Cherts	Wadge *et al.* (1982)
Hispaniola	North Coast	Harzburgite* Dunite, Chromite*	Extensive (Antig.-Chry.)	Inter-layered mela-, leuco-	Diorite Plagio-granite	—	Sheared (Pillows in olistostromes)	Chert. lmst., Red clay	Lewis (1980) Nagle (1966, 1974) Bowin & Nagle (1980) Eberle *et al.* (1980)
Hispaniola	Cordillera Central	Harzburgite* Dunite dykes, Websterite and Cortlandite in Haiti	Extensive	Tectonic inclusions (Pl-opx-cpx) (Pl-Amp in Haiti)	— (Diorites in Haiti)	—	Associated metabasalts?	Cherts Radiolarites Black shales	Bowin (1966) Lewis (1980) Nicolini (1977)
Hispaniola	Southern Peninsula	—	—	Minor dykes	—	Sills	Massive, pillowed	Limestones Cherts, clastics	Maurasse *et al.* (1979)
Puerto Rico	Bermeja	Harzburgite* Dunite	Variable	—	—	—	Metatholeiites (some pillows)	Cherts	Mattson (1960 1974)

*Tectonized

TABLE 2. *Emplacement histories of northern Caribbean ophiolites*

		Minimum age of formation	Latest period of emplacement	Lithologic association	Tectonic association	Selected refs
Guatemala	North	Upper Valanginian –Lower Cenomanian	Campanian–Maestrichtian	Amphibolite sole. Rests on Lower Campanian flysch. Overlying seds. increase proportion of calcalkaline volcs. upsection	Foreland carbonate platform to north. Cretaceous arc to south. Present-day plate boundary.	Rosenfeld (1981), Williams (1975)
Guatemala	South	Upper Valanginian –Aptian	Lower Santonian–Maestrichtian	Thick underlying amphibolites. Blueschist/eclogite exotics. Overlying flysch, island-arc volcs.	Present-day plate boundary. Paleozoic metamorphic basement to north and south.	Muller (1979) Roper (1978) Sutter (1979)
Cuba	West	Pre-Albian	Paleocene–early M. Eocene	Andesitic volcanics and clastics. Limestones and olistostromes. Blueschist exotics.	Multiple (5–6) recumbent nappes thrust S–N over Bahama Platform.	Soto (1978a) Mossakovskiy & Albear (1978)
Cuba	Las Villas	Pre-Aptian	Lower–Middle Eocene	Overlain by submarine andesite/2 px basalts, clastics, pelagics. Underlying thrust has sole of granodiorite/ amphibolite/exotics	Major thrust sheet underlain by dismembered ? gabbro sheet. Mélange at base. All thrusts S–N over Bahama Platform	Meyerhoff & Hatten (1968) Pardo (1975), Mossakovskiy & Albear (1978)
Cuba	Camaguey	Pre-Aptian	Early Middle Eocene	Overlain by submarine andesites, basalts and carbonates. Detrital serpentinites in carbonates.	Cretaceous volcanics/ lmst. folded with ophiòlite prior to Middle Eocene over-thrusts onto Bahamas. 2 major mélange-bearing thrust planes below carbonate thrust sheets.	Flint *et al.* (1948) Thayer & Guild (1947)
Cuba	Gibara/ Holguin	Pre-Aptian	Early Middle Eocene	Pelagic limestones, cherts, wackes, reefal limestones	Extreme imbrication with limestones in recumbent thrusts. S–N over Bahama Platform	Kozary (1968), Knipper & Cabrera (1974)

Cuba	Nipe/ Cristal	?	Late Maestrichtian–Early Paleocene	Underlain by L. Cretaceous tuffs, greywackes (cut by 119 Ma pegmatite)	Diabase/Serpentinite submarine mélange underlies serpentinite. S–N thrust. Not underlain by Bahama Platform, Dome-like uplift.	Cobiella (1978) Lewis & Straczek (1955)
Cuba	Purial	?	Paleocene	Metabasalts (blueschist greenschist amphibolite facies), tuffs, clastics	Dome uplift of thrust sheet from South. Underlying metamorphics folded on NE–SW axes. Not underlain by Bahama Platform.	Hernandez (1979) Cobiella *et al.* (1977) Boiteau & Michard (1976)
Jamaica	Blue Mts.	Campanian	Maestrichtian–? Paleocene	Overlain by Maestrichtian flysch. Fault contact blueschist-facies metabasalt. Greenschist, amphibolite	Base not seen. No thrusts identified. High-angle strike-slip and normal faulting of Cenozoic age dominates	Wadge *et al.* (1982)
Hispaniola	North Coast	Turonian?	Paleocene	Overlain by Paleocene andesite volcanics, Tertiary turbidites. Faulted against blueschists, marbles, amphibolites	Allochthonous slices separated by high-pressure metamorphics in subduction complex. Superposed Cenozoic strike-slip faulting	Eberle *et al.* (1980) Nagle (1966) Perfit & McCulloch (1982)
Hispaniola	Cordillera Central	Valanginian?	Albian?	Greenschist-amphibolite-facies metabasalts, metasediments, keratophyres, tonalites, rhyolites, Hb-granites	Base not seen. Major faults are high angle including peridotite margins. Thrusts in Haiti	Bowin (1966) Palmer (1963) Haldemann *et al.* (1980) Nicolini (1977)
Hispaniola	Southern Peninsula	Turonian–Campanian	Eocene–Miocene	Pelagic limestones, cherts some clastic sediments	Late Cretaceous S–N thrusts? Late Cenozoic strike-slip tectonics	Maurasse *et al.* (1979) Mercier de Lepinay *et al.* (1979)
Puerto Rico	Bermeja	Portlandian	Albian	Overlain by Upper Cret. volcanics. Fault contact with amphibolite-facies metabasalts	Northward thrusting from Albian to Maestrichtian and domal uplift. Strike-slip in Cenozoic	Mattson (1960, 1974) Mattson & Pessagno (1979)

by Kozary (1968) are probably accretionary-wedge structures rather than collision structures. In Nipe/Cristal and Purial of the Oriente province, Cuba the imbricate stacking of thrust sheets typical of further west in Cuba is lacking and there is no borehole evidence for under-thrusting by the Bahama platform. Mattson (1974) suggested that the Puerto Rican ophiolite was emplaced as a gravity nappe from the south and that movement must have taken place before the Campanian and probably in Albian or earlier times.

Crustal structure

Guatemala is the only region where there is undisputed continental crust underlying the Caribbean ophiolites. Both the Maya block (Yucatan) to the north of the Polochic fault zone and the Chortis block to the south of the Motagua fault zone contain Paleozoic and ? Precambrian rocks (Dengo 1972). A wide-angle seismic-reflection study just south of the ophiolites (Kim *et al.* 1982) revealed a three-layer crust with the Moho at 37.4 km. Case (1980) reviewed the gravity data for the area. The regional negative Bouguer anomalies are consistent with continental crust. The northern-belt ophiolites have very weak associated anomalies consistent with their emplacement as detached allochthonous sheets (Williams 1975). Local positive Bouguer anomalies of 30–50 mgal are associated with the southern belt suggesting a deeper-rooted character.

The crustal structure of Cuba has been determined from regional seismic profiling by Scherbakova *et al.* (1980) and Bovenko *et al.* (1979, 1980). The mean depth to the Moho beneath western and central Cuba is 27 km (minimum 21 km, maximum 36 km). In the north, the Bahama Platform has a three-layer structure and a thickness of about 30 km. The trace of the ophiolite outcrop is approximately marked by a belt of thinner crust bounded in some places to the south by a discontinuity. This discontinuity marks an abrupt thickening of the crust by 5–7 km (Scherbakova *et al.* 1980). The crustal structure of easternmost Cuba is very different from the rest of Cuba. Beneath the Cauto Basin the crust thins dramatically (minimum of 12 km) with a two-layer structure beneath the Nipe/Cristal and Purial ophiolites. The crust remains thin (~20 km) further south in the Sierra Maestra but the three-layer structure returns (Fig. 2). The distinct character of Cuba SE of the Cauto Basin is reinforced by the large positive Bouguer anomalies over the ophiolites there (Bowin 1976). Except for the Sierra Maestra, in southernmost Oriente, where Paleocene/Eocene magmatism may have modified the crust, we interpret the present-day crustal structure of Cuba in terms of the configuration of the crust immediately after ophiolite emplacement. The northward sense of thrusting noted in the surface geology is also seen in the seismic structure beneath the western and central Cuban ophiolites. The oceanic-type crustal structure beneath Nipe/Cristal and Purial implies that these ophiolites have a different provenance from the other Cuban examples.

Crust of a transitional character and a thickness of about 20 km underlies Jamaica (Arden 1975). Gravity studies (Wadge *et al.* 1980) suggest that the Jamaican ophiolite is the surface expression of a dense, upfaulted block. The tectonism that produced this feature is Cenozoic in age and the original late Cretaceous structure is masked. Moreover, on the basis of the anomalous juxtaposition of trench-generated blueschists and arc-generated volcanics within the Blue Mountains, Draper (1979) suggested that the ophiolite-bearing eastern block of the island was not originally contiguous with the rest of the island.

The gravity field of Hispaniola, which has been described by Reblin (1973) and Bowin (1975, 1976), indicates an average crustal thickness of 25 km. Most of the island is characterized by positive Bouguer anomalies which would be much greater but for the low-density sediments in the Cenozoic basins (Case 1980). The high positive anomalies on the Southern Peninsula of Haiti appear to be part of the regional gradient from the ocean basin to the south. A gravity high extends across the central part of the island following the Cordillera Central belt. Recent drilling has confirmed that the peridotite belt continues eastwards along this trend beneath the Tertiary limestone cover into eastern Hispaniola. In contrast, over the North Coast belt there are much smaller and more localized, positive anomalies. High positive Bouguer gravity anomalies over central Puerto Rico (Bowin 1976) indicate the presence of dense rock beneath much of the island, not simply over the ophiolite exposed in the SW. Mattson & Schwartz (1971) and Mattson (1974) suggest that the pre-Albian units associated with the ophiolitic rocks have a shallow dip northwards. They may underlie the entire island and be the cause of the positive anomalies.

Plate tectonic evolution

In recent years, several models of the plate tectonic evolution of the northern Caribbean have been proposed (e.g. Burke *et al.* 1978; Mattson 1979; Maurasse 1981; Pindell & Dewey

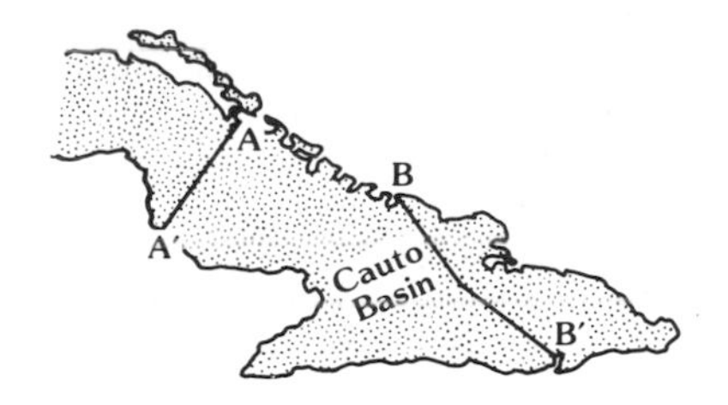

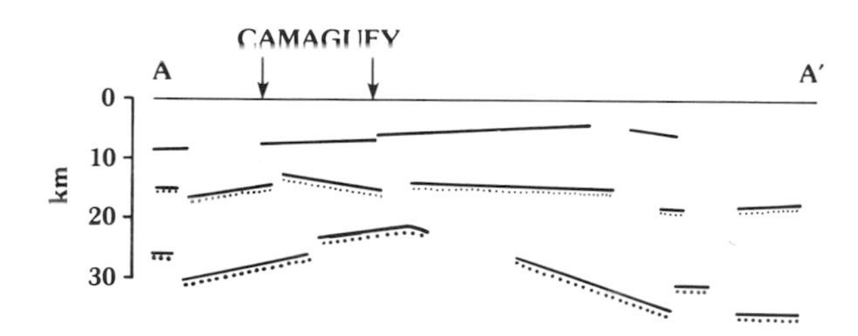

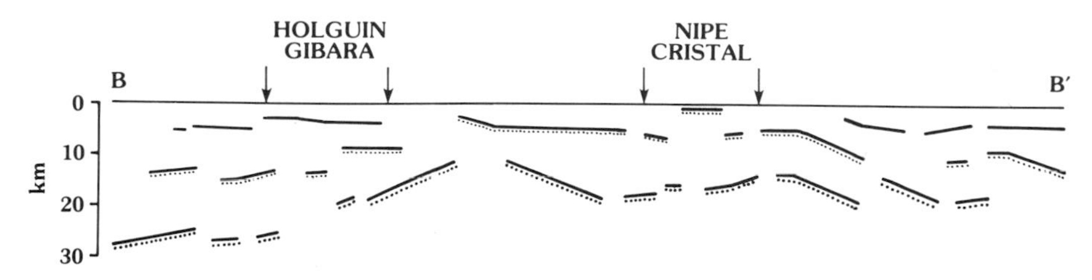

FIG. 2. Two seismic profiles across eastern Cuba from Bovenko *et al.* (1980). Three crustal layers are defined. The single line separates an upper layer with V_P seismic velocities from 2 to 4.6 km s^{-1} from an intermediate layer with $V_P = 6.2$ km s^{-1}. This intermediate layer is separated from a deeper layer, $V_P = 7.2$ km s^{-1}, by the continuous small dot line pair. The continuous large dot line pair is the Moho. The outcrop locations of the ophiolites are shown. Note the shoaling of the Moho beneath Camaguey and the two-layer crustal structure in the Cauto Basin and underlying Nipe-Cristal.

1982). In the following discussion we have paid particular attention to the model of Pindell & Dewey since it incorporates and refines many of the features of earlier models. We consider the following elements from these models to be pertinent to the origin and emplacement of the northern Caribbean ophiolites:

(i) The Greater Antilles began as an island arc in early Cretaceous times, formed by subduction of Atlantic crust beneath a NE-migrating plate which originated in the Pacific. This subduction also consumed the spreading ridge which was formed during the original separation of North and South America. In contrast to this idea, Mattson (1979) proposed that the early Cretaceous subduction zone dipped northward and consumed a proto-Caribbean crust.

(ii) In the late Cretaceous, the western end of this arc collided with the continental crust of the Yucatan peninsula. The rest of the arc continued to move northwards but the western part (Cuba) was split by back-arc spreading to form the Yucatan Basin and the eastern half (Hispaniola, Puerto Rico) moved independently. Also during the late Cretaceous (~90–70 Ma) the Caribbean crust was apparently thickened by regional tholeiitic magmatism (B″ magmatism). Maurasse (1981) suggested that this also resulted from a form of back-arc spreading behind the Greater Antillean arc.

(iii) The Cuban arc collided with the Florida/Bahamas platform in the Eocene.

(iv) Left-lateral transform motion began during the Oligocene transferring Cuba and the Yucatan Basin to the North American plate.

(v) Continued transform motion and spreading in the Cayman Trough to the present-day translated the Caribbean plate to the east relative to North America probably by at least 1100 km (Wadge & Burke, in press). This motion also juxtaposed the continental Chortis block, formerly part of SW Mexico, against Yucatan.

These models predict that the northern Caribbean ophiolites originated from the following sources: Atlantic ocean crust occupying the present site of thc Caribbcan; Caribbean crust originating in the Pacific, or, Caribbean crust resulting from back-arc spreading within the Caribbean. The age range of the consumed Atlantic crust must have been late Jurassic/early Cretaceous. DSDP drilling has demonstrated that much of the Caribbean crust is older than late Cretaceous and Ghosh *et al.* (in prep.) have proposed a late Jurassic/early Cretaceous spreading history for the Venezuelan Basin based on magnetic anomalies. Thus, the age of ocean crust is a poor discriminator between an Atlantic or a Caribbean (Pacific) origin for those ophiolites formed from late Jurassic and early Cretaceous ocean crust. Although the Yucatan Basin crust is of suitable age, the Pindell & Dewey model does not provide an explicit mechanism for obduction of this crust.

The concept of a northward-moving island-arc system colliding with a passive continental margin in southern Yucatan during the Campanian (79–70 Ma; Sutter 1979) is consistent with the geology of late Cretaceous rocks in eastern Guatemala (Rosenfeld 1981) (Fig. 3). A reconstruction of the pre-Cenozoic position of the Chortis block farther west exposes southern Yucatan (Wadge & Burke, in press) and there is no necessity, if this reconstruction is proven to be correct, to invoke a continent–continent collision (Yucatan–Chortis) as in Wadge *et al.* (1982, Fig. 6). Cenozoic strike-slip motion along the suture has left intact only the northernmost parts of the allochthonous ophiolite sheets on the Yucatan continental margin (Sierra de Santa Cruz). The southern belt, along the Motagua fault zone, must be regarded as the root zone for the ophiolite allochthons.

Although complex in local detail, the general evidence from the Cuban ophiolites (excepting Nipe/Cristal and Purial) is consistent with another island arc–continent (Florida–Bahamas) collision, occurring later than in Guatemala. However, unlike the Guatemalan ophiolite belts, most of Cuba was not involved in later transform motion, and hence much of the original obduction structure is preserved. Deep seismic structure and shallow structure from boreholes (Shaposhnikova 1974) are both compatible with repeated northward thrusting of ophiolitic crust onto the Florida–Bahamas carbonate foreland. The outcrop pattern of the central Cuban ophiolites has been interpreted in terms of an overthrust accretionary prism (Gealey 1980), the ocean crust having been trapped behind the trench and beneath the forearc basin. In central and western Cuba, however, large outcrops of high-pressure metamorphic rocks are found only in Escambray south of Las Villas. Millan & Somin (1981) showed that the ophiolite sheets overrode the high-pressure metamorphic rocks from the south, a region now off the south coast of Cuba.

Along strike to the east, the continuation of the Cretaceous subduction zone appears to pass through northern Hispaniola and into the Puerto Rico Trench. There is little doubt that the North Coast belt ophiolites are the subaerial equivalents of the rocks exposed on the inner wall of the Puerto Rico trench (Perfit *et al.* 1980), and these were uplifted by the entry of the southeastern end of the Bahama platform into the trench during the early Cenozoic. The consequences of this event were very different from those in Cuba since no large-scale thrusting resulted in northern Hispaniola. This may simply have been the result of a northwardly convex Greater Antilles arc colliding with a southwardly convex Bahama platform. Intense overthrusting would have taken place at the initial contact zone, but lesser, later convergence farther east resulted in upthrusting of the forearc region in Hispaniola.

The other northern Caribbean ophiolites are more difficult to explain by the Pindell & Dewey model. The most difficult problem is that of the locations and amounts of E–W-trending Cenozoic strike-slip displacements. The original positions of the ophiolites and hence their tectonic interpretations are dependent upon these subsequent left-lateral movements. This is particularly true for the three Hispaniolan belts. More westerly positions for both Oriente and central Hispaniola during the Cretaceous are called for by the Pindell & Dewey model. Oriente, which is now sutured along the Cauto Basin to the North American plate, presumably moved less far than central Hispaniola to the south. One such reconstruction would put both central Hispaniola and Oriente as part of the remnant arc, left behind by the opening Yucatan Basin (Fig. 3). The evidence for oceanic crust beneath the Cauto Basin indicates that this may be the eastward extension of the Yucatan Basin, and hence the Nipe/Cristal and Purial ophiolites may be Yucatan Basin crust rather than Atlantic crust. In this case, back-arc spreading could have com-

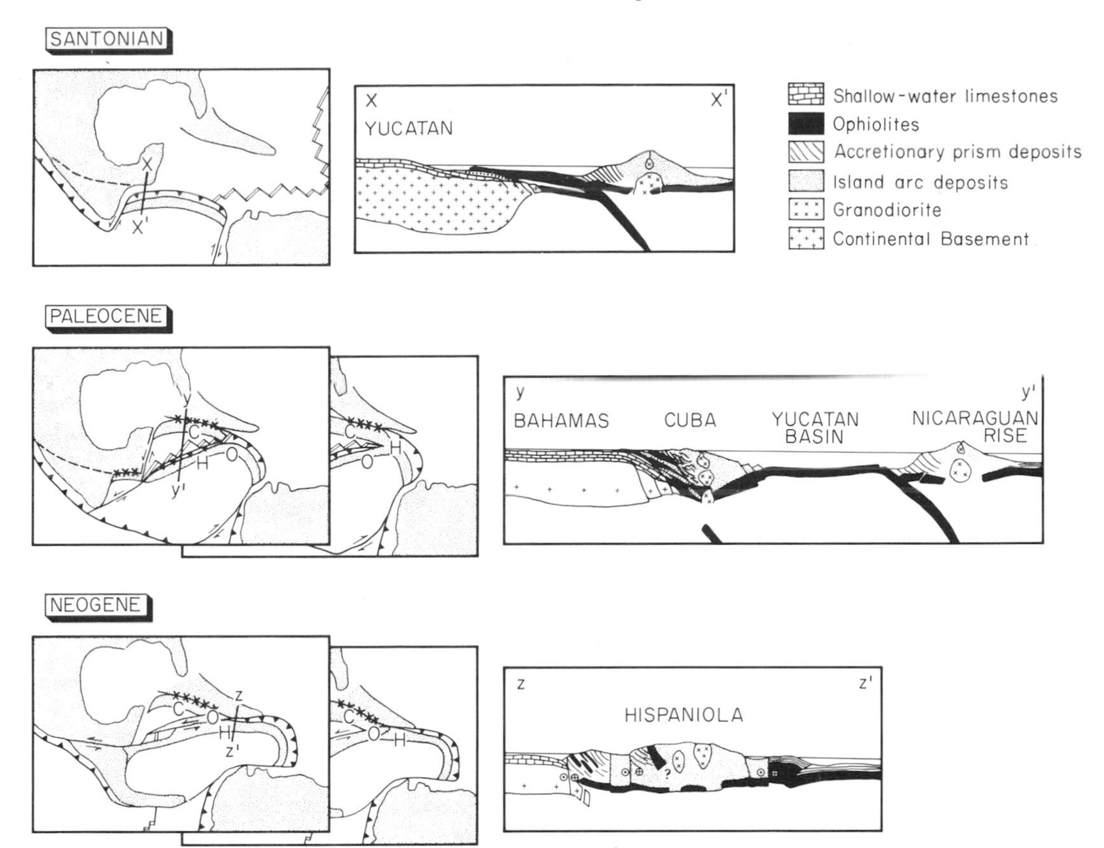

FIG. 3. Schematic interpretations of the roles played by ophiolites at three periods. The Santonian map shows the approach of the Greater Antilles arc from the SW prior to the collision with Yucatan . In the section the allochthonous ophiolite sheet is shown being emplaced ahead of the arrival of the subduction zone. By the Paleocene the Cuban arc has collided with the Bahamas, which underthrusts the northern half of the arc. Back-arc spreading in the Yucatan basin is partly consumed to the south beneath the Nicaraguan Rise. Two alternative maps for the positions of Oriente (O) and Hispaniola (H) relative to the main Cuban arc (C) are shown and their subsequent development continued in the Neogene. The section through Hispaniola during the Neogene emphasizes the different settings of the three Hispaniolan belts and the intervening strike-slip fault systems. The North Coast belt is part of a subduction complex sealed when the Bahamas platform entered the trench. The Cordillera Central belt is shown as an older subduction complex. To the south, the Southern Peninsula is an uplifted block of thickened Caribbean crust perhaps associated with some minor underthrusting.

pletely cut the arc after the Yucatan collision rather than partially splitting it with a pivotal opening in Oriente (Fig. 3). Central Hispaniola would have occupied a position south or west of Oriente. Alternatively, central Hispaniola may have been the extension of the central Cuban arc until the Paleogene collision event. Subsequent left-lateral displacements would move central Hispaniola only a short distance eastwards adjacent to the North Coast belt to be replaced by the back-arc/remnant arc fragment of Oriente.

The origin of the Cordillera Central belt is still not clear. The present position of the early Cretaceous rocks of the Los Ranchos Formation, and other volcanic arc rocks of the Cordillera Central is NE of, and probably contemporaneous with, the ophiolite belt (Kesler *et al.* 1981). Bowin (1975) and Mattson (1979) interpreted this as resulting from northward subduction of Caribbean crust, with the Duarte Complex as a metamorphosed fragment of this crust. Draper & Lewis (1980, 1982), however, have suggested that the Duarte Complex is the lower part of a volcanic arc edifice and that the peridotite belt and the low-grade Amina-Maimon schists are part of an accretionary prism formed by a southwestward-dipping subduction zone. In this interpretation, the ophiolite originated in a small basin between two volcanic arcs which collided in the middle Cretaceous, and are now represented by the southwestern Cordillera Central and the Cordillera Oriental.

The Puerto Rican, Southern Peninsula and Jamaican ophiolites are furthest removed from the present Atlantic and hence are the most likely candidates for origins as Caribbean crust. Of these, the case for the Southern Peninsula ophiolite as Caribbean crust thickened during the late Cretaceous is easily the strongest (Maurasse *et al.* 1979). The Pindell & Dewey model cannot adequately explain the Puerto Rican and Jamaican ophiolites. Other explanations for these have included northward subduction of Caribbean crust (e.g. Burke *et al.* 1978), particularly during the early Cretaceous (Mattson 1979), and southward subduction of Yucatan Basin crust (Wadge *et al.* 1982). This latter scenario would account for the apparent youthfulness of the Jamaican ophiolite. The early Cretaceous volcanic rocks associated with the Bermeja complex are not exposed and hence the question of subduction polarity is difficult to resolve in Puerto Rico (Mattson 1979).

Conclusions

1. Most of the northern Caribbean ophiolites consist of latest Jurassic to mid-Cretaceous ocean crust which was obducted during the late Cretaceous.
2. The provenance of these ocean-crust fragments can be one of the following: Pacific plate; crust formed during the opening of the North Atlantic; back-arc basin crust formed within the Caribbean. Both Pacific and Atlantic ocean crust should give early Cretaceous ages. However, late Cretaceous ophiolites (e.g. Jamaica) indicate a Caribbean origin.
3. The concept of a northward-moving island arc which subducted Atlantic crust to the south and collided with continental crust in Yucatan and the Bahamas in late Cretaceous–Paleogene times, is consistent with the histories of emplacement of the Guatemalan and the western and central Cuban ophiolites.
4. With the exception of parts of the northern belt of Guatemala, all the northern Caribbean ophiolites have been dismembered tectonically. Dismemberment has occurred by: obduction within an accretionary prism (e.g. North Coast belt, Hispaniola); subsequent imbricate overthrusting during continental collision (e.g. central Cuba); and most ubiquitously, by Cenozoic strike-slip faulting (e.g. Jamaica).
5. None of the existing plate tectonic models satisfactorily explains all the ophiolites. The older ophiolites (e.g. Cordillera Central belt, Hispaniola) are particularly difficult to accommodate in these models.

ACKNOWLEDGMENTS: J. Rosenfeld and T.W. Donnelly made helpful comments concerning the Guatemalan ophiolites. G. Wadge would like to thank Lunar and Planetary Institute for support during this work. G. Draper was partly supported by the Latin American–Caribbean Center and the International Affairs Center of Florida International University. J.F. Lewis was supported by grants NSF-INT78-06265 and NSF-INT81-16703. This is Lunar and Planetary Institute Contribution No. 508.

References

ARDEN, D. D. JR 1975. The geology of Jamaica and the Nicaraguan Rise. *In*: NAIRN, A. E. M. & STEHLI, F. (eds) *The Ocean Basin and Margins. Vol. 3.* Plenum Press, New York, pp. 617–41.

BERTRAND, J. & VUAGNAT, M. 1975. Sur la présence de basaltes en coussine dans la zone ophiolitique méridionale de la Cordillère centrale du Guatémala. *Bull. suisse Minéral. Pétrogr.* **55**, 136–42.

—— & —— 1976. Etude pétrographique de diverse ultrabasites ophiolitiques du Guatémala et de leurs inclusions. *Bull. suisse Minéral. Pétrogr.* **56**, 527–40.

—— & —— 1980. Inclusions in the serpentinite mélange of the Motagua Fault Zone, Guatemala. *Arch. Sci. (Genève),* **33**, 321–35.

BOITEAU, A. & MICHARD, A. 1976. Données nouvelles sur le socle métamorphique de Cuba. Problèmes d'application de la tectoniques des plaques. *Trans. 7th. Caribb. Geol. Conf., Guadeloupe*, 221–6.

——, —— & SALIOT, P. 1972. Métamorphisme de haute pression dans le complexe ophiolitique du Purial (Oriente, Cuba). *C.r. Acad. Sci. Paris, Ser. D, 274*, 2137–40.

BOVENKO, V. G., SCHERBAKOVA, B. E. & ERNANDES, G. 1980. New geophysical data on the deep structure of eastern Cuba. *Soviet Geology* **9**, 101–9.

——, —— & —— 1979. Deep geological structure of western Cuba. *In*: *The Tectonic and Geodynamic of the Caribbean Region.* Nauka, Moscow, pp. 130–42.

BOWIN, C. 1966. Geology of central Dominican Republic: A case history of part of an island arc. *In*: HESS, H. H. *Caribbean Geological Investigations.* Geol. Soc. Am. Mem. **98**, 11–84.

—— 1975. The geology of Hispaniola. *In*: NAIRN, A. E. M. & STEHLI, F. (eds) *The Ocean Basins and Margins, Vol. 3,* Plenum Press, New York, pp. 501–552.

—— 1976. Caribbean gravity field and plate tectonics. *Geol. Soc. Am. Special paper* **169**, 79 pp.

—— & NAGLE, F. 1980. Igneous and metamorphic rocks of northern Dominican Republic. *Trans. 9th. Caribb. Geol. Conf., Santo Domingo*, 1980, 39–45.

BURKE, K., FOX, P. J. & SENGÖR, A. M. C. 1978. Buoyant ocean floor and the evolution of the Caribbean. *J. Geophys. Res.* **83**, 3949–54.

CASE, J. E. 1980. Crustal setting of mafic and ultramafic rocks and associated ore deposits of the Caribbean region. *U.S. Geological Survey Open-File Report,* **80-304**, 95 pp.

COBIELLA, J. L. 1978. Una melange en Cuba Oriental. *Mineria en Cuba,* **4**, (4) 46–51.

——, CAMPOS, M., BOITEAU, A. & QUINTAS, F. 1977. Geologia del flanco sur de la Sierra del Purial. *Mineria en Cuba,* **3**, (1) 55–62; (2) 44–53.

COLEMAN, R. G. 1977. *Ophiolites*. Springer-Verlag, Berlin, 229 pp.

DENGO, G. 1972. Review of Caribbean serpentinites and their tectonic implications. *Mem. Geol. Soc. Am.* **132**, 303–12.

DRAPER, G. 1979. *The regionally metamorphosed rocks of Eastern Jamaica.* Thesis, PhD, Univ. West Indies, Kingston, Jamaica, (unpubl.) 277 pp.

—— & LEWIS, J. F. 1980. Petrology, deformation and tectonic history of the Amina Schists, northern Dominican Republic. *Trans. 9th. Caribb. Geol. Conf., Santo Domingo,* 1980, 53–64

—— & —— 1982. Metamorphic and plutonic belts in the Dominican Republic: Implications for island arc tectonics in the northern Caribbean plate boundary. *Abstr. Geol. Soc. Am.* **14**, 477.

DUCLOZ, C. & VUAGNAT, M. 1962. À propos de l'age des serpentinites de Cuba. *Arch. Sci. (Genève)* **16**, 351–402.

EBERLE, W., HIRDES, W., MUFF, R. & PALAEZ, M. 1980. The geology of the Cordillera Septentrional (Dominican Republic). *Trans. 9th. Caribb. Geol. Conf., Santo Domingo,* 1980, 619–32.

FLINT, D. E., ALBEAR, J. F. & GUILD, P. W. 1948. Geology and chromite deposits of the Camagüey Province, Cuba. *U.S. Geol. Survey Bull.* **954-B**, 39–63.

FONSECA, E., PHILIPOV, C. & PEREZ, M. 1980. Caracteristicas petrologicas y petroquimicas generales de las rocas vulcanogenas e intrusivas basicas de la zona norte, region Bahia Honda. *Ser. Geol. (Havana)* **3**, 15 pp.

GEALEY, W. K. 1980. Ophiolite obduction mechanism. *In*: PANAYIOTOU, A. (ed.) *Ophiolites, Proc. Int. Ophiolite Symp., Cyprus 1979,* 228–43.

GHOSH, N., HALL, S. A. & CASEY, J. F. (in prep.) Seafloor spreading and magnetic anomalies in the Venezuela Basin.

GUILD, P. W. 1947. Petrology and structure of the Moa chromite district, Oriente Province, Cuba. *Trans. Am. Geophys. Union* **28**, 218–46.

HALDEMANN, E. G., BROUWER, S. B., BLOWES, J. H. & SNOW, W. E. 1980. Lateritic nickel deposits at Bonao Falconridge Dominicana C. por A., *Field Guide, 9th. Caribb. Geol. Conf., Santo Domingo,* 69–78.

HERNANDEZ, M. 1979. Datos preliminaros sobre las caracteristicas petrograficas de las rocas del macizo Sierra Purial. *Mineria en Cuba,* **5**, No. 2, 2–7.

KEIJZER, F. G. 1945–7. Outline of the geology of the eastern part of the province of Oriente, Cuba. *Geog. en Geol. Mededeelingen,* **2**, (6) 239 pp.

KESLER, S. E., RUSSEL, N., SEAWARD, M., RIVERA, J., MCCURDY, K., CUMMING, G. L. & SUTTER, J. F. 1981. Geology and geochemistry of sulfide mineralisation underlying the Pueblo Viejo gold-silver oxide deposit, Dominican Republic. *Econ. Geol.* **76**, 1096–117.

KHUDOLEY, K. M. & MEYERHOFF, A. A. 1971. Paleogeography and geological history of Greater Antilles. *Mem. Geol. Soc. Am.* **129**, 199 pp.

KIM, J. J., MATUMOTO, T. & LATHAM, G. V. 1982. A crustal section of northern Central America as inferred from wide-angle reflections from shallow earthquakes. *Bull. Seismol. Soc. Am.* **72**, 925–40.

KNIPPER, A. L. & CABRERA, R. 1974. Tectonica y geologica historica de la zona de articulacion entre el mio- y eugeosynclinal y del cinturon hiperbasico de Cuba. *In*: *Contribucion a la Geologia de Cuba, Publ. Especial No. 2* Instituto de Geologia y Paleontologia, Havana, pp. 15–77.

KOZARY, M. T. 1968. Ultramafic rocks in thrust zones of northwestern Oriente province, Cuba. *Bull. Am. Assoc. Petrol. Geol.* **52**, 2298–317.

LAWRENCE, D. P. 1976. Tectonic implications of the geochemistry and petrology of the El Tambor Formation: probable oceanic crust in central Guatemala. *Abstr. Geol. Soc. Am.* **8**, 973–4.

LEWIS, G. E. & STRACZEK, J. A. 1955. Geology of south-central Oriente, Cuba. *Bull. U.S. Geol. Surv.* **975D**, 171–336.

LEWIS, J. F. 1980. Ultrabasic and associated rocks in Hispaniola. *Trans. 9th. Caribb. Geol. Conf., Santo Domingo*, 1980, 403–8.

MATTSON, P. H. 1960. Geology of the Mayaguez area, Puerto Rico. *Bull. Geol. Soc. Am.* **71**, 319–62.

—— 1974. Middle Cretaceous nappe structures in Puerto Rico ophiolites and their relation to the tectonic history of the Greater Antilles. *Bull. Geol. Soc. Am.* **84**, 21–38.

—— 1979. Subduction, buoyant braking, flipping and strike-slip faulting in the northern Caribbean. *J. Geol.* **87**, 293–304.

—— & PESSAGNO, E. A., JR 1979. Jurassic and early Cretaceous radiolarians in Puerto Rican ophiolite—tectonic implications. *Geology* **7**, 440–4.

—— & SCHWARTZ, D. P. 1971. Control of intensity of deformation in Puerto Rico by mobile serpentinized peridotite basement. *Mem. Geol. Soc. Am.* **130**, 97–106.

MAURASSE, F. J.-M. 1981. Relations between the geologic setting of Hispaniola and the evolution of the Caribbean. *Trans. 1st. Colloque Géol. d'Haiti, Port au Prince, Haiti,* 1980, 246–64.

——, HUSLER, J., GEORGES, G., SCHMITT, R. & DAMOND, P. 1979. Upraised Caribbean sea-floor below acoustic reflection B″ at the southern peninsula of Haiti. *Geol. Mijnb.* **58**, 71–83.

MCBIRNEY, A. R. 1963. Geology of a part of the Central Guatemalan Cordillera. *Univ. California Publ. in Geol. Sci.* **38**, 177–242.

MERCIER DE LEPINAY, B., LABESSE, B., SIGAL, J. & VILA, J.-M. 1979. Sedimentation chaotique et tectonique tangentielle Maestrichtiennes dans la presqu'île du sud d'Haiti (île d'Hispaniola, Grandes Antilles). *C. r. Acad. Sc. Paris, Ser. D.* **289**, 887–90.

MEYERHOFF, A. A. & HATTEN, C. W. 1968. Diapiric

structures in central Cuba. *In*: BRAUNSTEIN, J. & O'BRIEN, G. E. (eds) *Diapirism and Diapirs.* Mem. Am. Assoc. Petrol. Geol. **8**, 315–57.

MILLAN, G. & SOMIN, M. L. 1981. Litologia, estratigrafia, tectonica, y metamorfismo del macizo de Escambray. *Acad. Ciencias de Cuba Industria* **452**, 104 pp.

MOSSAKOVSKIY, A. A. & ALBEAR, J. F. 1978. Nappe structure of western and northern Cuba and history of its emplacement in the light of a study of olistostromes and molasse. *Geotectonics* **12**, 225–36.

MULLER, P. D. 1979. *Geology of the Los Amates quadrangle and vicinity, Guatemala, Central America.* Thesis, PhD, State Univ. New York at Binghamton, (unpubl.) 326 pp.

NAGLE, F. 1966. *Geology of the Puerto Plata area, Dominican Republic.* Thesis, PhD, Princeton University (unpubl.) 171 pp.

—— 1974. Blueschists, eclogite, paired metamorphic belts and the early tectonic history of Hispaniola. *Bull. Geol. Soc. Am.* **85**, 1461–6.

NICOLINI, P. 1977. *Les porphyres cuprifères et les complexes ultra-basiques du Nord-est d'Haiti. Essai de gitologie provisionelle.* Thesis, L'Université Pierre et Marie Curie, Paris, 203 pp.

PALMER, H. C. 1963. *Geology of the Moncion-Jarabacoa area, Dominican Republic.* Thesis, PhD, Princeton University (unpubl.) 256 pp.

PARDO, G. 1975. Geology of Cuba. *In*: NAIRN, A. E. M. & STEHLI, F. (eds) *The Ocean Basins and Margins. Vol. 3*, Plenum Press, New York, 553–615.

PERFIT, M. R. & MCCULLOCH, M. T. 1982. Nd- and Sr-isotope geochemistry of eclogites and blue-schists from the Hispaniola-Puerto Rico subduction zone. *Terra Cognita* **2**, 325.

——, HEEZEN, B. C., RAWSON, M. & DONNELLY, T. W. 1980. Chemistry, origin and tectonic significance of metamorphic rocks from the Puerto Rico trench. *Mar. Geol.* **34**, 125–56.

PETERSEN, E. U. & ZANTOP, H. 1980. The Oxec deposit, Guatemala: an ophiolite copper occurrence. *Econ. Geol.* **75**, 1053–65.

PINDELL, J. & DEWEY, J. F. 1982. Permo-Triassic reconstruction of western Pangea and the evolution of the Gulf of Mexico/Caribbean region. *Tectonics* **1**, 179–212.

REBLIN, M. T. 1973. *Regional gravity survey of the Dominican Republic.* Thesis, M.Sc. University of Utah. (unpubl.) 95 pp.

ROPER, P. J. 1978. Structural fabric of serpentinite and amphibolite along the Motagua fault zone in El Progresso quadrangle, Guatemala. *Trans. Gulf Coast Assoc. Geol. Socs.* **28**, 449–58.

ROSENFELD, J. H. 1981. *Geology of the western Sierra de Santa Cruz, Guatemala, Central America. An ophiolite sequence.* Thesis, PhD, State University of New York at Binghamton (unpubl.) 313 pp.

SCHERBAKOVA, B. E., BOVENKO, B. G. & HERNANDEZ, H. 1980. Relief of the Mohorovicić Discontinuity surface under western Cuba. *Doklady, Earth Sciences* **238**, 7–9.

SHAPOSHNIKOVA, K. I. 1974. Tectonics of central Cuba. *Geotectonics* **1**, 14–20.

SOTO, R. S. 1978a. La secuencia espilitico diabasica perforada por el pozo 'Mercedes 2', *Geol. Mijnb.* **57**, 382.

—— 1978b. Algunos aspectos petrologicos de las rocas ultramaficas de Cuba. *Geol. Mijnb.* **57**, 382.

SUTTER, J. F. 1979. Late Cretaceous collisional tectonics along the Motagua Fault Zone, Guatemala. *Abstr. Geol. Soc. Am.* **11**, 525–6.

THAYER, T. P. & GUILD, P. W. 1947. Thrust faults and related structures in eastern Cuba. *Trans. Am. Geophys. Union* **28**, 919–30.

WADGE, G. & BURKE, K. (in press) Neogene Caribbean plate rotation and associated Central American tectonic evolution. *Tectonics.*

——, DRAPER, G. & ROBINSON, E. 1980. Gravity anomalies in the Blue Mountains, eastern Jamaica. *Trans. 9th. Caribb. Geol. Conf., Santo Domingo*, 1980, 467–74.

——, JACKSON, T. A., ISAACS, M. C. & SMITH, T. E. 1982. The ophiolitic Bath-Dunrobin Formation, Jamaica: significance for Cretaceous plate margin evolution in the north-western Caribbean. *J. Geol. Soc. London* **139**, 321–33.

WILLIAMS, M. D. 1975. Emplacement of Sierra de Santa Cruz, Eastern Guatemala. *Bull. Am. Assoc. Petrol. Geol.* **59**, 1211–6.

G. WADGE, Lunar and Planetary Institute, 3303 NASA Road One, Houston, Texas 77058, USA. Present address: Seismic Research Unit, University of the West Indies, St Augustine, Trinidad.

G. DRAPER, Department of Physical Sciences, Florida International University, Tamiami Trail, Miami, Florida 33134, USA.

J. F. LEWIS, Department of Geology, George Washington University, Washington DC, 20006, USA.

A conspectus of Scandinavian Caledonian ophiolites

B.A. Sturt, D. Roberts & H. Furnes

SUMMARY: Dismembered and fragmented ophiolite assemblages constitute important elements of the metamorphic allochthon of the Scandinavian Caledonides. Ophiolites recognized to date range in age from probable Vendian to Middle Ordovician and occur at two principal tectonostratigraphic levels. Field criteria, with support from geochemical data, permit a classification of many of the ophiolite complexes into two fundamental groups. Group I is characterized by a generally well-developed pseudostratigraphy, MORB as well as IAT and WPB petrochemistry, plagiogranites at higher levels, and a conformable cap of oceanic sediments with local ocean or immature arc volcanism. The group is representative of either a major ocean or evolved marginal-basin regime. Evidence points to their obduction, and initial internal deformation, in the Finnmarkian orogenic event in pre-Middle Arenig time. Group II complexes are younger, of late Arenig to Llanvirn and possibly Llandeilo age, and show a poorer-defined pseudostratigraphy. They are laterally and vertically intercalated with either siliciclastic sediments from a contemporary magmatic arc or arcs, and are considered as having been generated in a restricted marginal-basin setting. Their prime tectonic deformation, and translation within nappes, is essentially of Silurian age. Group I ophiolites were also further dissected and deformed during this main Caledonian event.

Dismembered and fragmented ophiolite assemblages are now known to constitute significant elements of the metamorphic allochthon of the Scandinavian Caledonides. Occurring almost exclusively in the Norwegian sector of the orogen and at higher levels in the tectonostratigraphy, the ophiolite remnants vary widely in character depending on the state of dismemberment and fragmentation of the original pseudostratigraphy and the degree of tectonic strain to which they have been subjected. They range in size from complexes up to 100 km in outcrop length to lensoid megaboudins measurable on the deci metre scale.

Ophiolites recognized to date show an age range from probable Vendian to Middle Ordovician, though with an important bipartite grouping discerned from the combined criteria of field relationships, faunal constraints and radiometric dating, aided by comprehensive geochemical studies. The two fundamental groups, I and II (Furnes *et al.* in press), equate in broad terms with Vendian–Cambrian and Ordovician ocean-floor assemblages respectively, with the late Cambrian–earliest Ordovician Finnmarkian orogenic event (Sturt *et al.* 1978) providing a natural and convenient line of separation between the two. Group I ophiolite complexes appear to be derivatives of either a major ocean or a mature, arc-remote, marginal-basin setting, and evidence points to their eastward obduction and initial internal deformation in the Finnmarkian, in pre-Middle Arenig time. Group II fragmented ophiolites in central Norway are of late Arenig to Llanvirn and possibly Llandeilo age and bear geochemical imprints of generation in a restricted back-arc or inter-arc basinal regime. Their prime tectonic deformation, and incorporation within easterly transported nappes (Roberts & Sturt 1980), is essentially of Silurian age. Group I ophiolites were further dissected and deformed during this main Caledonian (Scandian) event.

Although the ophiolites of both groups occur in the two highest principal tectonostratigraphic levels in the metamorphic allochthon (Fig. 1) (Roberts *et al.* 1983a), their actual location within the nappe pile has been complicated in some areas by multicomponental thrusting and slicing with resultant repetition of stratigraphies. In this contribution we outline the basic divisions of the ophiolite types and their distribution, and note the character and importance of major unconformities and tectono-metamorphic breaks throughout the mountain belt.

Group I ophiolites

These are characterized by having been subjected to two distinctive tectono-metamorphic cycles during Caledonian orogenic evolution—i.e. both Finnmarkian and Scandian. They represent material derived from an oceanic crust which was essentially pre-Ordovician in age, thus providing a record from Iapetus in its early development stage (Fig. 2). The evidence for a pre-Ordovician age for this oceanic crust is varied and ranges from geochronology, faunal/stratigraphic control to event correlation (Table 1). There is also evidence for the existence of an early primitive ensimatic island arc, probably representing an initial stage of subduction. These island-arc products were eventually obducted together with the subjacent oceanic crust during the Finnmarkian orogenic phase.

The ophiolite fragments in SW Norway show a series of features relevant to this sequence of

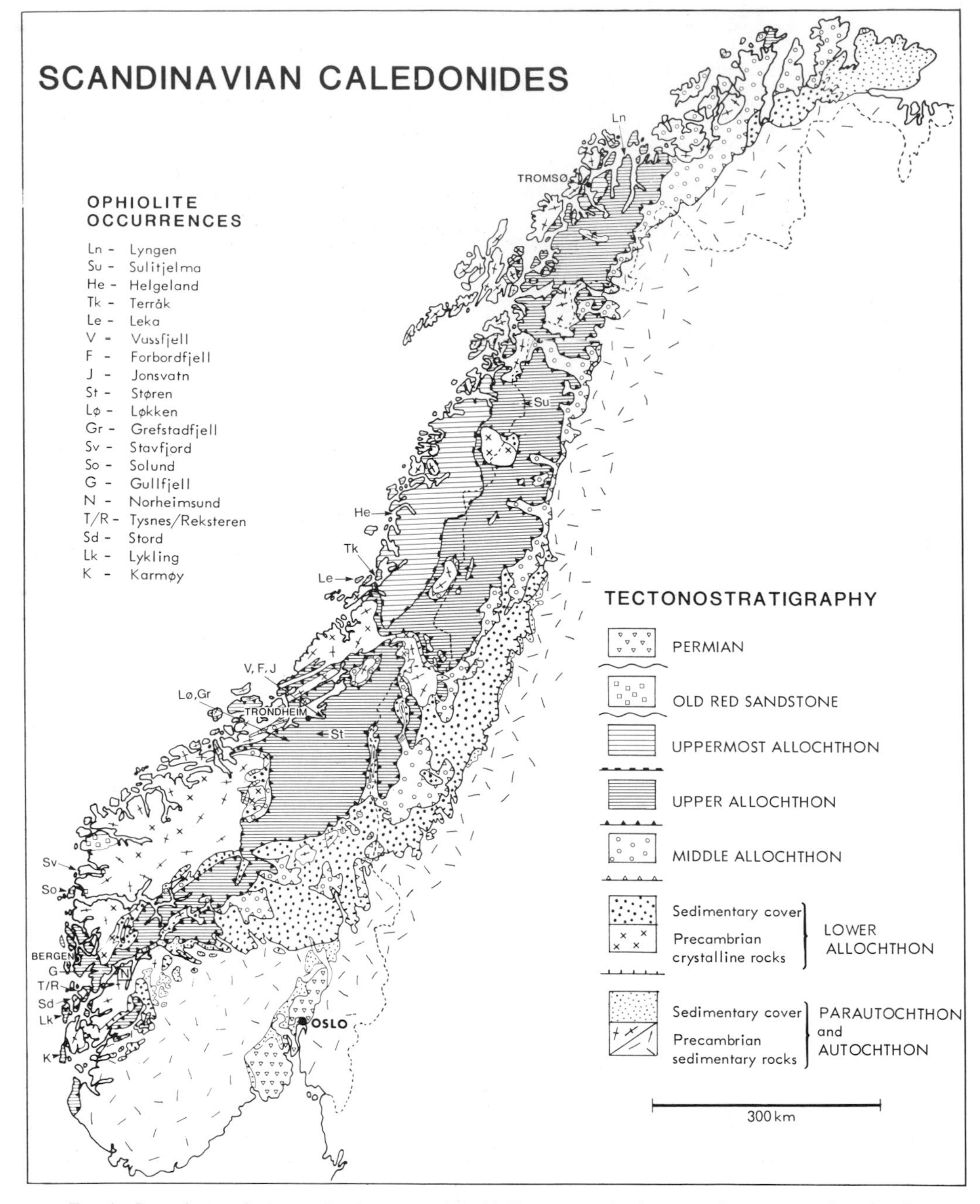

FIG. 1. Locations of the main fragmented ophiolites recognised up to the present time in the Scandinavian Caledonides. The background Caledonian tectonostratigraphy is simplified from an IGCP Project 27 1:1,000,000 map compilation, modified from Roberts & Gee (1983).

events (Table 1). The Karmøy Ophiolite has a virtually complete pseudostratigraphy (Sturt & Thon 1978a; Sturt *et al.* 1979, 1980) with a thick cap-rock sequence of essentially pelagics, hemipelagics, off-axis alkaline basaltic lavas and volcaniclastic sediments (Sturt *et al.* 1980; Solli 1981). The cap-rock succession (Torvastad Group) has been affected by polyphasal deformation and metamorphism in upper greenschist facies prior to the emplacement of the pre-Upper

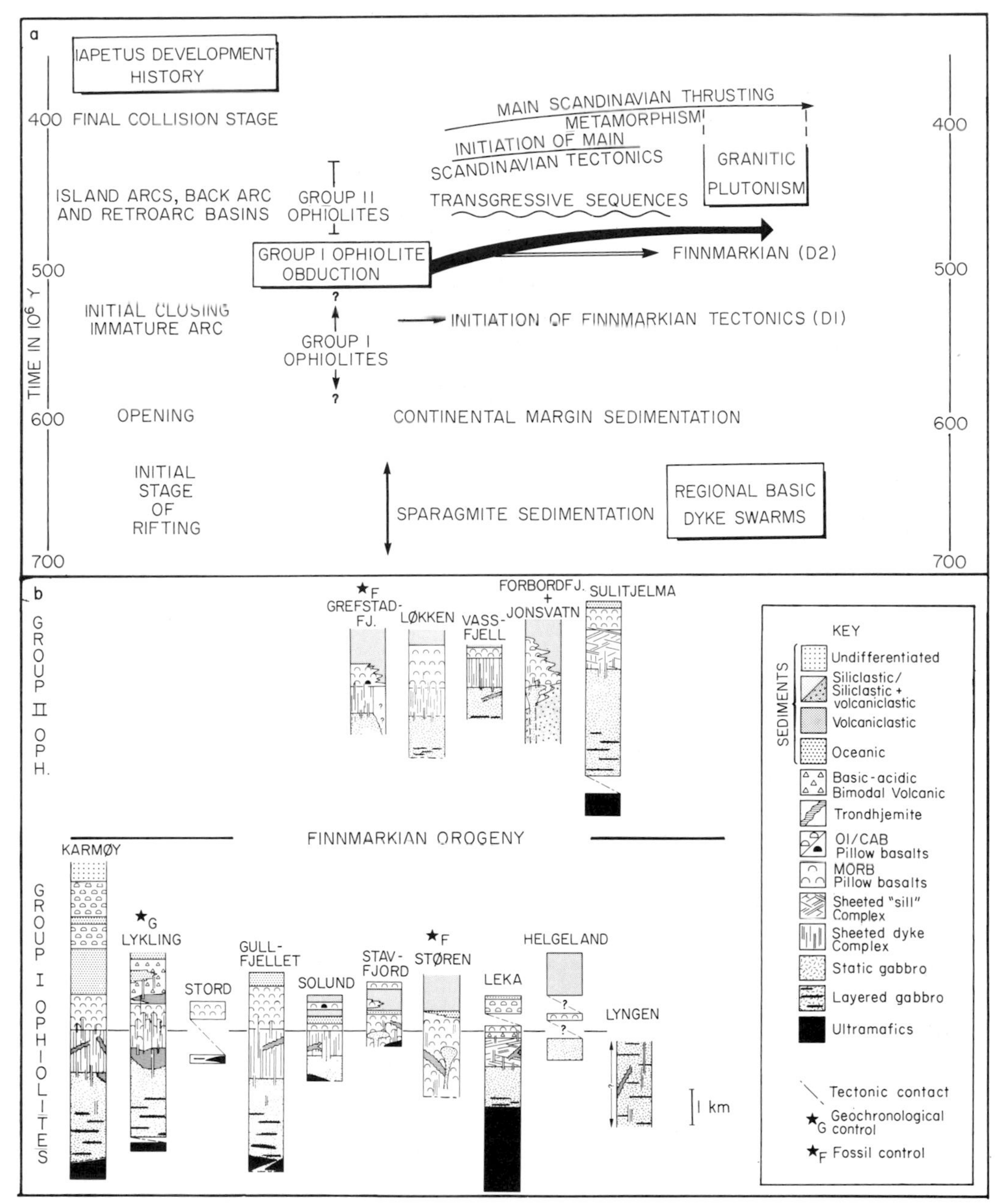

FIG. 2. (a) Iapetus development history, and the time of generation of the two groups of Norwegian ophiolites in the context of the evolving orogen. (b) Simplified pseudostratigraphic columns of the Group I and Group II ophiolite fragments, with the Finnmarkian orogenic phase indicated as the line of separation of the two groups.

Ordovician granitoids of the West Karmøy Igneous Complex (WKIC). This latter provides evidence, in its extensive xenolith population, for the prior obduction of the ophiolite. This is seen in the abundant xenoliths of continental basement rocks brought up through the ophiolite (Sturt & Thon 1978b; Ledru 1980), demonstrating that the Karmøy Ophiolite had already been obducted onto a westward extension of the Baltoscandian craton prior to intrusion of the granitoids. The geochronology of the WKIC is not yet refined and only shows that the late intrusive phases are

TABLE 1. *Summary of available information from the early, Group I ophiolites in the Scandinavian Caledonies*

Ophiolite fragment	Pseudostratigraphy	Cap-rock	Associated off-axis lavas	Evidence for transform fault	Evidence for primitive island arc	Transgressive unconformity and overlying sediments	Evidence for pre-unconformity deformation and metamorphism
Karmøy	Pillow lava Sheeted dykes Varied-textured and cumulate gabbro Ultramafic cumulates	Very thick Pelagics/ hemi-pelagics No fauna	Alkaline within plate basalts and minor trachytic tuffs	Dyke rotation Basal cap-rock sediments	Late low-K tholeiite dykes Cut deformed gabbro	Marked. Weathering profile. Cont./ shallow marine Sub-Ashgill	Polyphasal def. Upper greensch. facies met. Cut by W. Karmøy Igneous Comp. prior to deposition of U. Ord.
Bømlo (Lykling)	Pillow lava Sheeted dykes Varied-textured and cumulate gabbro Serpentinites	Pelagics/ hemi-pelagics Chitinzoa (probably Tremadoc)	Not known	Deep marine fault-scarp breccias (sedimentary)	Low-K tholeiites and q-keratophyres Rb/Sr w.r. 535±46 Ma	Marked Continental basaltic to rhyolitic volc. Rb/Sr w.r. 464±16 Ma 468±23 Ma	Overfolding, with associated cleavage Low greensch. facies met.
Bergen arcs (Gullfjellet)	Pillow lava Sheeted dykes Gabbros Few ultramafic cumulates	Pelagics/ hemi-pelagics Mélanges No fauna	Not known	Not known	Low-K tholeiites	Marked Continental shallow marine Sub-Ashgill	Polyphasal folding Intr. of q-diorite (def. pre-unconf.) Upper greenschist/ amph. facies
Solund-Stavfjord	Pillow lava Sheeted dykes Varie-textured and minor layered gabbro Minor serpentinites	Pelagics/ hemi-pelagics (but little known) No fauna	Not known	Not known	Not known	Yes Continental shallow marine Sub-Ashgill	Yes, but little data available

Støren (+ Fundsjø arc)	Pillow lava Local dykes Gabbros	Pelagics/ hemi-pelagics Contains Dictyonema (Tremadoc)	Not known	Not known	Low-K and -Ti tholeiites Keratophyres Alb. granites	Yes Continental shallow marine Sub-upper-middle Arenig	Two phases of deformation pre-date trondhjemites dated to 478–447 Ma
Leka	Pillow lava (?) Sheeted dykes Cumulate gabbros Ultramafic cumulates Harzburgite (depleted mantle ?)	Minor pelagics/ hemi-pelagics No fauna	Alkaline within plate basalts	Not known	Not known	Marked Weathering profile. Cont./ shallow marine Age unknown	Earlier cleavage in sheeted dykes
Helgeland nappe complex complex (Terråk, Velfjord, Rødøy etc)	Different parts in different fragments Deformed harzburgite at Terråk	Little yet recorded	Not known	Not known	Not known	Marked Weathering profile. Cont./ shallow marine Age unknown	Folded gabbro and ultramafic cumulate layering
Lyngen	Lyngen Massif of mainly gabbro Kjosen Fm contains also pillow lava and dykes (G. Oliver, pers. comm.)	Little yet recorded	Not known	Not known	Not known	Marked Weathering profile. Cont./ shallow marine Sub-Ashgill	Strong folding of gabbro

of general mid-Ordovician age (Sturt *et al.* 1979). The Torvastad Group metasediments also carry evidence indicating that prior to the emplacement of the WKIC, the ophiolite had been involved in orogenic deformation and metamorphism.

Turning to the nearby island of Bømlo, the Lykling Ophiolite was strongly folded, and in places inverted, before being unconformably overlain by bimodal continental volcanics (Nordås *et al.* 1983) now reliably dated as Middle Ordovician in age (Furnes *et al.* 1983a). An ensimatic island arc was developed on the edifice of oceanic crust with extensive eruption of low-K tholeiites and keratophyres (Nordås *et al.* 1983) which have recently been dated at 535 ± 46 Ma (Furnes *et al.* 1983a). This latter dating shows that the ophiolite itself must represent older oceanic crust. In the pelagic sediments directly above the ophiolite *Chitinozoa* have been recovered (C. Magnus & S. Alpaslan, pers. comm. 1982) which indicate a probable Tremadoc age.

In the Bergen Arc region, ophiolitic mélanges are associated with the Gullfjellet Ophiolite in the Os district (Thon in press; S.E. Ingdahl pers. comm. 1983). These, together with cap-rock sediments, have been affected by polyphasal folding and metamorphism in upper greenschist/amphibolite facies prior to deposition of the sub-Ashgill sequence of the Moberg Conglomerate. The ophiolite was also cut by quartz-diorite intrusions which were strongly deformed before deposition of the Moberg Conglomerate. This shows a similarity to Karmøy, again with clear evidence for the involvement of the ophiolite in a pre-upper Ordovician tectono-thermal orogenic event.

In the Trondheim region (Fig. 1) there is clear evidence for the deep erosion of the Støren ophiolite slice and subjacent mélange (Horne 1979), before the deposition of an overlying unconformable sequence of Lower Hovin Group rocks containing a middle to late (Upper) Arenig fauna (Bruton & Bockelie 1980; Ryan & Sturt in press; Grenne & Lagerblad in press). In the eastern part of the same region, the Fundsjø group represents an ensimatic arc with volcanics in part interbedded with and overlain by sediments and volcaniclastics; these are considered by correlation to be at least in part of Tremadoc age (Grenne & Lagerblad in press). The rocks of the Fundsjø Group have suffered several phases of folding and metamorphism prior to the deposition of the overlying Sulåmo Group (Grenne & Lagerblad in press; Hardenby *et al.* in press). Morever, the Fundsjø and Støren Groups and their Gula Complex substrate (but not the overlying sediments) are cut by many trondhjemite dykes with maximum Rb/Sr and zircon ages of 478 Ma (Klingspor & Gee in press).

The remaining Group I ophiolites (Figs 1 & 2) do not yet have such well-defined stratigraphic or radiometric constraints as those outlined above; in some cases critical contacts are not exposed, while in others (detailed) mapping has still to be carried out. Islands adjacent to Leka carry amphibolite-facies rocks similar to Gula Complex lithologies, yet the critical boundary with the ophiolite is hidden by sea. At Lyngen in north Norway, conglomerates overlying the ophiolite contain clasts derived from the Finnmarkian-deformed metasedimentary successions, all members of the immediately subjacent ophiolite pseudostratigraphy, as well as exotic island-arc debris (Minsaas & Sturt 1983).

There are some striking differences between the different ophiolite fragments with regard to palaeogeographical setting. Firstly, evidence for transform faulting is restricted to the southernmost localities (Karmøy, Lykling) where features of the sedimentary cap-rocks, and rotational patterns of sheeted dykes are strongly suggestive of transform activity (Amaliksen 1980; Solli 1981). Thick developments of ultramafic cumulates, often with chromite concentrations, and the only instances yet recorded of upper-mantle tectonites (Table 1) are confined to Leka and other occurrences in the Helgeland Nappe Complex (Furnes *et al.* 1980b; Prestvik 1980). As shown by Furnes *et al.* (1982) there are also regional differences in the geochemical patterns which are summarized below.

The major- and trace-element characteristics of the pillow lavas and sheeted dykes from the Group I ophiolite complexes have been reported in a number of recent papers (e.g. Furnes *et al.* 1980a, b, 1982, in press; Prestvik 1980, Stephens *et al.* in press). Only the main features, therefore, are briefly mentioned here.

From Bergen southwards, the Karmøy, Lykling and Gullfjellet ophiolites are characterized by showing profound variations in the total abundances as well as the ratios between incompatible elements. In discriminant diagrams, such as TiO_2–FeO^t/MgO, Ti–Cr, TiO_2–Zr/P_2O_5 and Ti-Y-Zr, the data show both alkaline and subalkaline character and they plot across the WPB, MORB and IAT fields. Their REE patterns, which vary from nearly flat, LREE-depleted to slightly LREE-enriched, are characterized by between-sequence variations, as well as marked within-sequence variations from the Gullfjellet ophiolite which, in addition to the above-mentioned patterns, also show an upward-convex shape (Furnes *et al.* 1980a, b, 1982, in press). Those immediately north of Bergen—i.e. the Solund and Stavfjorden ophiolites—are in all geochemical respects similar to MORB, though their REE patterns are characteristically upward-convex in form (Furnes *et al.* 1982). The metabasalts from the Støren ophiolite in the Trondheim region are exclusively subalkaline but on the Ti-Y-Zr diagram they cluster around the WPB and MORB boundary. Their REE patterns vary from flat to slightly LREE-enriched (Loeschke & Schock 1980).

Geochemical data from the many fragmented-ophiolite occurrences north of Trondheim, i.e. those within the Uppermost Allochthon (the Helgeland Nappe) (Fig. 1), are limited, though the Leka ophiolite and some isolated greenstone occurrences at Skålvær in Helgeland (Fig. 1) show normal MORB geochemistry with respect to the Ti-Y-Zr and Ti-Cr diagrams, and their REE patterns (Prestvik & Roaldset 1978; Gustavson 1978; Stephens *et al.* in press). The REE patterns, however, are also here characterized by both between-sequence variations and for the Skålvær occurrence, within-sequence variations (Furnes *et al.* 1982, in press). Geochemical data from the northernmost ophiolite, the Lyngen gabbro and the Kjosen Formation greenstones (Figs 1 & 2, Table 1), are not yet available.

Group II ophiolites

Ophiolite fragments of this category have been subjected to only one major Caledonian tectonometamorphic cycle, that of the middle to late Silurian, Scandian phase. With the exception of the Sulitjelma occurrence, all examples so far recorded occur within the Upper Allochthon of the Trondheim region (Grenne & Roberts 1980, 1981; Roberts *et al.* 1983, in press). These ophiolitic assemblages differ from those of Group I in being intra-Ordovician in age, thus post-dating the Finnmarkian orogenesis. They also occur in association with arc-derived volcaniclastic as well as siliciclastic sediments (Lower Hovin Group) from two completely different sources, and show certain differences in geochemical signature when compared with Group I ophiolites. The regional geological picture denotes that these Ordovician magmato-sedimentary associations, floored by a major unconformity and tectono-metamorphic break, accumulated in a fault-dissected, back-arc, marginal ocean basin milieu (Gale & Roberts 1974, Roberts *et al.* 1983), with the Støren ophiolite slice and Gula Complex as part of the immediate substrate. These Group II ophiolites are also host to some important Cu-Zn stratiform ore deposits (Grenne *et al.* 1980; Grenne & Roberts 1981).

The Løkken, Grefstadfjell and Vassfjell ophiolite complexes SW of Trondheim (Figs 1 & 2) all carry well-developed gabbro-sheeted dyke-pillowed volcanite pseudostratigraphies, and vary in structural disposition from right-way-up to inverted. An olistostromal mega-breccia occurs in fault contact with the Vassfjell ophiolite. Faunal control is best in association with the Grefstadfjell occurrence, denoting a late Arenig age for this particular ophiolite fragment (Ryan *et al.* 1980). Overall, these three complexes comprise c. 1 km of gabbro with small bodies of plagiogranite, which passes up into a 100% sheeted dolerite dyke member up to 1 km thick, and then a basaltic pillow lava pile with intercalated chert, jasper and Mn-pelites. Pervasive sea-floor hydrothermal activity is indicated by the widespread occurrence of sulphide disseminations, and massive stratiform Cu-Zn pyrite orebodies either within the lava pile or in the dyke-to-lava transition zone (Grenne *et al.* 1980; Grenne & Roberts 1981).

The Forbordfjell and Jonsvatn fragments (Figs 1 & 2) are composed dominantly of pillowed, massive and brecciated tholeiitic lavas (Grenne & Roberts 1980) with only a minor proportion of dykes and gabbro. These particular ophiolite fragments appear to be slightly younger than those described above, although the faunal evidence here offers poor constraints. In the same district, similar lava sequences occur at Frosta and Snåsa. Further north the Sulitjelma magmatic complex (Fig. 1), although strongly deformed, displays the characteristic ophiolite pseudostratigraphy (Boyle 1980) and hosts a Cu-Zn sulphide orebody. Gabbro-metasediment relationships (Mason 1980) suggest generation of the complex in a marginal basin along or adjacent to a continental margin (Boyle *et al.* in press). In northern Norway, geochemical data indicates the presence of oceanic tholeiites within the Ordovician rocks of the Upper Allochthon in some areas (Gayer *et al.* in press). At Vaddas, *c.* 50 km east of Lyngen, a mafic magmatic complex is present with local Cu-Zn sulphide mineralization, the members of which show ocean-floor geochemical features (I. Lindahl & B. Stevens, pers. comm. 1982). Tectonostratigraphic problems in the area are such, however, that in Lindahl & Stevens' view no decision can yet be reached on the genesis or palaeotectonic setting of the Vaddas complex. Some of the geological and ore-mineralogical similarities with Sulitjelma are nevertheless striking, as first pointed out by Vogt (1927).

Geochemical studies on the Group II ophiolitic lavas and dykes have revealed consistent, normal MORB signatures with slight LREE depletion.

Some complexes, however, notably the slightly younger Forbordfjell and Jonsvatn occurrences, show LREE enrichment and within-plate tholeiite tendencies in the oldest and youngest members of the lava pile, whereas the bulk of the volcanites are of OFB character.

Tectonostratigraphic significance and distribution of the ophiolites

Our general thesis regarding relationships between Iapetus development and orogenic stage is outlined in Fig. 2a. According to this concept there was, in late Precambrian and Cambrian times, an extremely wide shelf to the west of the existing margin of the Baltoscandian continental plate. Along this margin and shelf, thick sequences of Riphean–Vendian sparagmitic sandstones and Cambrian sediments accumulated as a partly continental, partly shallow-marine succession. The tectonic regime was one of distension and rifting with the development of grabens and half-grabens, and associated volcanicity of initial rifting character (Solyom *et al.* 1979; Furnes *et al.* 1983b). To appreciate the extent of this wide shelf area requires the palinspastic restoration of the stacked sequence of Finnmarkian nappes, with their distinctive basement/cover pairings (Sturt *et al.* 1979; Ramsay *et al.* in press), which reveals a shelf >400 km in width; and the restoration of the Särv Nappe to the west of the present coastline demands a similar allochthoneity for the central sparagmite basin (Nystuen 1981). We are thus presented with a rifted passive margin such as that which borders much of the Atlantic Ocean today.

The next stage in the development history involved a continuation of rifting which ultimately led to the production of oceanic crust and the inception of a major period of sea-floor spreading (Roberts & Gale 1978). It is extremely difficult to pin down the start of this stage; attempts have been made—e.g. by Furnes *et al.* (1980b)—in considering the dyking of the sparagmite basinal areas to define this inception, but this presupposes that rift-related magmatism did not continue on the rifted margin. There are as yet no reliable data which allow us to date the oceanic crust, and even when this becomes available the problem of dating the initial stage of Iapetus crustal accretion is fraught with the difficulty of estimating the position that any one fragment of ocean-floor had in relation to the location of the initial spreading axis.

One feature of potential value here is that of the age of the pelagic/hemi-pelagic sedimentary cap-rock sequences. It has been claimed by Sturt *et al.* (1980) that the thick cap-rock sequence on Karmøy reflects a considerable time span and hence a long period of ocean-floor development. These initial estimates would now appear to be invalid, as recent fieldwork (Sturt & Ramsay, in prep.) indicates the presence of repetitions within the previously assumed continuous stratigraphy of the Torvastad Group. Unfortunately no fauna has yet been recovered from the Torvastad Group, thus making time estimates especially difficult.

A new development in SW Norway is the recognition of primitive ensimatic island-arc activity on this old oceanic crust. On Bømlo this activity is dated to 535 ± 46 Ma. (Furnes *et al.* 1983a) and for the Fundsjø Group of the Trondheim region this is shown to be either Tremadocian or pre-Tremadocian in age. Other indications of this initial arc volcanics (marked by low-Ti tholeiites) are seen above the Karmøy and Gullfjellet ophiolites, but no reliable dating is yet available. In West Finnmark, the highest part of the Finnmarkian stratigraphy is represented by the thick distal turbidites of the Hellefjord Group preserved only in the uppermost nappes (Ramsay *et al.* in press). The Hellefjord is the only unit in this thick sequence which bears evidence of a volcanic component, albeit minor (Roberts 1968); and although it cannot be precisely dated it is known to be post-lower Middle Cambrian in age (Holland & Sturt 1970). This means that the age of the Hellefjord Group cannot be far removed from that obtained for the ensimatic arc stage in the central and southern parts of Scandinavia. It is thus tempting to suggest that these turbidites were developed in an arc-related foredeep in front of an immature arc further to the west. This would in turn imply that the initial subduction in late Cambrian times had a westward polarity, and not eastwards as has earlier been assumed (Ramsay 1973; Robins & Gardner 1975; Roberts *et al.* in press). If westward subduction did obtain, then a number of anomalies, inherent in an eastward polarity hypothesis, would disappear. These include the orogen-parallel extent of wide, laterally continuous, facies belts developed on an essentially stable though rifted shelf, and the total absence of subduction-related volcanic products in late Precambrian and Cambrian sequences. Here it must be remarked that the Baltoscandian craton in Cambrian times was a relatively elevated stable platform covered by only a thin sedimentary veneer, again emphasizing the passive nature of the continental margin.

The formation of the ensimatic arc would

obviously coincide with a stage of oceanic contraction (Fig. 2a) which represents a prelude to Finnmarkian orogenesis. It is of interest that the dating of the initial stage of Finnmarkian deformation (D1) has been shown to have occurred around 535 Ma (Sturt *et al.* 1978) which closely coincides with the age obtained for the ensimatic arc development on Bømlo. The evidence for orogenic deformation of the Group I ophiolites, i.e. pre-478 Ma (Klingspor & Gee in press), implies that these ophiolites must have comprised an integral part of the Finnmarkian orogenic belt during at least part of the orogenic structuring, thus placing a further constraint on the age of obduction. The scant faunal evidence, from the cap-rocks of the Group I ophiolites, indicates that sedimentation continued at least into the Tremadoc. Thus the ophiolites could not have been part of the deforming belt until Finnmarkian D2 (500–485 Ma) at the very beginning of the Ordovician. Here it should be recalled that the foreland autochthon sediments of East Finnmark extend up into the Tremadoc, but the geochronology of slates (ranging in sedimentation age from Upper Riphean to Vendian) has yielded slaty cleavage ages in the time range equivalent to Finnmarkian D2 (Sturt *et al.* 1978; Ryan & Sturt in press).

A particularly marked feature of Early-Middle Ordovician history in western Scandinavia is the presence of a major belt-length unconformity where coarse Ordovician clastics overlie a substrate consisting essentially of Group I ophiolites (Table 1) though in N. Troms spreading across on to subjacent Finnmarkian metamorphics (Ramsay *et al.* in press). Thus, Finnmarkian orogenesis and Group I ophiolite obduction preceeded a major orogenic uplift which probably coincided with or was somewhat later than a changeover to eastward subduction polarity. This latter was in strong evidence in mid-Ordovician times, as witnessed by the presence of remnants of evolved island-arc complexes including volcanic/plutonic suites of marked calc-alkaline affinity (Roberts 1980; Nordås *et al.* in press; Roberts *et al.* in press). Evidence strongly indicates (Roberts *et al.* in press) that, in the Trondheim district, basic to andesitic volcanics and sediments of comparable age accumulated in a fault-dissected, back-arc, marginal basin (Gale & Roberts 1974); Bruton & Bockelie 1980). The sediment composition reflects the nature of the synchronous volcanism, and there was also an input of metamorphic material from a continental source. These rock sequences are characterized by rapidly changing facies and bimodal sediment-dispersal patterns. A similar pattern of evolved island-arc volcanic/plutonic complexes is seen in western Norway, especially in the Sunnhordland region. Here, such complexes extend in age from middle Ordovician into at least early Silurian times. Reconnaissance investigations in the Varaldsøy-Norheimsund region (Sturt, Naterstad, Furnes & Andersen, in prep.) indicate the possibility that Lower Palaeozoic assemblages post-dating Group I ophiolites may also have been deposited in a back-arc marginal basin. Thus, certainly by middle Ordovician times the Balto-scandian plate margin was of destructive type, most likely related to eastward subduction of the Iapetus oceanic crust.

It is within this apparent marginal-basin setting, and in the Trondheim Region, that the majority of the Group II ophiolites occur. As noted earlier these particular ophiolite assemblages appear to occur at two stratigraphic levels. These are considered by Roberts *et al.* (1983) as reflecting two slightly separate phases of crustal thinning and oceanic lithosphere accretion. Thus the palaeogeography of the central Norwegian Ordovician marginal basin bears many resemblances to those of parts of SE Asia or southern Chile.

Major granitic batholiths cut the Group I ophilites and their transgressive Ordovician–Silurian cover sequences, being particularly extensively exposed in Nordland (e.g. the Bindal Batholith) and in Sunnhordland. The batholith-emplacement cycle extends in time through into the Scandian tectono-thermal stage towards the end of the Silurian. The Group I ophiolites are thus now highly disrupted and fragmented by obduction, involvement in Finnmarkian tectonics, deep erosion and burial by transgressive Ordovician/Silurian cover sequences. They were further disrupted by folding and large-scale thrusting during Scandian orogenesis and partly by renewed local thrusting in post-mid-Devonian times. Ophiolite fragments in Helgeland, for example, were repeatedly sliced and now occur at several tectonic levels within the Uppermost Allochthon (Sturt & Ramsay, unpubl. data). Similar dissection obtains within the Upper Allochthon. It is thus difficult to state with any certainty whether or not these oceanic remnants once formed part of a coherent obducted ophiolite slab; their similarities in post-obduction development, however, do make this an interesting and appealing possibility.

ACKNOWLEDGMENTS: We thank E. Irgens for making the illustrations. This is a Norwegian contribution to the IGCP Project 27 'The Caledonian Orogen'.

References

AMALIKSEN, K. G. 1980. Lykling-ophiolittens breksjer —en indikasjon på en fossil 'Fracture Zone'. *Geologinytt* **13**, 4.

BOYLE, A. P. 1980. The Sulitjelma amphibolites, Norway: Part of a Lower Palaeozoic ophiolite complex? *In*: PANAYIOTOU, A. (ed.) *Proc. Int. Ophiolite Symp., Nicosia, 1979*, 567–75.

——, HANSEN, T. S., KOLLUNG, G. & MASON, R. 1983. A new tectonic perspective of the Sulitjelma region. *In*: GEE, D. G. & STURT, B. A. (eds) *The Caledonide Orogen—Scandinavia and Related Areas*. John Wiley, New York (in press).

BRUTON, D. & BOCKELIE, J. F. 1980. Geology and paleontology of the Hølonda area, western Norway —a fragment of North America? *In*: WONES, D. R. (ed.) *The Caledonides in the USA*. Virginia Polytechnic Inst. and State Univ., Dept. of Geol. Sci. Mem. **2**, pp. 41–7.

FURNES, H., AUSTRHEIM, H., AMALIKSEN, K. G. & NORDÅS, J. 1983a. Evidence for an incipient early Caledonian (Cambrian) orogenic phase in southwest Norway. *Geol. Mag.* (in press).

——, NYSTUEN, J. P., BRUNFELT, A. & SOLHEIM, S. 1983b. Geochemistry of Upper Riphean–Vendian basalts associated with the 'sparagmites' of southern Norway. *Geol. Mag.* **120**, 349–61.

——, ROBERTS, D., STURT, B. A. THON, A. & GALE, G. H. 1980b. Ophiolite fragments in the Scandinavian Caledonides. *In*: PANAYIOTOU, A. (ed.) *Proc. Int. Ophiolite Symp., Nicosia, 1979*, 582–99.

——, RYAN, P. D., GRENNE, T., ROBERTS, D., STURT, B. A. & PRESTVIK, T. Geological and geochemical classification of ophiolite fragments in the Scandinavian Caledonides. *In*: GEE, D. G. & STURT, B. A. (eds) *The Caledonide Orogen—Scandinavia and Related Areas*. John Wiley, New York (in press).

——, STURT, B. A. & GRIFFIN, W. L. 1980a. Trace element geochemistry of metabasalts from the Karmøy ophiolite, southwest Norwegian Caledonides. *Earth planet. Sci. Lett.* **50**, 75–91.

——, THON, A., NORDÅS, J. & GARMANN, L. B. 1982. Geochemistry of Caledonian metabasalts from some Norwegian ophiolite fragments. *Contrib. Mineral. Petrol.* **79**, 295–307.

GALE, G. H. & ROBERTS, D. 1974. Trace element geochemistry of Norwegian Lower Palaeozoic basic volcanics and its tectonic implications. *Earth planet. Sci. Lett.* **22**, 380–90.

GAYER, R. A., HUMPHREY, R. J., BINNS, R. E. & CHAPMAN, T. 1983. Tectonic modelling of the Finnmark and Troms Caledonides based on high level igneous rock geochemistry. *In*: GEE, D. G. & STURT, B. A. (eds) *The Caledonide Orogen—Scandinavia and Related Areas*. John Wiley, New York (in press).

GRENNE, T. & LAGERBLAD, B. 1983. The Fundsjø Group, central Norway—A Lower Palaeozoic island arc sequence: geochemistry and regional implications. *In*: GEE, D. G. & STURT, B. A. (eds) *The Caledonide Orogen—Scandinavia and Related Areas*, John Wiley, New York (in press).

—— & ROBERTS, D. 1980. Geochemistry and volcanic setting of the Ordovician Forbordfjell and Jonsvatn greenstones, Trondheim region, central Norwegian Caledonides. *Contrib. Mineral. Petrol.* **74**, 374–386.

—— & —— 1981. Fragmented ophiolite sequences in Trøndelag, central Norway. *Excursion guide B12, Uppsala, IGCP Symp.* 1981.

——, GRAMMELTVEDT, G. & VOKES, F. M. 1980. Cyprus-type sulphide deposits in the western Trondheim district, central Norwegian Caledonites. *In*: PANAYIOTOU, A. (ed.) *Proc. Int. Ophiolite Symp., Nicosia 1979*, 727–743.

GUSTAVSON, M. 1978. Geochemistry of the Skålvær greenstone, and a geotectonic model for the Caledonides of Helgeland, north Norway. *Norsk geol. Tidsskr.* **58**, 161–74.

HARDENBY, C., LAGERBLAD, B. & ANDREASSON, P. G. Structural development of the northern Trondheim Nappe Complex, central Scandinavian Caledonides. *In*: GEE, D. G. & STURT, B. A. (eds) *The Caledonide Orogen—Scandinavia and Related Areas*, John Wiley, New York (in press).

HOLLAND, C. H. & STURT, B. A. 1970. On the occurrence of archaeocyathids in the Caledonian metamorphic rocks of Sørøy, and their stratigraphical significance. *Norsk. geol. Tidsskr.* **50**, 345–55.

HORNE, G. S. 1979. Mélange in the Trondheim Nappe suggests a new tectonic model for central Norwegian Caledonides. *Nature, Lond.* **281**, 267–70.

KLINGSPOR, I. & GEE, D. G. Isotope age-determination studies of the Trøndelag trondhjemites, *In*: GEE, D. G. & STURT, B. A. (eds) *The Caledonide Orogen—Scandinavia and Related Areas*. John Wiley, New York (in press).

LEDRU, P. 1980. Evolution structurale et magmatique du complexe plutonique de Karmøy (Sud-Ouest des Calédonides norvegiennes). *Bull. Soc. geol. mineral. Bretange,* **XII**, (2), 1–106.

LOESCHKE, J. & SCHOCK, H. H. 1980. Rare earth element contents in Norwegian greenstones and their geotectonic implications. *Norsk. geol. Tidsskr.* **60**, 29–37.

MASON, R. 1980. Temperature and pressure estimates in the contact aureole of the Sulitjelma gabbro, Norway: implications for an ophiolite origin. *In*: PANAYIOTOU, A. (ed.) *Proc. Int. Ophiolite Symp. Nicosia, 1979*, pp. 576–581.

MINSAAS, O. & STURT, B. A. The Ordovician clastic sequence immediately overlying the Lyngen gabbro complex, and its environmental significance. *In*: GEE, D. G. & STURT, B. A. (eds) *The Caledonide Orogen—Scandinavia and Related Areas*. John Wiley, New York (in press).

NORDÅS, J., AMALIKSEN, K. G., BREKKE, H., SUTHREN, R. FURNES, H., STURT, B. A. & ROBINS, B. Lithostratigraphy and petrochemistry of Caledonian rocks on Bømlo, SW Norway. *In*: GEE, D. G. & STURT, B. A. (eds) *The Caledonide Orogen—Scandinavia and Related Areas*. John Wiley, New York (in press).

NYSTUEN, J. P. 1981. The Late Precambrian

'sparagmite' of southern Norway: A major Caledonian allochthon—The Osen-Røa Nappe complex. *Am. J. Sci.* **281**, 69–94.

PRESTVIK, T. 1980. The Caledonian ophiolite complex of Leka, north central Norway. *In*: PANAYIOTOU, A. (ed.) *Proc. Int. Ophiolite Symp., Nicosia, 1979*, pp. 555–66.

—— & ROALDSET, E. 1978. Rare earth element abundances in Caledonian metavolcanics from the island of Leka, Norway. *Geochem. J.* **12**, 89–100.

RAMSAY, D. M. 1973. Possible existence of a stillborn marginal ocean in the Caledonian orogenic belt of northwest Norway. *Nature, Phys. Sci.* **245**, 107–109.

——, STURT, B. A. ZWAAN, K. B. & ROBERTS, D. Caledonides of northern Norway. *In*: GEE, D. G. & STURT, B. A. (eds) *The Caledonide Orogen—Scandinavia and Related Areas*. John Wiley, New York (in press).

ROBERTS, D. 1968. The structural and metamorphic history of the Langstrand-Finfjord area, Sørøy, northern Norway. *Norges geol. Unders.* **253**, 1–160.

—— 1980. Petrochemistry and palaeogeographic setting of Ordovician volcanic rocks of Smøla, Central Norway. *Norges geol. Unders.* **359**, 43–60.

—— & GALE, G. H. 1978. The Caledonian-Appalachian Iapetus Ocean. *In*: TARLING, D. H. (ed.) *Evolution of the Earth's Crust*. Academic Press, London, New York and San Francisco, pp. 255–342.

—— & GEE, D. G. Caledonian tectonics in Scandinavia. *In*: GEE, D. G. & STURT, B. A. (eds) *The Caledonide Orogen—Scandinavia and Related Areas*. John Wiley, New York (in press).

—— & STURT, B. A. 1980. Caledonian deformation in Norway. *J. geol. Soc. London* **137**, 241–50.

——, GRENNE, T. & RYAN, P. D. Ordovician marginal basin development in the central Norwegian Caledonides. *J. geol. Soc. London* (in press).

——, STURT, B. A. & FURNES, H. Volcanite assemblages and environments in the Scandinavian Caledonides and the sequential development history of the mountain belt. *In*: GEE, D. G. & STURT, B. A. (eds) *The Caledonide Orogen—Scandinavia and Related Areas*. John Wiley, New York (in press).

ROBINS, B. & GARDNER, P. M. 1975. The magmatic evolution of the Seiland Province, and the Caledonian plate boundaries in northern Norway. *Earth planet. Sci. Lett.* **26**, 167–78.

RYAN, P. D. & STURT, B. A. Early Caledonian orogenesis in northwestern Europe. *In*: GEE, D. G. & STURT, B. A. (eds) *The Caledonide Orogen—Scandinavia and Related Areas*. John Wiley, New York (in press).

——, WILLIAMS, D. M. & SKEVINGTON, D. 1980. A revised interpretation of the Ordovician stratigraphy of sør-Trøndelag, and its implications for the evolution of the Scandinavian Caledonides. *In*: WONES, D. R. (ed.) *The Caledonides in the USA*. Virginia Polytechnic Inst. and State Univ., Dept. of Geol. Sci., Mem. **2**, pp. 99–106.

SOLLI, T. 1981. *The geology of the Torvastad Group, the cap rocks to the Karmøy ophiolite*. Unpubl. Cand. real. thesis Univ. of Bergen.

SOLYOM, Z., GORBATCHEV, R. & JOHANSSON, I. 1979. The Ottfjället Dolerites. Geochemistry of the dyke swarm in relation to the geodynamics of the Caledonide orogen in central Scandinavia. *Sveriges geol. Unders.* **C756**, 38 pp.

STEPHENS, M. B., FURNES, H., ROBINS, B. & STURT, B. A. Igneous activity within the Scandinavian Caledonides. *In*: GEE, D. G. & STURT, B. A. (eds) *The Caledonide Orogen—Scandinavia and Related Areas*. John Wiley, New York (in press).

STURT, B. A. & THON, A. 1978a. An ophiolite complex of probable early Caledonian age discovered on Karmøy. *Nature, Lond.* **275**, 538–9.

—— & —— 1978b. A major early Caledonian igneous complex and a profound unconformity in the Lower Palaeozoic sequence of Karmøy, southwest Norway. *Norsk. geol. Tidssk.* **58**, 221–8.

——, PRINGLE, I. R. & RAMSAY, D. M. 1978. The Finnmarkian phase of the Caledonian orogeny. *J. geol. Soc. London* **135**, 597–610.

——, THON, A. & FURNES, H. 1979. The Karmøy ophiolite, southwest Norway. *Geology* **7**, 316–20.

——, —— & —— 1980. The geology and preliminary geochemistry of the Karmøy ophiolite, SW Norway. *In*: PANAYIOTOU, A. (ed.) *Proc. Int. Ophiolite Symp. Nicosia, 1979*, pp. 538–54.

THON, A. The Gullfjellet ophiolite complex and the structural evolution of the Major Bergen Arc, west Norwegian Caledonides. *In*: GEE, D. G. & STURT, B. A. (eds) *The Caledonide Orogen—Scandinavia and Related Areas*. John Wiley, New York (in press).

VOGT, TH. 1927. Sulitjelmafeltets geologi og petrografi. *Norges geol. Unders.* **121**, 560 pp.

B. A. STURT & H. FURNES, Geologisk Institutt, avd. A. Allégt 41, Universitetet i Bergen, 5014 Bergen-Univ., Norway.

D. ROBERTS, Norges geologiske undersøkelse, Postboks 3006, 7000 Trondheim, Norway.

Ophiolites: Figments of Oceanic Lithosphere?

R. Hall

SUMMARY: To resolve the problem of what a particular ophiolite represents it is necessary to look both at the ophiolite and at the rocks surrounding it: those immediately adjacent to the ophiolite and those of the region. Detailed geochemical and petrological studies are unlikely, on their own, to contribute to an understanding of the tectonic setting of ophiolite formation, nor of emplacement. Evidence from ophiolite complexes of the Mesozoic Arabian margin is inconsistent with the view that they originated at the spreading centre of a major ocean, or in back-arc basins. Instead, it is argued that these ophiolites represent discontinuous 'oceanic' rifts formed *within* the Arabian passive margin. Metamorphism of rocks beneath the ophiolite is considered to mark extension of the margin associated with the formation of these oceanic rifts in the late Cretaceous. These ophiolites do not mark a major suture and the true suture lies to the north of the ophiolite zone.

'Rules' for the Arabian margin

A discontinuous zone of ophiolites, referred to by Ricou (1971a) as the 'croissant ophiolitique' extends through southern Turkey to Oman (Fig. 1). This is the southernmost zone of ophiolites in the Middle-Eastern sector of Tethys. The ophiolites vary in character along the belt from ophiolitic olistostromes and mélanges to the almost intact stratiform complex of Oman. Despite their discontinuity and differences in their style of occurrence and deformation these ophiolites share a number of features which can be considered as 'rules' in the ophiolite-emplacement 'game'. All were emplaced on the Arabian *passive* continental margin (rule 1), where evidence exists it appears that they were emplaced soon after their formation (rule 2), and all were emplaced in a rather short period in the late Cretaceous (rule 3). I consider that any model should be able to account for the emplacement of *all* of the ophiolites, simply because ophiolite emplacement is a unique and unusual event in the history of the Arabian margin, and occurs in such a limited interval of time. Furthermore, emplacement models must also take into account the subsequent history of the region: continent–continent collision did not occur in southern Turkey and Iran until the Miocene, and in Oman has not yet occurred (rule 4). These rules immediately exclude a whole range of models which have been proposed for emplacement of other ophiolites, e.g. those involving active continental margins, and those related to continent–continent collision.

The Neyriz complex of southern Iran is one of the ophiolites of the 'croissant' and several authors have drawn attention to its similarities to Oman (e.g. Ricou 1971a; Hallam 1976; Coleman 1981). However, Neyriz has some unusual features: no Triassic or Jurassic volcanic rocks have so far been found, while sedimentary rocks associated with the volcanics are of Senonian age, and include shallow marine limestones. Most of the volcanic rocks are tholeiitic, although some alkali basalts similar to those of the Haybi complex of Oman have been reported (Arvin 1982). Vesicularity of these rocks suggests relatively shallow eruptive depths. Rhyolitic ignimbrites occur locally at the top of the volcanic sequence (Arvin 1982). Such rocks are rare, or even entirely absent, in island arcs built on oceanic crust, although they are commonly associated with volcanic mountain belts on continental crust (e.g. the Andes) or continental rifts such as the East African rift (Williams & McBirney 1979). Unusual skarns are found at the contact between the ophiolite peridotite and folded marbles and indicate high-temperature-contact metamorphism (Ricou 1971b; Hall 1980). A general ophiolite model for the Arabian margin should also be able to incorporate these anomalies.

That the Arabian continental margin was a passive margin is undisputed; it has a relatively continuous Mesozoic record of marine sedimentation without active margin-type volcanism. Many of the Tethyan passive margins are considered to have been formed by rifting in the Triassic and the Arabian margin conforms to this pattern; until the Triassic, central Iran and northern Arabia have a similar history but after the Triassic their stratigraphy is quite different. The Southern Tethys ocean formed between Arabia and central Iran-Lut should therefore have been floored by oceanic crust of Triassic age in the region adjacent to the Arabian passive

This paper was selected for conference presentation as it is controversial and would stimulate discussion; this indeed proved to be the case. However, in accepting the paper for publication we do so against the advice of the referees who, whilst accepting that some of the authors' criticisms of the concensus ophiolite model are valid, considered the author's own model to be unrealistic. Dr Hall rejected their recommendation that the alternative model be excluded, thus in order to present the critically acceptable part of the paper, we are obliged, albeit reluctantly, to publish the paper unmodified.

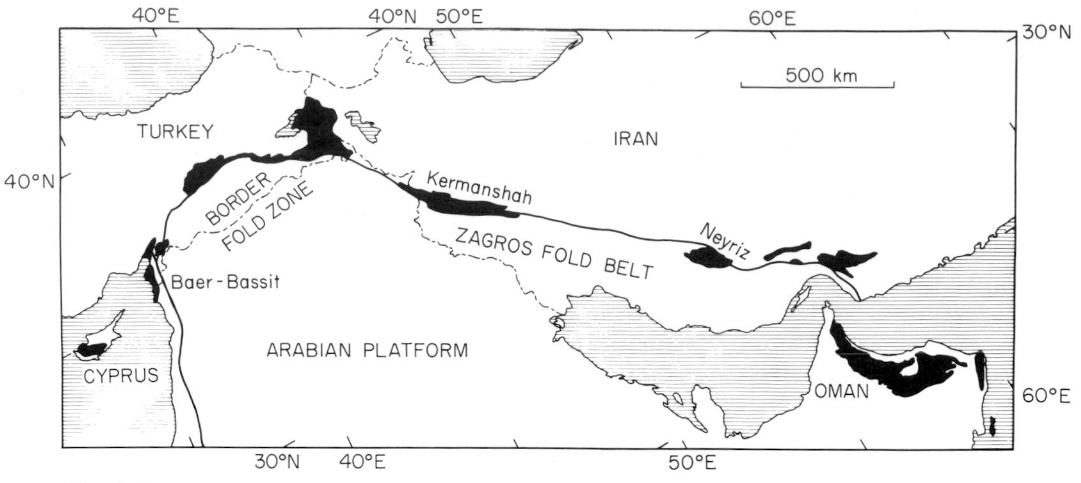

FIG. 1. Late Cretaceous ophiolite complexes of the Arabian Margin.

margin. Rifting is considered to be marked by late Triassic, largely alkaline, basic volcanism. However, although this evidence is consistent with what is known and expected during the early stages of continental rifting we need not assume that the occurrence of Triassic volcanics necessarily marks the exact margin between continental and new oceanic crust. Scandone (1975) observed that the Triassic rift zones in the western part of Tethys were not the zones of later ocean-crust formation, and the present Atlantic and Red Sea margins are marked by discontinuous zones of basic igneous rocks up to about 200 km from the present continent–ocean crust boundary which were formed at the same time as ocean crust in the main ocean. These zones are often referred to as 'failed rifts', an unfortunate term since it implies that rifting has ceased and is unlikely to be renewed. Comparison with the present margins of the Red Sea and Atlantic (Hall 1982) suggests that the Mesozoic Arabian passive margin was likely to have been up to several hundred kilometres wide, underlain by thin extended continental crust, and included large volumes of basic igneous rocks of Triassic age in 'failed rifts'.

Emplacement models

Those of us that have worked on ophiolites along this 'croissant' have tended to look in particular to the Oman ophiolite for a model of ophiolite emplacement. It is probably the best-studied ophiolite complex in the belt, and it is certainly the most complete and least deformed. Continent–continent collision has not yet occurred in Oman, and thus the Oman ophiolite should be able to provide some insights into the pre-Miocene history of the other ophiolites. Three types of models have been proposed that attempt to obey the rules specified above. The first of these, that of Coleman (1981), is a modification of the original proposal of Glennie *et al.* (1973), and is the simplest in suggesting detachment of the ophiolite from the northern side of a newly inactive spreading ridge in the Southern Tethys between Oman and Iran (Fig. 2). The ophiolite must be derived from close to a recently spreading centre to satisfy rule 2, and therefore must have crossed the older parts of the ocean during its emplacement to arrive on the continental margin. Glennie *et al.* (1973) suggest that the original width of the sub-ophiolite nappes was at least 400 km, although they do not estimate the width of oceanic crust in the ocean itself, while Coleman (1981) provides no figure beyond the minimum of 165 km indicated by the exposure width of the ophiolite. However, unless the pattern of spreading was unusual and highly asymmetrical the minimum half-width of the ocean can be estimated to satisfy rule 4 from the width of the ophiolite and the amount of oceanic lithosphere currently beneath the Makran active margin. In addition to the 165 km of ophiolite on land, about 500 km of oceanic crust is required beneath the Makran accretionary wedge (Farhoudi & Karig 1977) to the active volcanic arc of Djaz Murian (Fig. 2c). This is an absolute minimum since arc volcanism began at least as early as late Cretaceous–Paleocene (Berberian *et al.* 1982). Thus a half-width of at least 665 km is indicated. Since the ophiolite is now underlain by 165 km of continental crust it must have travelled at least 830 km to its present position as a slab about 12 km in thickness without significant disruption. Although this might be mechanically feasible

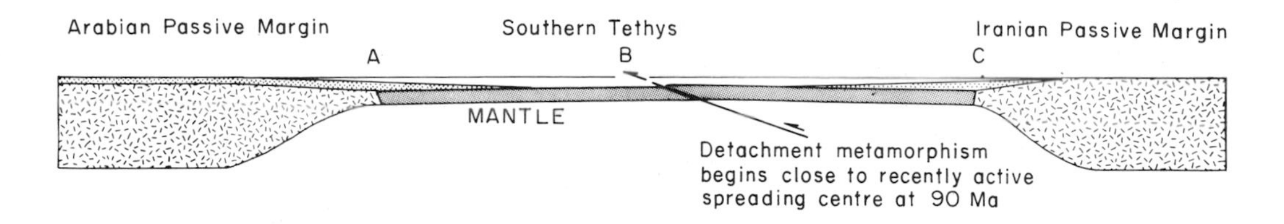

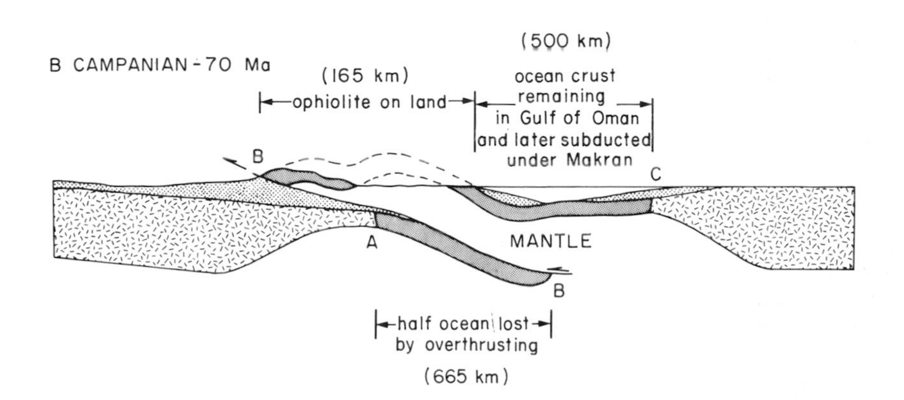

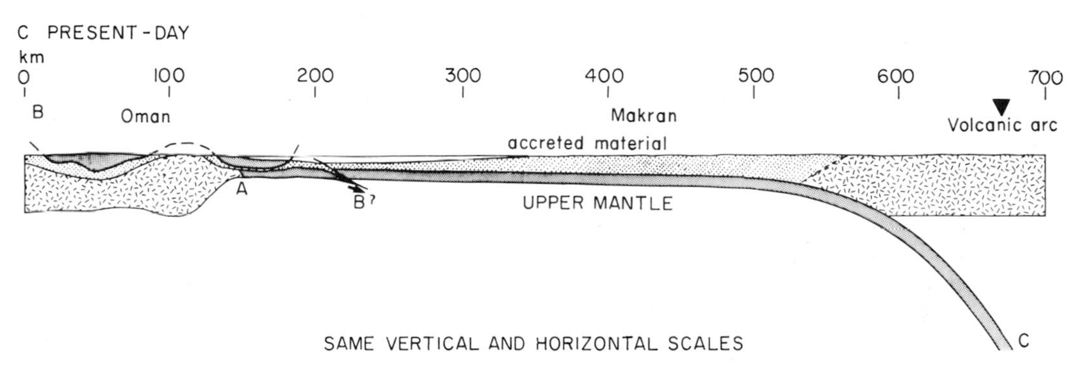

FIG. 2. A and B show detachment and emplacement of the Oman ophiolite according to Coleman (1981). C shows the present configuration of the Oman–Makran region along 59°E based on Manghnani & Coleman (1981) and Farhoudi & Karig (1977).

under the most favourable conditions (see Hubbert & Rubey 1959, but cf. Hsü 1969), it seems unlikely. The dimensions of the Oman ophiolite are now about 450 × 90 × 4–10 km (Searle & Malpas 1980) and one of the conditions would have to be the virtual absence of friction at the base of the sheet which would preclude significant frictional heating. Furthermore, the problem of the missing southern half of the ocean remains: a minimum of 665 km of oceanic lithosphere has disappeared without trace. I therefore conclude that this model is implausible.

Gealey (1977) neatly avoids some of these problems by proposing collision between the Arabian passive margin and an intra-oceanic island arc. He suggests that the Oman ophiolite represents the forearc limb of the island arc. This model can satisfy all the rules, but only if rather special circumstances are invoked. To satisfy rules 2 and 3 subduction must be initiated at, or very near to, a spreading centre active immediately before the initiation of subduction, and the subduction zone must have been parallel to the Arabian margin for a distance of about 3000 km

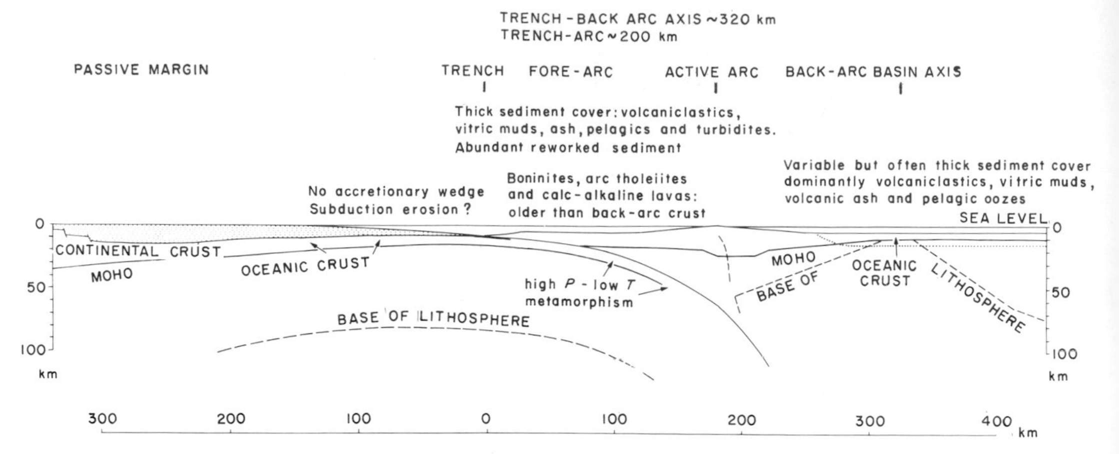

FIG. 3. Scaled cross-section across an island-arc–passive continental-margin convergence zone shortly before collision. This was the situation of the Oman ophiolite in the late Cretaceous according to the models of Gealey (1977) and Pearce *et al.* (1981). The arc is based on information from the Marianas from Sager (1980) and Hussong *et al.* (1981), and the passive margin is based on the Atlantic margin from information in Sheridan *et al.* (1979). The area of the back-arc basin enclosed by the dotted line corresponds approximately to the dimensions of the Semail ophiolite.

so that collision between the continental margin and the arc occurred synchronously along the belt. As Gealey himself observes (1977, p. 1190), the evidence for the island arc is weak. In the Oman region it must be located in the present Gulf of Oman, whereas elsewhere it has presumably been lost during Miocene continent–continent collision. Gealey's model has been modified by Pearce *et al.* (1981) who derive the ophiolite on geochemical grounds from a back-arc basin rather than the forearc limb. Comparison with modern arcs suggests that the whole of the forearc region and the island arc itself have been lost during collision, although the back-arc basin has survived remarkably well. Fig. 3 shows the scale of the loss; one feels that Pearce *et al.* are in a somewhat similar position to Mr Worthing—losing the forearc may be regarded as misfortune, but to lose both the forearc and the arc. . . . The sedimentary cover of the ophiolite should indicate whether either of these island-arc models is plausible or not. The sediments drilled in the Marianas arc (Hussong *et al.* 1981) in both forearc and back-arc regions, even quite close to the active spreading centre, include a considerable volcanic component (Fig. 3). No such sediments have yet been found in the Oman region, nor associated with other ophiolites of the belt. On the contrary, the sediments preserved are entirely pelagic and lack any volcanogenic material (e.g. Tippit *et al.* 1981).

Stoneley (1974, 1975) has suggested that the Southern Tethyan ophiolites originated within the Arabian passive margin during the late Cretaceous. This suggestion avoids some of the problems of the previous two models and the anomalous features of Neyriz can be incorporated in a general model (Hall 1982) which is summarized in Fig. 4. The form of the Arabian margin in the Triassic is based on cross-sections of the Red Sea margins constructed from geological and geophysical data by Lowell & Genik (1972). Continental lithosphere becomes thinner in an irregular way as the continent–ocean crust boundary is approached and crustal extension is assumed to occur by listric normal faulting. Parallel to the Southern Tethys is a 'failed rift' (cf. the Danakil rift) which is underlain by thinned lithosphere. This 'failed rift' is reactivated in the Cenomanian to form a narrow discontinuous ocean which may have been up to 2–300 km wide in the Oman region, but rather less to the north. I suggest that in about the late Santonian there was a change to a convergent regime and an attempt to begin subduction at the passive margins of the Southern Tethys. McKenzie (1977) has argued that the creation of new subduction zones is not an easy process and considerable work is required to produce a sinking slab about 180 km long before subduction becomes self-sustaining. I suggest that during the change from an extensional to a convergent regime in the Southern Tethys less work was required to compress the over-extended hot and weak Arabian passive margin than to initiate subduction. It was therefore not until after compression of the continental margins that subduction began in the Southern Tethys, at its

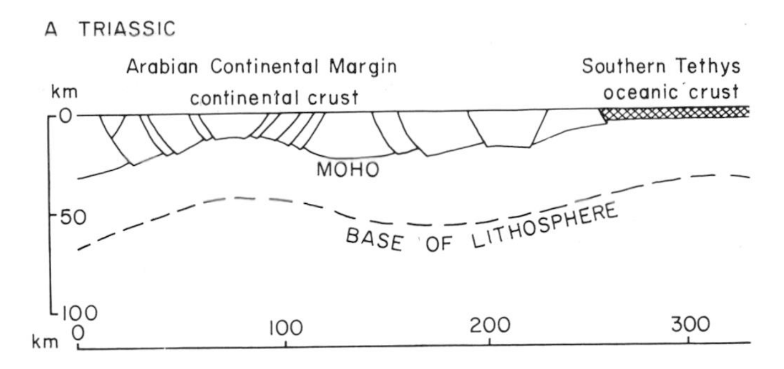

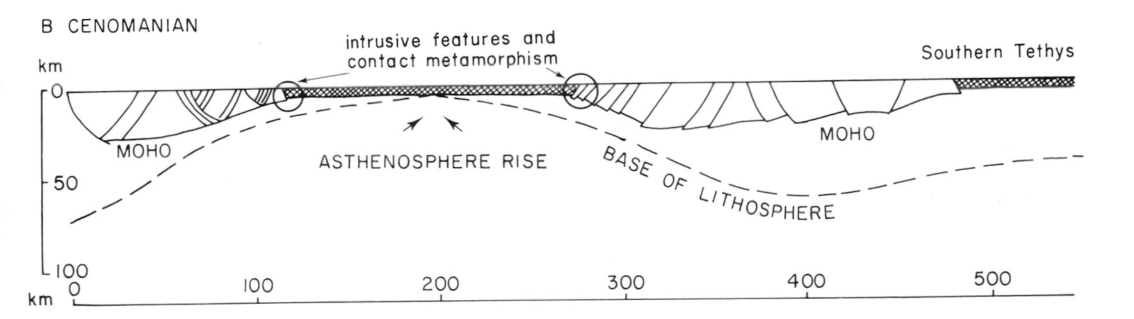

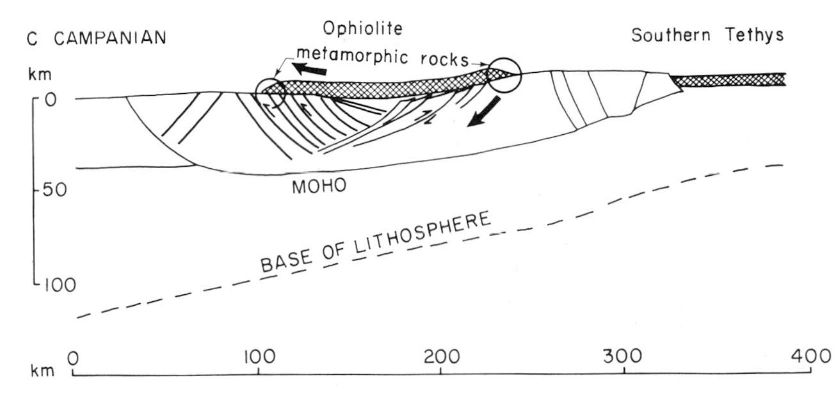

FIG. 4. Suggested Mesozoic evolution of the Arabian margin. No sedimentary cover is shown and the sections are drawn with no vertical exaggeration. Lithosphere thicknesses are speculative but note in particular the thick lithosphere under the Southern Tethys, consistent with its age, and very thin lithosphere beneath the 'failed rift' and ophiolite. Extension is assumed to occur by listric normal faults reactivated as listric thrust faults in the Campanian. (The diagram is discussed in the text and more fully in Hall (1982).)

northern margin. Subduction-related volcanism at the northern margin should therefore have begun later than ophiolite emplacement, the time lag depending on subduction rates and dip of the Benioff zone. In fact, in central Iran and north of the Djaz Murian depression volcanic activity is recorded from the late Cretaceous–Paleocene (Berberian *et al.* 1982).

Testing the models

Ophiolites and geochemical evidence

Most of the evidence from ophiolites themselves does not tell us very much about their environment of formation or means of emplacement. Since so many ophiolites seem to have been emplaced soon after their formation and are considered to have been hot during their emplacement it appears that they were not formed in major ocean basins like the Atlantic or Pacific, and most authors appeal to narrow Red Sea-type basins or marginal basins to explain their existence. Their structure, metamorphism and sedimentary cover seem only to require marine conditions with extrusion and intrusion of basic magma in relatively narrow zones, and such conditions might be found in a variety of tectonic settings. However, geochemical evidence, in particular arguments based on 'immobile' elements, have been proposed to resolve the problem of tectonic setting. Clearly, if geochemical methods are to be used to determine tectonic setting they must work in *all* modern environments and be capable of distinguishing tectonic settings that have no analogues at present. Geochemical differences can presumably be used as diagnostic of tectonic setting insofar as they result from processes that are restricted to those settings. These processes, which include melting behaviour in a heterogeneous mantle under a wide range of P–T conditions, variations in the amount of melting, fractional crystallization and differentiation at different levels of the lithosphere during magma ascent, and possibilities of magma mixing and crustal contamination, are still incompletely understood. To complicate matters, 'immobile' elements not infrequently become mobile during alteration and metamorphism. Furthermore, each of the three models of ophiolite genesis and emplacement appeals at one stage or another to some non-actualistic variant of present-day processes. Can we be sure that currently unknown tectonic environments obey empirically derived discriminant rules? Figure 5 emphasizes the need for caution. Using the same discriminant methods chosen by Pearce *et al.* (1981) to deduce a back-arc setting for the Oman ophiolite, two of three groups of Tertiary basaltic rocks recognized from Skye, whose tectonic setting is presumably well known, fail to be correctly classified. The Fairy Bridge basalts are clearly classified as ocean-floor tholeiites while the Preshal Mhor basalts are arguably ocean-floor or island-arc tholeiites. I suggest that the Skye basalts have a comparable tectonic setting to many of the ophiolites of the Tethyan belt, i.e. within a passive continental margin.

Metamorphic rocks associated with ophiolites

Until a few years ago, one of the major problems associated with emplacement of 'alpine-type' ultramafics was their lack of high-temperature metamorphic aureoles. Now that many of these ultramafic bodies are recognized as obducted ophiolites their high-temperature basal metamorphic rocks have also been discovered. There now seems to be a great enthusiasm for identifying metamorphic rocks associated with ophiolites as part of their sub-ophiolite sole despite the observations that many of them are too thick, compositionally unlikely, structurally in the wrong position or the wrong age to be produced by ophiolite obduction (e.g. Crete, south-east Turkey, Neyriz). Even in Oman where an inverted metamorphic sequence is well preserved at the base of the ophiolite their interpretation as due to obduction has a number of problems.

Radiometric dating is showing that many sub-ophiolite metamorphic rocks are considerably older than ophiolite emplacement (e.g. Spray & Roddick 1980; Thuizat *et al.* 1981). Therefore, they are considered to pre-date the obduction event and instead date the initiation of thrusting within the ocean basin. In Oman, ages of the amphibolites immediately beneath the ophiolite are only a few Ma younger than ages of igneous rocks at the spreading centre. Since the igneous rocks are considered to have moved into an off-axis position, indicating continued spreading after their formation, and metamorphism of ocean crust to amphibolite facies requires thrusting, heating to approximately 750°C and cooling to a temperature at which the K-Ar system closes (?500°C), spreading and thrusting must have been almost contemporaneous. Lanphere (1981) considers that overthrusting occurred 3–7 Ma after formation of the ophiolite at the spreading centre, but modifying the assumptions used in his argument (e.g. increasing temperature of hornblende formation, lowering blocking temperature, changing conduction rates), all of which are poorly constrained, could lead to the conclusion that metamorphism is as old, or possibly older, than spreading ages. There is then a gap of about 10 Ma between the oldest and highest-grade rocks and the younger, structurally lower rocks in the metamorphic sole (Lanphere 1981).

Unusual alkali peridotites and jacupirangites occur in the Haybi complex immediately beneath the metamorphic sole in Oman, and metamorphosed jacupirangites occur with amphibolites in the metamorphic sole. These rocks are late Cretaceous in age (92.5 ± 4 Ma, Searle *et al.* 1980) and occur as dykes and sills indicating intrusion in an extensional environment. Searle *et al.* compare them to a jacupirangite–syenite association now found in a similar tectonic situation beneath the Bay of Islands ophiolite where Jamieson & Talkington (1980) interpret them as

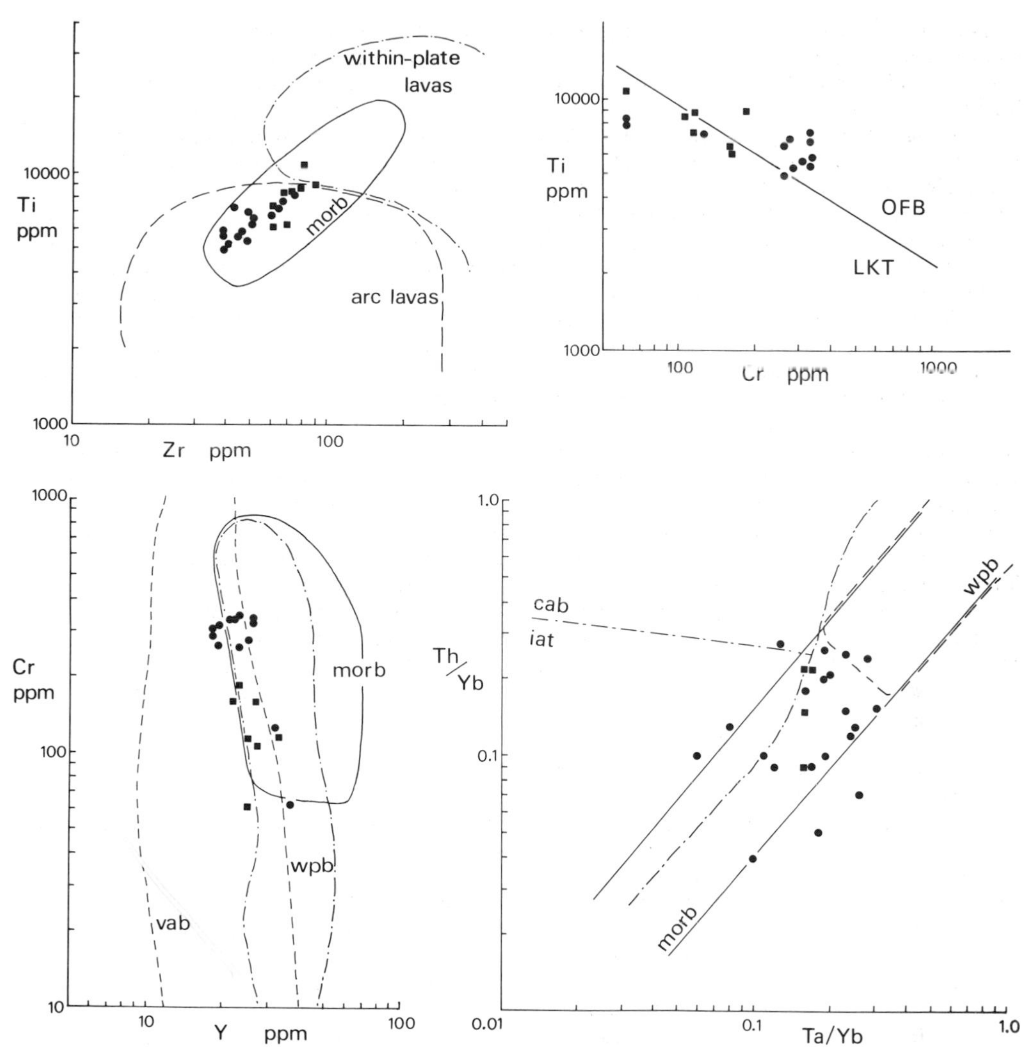

FIG. 5. Discriminant plots used by Pearce *et al.* (1981) to deduce a back-arc setting for the Oman ophiolite applied to Tertiary basalts from Skye: dots are Preshal Mhor basalts, squares are Fairy Bridge basalts. Data from Thompson (1982), Thompson *et al.* (1972, 1980) and Mattey (1980).

part of an ocean-island sequence although conceding that 'the overwhelming majority of similar alkaline rock suites has been described from continental environments where they are generally associated with rifting'. (Jamieson & Talkington 1980, p. 474). In the Neyriz area the high-temperature skarns at the contact of the ophiolite peridotite are interpreted as melts produced at the contact of hot intrusive peridotite (Hall 1980). These rocks are not easily explained by models postulating formation of these ophiolites at a mid-ocean ridge or somewhere in an island arc, although they fit well into the rifted continental-margin model.

Lithologies of the Oman sub-ophiolite sequence are perfectly consistent with an origin by metamorphism of igneous and sedimentary rocks formed at the margin of an ocean basin. On the other hand, the high-temperature metamorphism that they have suffered seems inconsistent with the subduction-zone origin suggested by Searle & Malpas (1980, 1982). Subduction-zone metamorphism is normally considered to be typically of a high-pressure–low-temperature type, as suggested by the low heat flow of these regions at the present, and thermal modelling of subducted cold slabs. It is odd that no high-pressure metamorphic rocks have been reported from the mélanges beneath the ophiolite if the continent–arc collision model is correct, since the model involves subducting the margin for a distance of at least 165 km to emplace the ophiolite.

I suggest that an alternative to the proposal that many of the metamorphic rocks associated with ophiolites are directly related to obduction might be the following (Fig. 6). Rifting of the Arabian continental margin in the Triassic resulted in formation of failed rifts within the margin containing Triassic igneous rocks. Renewed extension of the continental margin in the mid-Cretaceous resulted in formation of new oceanic lithosphere. Prograde metamorphism of crust of the continent–ocean margin should have culminated at this stage since a thermal peak occurs at the time of break-up, after which the lithosphere cools, and this type of metamorphism is considered to contribute to the post-break-up subsidence of continental margins (e.g. Falvey & Middleton 1981; Royden *et al.* 1980). Extension on listric faults produced mylonites and folds in ductile shear zones at depth. Mineral growth and recrystallization occurred during shearing due to frictional heating in the shear zones and partly as a consequence of rise of isothermal surfaces during lithosphere thinning. Mineral formation would therefore be syn- or post-tectonic. Subsequently, cessation of movement on faults, and subsidence of isothermal surfaces during lithosphere cooling and thickening, caused these minerals to cool below their blocking temperatures. Thus, metamorphic ages would post-date break-up, although by how much would depend on the contribution made by frictional heating and rate of lithosphere cooling. Furthermore, this is the zone in which one would expect to see intrusive relationships and formation of rocks such as the Neyriz skarns and Oman jacupirangites and alkali peridotites associated with the early stages of continental rifting. Reactivation of listric normal faults as listric reverse faults during compression of the continental margin and emplacement of the ophiolite formed in this narrow rift would result in imbrication of these mylonites, renewed dynamothermal metamorphism, and younger metamorphic ages as the rocks moved upwards. The metamorphic rocks would therefore have a polyphase history of deformation and mineral growth recording extension, static cooling and thrusting. The suggestion that the older metamorphic ages are indicative of extension is also more consistent with sedimentary evidence indicating extension of the continental margin at the same period (see below), evidence of dykes cutting the metamorphics, and would explain the observation that spreading in Oman is almost as young as, and elsewhere much younger than, high-temperature metamorphic cooling ages. Interestingly, this model is also consistent with the suggestion of Spray (1982) that many ophiolites represent the edges of ocean basins adjacent to continental margins, based on the occurrence of unusual mafic segregations in ophiolite-mantle sequences.

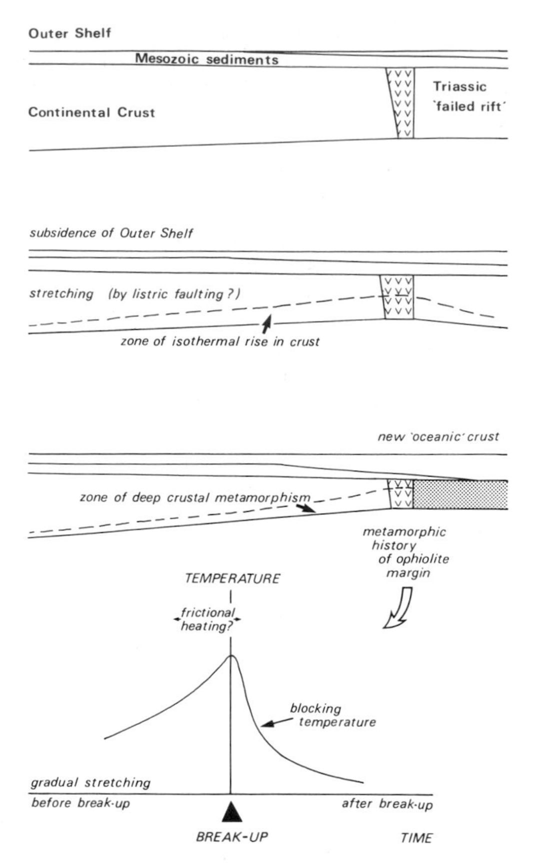

FIG. 6. Schematic thermal evolution of Triassic 'failed rift' after renewed extension leading to break-up in the late Cretaceous. Extensional faults at the margin of the ophiolite should be marked by mylonites at depth (cf. Sibson 1977) and shear heating may contribute to the maximum temperature reached. This is also the zone in which intrusive features and contact metamorphism should be observed. Note that the thermal maximum occurs at the time of break-up but metamorphic ages are younger, depending on blocking temperatures.

Sedimentary evidence

The sedimentary rocks associated with ophiolites, both above and below, should also be able to provide some tests for models of ophiolite formation and emplacement. I have pointed out above that the lack of volcanogenic material in the sedimentary cover of the ophiolites in the 'croissant' is inconsistent with an island-arc origin. If these ophiolites did originate within the Arabian continental margin we might expect to see some effects of renewed rifting and extension in the sedimentary sequences over a wide area. There are some tantalizing indications of this. In the Neyriz area, the stratigraphy of the sedimentary rocks in nappes beneath the ophiolite indicate a tensional regime throughout most of the late Cretaceous (Hallam 1976; Stoneley 1981). On the outer part of the shelf there was a change to deeper-water sedimentation associated with slumping and normal faulting in the Cenomanian (Stoneley 1981). To the NE, in the basin represented by the Pichakun nappes, there was a change in the direction of derivation of sediment in the Aptian: earlier detritus was derived from the SW, but later from the NE (Ricou 1968). This cannot represent the beginning of compression of the continental margin and formation of a foreland basin in front of the advancing nappes, since sediments associated with the Neyriz pillow basalts indicate that volcanic activity continued into the Senonian (Arvin 1982). I therefore prefer to interpret this as evidence of rifting within the deeper parts of the continental margin, followed by subsidence and 'oceanic' crust formation. Not until the late Santonian or early Campanian were sedimentary nappes emplaced from the east followed soon after by the ophiolite. A similar sequence of events is observed in SE Turkey where a late Turonian unconformity marks a change in sedimentation on the outer shelf to deeper-water conditions terminated by ophiolitic gravity slides in the late Campanian–early Maastrichtian (Rigo de Righi & Cortesini 1964). In Oman, the history of sedimentation in the Hawasina margin becomes less clear in the early-mid Cretaceous. Instability on the shelf edge may be indicated by the deposition of mid-Cretaceous thick debris flows and the later Cretaceous history is thought to be represented by mélanges (Graham 1980). Although previous interpretations have been different (e.g. Glennie *et al.* 1973; Graham 1980) there appears to be nothing to contradict the same interpretation as in the Neyriz area: rifting within a continental margin, perhaps beginning a little earlier (?Albian).

The biggest problem with this type of interpretation is separating the effects of regional tectonics and eustatic changes in sea level. Global eustatic changes in the late Cretaceous will have affected these stratigraphic sequences, and equally, rapid spreading in the late Cretaceous must have contributed to sea-level changes.

Conclusions

Emplacement of the ophiolites of the 'croissant ophiolitique' is a unique event in which young 'oceanic' lithosphere was emplaced very soon after its formation. I suggest that the almost simultaneous emplacement of ophiolites of similar age along a belt about 3000 km in length is consistent with their origin by rifting within the Arabian passive continental margin in the early late Cretaceous. The present discontinuous nature of the zone of ophiolites is interpreted as a primary feature reflecting an originally discontinuous rift. Metamorphic rocks now associated with the ophiolites were formed as a result of intrusion in the initial stages of rifting and by dynamothermal metamorphism associated with extension of the margin. The older metamorphic rocks do not therefore date the initiation of overthrusting but break-up initiating spreading. A change from an extensional to compressional regime in the margin in the late Cretaceous caused detachment and emplacement of the ophiolites within the passive margin. The ophiolites are therefore essentially parautochthonous; the relatively small distances of overthrusting (less than 100 km on each side of the Oman ophiolite) can explain why very large but extremely thin slabs can be emplaced without major disruption. I suggest that the compression of the Arabian passive margin reflects a major change in plate motions at the end of the Cretaceous associated with the initiation of subduction of the Southern Tethys. This was followed by subduction-related volcanism at the northern margin and Miocene continent–continent collision in SE Turkey and Iran. Continent–continent collision has not yet occurred in the Oman region and subduction of the last remnants of the Southern Tethys continues under the Makran. The ophiolites do not mark what is normally interpreted as a suture, but a discontinuous zone within the Arabian passive margin. The true suture lies to the north of the ophiolite zone and is marked by a zone of mélanges including Tertiary flysch and disrupted ophiolites (Hall 1976; Stoneley 1981). The closest present-day analogue of the Arabian margin ophiolites is the Red Sea–Dead Sea system which appears to be underlain discontinuously by oceanic lithosphere in the Red Sea and Dead Sea (Ginzburg *et al.* 1981) and is close to a continental margin for at least part of its length.

ACKNOWLEDGMENTS: I thank Mohsen Arvin, Mike Audley-Charles, Curtis Cohen, Ernie Rutter and Bob Stoneley for discussion.

References

ARVIN, M. 1982. *Petrology and geochemistry of ophiolites and associated rocks from the Zagros suture, Neyriz, Iran.* Ph.D. Thesis, London University.

BERBERIAN, F., MUIR, I. D., PANKHURST, R. J. & BERBERIAN, M. 1982. Late Cretaceous and early Miocene Andean-type plutonic activity in northern Makran and Central Iran. *J. Geol. Soc. London* **139**, 605–14.

COLEMAN, R. G. 1981. Tectonic setting for ophiolite obduction in Oman. *J. geophys. Res.* **86**, 2497–508.

FALVEY, D. A. & MIDDLETON, M. F. 1981. Passive continental margins: evidence for prebreakup deep crustal metamorphic subsidence mechanism. *Oceanol. Act. Proc. 26th Int. geol. Congr. Paris 1980* 103–14.

FARHOUDI, G. & KARIG, D. E. 1977. Makran of Iran and Pakistan as an active arc system. *Geology* **5**, 664–8.

GEALEY, W. K. 1977. Ophiolite obduction and geologic evolution of the Oman Mountains and adjacent areas. *Bull. geol. Soc. Am.* **88**, 1183–91.

GINZBURG, A., MAKRIS, J., FUCHS, K. & PRODEHL, C. 1981. The structure of the crust and upper mantle in the Dead Sea rift. *Tectonophysics* **80**, 109–19.

GLENNIE, K. W., BOEUF, M. G. A., HUGHES-CLARKE, M. W., MOODY-STUART, M., PILAAR, W. F. H. & REINHARDT, B. M. 1973. Late Cretaceous nappes in Oman Mountains and their geologic evolution. *Bull. Am. Assoc. Petrol. Geol.* **57**, 5–27.

GRAHAM, G. 1980. Evolution of a passive margin and nappe emplacement in the Oman Mountains. *In*: PANAYIOTOU, A. (ed.) *Ophiolites: Proc. International Ophiolite Symposium Cyprus 1979*, pp. 414–23.

HALL, R. 1976. Ophiolite emplacement and the evolution of the Taurus suture zone, southeastern Turkey. *Bull. geol. Soc. Am.* **87**, 1978–88.

—— 1980. Contact metamorphism by an ophiolite peridotite from Neyriz, Iran. *Science* **208**, 1259–62.

—— 1982. Ophiolites and passive continental margins. *Ofioliti* **7**, 279–98.

HALLAM, A. 1976. Geology and plate tectonics interpretation of the sediments of the Mesozoic radiolarite-ophiolite complex in the Neyriz region, southern Iran. *Bull. geol. Soc. Am.* **87**, 47–52.

HSÜ, K. J. 1969. Role of cohesive strength in the mechanics of overthrust faulting and of landsliding. *Bull. geol. Soc. Am.* **80**, 927–52.

HUBBERT, M. K. & RUBEY, W. W. 1959. Role of fluid pressure in mechanics of overthrust faulting. *Bull. geol. Soc. Am.* **70**, 115–66.

HUSSONG, D. M., UYEDA, S. *et al.* 1981. *Initial Reports DSDP Leg 60.* Washington (U.S. Government Printing Office).

JAMIESON, R. J. & TALKINGTON, R. W. 1980. A jacupirangite-syenite assemblage beneath the White Hills peridotite, northwestern Newfoundland. *Am. J. Sci.* **280**, 459–77.

LANPHERE, M. A. 1981. K-Ar ages of metamorphic rocks at the base of the Samail ophiolite, Oman. *J. geophys. Res.* **86**, 2777–82.

LOWELL, J. D. & GENIK, G. J. 1972. Sea-floor spreading and structural evolution of southern Red Sea. *Bull. Am. Assoc. Petrol. Geol.* **56**, 247–59.

MATTEY, D. P. 1980. *The petrology of high-calcium, low-alkali tholeiite dykes from the Isle of Skye regional swarm.* Ph.D. Thesis, London University.

MANGHNANI, M. L. & COLEMAN, R. G. (1981) Gravity profiles across the Samail ophiolite. *J. geophys. Res.* **86**, 2509–25.

MCKENZIE, D. P. 1977. The initiation of trenches: a finite amplitude instability. *In*: TALWANI, M. & PITMAN, W. C. (eds) *Island Arcs, Deep Sea Trenches and Back Arc Basins.* American Geophysical Union, Washington D.C. pp. 57–61.

PEARCE, J. A., ALABASTER, T., SHELTON, A. W. & SEARLE, M. P. 1981. The Oman ophiolite as a Cretaceous arc-basin complex: evidence and implications. *Philos. Trans. R. Soc. London* **A300**, 299–317.

RICOU, L.-E. 1968. Sur la mise en place au Crétacé supérieur d'importantes nappes à radiolarites et ophiolites dans les Monts Zagros (Iran). *C. R. Acad. Sci. Paris* **267**, 2272–5.

—— 1971a. Le croissant ophiolitique peri-arabe: une ceinture de nappes mises en place au Crétacé supérieur. *Rev. Géogr. phys. Géol. dyn. Paris* **13**, 327–49.

—— 1971b. Le métamorphisme au contact des péridotites de Neyriz (Zagros interne, Iran): développement de skarns à pyroxène. *Bull. Soc. géol. Fr.* **13**, 146–55.

RIGO DE RIGHI, M. & CORTESINI, A. 1964. Gravity tectonics in Foothills Structure Belt of southeast Turkey. *Bull. Am. Assoc. Petrol. Geol.* **48**, 1911–37.

ROYDEN, L., SCLATER, J. G. & VON HERZEN, R. P. 1980. Continental margin subsidence and heat flow: important parameters in formation of petroleum hydrocarbons. *Bull. Am. Assoc. Petrol. Geol* **64**, 173–87.

SAGER, W. 1980. Structure of the Mariana Arc inferred from seismic and gravity data. *J. geophys. Res.* **85**, 5382–8.

SCANDONE, P. 1975. Triassic seaways and the Jurassic Tethys ocean in the central Mediterranean area. *Nature, Lond.* **256**, 117–9.

SEARLE, M. P. & MALPAS, J. 1980. Structure and metamorphism of rocks beneath the Semail ophiolite of Oman and their significance in ophiolite obduction. *Trans. R. Soc. Edinburgh* **71**, 247–262.

—— & MALPAS, J. 1982. Petrochemistry and origin of sub-ophiolitic metamorphic and related rocks in the Oman Mountains. *J. geol. Soc. London* **139**, 235–48.

——, LIPPARD, S. J., SMEWING, J. D. & REX, D. C. 1980. Volcanic rocks beneath the Semail ophiolite nappe in the northern Oman mountains and their significance in the Mesozoic evolution of Tethys. *J. geol. Soc. London* **137**, 589–604.

SHERIDAN, R. E., GROW, J. A., BEHRENDT, J. C. & BAYER, K. C. 1979. Seismic refraction study of the continental edge off the eastern United States. *Tectonophysics* **59**, 1–26.

SIBSON, R. H. 1977. Fault rocks and fault mechanisms. *J. geol. Soc. London.* **133**, 191–213.

SPRAY, J. G. 1982. Mafic segregations in ophiolite mantle sequences. *Nature, Lond.* **299**, 524–528.

—— & RODDICK, J. C. 1980. Petrology and $^{40}Ar/^{39}Ar$ geochronology of some Hellenic sub-ophiolite metamorphic rocks. *Contrib. Mineral. Petrol.* **72**, 43–55.

STONELEY, R. 1974. Evolution of continental margins bounding a former Southern Tethys. *In*: BURK, C. A. & DRAKE, C. L. (eds) *The geology of Continental Margins.* Springer-Verlag, New York, pp. 889–903.

—— 1975. On the origin of ophiolite complexes in the Southern Tethys region. *Tectonophysics* **25**, 303–22.

—— 1981. The geology of the Kuh-e Dalneshin area of southern Iran, and its bearing on the evolution of Southern Tethys. *J. geol. Soc. Lond* **138**, 509–26.

THOMPSON, R. N. 1982. Magmatism of the British Tertiary volcanic province. *Scott. J. Geol.* **18**, 49–107.

——, ESSON, J. & DUNHAM, A. C. 1972. Major element chemical variation in the Eocene lavas of the Isle of Skye, Scotland. *J. Petrol.* **13**, 219–53.

——, GIBSON, I. L., MARRINER, G. F., MATTEY, D. P. & MORRISON, M. A. 1980. Trace element evidence of multistage mantle fusion and polybaric fractional crystallisation in the Paleocene lavas of Skye, NW Scotland. *J. Petrol.* **21**, 265–93.

THUIZAT, R., WHITECHURCH, H., MONTIGNY, R. & JUTEAU, T. 1981. K-Ar dating of some infra-ophiolitic metamorphic soles from the eastern Mediterranean: new evidence for oceanic thrustings before obduction. *Earth planet. Sci. Lett.* **52**, 302–10.

TIPPIT, P. R., PESSAGNO, E. A. & SMEWING, J. D. 1981. The biostratigraphy of sediments in the volcanic unit of the Samail ophiolite. *J. geophys. Res.* **86**, 2756–62.

WILLIAMS, H. & MCBIRNEY, A. R. 1979. *Volcanology.* Freeman, Cooper & Co., San Francisco.

R. HALL, Department of Geology, University College, London WC1E 6BT.

The role of Landsat Multispectral Scanner (MSS) imagery in mapping the Oman ophiolite

D.A. Rothery

SUMMARY: The Oman ophiolite is well exposed, and all the major structural and lithological features are visible on suitably processed Landsat images. Visual interpretation of such images provides a synoptic view of the Oman mountains which is in many aspects as useful as inspection of published maps at scales of 1:500,000 to 1:100,000. This study is an investigation of image processing for semi-automated lithological mapping. The differences between the visible and near-infrared multispectral signatures of ophiolite rock types in the Oman mountains have been shown by *in-situ* measurements to be considerably less than for fresh, particulate samples of similar rocks, in the Landsat MSS spectral bands. The muting of their reflectance spectra is due to a combination of alteration and weathering, and this hinders discrimination between rock types on MSS images. The problem is further compounded by topographically induced variations in brightness (BV), and the presence of shadows containing no valid data. The first-order topographic effects can be partly overcome by the generation of intensity-normalized bands. By using training sets of less than 2 km^2 from within an area of less than 2000 km^2; a thematic map can be produced which shows the approximate distribution of major rock types over virtually an entire MSS scene (30,000 km^2). The problem of misclassification due to topography and other effects can be largely side-stepped by the use of a suitable post-classification filter.

The importance of remote sensing in various aspects of geological exploration and interpretation has increased in recent years. The growth of the space programme has enabled the acquisition of images from orbital altitudes, offering synoptic coverage of large areas, often with little geometric distortion. NASA's policy of making images acquired by earth satellites openly available at nominal prices has encouraged a good deal of research into the geological applications of these images. The most useful images have proved to be those obtained by the Multispectral Scanner (MSS) onboard the Landsat series of satellites. Examination of MSS images has led to significant refinements in geological mapping, such as the subdivision of the Transvaal dolomite achieved by Grootenboer *et al.* (1973), has enabled regional studies of the Cenozoic and Recent faulting resulting from the India–Eurasia collision (Molnar & Tapponnier 1975, 1978) and has provoked revised interpretations of major structural features such as the Jordan rift (Brown & Huffman 1976). In addition, the association of mineral deposits with multispectral reflectance anomalies and structural features detectable by the MSS has become widely recognized (Rowan *et al.* 1974; Goetz & Rowan 1981).

The new generation of satellites with more suitable band-passes and improved resolution, such as Landsat 4 (launched July 1982) and the French SPOT (1985), will increase the potential usefulness of remote sensing in geology. However, large price rises to adjust the cost of images to commercially viable levels are likely to diminish the extent to which they can be used in low-budget projects. In order to justify the expenditure, it will therefore become even more important that a realistic judgement be made of the extent to which various image processing and interpretation techniques could contribute to each particular study. This remote-sensing study of the Oman ophiolite was largely retrospective, being performed in conjunction with an established field-based mapping project. It is thus possible to assess the limitations of a purely remote-sensing-based study and to recognize the extent to which the geological mapping of this region could have been achieved if access on the ground had been more restricted.

Landsat MSS images and image processing

The Landsat series and MSS imagery have been reviewed by several authors (Horan *et al.* 1974; NASA 1977; Slater 1979). In brief, the system acquires coincident images of the ground in four spectral bands: band 4 (500–600 nm), band 5 (600–700 nm), band 6 (700–800 nm) and band 7 (800–1100 nm). The scene recorded represents an approximately square area about 185 km across, and each image consists of a set of scan lines on average 79 m apart broken into picture elements, or pixels, sampled every 58 m. Data is transmitted in digital form, every pixel being assigned a brightness value (BV) over the range 0–127 (7 bits) for bands 4–6, and 0–63 (6 bits) for band 7.

The image can be reconstructed either in photographic form ('hard copy') or displayed on a television screen. In this study, interactive processing of an MSS image of the Oman mountains was performed on a Dipix LCT-11 image-analysis system, which has a PDP 11/23 computer to perform the data manipulation and a high-quality colour television monitor to display the image.

Geological maps of the Oman mountains

The most comprehensive map of the Oman mountains is that produced by Glennie *et al.* (1974), which covers the whole of the mountain belt at a scale of 1:500,000. This was produced by interpretation of vertical aerial photographs, backed up by a limited number of field surveys and helicopter traverses. In conjunction with detailed petrological, geochemical and structural studies, the Open University Oman Ophiolite Project has produced for 1:100 000 scale maps covering the central portion of the mountains, based on extensive field work together with photointerpretation (Smewing 1979; Lippard 1980; Lippard & Rothery 1981; Browning & Lippard 1982). These maps show the lithological mapping of Glennie *et al.* (1974) to be substantially correct, but several refinements were made, including the introduction of new mappable units within the ophiolite and the structural reinterpretation of some areas. A complete ophiolite sequence is recognized, comprising a 'mantle sequence' of serpentinized harzburgite and an overlying 'crustal sequence' of layered and unlayered plutonic rocks, sheeted dykes, pillow lavas and pelagic sediments (see Browning, this volume; Smewing *et al.*, this volume). A further 1:100,000 scale map was produced by an American group (Bailey 1981) covering a 35 km wide strip running south from Muscat, based mainly on field mapping in the ophiolite and photointerpretation of the adjacent terrain.

Justification for this remote-sensing study

The excellent exposure, good geological maps and the personal experience of the terrain gained during conventional mapping combine to make the Oman ophiolite a suitable area in which to test the validity of remote-sensing and image-processing techniques for mapping. Previous remote-sensing studies of the Oman mountains, comprising a thermal-inertia study by Pohn *et al.* (1974) and Pohn (1976) and an attempt at thematic mapping using MSS data by Carlson & Stoiber (1977) were not associated with any field-based studies and used poorer geological maps for ground truth than are now available.

Black-and-white prints from standard-product MSS negatives provide a synoptic view of the Oman mountains, on which the structural division of the Semail Nappe is clear, and several lithological units may be recognized. Visual interpretation of such images in conjunction with limited field traverses would probably have enabled production of geological maps of the ophiolite equivalent in standard to the 1:500,000 scale maps by Glennie *et al.* (1974). Interactive digital processing of MSS data using all four bands enables distinctions between rock units to be enhanced in a variety of ways not possible by photographic techniques. In this paper, use of the data for thematic mapping, i.e. semi-automated lithological discrimination, is discussed.

Image processing for lithological discrimination

Rationale

In order to understand the multispectral information content in MSS images of this terrain, the state of the rocks must be considered. The rocks in the Oman ophiolite are neither completely fresh on their surfaces, nor in most cases have they retained their primary igneous mineralogy. Thus the reflected light detected by the MSS comes from weathered surfaces of rocks which have usually been metamorphosed to some extent. *In-situ* measurements of bidirectional reflectance factor (BRF) on a variety of rock types in the Oman mountains (Rothery & Milton 1981; Rothery 1983b, 1984) show that weathering and alteration severely reduce the differences between the average spectral reflectances of the Oman rock types, on the basis of BRFs and BRF ratios in the MSS band-passes, and that there is considerable overlap between the fields occupied by many rock types. On MSS images these overlaps tend to be increased by topographic slopes which introduce brightness variations within a single rock unit. Thus, any technique for lithological discrimination or mapping on MSS images needs to be capable of coping with, and preferably compensating for, overlaps between classes. In this study, supervised maximum-likelihood classification in conjunction with post-classification filtering was found to produce satisfactory thematic maps.

The test area

The thematic mapping procedure was developed using a subscene of 54 × 37 km, which was geometrically corrected and resampled from the original MSS data by cubic convolution to give 100 × 100 m square pixels. Etheridge & Nelson (1979) have indicated that cubic convolution resampling does not significantly alter the accuracy of subsequent maximum-likelihood classification. A simplified geological map of the

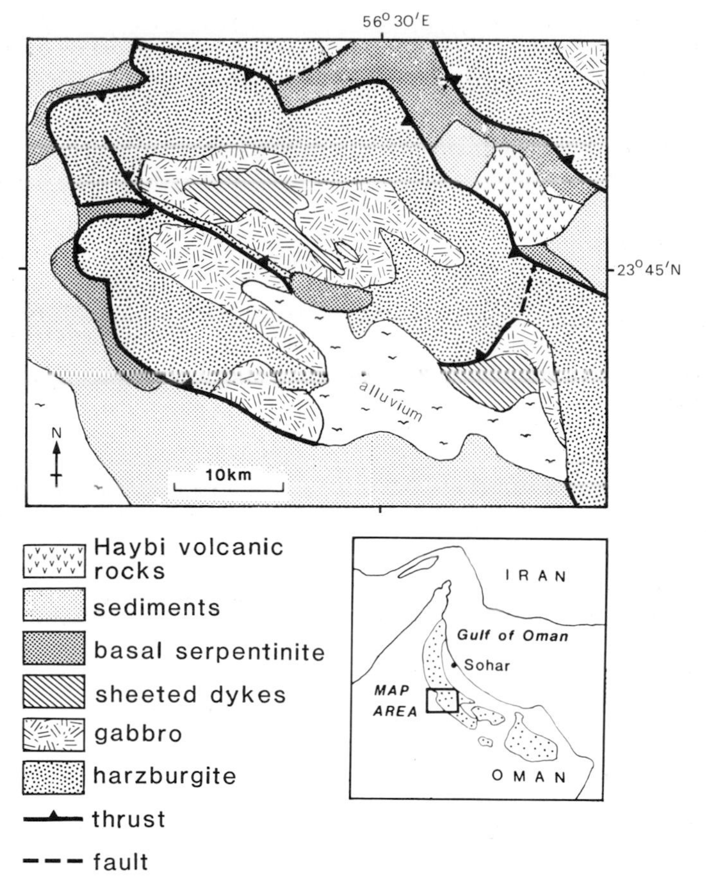

FIG. 1. Simplified geological map of the test area.

test area is shown in Fig. 1 and a band 7 image of it is shown in Fig. 2. The area includes the major ophiolite rock types in a synclinal area near the leading edge of the ophiolite nappe, the serpentinite sheet below the nappe, and a variety of rock types in the adjacent terrain.

Training areas were selected from known outcrops typical of each rock type, in flat areas where possible, or, in the more rugged terrain, on sunlit slopes. Several areas within each rock type were combined to form the training set for that lithology (40–180 pixels per set), so that a representative range of variation was included within each set.

Maximum-likelihood classification and post-classification filtering

The four-dimensional distributions of the BV data in each training set (based on the four MSS bands) were used as a basis for maximum-likelihood classification of the subscene. The maximum-likelihood classification procedure determines a four-dimensional ellipsoid corresponding to a user-specified percentage of the nominally normal probability distribution function of the data in each training set, and when the possibility arises that a pixel belongs to more than one class it is assigned to the most likely theme on the basis of Bayesian decision rules (Sprecht 1967; Andrews 1972). The results were displayed by colour coding the pixels assigned to each theme and compared with the 'correct' geological map based on field studies. It was not possible to distinguish ophiolite basaltic pillow lavas from the mantle-sequence harzburgite, because of their almost identical BV ranges in the training sets, probably due to similar desert varnishing. Seven major lithological units were successfully discriminated, being harzburgite, gabbro, sheeted dykes, serpentinite, basaltic tholeiitic lavas from the Haybi Complex (Searle *et al.* 1980) and two sedimentary-rock classes (mostly limestone). The acidic rocks in the ophiolite have a very restricted occurrence, and were not considered, and the mineralized areas

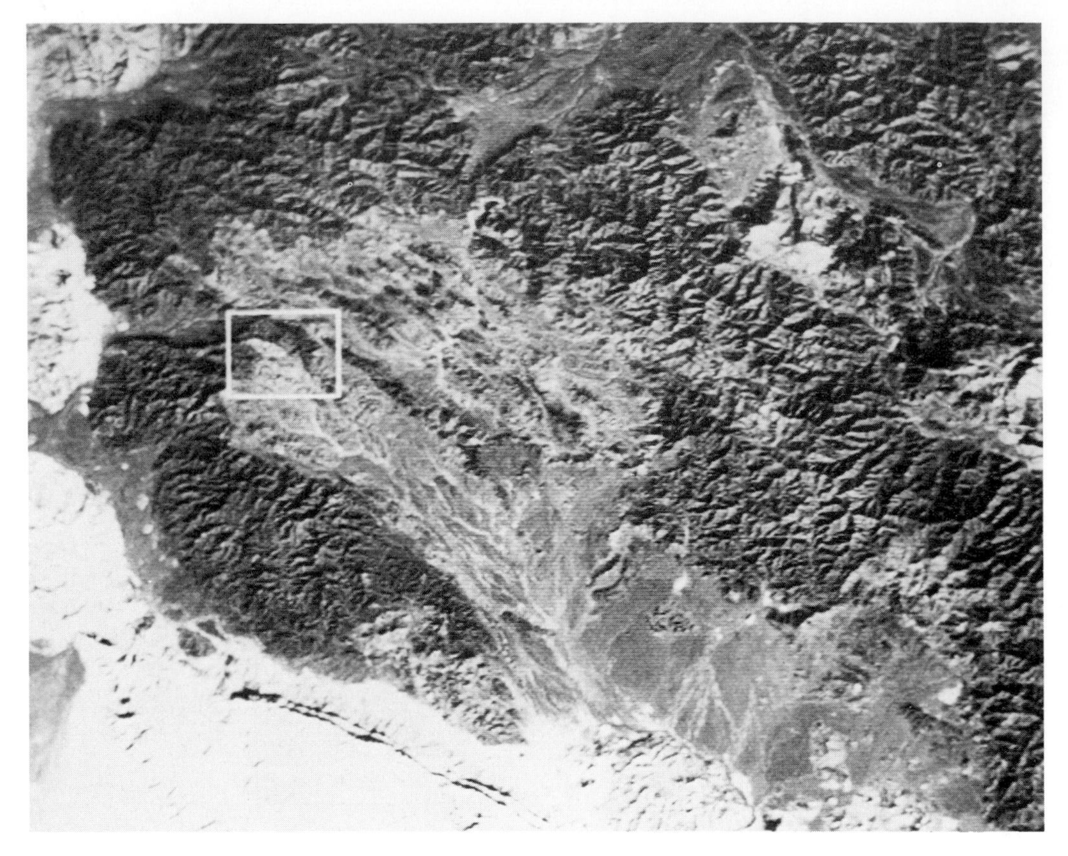

FIG. 2. Interactively contrast stretched band 7 image of the test area, which has been geometrically corrected and resampled to 100 m square pixels. The small rectangle indicates the area shown enlarged in Fig. 3.

(gossans and chromitites) are too small to be uniquely detected.

The most successful classification was achieved using 50% of the probability distribution function for the sheeted dyke and Haybi volcanic training sets and 95% of the probability distribution function for the others. Even in this best case there was a large proportion of misclassified pixels scattered throughout the image; about 30% of all classified pixels were judged to have been assigned to an incorrect theme. This is easy to appreciate when each theme is assigned to a distinctive colour on the display, but cannot be shown so well when themes have to be shown as shades of grey (as for publication). An enlarged portion of the image is reproduced in Fig. 3. This shows the themes in distinct shades of grey and the unclassified areas (corresponding mostly with the shadows) shown in white.

There is clearly a lot of misclassification in Fig. 3, and the intimate mixture of pixels makes this thematic map difficult to interpret. Misclassification occurs either in small areas receiving atypically full or oblique illumination, and which therefore appear either brighter or darker than the bulk of the pixels in the same rock type, or in areas which naturally lie near the extremes of the multispectral reflectance overlaps, and were assigned to the wrong theme by the maximum-likelihood decision. Misclassified pixels therefore tend to occur singly or in small groups. This being so, the quality of the thematic map can be greatly improved by post-classification filtering, otherwise known as 'small-area replacement' (Letts 1979).

In this study, it was found suitable to delete homogeneously classified areas consisting of less than eight pixels, and then allow the surviving, larger, classified areas to expand to fill the vacated or unclassified spaces. The effect of this is filtering on the image is shown in Fig. 4. The filtered thematic map is much clearer than the unfiltered version. Only four themes remain within the outlined area; harzburgite, serpentinite,

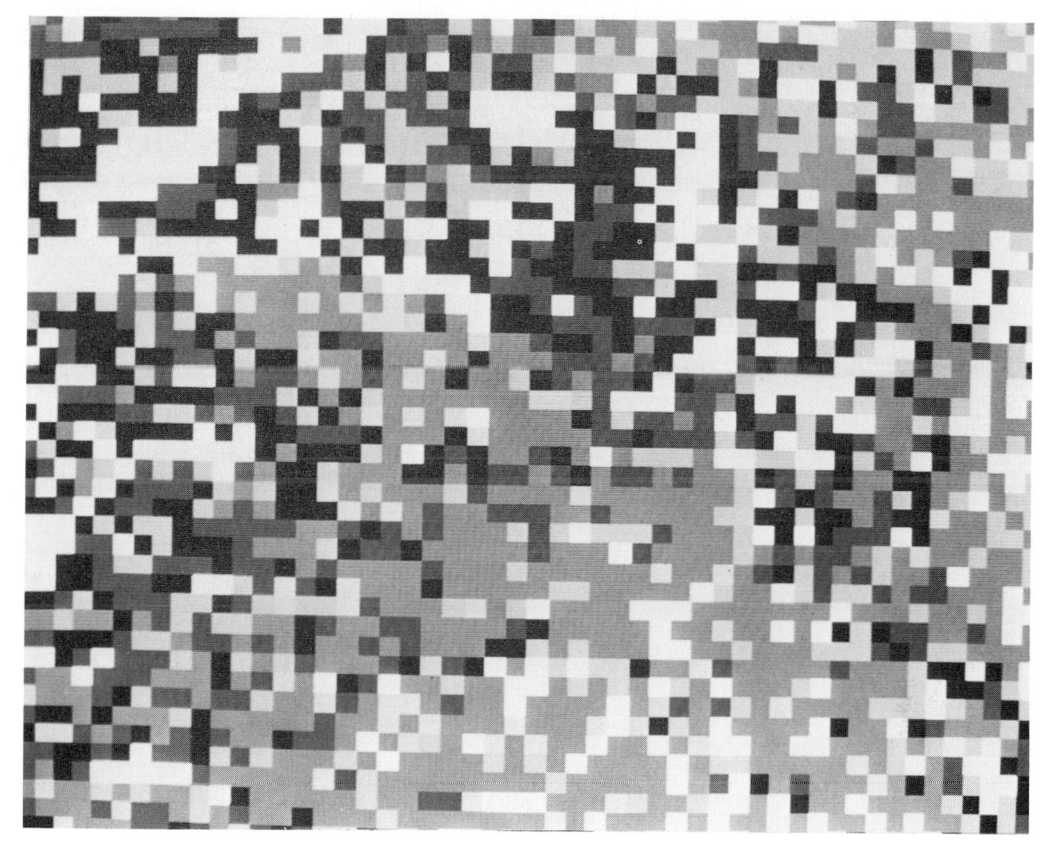

FIG. 3. Enlarged portion of outlined area in Fig. 2 with unfiltered themes superimposed, produced by supervised maximum-likelihood classification using the four MSS bands. Unclassified areas are shown in white, themes are in various shades of grey (for key see Fig. 4 caption).

gabbro and sheeted dykes (which is entirely consistent with the known geology), small areas previously misclassified as sediments and Haybi volcanic rocks having been removed.

The overall success of the thematic mapping is readily apparent. The distribution of the harzburgite theme coincides almost exactly with the outcrop of the mantle sequence, and the gabbro and sheeted-dyke themes are also mostly correct. The serpentinite theme picks out all the areas of basal serpentinite, and also occurs along a few linear faults and in other areas within the mantle sequence (as discussed later). This theme also occurs around the junction of the mantle-sequence harzburgite and the gabbros, reflecting partly the occurrence of layered peridotites near the base of the crustal sequence and partly the mixture of harzburgite-derived and gabbro-derived scree on the surface. The single area of Haybi volcanic rocks is shown correctly, and the sedimentary rocks are identified very well, the two sediment themes distinguishing the deeper-water succession below the ophiolite from the shallower marine limestones unconformably above the thrust belt in the SW of the area.

Alluvium is classified according to the main source rock type, except that alluvium derived from harzburgite is usually shown as serpentinite, because of serpentinization and weathering of the pebbles. Areas of shadow on the original image are assigned to the theme of the dominant surrounding rock type. The major misclassification is patches of mantle sequence wrongly shown as the Haybi volcanic or sheeted-dyke themes and a few areas of sheeted dykes shown wrongly as harzburgite or serpentinite. It is understandable that some misclassification should remain, since these are spectrally similar rock types (Rothery & Milton 1981; Rothery 1984), but misclassification is much less than before filtering. In defining the size of the minimum theme area allowed to pass through the post-classification filter, a trade-off is reached between deleting small correctly classified areas

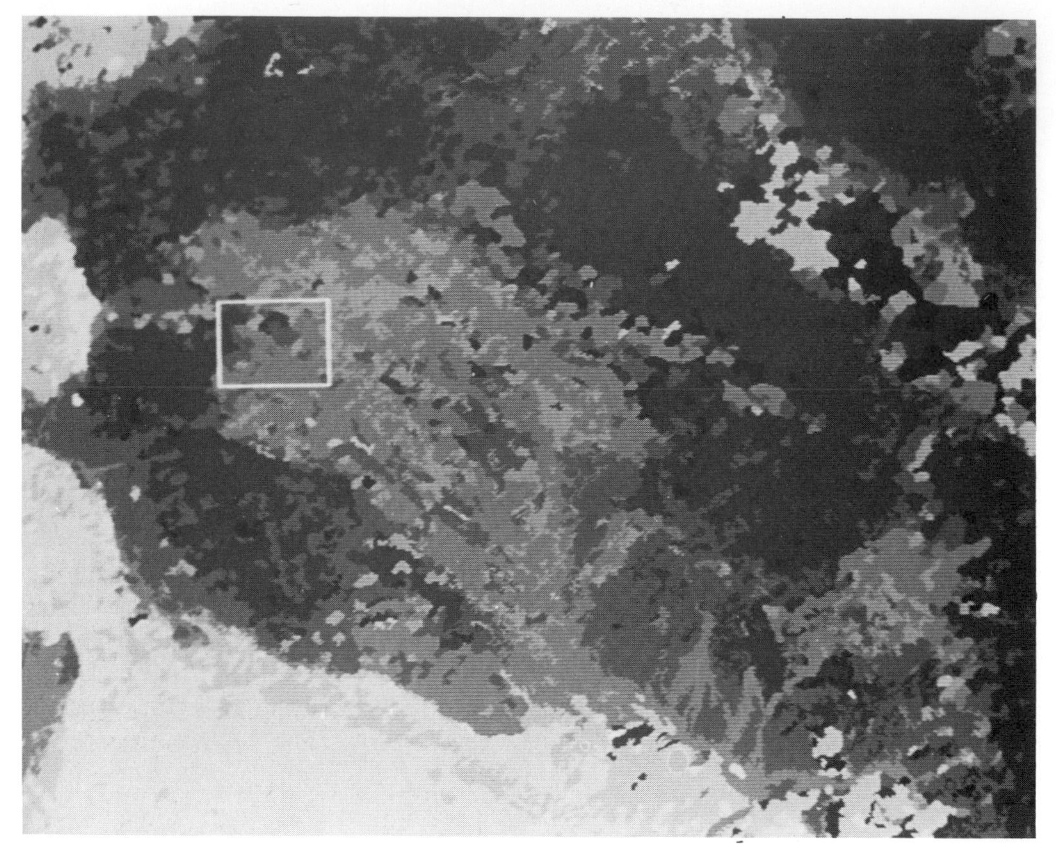

FIG. 4. The whole test-area image, shown as a thematic map after filtering. There are seven themes, and no areas remain unclassified. Themes are, from darkest to lightest: Haybi volcanic rocks, harzburgite, serpentinite, gabbro, sheeted dykes, deep-water sediments, shallow water limestone.

and leaving in too many incorrect patches, and the best choice has to be determined interactively.

The presentation of the image in the form of a conventional thematic map necessitates the loss of textural information, which can lead to incomplete or misleading interpretation. It was found possible to overcome this by displaying the themes not as areas of uniform hue and brightness (i.e. solid colours) but by displaying each theme in a particular hue (fixed proportions of red, green and blue) and modulating its brightness in response to the brightness of a feature showing the topography of the terrain, such as one of the four MSS bands or the first principal component. The resulting image has apparently 'transparent' themes, enabling structural features, such as faults, and the drainage pattern to show through. In this way the distinction between rugged outcrop and flat alluvial plains can be easily made, and some misclassified areas can be recognized as such because their structural position and drainage pattern are characteristic of the true rock type.

Classification using intensity-normalized bands

Several authors have suggested that the problem of topographically induced variations in BV may be circumvented by using band to band ratios, thus using one band to calibrate another. In this study, 'intensity-normalized' (Plessey Radar 1979) bands were generated in which the BV of each pixel was normalized by dividing it by its average BV in all four bands. The importance of performing a correction for atmospheric scattering as a preliminary step is stressed in Rothery (1984).

The effect of intensity-normalization is to suppress brightness variations but to accentuate colour differences, and noise is minimized by the use of a four-band average, rather than a single band, in the denominator. Maximum-likelihood classification was performed on the intensity-normalized image, using the same training areas as before. Post-classification filtering was again desirable to remove small misclassified areas, and the resulting thematic map is shown in Fig. 5.

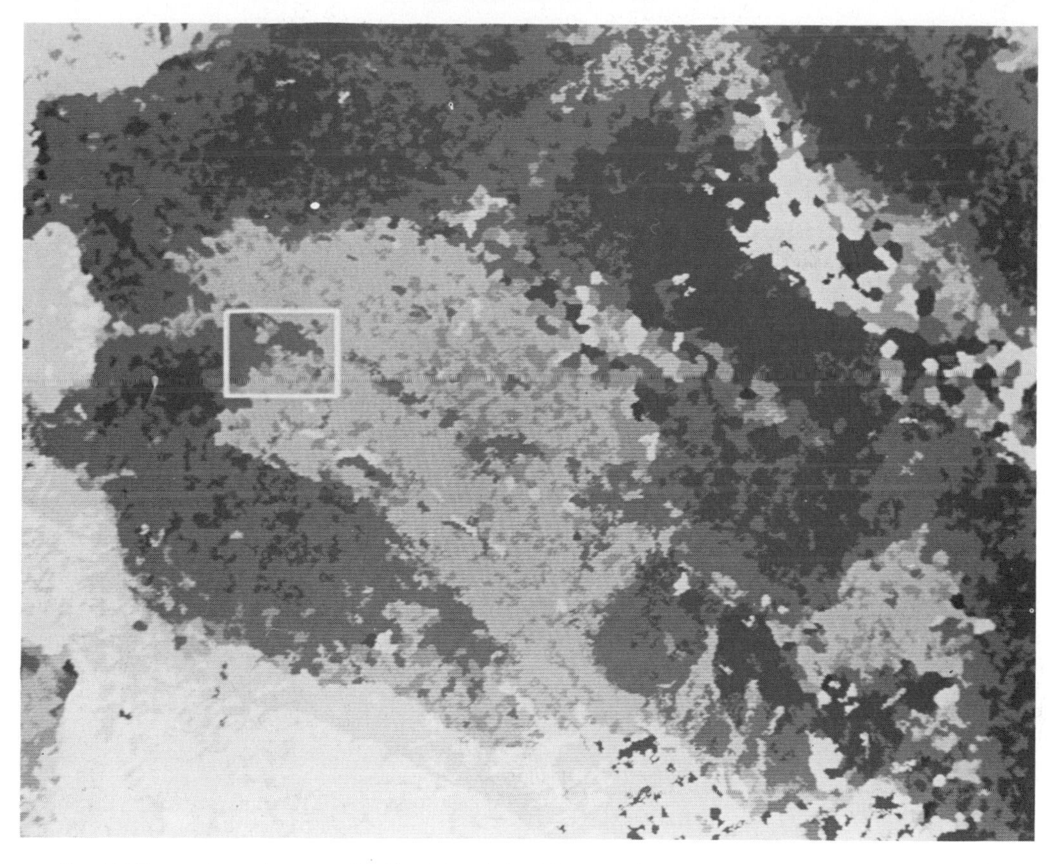

FIG. 5. Filtered thematic map as for Fig. 4, but derived from maximum-likelihood classification based on intensity-normalized bands.

There is less misclassification of the harzburgite as sheeted dykes (and *vice versa*) in both the filtered and unfiltered thematic maps derived in this way, which is predictable from *in-situ* reflectance measurements (Rothery 1984). Harzburgite and ophiolite lavas can also be discriminated better on intensity-normalized images, but this is not illustrated here. A large part of the area of gabbro is now assigned to the sheeted-dyke theme, partly because of the loss of albedo information involved in intensity-normalization (Rothery 1984). The extent of the sheeted-dyke theme within the gabbro might be overcome by reducing the percentage of the normal probability distribution function used for the sheeted-dyke theme to below 50%, but it does reflect, to some extent, the widespread occurrence of dykes from the sheeted-dyke formation throughout much of the gabbroic part of the crustal sequence in this area (Rothery 1983a). The two sediment classes are slightly better discriminated than before, but the Haybi volcanic area is almost completely missing.

The most significant difference between the two filtered thematic maps is that the intensity-normalized version picks out much more serpentinite within the dominantly harzburgite mantle sequence. The serpentinite shown within the mantle sequence is not distributed randomly, but occurs along major thrusts and faults (which can also be seen to some extent on the first thematic map) and also within the SW portion of the area. This was part of the leading edge of the ophiolite nappe during emplacement, where deformation would be expected to be the most intense, and it appears that it is this deformation which has encouraged subsequent serpentinization of the harzburgite to a greater-than-average extent. This observation of serpentinization related to emplacement features supports the claim of Wenner & Taylor (1973), based on oxygen and hydrogen isotope studies, that serpentinization in ophiolites occurs either during or after emplacement, rather than within the oceanic realm. The distribution of the most serpentinized areas of the mantle sequence is not easy to define in the field, and this technique provides a useful means of assessing it, and, by inference, the areas of highest strain, over large areas of terrain.

Extending the thematic mapping over a larger area

A thematic map covering most of the MSS scene has been produced using the identical classification process as above, on the sole basis of the data in the original training areas. The quality of the thematic map seems to be uniform throughout. The tripartite division of the ophiolite into harzburgite, gabbro and sheeted dykes is just as effectively shown as in the test area, and agrees broadly with the Open University 1:100,000 scale geological maps. The increased serpentinization within the mantle sequence near the leading edge of the nappe and along major faults can be seen throughout. It thus appears that, having established interactively a satisfactory thematic mapping procedure in a small test area, the same routine can be applied with equal success over the entire scene, provided that (as in this area) the nature of the terrain is uniform throughout.

Conclusions

Had the rest of the mountain belt within this image not already been mapped at 1:100,000 scale, this study would have been a useful preliminary exercise in the mapping programme. This type of work is likely to be valuable near the start of any similar study. Intensity-normalization improves discrimination between some lithologies, but concomitant with the suppression of topographically induced brightness variations is the loss of information on true differences in brightness between rock types. The relative merits of using intensity-normalized as opposed to ordinary MSS bands for classification depend therefore on the reflectance spectra of the rock types concerned and the nature of the topography. In either case, post-classification filtering is recommended to reduce misclassification, except when looking for pixel-sized targets.

Thematic maps should, however, be interpreted in conjunction with other images which preserve structural and textural information, such as principal components, single bands or colour composites. For ease of interpretation, modulation of coloured-theme brightness by a single feature is recommended as a way of combining most of the advantages of thematic and non-thematic information in a single image.

In addition to confirming most of the field-based mapping in this project, thematic mapping has highlighted features hitherto overlooked. The most significant such result is the detection of increased serpentinization of the mantle-sequence rocks along fault zones and near the leading edge of the nappe.

ACKNOWLEDGMENTS: I would like to thank particularly S.A. Drury and S.J. Lippard for reviewing various stages of the manuscript, and other members of the Open University Oman ophiolite group for help and advice throughout.

References

ALABASTER, T., PEARCE, J. A., MALLICK, D. I. J. & ELBOUSHI, I. 1980. The volcanic stratigraphy and location of massive sulphide deposits in the Oman ophiolite. *In*: PANAYIOTOU, A. (ed.) *Ophiolites Proceedings International Ophiolite Symposium Cyprus 1979*, pp. 751–7.

ANDREWS, H. C. 1972. *Introduction to Mathematical Techniques in Pattern Recognition.* Wiley-Interscience, New York, pp. 105–12.

BAILEY, E. H. 1981. Geologic Map of Muscat-Ibra Area, Sultanate of Oman. *J. Geophys. Res.* **86**, (Pocket map).

BROWN, G. F. & HUFFMAN, A. C. 1976. An interpretation of the Jordan Rift Valley, in *ERTS. I. A New Window on our Planet*, U.S. Geol. Surv. Prof. Paper 929, 362 pp.

BROWNING, P. & LIPPARD, S. J. (eds) 1982. Wadi Hawasina—Rustaq. *Oman Geological Ophiolite Project Map 4/5*, Open University, Directorate of Overseas Surveys.

CARLSON, C. G. & STOIBER, R. E. 1977. Geologic mapping by computer analysis of digital Landsat data. *Geol. Soc. Am. 1977 Annual Meeting, Seattle, WA*. Programme with abstracts, p. 1243.

ETHERIDGE, J. & NELSON, C. 1979. Some effects of Nearest Neighbour, Bilinear Interpolation, and Cubic Convolution Resampling on Landsat data. *Proc. 5th Ann. Symp. on Machine Processing of Remotely Sensed Data. Purdue Universty, 1979*, p. 84.

GLENNIE, K. W., BOEUF, M. G. A., HUGHES-CLARKE, M. W., MOODY-STUART, M., PILAAR, W. F. H. & REINHARDT, B. M. 1974. Geology of the Oman Mountains, *Kon. Ned. Geol. Mijnboukundig Gennot. Verh.* **31**, 423 pp.

GOETZ, A. F. H. & ROWAN, L. C. 1981. Geologic Remote Sensing. *Science* **211**, 781–91.

GROOTENBOER, J., ERIKSON, K. & TRUSWELL, J. 1973. Stratigraphic subdivision of the Transvaal dolomite from ERTS imagery. *Proc. 3rd ERTS symposium, Washington, Dec. 1973*, NASA SP-351 Vol. 1, Washington, pp. 657–64.

HORAN, J. J., SCHWARTZ, D. S. & LOVE, J. D. 1974. Partial performance degradation of a remote sensor in a space environment, and some probable causes. *Applied Optics* **13**, 1230–7.

LETTS, P. J. 1979. Small Area Replacement in Digital Thematic Maps, *Proc. 5th Ann. Symp. on Machine Processing of Remotely Sensed Data, Purdue University 1979*, p. 431.

LIPPARD, S. J. (ed.) 1980. Wadi Jizi *Oman Geological Ophiolite Project Map 2*, Open University, Directorate of Overseas Surveys.

—— & ROTHERY, D. A. (eds) 1981. Wadi Ahin—Yanqul. *Oman Geological Ophiolite Project Map 3*, Open University, Directorate of Overseas Surveys.

MOLNAR, P. & TAPPONNIER, P. 1975. Cenozoic Tectonics of Asia: Effects of a Continental Collision, *Science* **189**, 419–26.

—— & —— 1978. Active Tectonics of Tibet. *J. Geophys. Res.* **83**, 5361–75.

NASA 1977. *Landsat Data Users' Handbook*, NASA-TM-4722, Greenbelt, Md.

PLESSEY RADAR 1979. *Image Data Processor IDP-3000* Users' Guide, Plessey TP 3332.

POHN, H. A. 1976. A Comparison of Landsat Images and Nimbus Thermal-Inertia Mapping of Oman. *J. Res. US Geol. Survey* **4**, 661–5.

——, OFFIELD, T. W. & WATSON, K. 1974. Thermal Inertia Mapping from Satellite—Discrimination of Geologic Units in Oman. *J. Res. US Geol. Surv.* **2**, 147–158.

ROTHERY, 1983a. The base of a sheeted dyke complex, Oman ophiolite: implications for magma chamber configuration at oceanic spreading axes, *J. geol. Soc. Lond.* **140**, 287–96.

—— 1983b. Supervised maximum-likelihood classification and post-classification filtering using MSS imagery for lithological mapping in the Oman ophiolite. *Proc. Int. Symp. on Remote Sensing of Environment, 2nd Thematic Conference, 'Remote Sensing for Exploration Geology', Fort Worth, Texas 1982*, Vol. I, pp. 417–26.

——1984. Reflectances of ophiolite rocks in the Landsat MSS bands: relevance to lithological mapping by remote sensing, *J. geol. Soc. Lond.* thematic paper set 'Geological Applications of Remote Sensing', (in press).

—— & MILTON, J. E. 1981. Lithological discrimination in an ophiolite terrain: Landsat MSS imagery and reflectance measurements in Oman, *In*: ALLAN, J. A. & BRADSHAW, M. J. (eds) *Geological and Terrain Analysis Applications of Remote Sensing Proc. Eighth Annual Conference of the Remote Sensing Society*, pp. 3–23.

ROWAN, L. C., WETLAUFER, P. H. , GOETZ, A. F. H., BILLINGSLEY, F. C. & STEWART, J. H. 1974. Discrimination of rock types and detection of hydrothermally altered areas in south-central Nevada by the use of computer-enhanced ERTS imagery, *U.S. Geol. Surv. Prof. Paper* **883**, 35 pp.

SEARLE, M. P., LIPPARD, S. J., SMEWING, J. D. & REX, D. C. 1980. Volcanic rocks beneath the Semail Ophiolite Nappe in the Oman mountains and their significance in the Mesozoic evolution of Tethys, *J.geol. Soc. Lond.* **137**, 589–604.

SLATER, P. N. 1979. A re-examination of the Landsat MSS. *Photogramm. Eng. & Remote Sensing* **45**, 1479–85.

SMEWING, J. D. (ed.) 1979. The Sumeini—Shinas Area. *Oman Geological Ophiolite Project Map 1*, Open University, Directorate of Overseas Surveys.

—— 1980. Regional setting and petrological characteristics of the Oman ophiolite in North Oman. *In*: ROCCI, G. (ed) *Ofioliti* special issue "Tethyan ophiolites" Vol. 2 Eastern Area, pp. 335–78.

SPRECHT, D. F. 1967. Generation of Polynomial Discriminant Functions for Pattern Recognition. *IEEE Trans. on Electronic Computers*, 308–19.

WENNER, D. B. & TAYLOR, H. P. JR 1973. Oxygen and hydrogen isotope studies of the serpentinisation of ultramafic rocks in oceanic environments and continental ophiolite complexes. *Am. J. Sci.* **273**, 207–39.

D. A. ROTHERY, Department of Earth Sciences, The Open University, Milton Keynes, MK7 6AA.